The Physics of Instabilities in Solid State Electron Devices

The Physics of Instabilities in Solid State Electron Devices

Melvin P. Shaw
Wayne State University
Detroit, Michigan

Vladimir V. Mitin
Wayne State University
Detroit, Michigan

Eckehard Schöll
Technische Universität
Berlin, Germany

Harold L. Grubin
Scientific Research Associates, Inc.
Glastonbury, Connecticut

Plenum Press • New York and London

Library of Congress Cataloging-in-Publication Data

The Physics of instabilitites in solid state electron devices / Melvin P. Shaw ... [et al.].
p. cm.
Includes bibliographical references and index.
ISBN 0-306-43788-0
1. Semiconductors. 2. Gunn effect. 3. Diodes. I. Shaw, Melvin P.
QC611.P583 1991
537.6'22--dc20 91-32951
CIP

ISBN 0-306-43788-0

A Division of Plenum Publishing Corporation
233 Spring Street, New York, N.Y. 10013

Printed in the United States of America

To my late mother, Yetta Shaw, and my mentors: Tom Eck; Peter Solomon; the late Al Badain and Julius Portnoy

—MPS

To my wife Ludmilla and my son Oleg

—VVM

To my wife Viola and my children Claudia and Alexander

—ES

To my wife Ruth and my children Scott and Rachael

—HLG

Preface

The past three decades have been a period where useful current and voltage instabilities in solids have progressed from exciting research problems to a wide variety of commercially available devices. Materials and electronics research has led to devices such as the tunnel (Esaki) diode, transferred electron (Gunn) diode, avalanche diodes, real-space transfer devices, and the like. These structures have proven to be very important in the generation, amplification, switching, and processing of microwave signals up to frequencies exceeding 100 GHz. In this treatise we focus on a detailed theoretical understanding of devices of the kind that can be made unstable against circuit oscillations, large amplitude switching events, and in some cases, internal rearrangement of the electric field or current density distribution.

The book is aimed at the semiconductor device physicist, engineer, and graduate student. A knowledge of solid state physics on an elementary or introductory level is assumed. Furthermore, we have geared the book to device engineers and physicists desirous of obtaining an understanding substantially deeper than that associated with a small signal equivalent circuit approach. We focus on both analytical and numerical treatment of specific device problems, concerning ourselves with the mechanism that determines the constitutive relation governing the device, the boundary conditions (contact effects), and the effect of the local circuit environment.

Our introductory chapter describes the mathematics of instabilities including stability, self-sustained oscillations, solitons, and chaos, and introduces some basic mechanisms for producing negative differential conductivity (NDC). The chapter also provides outline descriptions of a wide variety of devices used in switching, oscillating, and amplifying devices such as the prototype p–n junction diode; tunnel and resonant tunneling diode; p–n–p–n diode; amorphous devices; Josephson junction; Gunn diode; real-space transfer devices, and others. Chapter 2 focuses on stability, and advances the view of a semiconductor device as a nonlinear dynamic system, with the details of phase portrait analysis and the role of the local circuit environment. The following six chapters deal with specific NDC problems: tunneling devices; avalanche and Gunn diodes; superconducting Josephson junctions; multilayer semiconductor structures; and thermal and electrothermal instabilities. A major thrust is to understand and predict the current–voltage characteristics of these devices, and their high-frequency properties.

We are grateful to our many colleagues, worldwide, for their interest and involvement in our work. Special thanks are due to J. T. Chen and Brian

Schwartz for their helpful comments on Chapter 6. E. S. is indebted to P. T. Landsberg and F. Schlögl and V. V. M. is indebted to V. A. Kochelap and Z. S. Gribnikov, for their stimulating influence over many years.

Much of the work discussed in the book was supported in part by grants from the National Science Foundation, the Office of Naval Research, the Deutsche Forschungsgemeinschaft, and the Academy of Sciences of UkrSSR.

M. P. Shaw
V. V. Mitin
E. Schöll
H. L. Grubin

Contents

1

Introduction

1.1. ELECTRONIC INSTABILITIES IN SOLID STATE MATERIALS AND DEVICES

1.1.1. Negative Differential Conductivity (NDC)

Many of the desirable features of a wide variety of solid state electronic devices result from situations where a sufficiently high bias is applied to the device so that it either switches from one conductive state to another or oscillates between different conductive states. These effects are classified as electronic instabilities. Among devices that behave this way are the tunnel diode (Esaki, 1958); Gunn diode (Gunn, 1964); avalanche diode (Shockley, 1954); Josephson junction device (Josephson, 1962); thyristor (Fulop, 1963); p–i–n diode (Prim, 1953); and electrothermally driven amorphous switch (Ovshinsky, 1968). In order to analytically understand the behavior of these different structures, it is most often, but not always, convenient to characterize them in terms of an effective region of negative differential conductivity (NDC) in their current density, **J**, – electric field, **F**, characteristic. In this region $dJ/dF < 0$ holds. Figure 1-1 shows the two major classes of NDC **J**(**F**) characteristics: SNDC; NNDC. Strictly speaking, the NNDC case has a single-valued **J**(**F**) curve and the SNDC case has a single-valued **F**(**J**) curve. [We will use the generic term **J**(**F**) when referring to an arbitrary NDC curve. Furthermore, we will often drop the vector notation since the two vectors will generally be parallel.] As can readily be seen from the curves, the S and N stand for the shape of the **J**(**F**) characteristic. All NDC characteristics are functions of frequency and this dependence must always be considered in determining the upper frequency limit of operation of a particular device.

The simplest circuit in which any of these devices operates is shown in the inset of Fig. 1-2. Here the NDC element is in series with a load resistor R and bias battery that provides a voltage, Φ_B. If I is the current in the circuit and Φ the voltage drop across the NDC element then

$$\Phi_B = IR + \Phi \tag{1-1}$$

is the equation of the dc *load line* (dcll). This line is plotted in Fig. 1-2; its slope is $-1/R$ and its intersection with the device characteristic, $I(\Phi)$, defines the steady state operating point. As we shall show in Chap. 2, intersections of the dcll with the $I(\Phi)$ characteristics are often stable so long as $dI/d\Phi > 0$ holds, which is the case depicted in Fig. 1-2 for one intersection. However, for situations where either $dI/d\Phi$ or dJ/dF have regions of NDC, operating points at intersections in

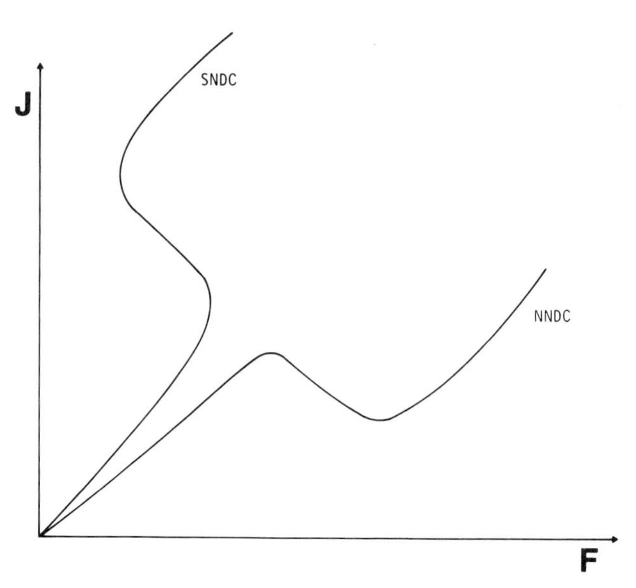

FIGURE 1-1. Current density **J** versus electric field **F** for two types of negative differential conductivity (NDC)—NNDC, SNDC (schematic).

these regimes (one of the three intersections) are often unstable against both the formation of inhomogeneous field and/or current density distributions (space charge nonuniformities) and/or circuit-controlled oscillatory effects.

Before we proceed, some caution is now in order. The electrical transport properties of a semiconductor are governed in a complex way by the microscopic

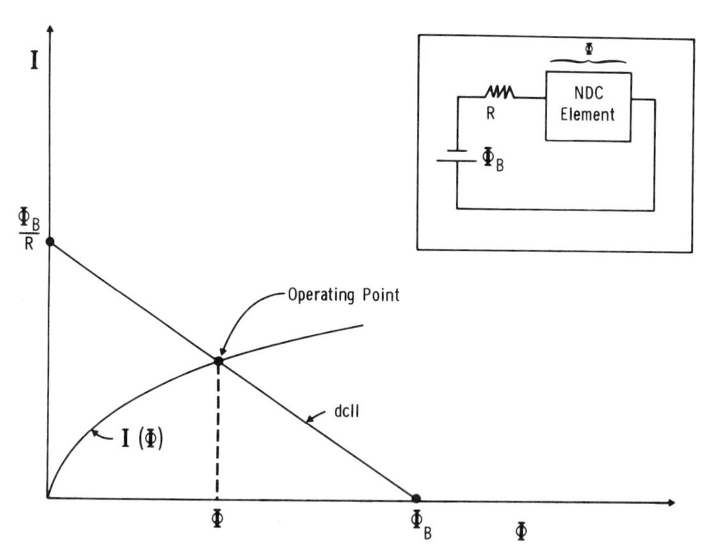

FIGURE 1-2. Schematic behavior of an NDC element in a resistive circuit with an applied voltage Φ_B and a load resistor R. The current I is plotted versus the voltage drop Φ across the NDC element. The steady state operating point is the intersection of the dc load line (dcll) and the $I(\Phi)$ characteristic of the NDC element. The inset shows the circuit.

bulk properties, which determine the local current density J as a function of the local electric field F, by the contacts, and by the external circuit components. Firstly, NDC does not always imply instability of the steady state, and vice versa. For example, SNDC states can be stabilized (and experimentally observed!) by a heavily loaded external circuit, and on the other hand, the Hopf bifurcation of a limit cycle oscillation can occur on a $J(F)$ characteristic with positive differential conductivity (Schöll, 1989b). Secondly, there is no one-to-one correspondence between SNDC and current filaments, or between NNDC and field domains (Schöll, 1987). Finally, it is important to distinguish between the local $J(F)$ characteristic and the global $I(\Phi)$ characteristic, which is determined by the total current

$$I = \int_S \mathbf{J}\, d\mathbf{S}$$

and the total voltage drop

$$\Phi = -\int_0^l F(z)\, dz$$

where S is the cross section of the current flow, and l is the contact-to-contact distance. The $J(F)$ and the $I(\Phi)$ characteristics are identical only for spatially homogeneous states. For spatially inhomogeneous states, negative differential conductivity of the $J(F)$ characteristic may result in either negative or positive differential conductance of the $I(\Phi)$ characteristic, and the same is true for positive differential conductivity of the $J(F)$ characteristic (Schöll, 1987).

In order to understand the detailed nature of the instabilities, we must ask ourselves two basic questions: *First, what is the mechanism responsible for the NDC region? Second, how do we analyze the resulting time- and space-dependent phenomena that often occur?*

The answer to the first question is usually different for the different devices discussed above. For example, we shall see that the NNDC characteristic of the tunnel diode results from the bias-induced onset of a region where the density of final available quantum mechanical tunneling states must vanish, so that the tunneling current experiences a precipitous decrease with increasing bias. We shall also find that the NNDC characteristic of the Gunn diode results from bias-induced intervalley transfer of hot electrons from a high mobility band to a lower mobility subband. In this particular case it is the average mobility of the charge carriers that is affected; since conductivity is proportional to mobility, NDC results from a region of NDM (negative differential mobility). As a final example, the SNDC characteristic exhibited by the p–i–n diode develops because as the number of injected carriers increases, more traps can be filled. Eventually, when all traps fill, an average carrier will transit the sample without being trapped; the voltage drop across the sample will decrease as the conductivity, and hence the current density, increases. In Sec. 1.3 we shall analyze four relatively simple models. The first three will show us how NNDC is generated from intervalley or real-space transfer of electrons. The fourth will show us how SNDC is generated from carrier generation and trapping.

The answer to the second question, on the other hand, we will find to be more generally applicable to a wide variety of devices. Here we will be concerned with the mathematics of instabilities, a subject that has been treated in considerable detail for the past 50 years, especially in systems where the constitutive curve [**J**(**F**) characteristic in our case] is independent of time, frequency, or internal dynamics.

There exists a large number of older monographs and review articles (see e.g., Volkov and Kogan, 1969; Hartnagel, 1969; Carroll, 1970; Thomas, 1973; Bonch-Bruevich et al., 1975; Shaw et al., 1979; Pozhela, 1981). Recently, interest in this field has been greatly revived by the striking connections with the modern field of *nonlinear dynamics,* with its bifurcations, deterministic chaos, and self-organized spatiotemporal structure formation. One purpose of this book is to provide an introduction to the theory of electronic instabilities in semiconductor devices within this framework. Semiconductors may be viewed as nonlinear dynamic systems that can exhibit electrical instabilities when they are driven far from thermodynamic equilibrium by strong external electric or magnetic fields, optical irradiation, or current injection. Such instabilities are known in a great number of very different dissipative systems occurring in physics, chemistry, and biology (Haken, 1983a; Nicolis and Prigogine, 1977) when a state far from thermodynamic equilibrium is maintained by a continuous influx and dissipation of energy. These instabilities bear a remarkable analogy with phase transitions. There are now powerful mathematical tools to deal with such nonequilibrium phase transitions and the nonlinear dynamics of charge transport in semiconductors (Schöll, 1987). In this book we shall apply some of these modern concepts to semiconductor devices. Physical aspects of electrical instabilities in semiconductors may also be found in reviews on nonlinear and chaotic transport in semiconductors (Abe, 1989), on a general dynamic systems approach to oscillatory current instabilities (Schöll, 1989b), on a derivation of macroscopic nonlinear and chaotic behavior from hot electron transport theory (Schöll, 1989c), on physical mechanisms and self-organized spatiotemporal structures in current instabilities (Schöll, 1991), and on microscopic theories of hot electron recombination and ionization (Reggiani and Mitin, 1989).

1.1.2. Solid State Switching, Oscillating, and Amplifying Devices

Before we embark on a development of the mathematical techniques required to understand instabilities in solid state devices, it is useful to provide a brief description of the wide variety of devices that are capable of switching, oscillating, or amplifying. We will discuss many of these in detail in Chaps. 3–8; they are listed alphabetically here.

Amorphous Thin Film Memory Switch (Chap. 8). This switch is similar to the following threshold switch except for the use of a bistable chalcogenide glass, for example $Ge_{14}Te_{83}Sb_3$, which can be reversibly altered by electrical pulses between a resistive noncrystalline state and a conductive microcrystalline state. Amorphous hydrogenated silicon thin film devices are also useful, but they cannot be reset conveniently and are therefore primarily used as PROM (programmable read only memory).

Amorphous Thin Film Threshold Switch (Chap. 8). These SNDC devices contain as their active part a noncrystalline (amorphous) semiconductor. In an Ovonic threshold switch a layer of chalcogenide (Se, Te, or S) alloy glass, for example, $Te_{50}As_{30}Si_{10}Ge_{10}$, is sputtered or evaporated between two nonreacting electrodes which are about 1 μm apart. The normal (OFF) resistance is of order 1 MΩ. If an applied voltage of either polarity exceeds the threshold voltage (2–200 V), the device switches in times as fast as 0.1 ns from the threshold point to its ON state. In this state the voltage drop is about 1 V, independent of current. As soon as the current falls below a holding current of about 1 mA, the device returns to its OFF state. Under pulse conditions the ON state is reached after a delay time of between 10 and 1000 ns, depending on the pulse height above threshold. The OFF state is the normal resistance of the semiconducting glass. In the ON state, injection from the contacts, trap filling, and the resultant increase in mobility produce a filamentary conducting region typically less than 10 μm in diameter. This device is useful as a surge suppressor.

The Avalanche Transistor. This is a transistor structure with alloyed step junctions and a base-layer resistivity that is larger than that of both the collector and emitter. With increasing Φ_{CE} avalanche breakdown of the collector occurs ($\Phi_{CE} \approx$ 60–80 V). The emitter current is first low, but with increasing current through the base the base potential increases and the emitter becomes forward biased, which causes a large emitter–collector current to flow at reduced $\Phi_{CE} \approx 30$ V. An SNDC region is traversed as the emitter becomes forward biased.

The Field Effect Transistor (FET). This is a majority-carrier (unipolar) device, in which the conductance between the source and drain electrodes along a thin semiconductor channel made of *n* or *p* material is controlled by the voltage difference between this channel and an adjacent gate electrode. Insulation between the channel and the gate is provided by a *p-n* junction (junction FET) or by a thin oxide layer yielding a metal-oxide-semiconductor (MOS)FET, also called an insulated gate (IG)FET. The gate voltage either enhances or depletes the carrier concentration in the channel and thus alters its conductance between low (OFF) and high (ON) values. The power consumption is very low, as it is essentially the change in energy stored on the device capacity. Because of device capacitances, the switching speed (particularly OFF to ON) is slower than that of fast bipolar devices. However, the small size of MOS devices, their self-isolation, and ease of manufacture permit large-scale integration of complex digital and memory circuits on one small silicon chip. The work function difference and hence the switching voltage (about 1.5 V) is decreased to 0.6 V by using *p*-doped polycrystalline Si instead of Al as a gate electrode. Such silicon-gate IGFETs have higher speed and use even less power.

The Gunn Diode (Chap. 5). This NNDC device is used mainly for microwave oscillation and amplification. However, it is also used as a switch when the contact/doping profile/length interrelated parameters are designed to produce a high electric field at the anode contact for sufficiently high values of bias.

Under these conditions the device switches from a low-voltage–high-current state to a high-voltage–reduced-current state. The mechanism for producing NNDC is the field-induced intervalley transfer of electrons from a high-mobility to low-mobility valley in the conduction band, and the manifestation of the current instability, e.g., oscillations or switching, is determined primarily by the cathode contact boundary condition.

Heterojunction Hot Electron Diode (Chap. 7). This SNDC device is fabricated from $GaAs/Ga_xAl_{1-x}As$ interfaces. Carriers are at first retarded from a heterojunction barrier. The barrier is subsequently thinned and reduced by the applied bias, and the field is increased in the narrow gap (GaAs) region. Eventually, the field in the region becomes large enough to heat carriers to sufficiently high temperatures so as to emit them thermionically over the barrier, thereby increasing the current, which in turn causes an even larger field in the narrow gap region. This produces the feedback mechanism required for stabilization of the SNDC curve.

IMPATT Devices (Chap. 4). These devices depend upon a combination of impact ionization of hot electrons and transit time effects. The IMPATT (impact ionization avalanche transit time) diodes can generate the highest cw (continuous wave) power output in the millimeter-wave regime (i.e., at frequencies $>$ 30 GHz). The originally proposed device (Read diode) involves a reverse biased n^+-p-i-p^+ structure, where n^+ and p^+ denote strongly n- or p-doped regions. In the n^+-p region (avalanche region) carriers are generated by impact ionization across the band gap; the generated holes are swept through the i region (drift region), and collected at the p^+ contact. When a periodic (ac) voltage is superimposed on the time-independent (dc) reverse bias, negative ac conductance can arise if the ac component of the carrier flow drifts opposite to the ac electric field. This corresponds to a π phase lag of the current behind the voltage. This phase lag is due to the finite build-up time of the avalanche current ("avalanche delay") and the finite time it takes the carriers to cross the drift region ("transit-time delay"). If the sum of these delay times is approximately one-half cycle of the operating frequency, negative conductance occurs. This can be achieved by properly matching the length of the drift region with the drift velocity and the frequency. Other transit-time devices are the BARITT (barrier injection and transit time) diode, the DOVETT (double velocity transit time) diode, and the TRAPATT (trapped plasma avalanche triggered transit) diode.

Integrated Logic Circuits. The switching elements in digital electronics are no longer discrete devices but instead integrated circuits: gates; clocks; counters; memory arrays, each containing numerous devices. These integrated elements are used together with drivers, encoders, and other circuitry to prepare the integrated circuits which comprise the random-access memories (RAM), electrically programmable memories (PROM), shift registers, and other peripherals needed to build, for example, a microprocessor or microcomputer. Even complex computing and data handling systems use only two basic logic operations, counting and gating. The basic logic elements of gating are the operations of coincidence starting "AND"; of anticoincidence or sorting out, "OR"; and of prohibition, "NOT."

Josephson Superconducting Junction (Chap. 6). When two superconductors are separated by a sufficiently thin nonsuperconducting region (less than about 50 Å), the tunneling of electron pairs produces a magnetic field-dependent current at zero voltage called the Josephson current. Switching can be achieved from this state to a voltage state by either current overdrive or a magnetic field. The switching time approaches several picoseconds and the switching event requires extremely low energy; it is our fastest known electronic switch, with power-delay products (less than about a femtojoule) well below those of either Si or GaAs devices. Because of the presence of the zero-voltage current, the Josephson element can be classified as an NNDC device. The recent breakthrough in high T_c superconductors may resuscitate this device as a possible competitor with conventional semiconductor devices.

The Junction Transistor. This consists of two *p-n* junctions in close proximity. Under normal conditions one junction is the forward-based emitter, the other the reverse-biased collector. The transistor is OFF when both collector and emitter are reverse biased (base current zero). The collector–emitter leakage current is about 10 nA for Si. In its ON state both junctions are forward biased ($\Phi_{CE} < 1$ V), injecting carriers into the base region, which reaches saturation as a result of space-charge and Ohmic effects. For fast turnoff speeds saturation must be avoided by, for example, a Schottky diode between collector and base, which becomes forward biased before saturation is reached. This device exhibits SNDC characteristics under breakdown conditions in the common emitter mode. However, it is used as a switch primarily between cutoff (OFF) and saturation (ON).

Metal Oxide Varistor. Here small metal particles (Sn, Ni, Nb, etc.) are surface oxidized and subsequently sintered. The transport mechanism involves an increase in current as the carriers tunnel through the oxide regions separating the particles. The current–voltage characteristics are generally rather soft and vary gradually, as opposed to structures that exhibit sharp discontinuous-type changes.

The MNOS Memory Switch. This is a MOS device with a silicon nitride dielectric layer (N) between the thin oxide and the gate electrode. Charges trapped in the silicon nitride layer hold the conductance state in the channel and thus provide the memory action: 5–10 V "write" and "erase" pulses of opposite polarity change the charge state of the traps. The memory access time is under 50 ns: the access time is at least twice as great in ferrite core memories. However, the MNOS memory is volatile, which means that it requires circuit voltage to keep its memory.

Optical Mass Memories. Generally made from chalcogenide materials, these thin film structures offer great promise for extremely high-density reversible memories. A focused laser beam is used to change the optical properties of the material in a microscopic region; the region is then read optically. Amorphous films are usually fabricated in disk form; they can be used as is, with the laser either changing the local region to crystalline, or vaporizing the region and creating a bubble in the material. The films are often heated after deposition to form crystallites and then induced to change regionally into the amorphous phase by elevating the local temperature optically and then rapidly removing the excitation (quenching). Several megabytes of information can be reversibly stored in a region of a few square inches.

Optoelectronic Devices. These may be classified as active or passive. Passive devices use light-emitting and -detecting diodes in close proximity. Light coupling achieves high electrical isolation between input and output. Active devices are self-oscillating semiconductor lasers, or laser amplifiers. These are used to generate or amplify short optical pulses in the picosecond regime, by either gain switching or active or passive mode-locking. Optical bistabilities with switching speeds on the order of nanoseconds are possible in both passive and active devices.

The p-i-n Diode (Chap. 7). This is a current-controlled microwave switch. It has a thin layer of undoped (intrinsic) region between strongly doped p and n layers. At microwave frequencies the intrinsic region acts like a capacitor. Forward bias (dc or low frequency) produces a carrier avalanche and makes the diode conductive. It sometimes shows SNDC characteristics.

The p-n Junction Diode. This is formed at the interface of an n- and p-type semiconductor. Electrons diffuse from the n to the p material and holes from the p to the n material until an internal electric field is built up which causes the resultant drift current to cancel the diffusion current. This prevents further increase of the equal and opposite space charges. The energy barrier in the high-resistance unbiased junction is about equal to the energy gap and nearly twice as big as the Φ_b of Schottky diodes. The current–voltage characteristic of the Schottky diode also holds for p-n junctions, but with much smaller values of I_r. This leads to a larger forward voltage drop Φ_F. For a 10-μA current $\Phi_F \approx 0.25$ V for a Schottky diode and $\Phi_F \approx 0.55$ V for a silicon p-n junction. For $\Phi > 0$, majority carriers flow across the junction. For $\Phi < 0$, they are pulled away from the junction and only the small minority-carrier density contributes to I_r until the breakdown voltage of the diode is reached, here $|I|$ increases rapidly due to Zener breakdown or avalanche multiplication of carriers in the high field of the reverse-biased junction. The turn-on time of a p-n diode is only a few nanoseconds. A turn-off time of microseconds is needed to sweep out the excess minority carriers which are stored in the junction during conduction.

The p-n-p-n Diode (Chap. 7). This has a reversed-biased collector junction, which yields a high OFF-state resistance ($>10^8\,\Omega$) between two forward-biased junctions. After onset of avalanche in the middle junction at the breakdown or switch-ON voltage, the two emitters saturate the two base regions and make the middle junction appear forward biased. A low-voltage ON state results. The device returns to its OFF state when the current falls below a holding current value. It is an SNDC device for one polarity of voltage.

The Real-Space Transfer Device (Chap. 1). This is the real-space analogue of the Gunn diode. It is based on the real-space transfer of electrons in a modulation-doped heterostructure. The conduction band edge of a GaAs/$Al_xGa_{1-x}As$ heterolayer, for instance, has a discontinuity ΔE_c due to the larger band gap of $Al_xGa_{1-x}As$ as compared to GaAs, e.g., $\Delta E_c = 250$ meV for $x = 0.3$. The AlGaAs layer is heavily doped, while the GaAs layer is undoped. Therefore, the ionized impurity scattering leads to a much smaller mobility in AlGaAs than in GaAs. Since the mobile carriers fall into the GaAs well, they are separated in real space from their parent donors, which are responsible for scattering. This results in very high current densities in the GaAs layer if an electric field is applied parallel to the layers. As the field is increased, the carriers are heated up strongly, resulting in a thermionic emission current from GaAs into AlGaAs. This real-space transfer reduces the average mobility and thus can lead to NNDC. The extremely high mobility in the GaAs (>5000 $cm^2/V\,s$ at 300 K) singles out this effect for applications in fast electronic switches and oscillators. A particularly important application is the high electron mobility transistor (HEMT), where the channel between source and drain contains a modulation-doped GaAs/AlGaAs heterojunction.

Resonant Tunneling Devices (Chap. 3). These NNDC devices are generally fabricated in the form of semiconductor heterostructures having two or more barriers, although single barrier heterostructures also produce NNDC effects. GaAs and $Ga_x\,Al_{1-x}As$ are generally employed. The basic principle of operation of these devices involves the creation of a resonant tunneling channel through a multiple quantum well structure, and the coupling of the channel to an emitter and collector. The resonance involves the tunneling transmission probability as a function of the energy of the incident electrons, which is related to the width and height of the quantum wells. The structure can often be thought of as multiwell, variably (in general) spaced superlattice energy filters.

The Schottky Diode. This is a rectifying metal–semiconductor (MS) contact. Interface states and to a small degree the difference in work functions between the metal (M) and the semiconductor (S) give rise to a Schottky barrier of height Φ_b and depletion of majority carriers in the barrier region. The observed current I is the difference between majority carrier flow in opposite directions across the barrier. The flow from M to S is impeded by $\exp(-\Phi_b/kT)$, the flow from S to M by $\exp[-(\Phi_b - q\Phi)/kT]$ because of the bias potential $q\Phi$ in S. This leads to the rectifier characteristic

$$I = I_r\left[\exp\left(\frac{q\Phi}{nkT}\right) - 1\right]$$

where n is an ideality factor equal to unity for a perfect diode, and I_r is the reverse leakage current. Switching times less than 0.1 ns are achieved for small (5-μm diameter) devices.

Silicon-Controlled Rectifier (SCR) (Chap. 7). Since switching of the *p-n-p-n* diode is initiated by the condition that the sum of the common base current gains α_{npn} and α_{pnp} becomes unity, rather than by the onset of avalanching, this condition can be achieved and the magnitude of the switch-ON voltage controlled by changing the current of either one of the two forward-biased junctions. This is done in the SCR by an additional terminal to one of the base regions. This SNDC device is a three-terminal *p-n-p-n* structure.

The Silicon-Controlled Switch (SCS) (Chap. 7). This has terminals to each of the two base regions of a *p-n-p-n* transistor. One permits control of ON switching; the other controls the holding current and thus OFF switching. This SNDC device is a four-terminal *p-n-p-n* structure.

Step-Recovery Diodes. These are *p-n* junctions whose graded doping profile minimizes the stored-minority-carrier transit time. They are used for fast and microwave switching.

Thyristors (Chap. 7). The SCR, SCS, and triac are a few of several solid state switches having thyratronlike SNDC characteristics. Some can also be light activated. Their low power consumption, trigger sensitivity, and reliability make them useful in many control circuits.

The Triac. This is an SCR device built such that operation with either voltage polarity yields a nearly identical SCR characteristic.

The Tunnel Diode (Chap. 3). This is a *p-n* junction in which both sides are so heavily doped that the Fermi level lies in the valence band on the *p* side and in the conduction band on the *n* side. At forward bias this overlap of the band edges allows electrons from the *n* side to tunnel quantum mechanically through the junction barrier into empty valence-band states without changing their energy. At higher bias the electron energies are too high for energy-conserving tunneling and the current drops through an NNDC range. The current rise at larger biases (>0.4 V) is the conventional forward-bias current of a *p-n* junction. The switching time is less than a nanosecond since only majority carriers are involved.

The Unijunction Transistor (UJT). This is an *n*-type semiconductor base with two Ohmic contacts, between which is placed a *p-n* junction (anode). Forward biasing of the anode injects minority carriers (holes) into the base, which decreases the resistance between the anode and the negative base contact. The resultant shift of the base potential below that of the anode increases its forward bias and thus the minority-carrier injection. This leads to an SNDC characteristic. The switching speed (ON to OFF) is orders of magnitude lower than that of bipolar transistors because of large charge storage effects in the high-resistivity base region.

The Zener Diode. This is a *p-n* junction with a sharp and well-controlled reverse-bias avalanche breakdown voltage (from 0.2 V to several hundred volts).

Its fast switching, voltage sensitivity, and low resistance in the ON state make it useful as an overload and transient suppressor and in control circuits.

The incredibly rapid pace of the development of integrated digital technology strives toward increasing speed and reducing cost, achieved by the higher packing density of very large-scale integration (VLSI). The fastest gating (less than 1 ns) is presently possible with bipolar junction transistors in the emitter-coupled logic (ECL) mode (less than 3 ns in the transistor–transistor logic (TTL) mode). However, the lowest cost is achievable with the slower MOS technology with its breakthroughs into LSI and VLSI during the past 20 years. MOS technology has limitations in that its gate propagation delay is greater than 10 ns. However, although ECL is the fastest logic switch, it also has the highest static power dissipation (about 40 mW per gate). The power dissipation for TTL is about 2 mW, and here is where the use of CMOS (complementary metal oxide semiconductor) becomes very desirable, since its power dissipation is about 10^{-3} mW. Small power dissipation is very important for future U (ultra) LSI applications involving millions of components and GSI (giga scale integration), with billions of drivers on the same chip. GSI will require feature sizes of about 0.1 μm. TTL gates require a chip area per gate of about $3 \times 10^4\ \mu m^2$, compared to $15 \times 10^2\ \mu m^2$ for CMOS. CMOS gates also have the lowest power-delay product of any semiconductor device, nearly 0.6 pJ.

Recent research indicates that vacuum microelectronic devices, which are based on micromachined on-chip miniature "vacuum tubes" with cold electron emitters, will have gate delays in the femtosecond range. At high temperatures, for high-speed applications it is expected that diamond semiconducting film will play a major role. For low-temperature, high-speed electronics, we expect that high-T_c superconducting devices as leads, flux containers, and/or Josephson junctions, could contribute substantially to the solid state switching technology.

The general mathematical techniques used for treating instabilities will be outlined in Sec. 1.2, and in Chap. 2 we shall analyze in detail the stability of systems germane to our needs. Chapters 3–8 will then cover particular devices, with whatever requisite physics and new analyses required outlined in each chapter. Chapter 3 covers important NNDC elements—the tunnel diode and resonant tunneling devices. Chapter 4 focuses on avalanche devices, which in many cases only exhibit SNDC at high frequencies. Chapter 5 covers Gunn diodes and Chap. 6 Josephson devices, both of which are NNDC elements. Chapters 7 and 8 cover the classic SNDC elements—the thyristor, *p-n-p-n* (Shockley) diode, *p-i-n* diode, and electrothermally driven devices.

1.2. THE MATHEMATICS OF INSTABILITIES

1.2.1. Stability and Bifurcation

One of the simplest ways of understanding the basic aspects of stability phenomena and nomenclature is to cite examples from mechanics (Haken, 1983a). The first is that of a particle of mass m, displaced a distance q from equilibrium, and simultaneously subjected to a driving force $F(q)$ and a frictional force $\gamma\, dq/dt$, where γ is a proportionality constant. The equation of motion of this system is

$$m\frac{d^2q}{dt^2} + \gamma\frac{dq}{dt} = F(q) \tag{1-2}$$

For purposes of illustrating stability questions, the overdamped case is considered, where $m\ddot{q} \ll \gamma\dot{q}$ and the second time derivative is ignored. [It is important to note that essential physics is lost in approximating a second-order differential equation by a first-order equation. (see, e.g., Cole, (1968).] Then scaling time through $t = \gamma t'$, Eq. (1-2) becomes

$$\frac{dq}{dt} = F(q) \tag{1-3}$$

where the prime has been dropped.

For conservative systems, by definition $F(q) = -d\Phi/dq$, where Φ is the potential. Hence, for the harmonic oscillator we have

$$F(q) = -kq \tag{1-4}$$

k being a constant, and

$$\Phi(q) = \tfrac{1}{2}kq^2 \tag{1-5}$$

plus an additive constant which has been arbitrarily set equal to zero. The potential function is sketched in Fig. 1-3. Note that $F(q)$, and hence dq/dt, vanish at $q = 0$. If a particle moves in such a potential it will be at rest, in equilibrium, at $q = 0$. If the particle is displaced from the point $q = 0$ it eventually returns to it; the position $q = 0$ is *stable.*

If the potential function had a different form, that of an anharmonic oscillator

$$F(q) = -kq - k_1q^3 \tag{1-6}$$

the equation of motion is

$$\frac{dq}{dt} = -kq - k_1q^3 \tag{1-7}$$

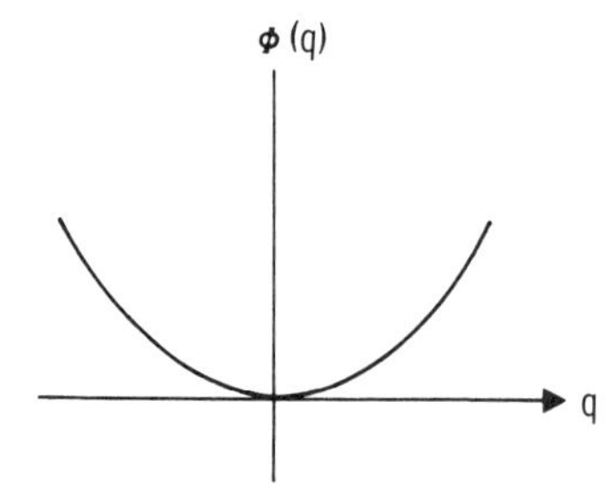

FIGURE 1-3. Potential Φ versus position q for a harmonic oscillator.

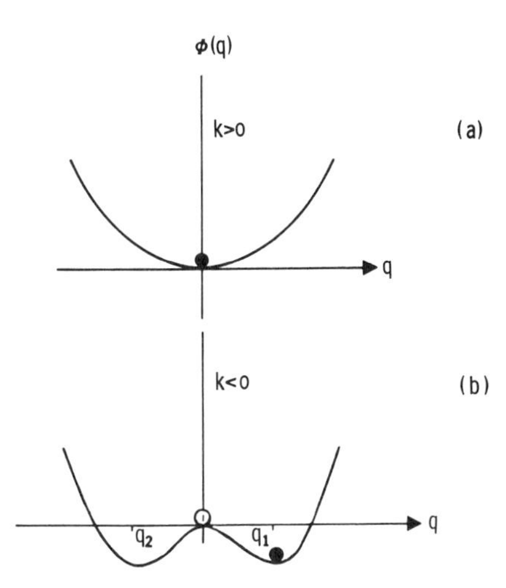

FIGURE 1-4. Potential Φ versus position q for an anharmonic oscillator $\Phi = kq^2/2 + k_1q^4/4$ with $k_1 > 0$ and (a) $k > 0$, (b) $k < 0$.

and the potential function is

$$\Phi(q) = \tfrac{1}{2}kq^2 + \tfrac{1}{4}k_1q^4 \tag{1-8}$$

which is plotted in Fig. 1-4 for two cases with $k_1 > 0$: (a) $k > 0$; (b) $k < 0$. Equilibrium points, q_e, occur when $dq/dt = 0$. For $k > 0$ there is only one stable solution, $q_e = 0$. For $k < 0$ there are three solutions, the two stable solutions $q_{e,1,2} = \pm\sqrt{|k|/k_1}$ and the unstable solution at $q_e = 0$. In Fig. 1-5 the equilibrium coordinate q_e is plotted as a function of k. For $k > 0$, the equilibrium coordinate is the straight line $q_e = 0$. The line represents a stable fixed point. For $k < 0$, the line $q_e = 0$ is unstable. The stable equilibrium is represented by the solid forked curve.

The graph of $q_e(k)$, the equilibrium position as a function of k (Fig. 1-5), has the form of a fork, and represents a *bifurcation*. Symmetry about the point $q = 0$ is maintained for $k < 0$ (as may be seen by replacing q and $\dot{q}$ by $-q$ and $-\dot{q}$), *but only in the sense that the positions available for equilibrium form a symmetrical pattern. In a physical problem right- or left-side equilibrium positions are outcomes determined by external influences, or internal statistical fluctuations, and once a choice is made symmetry is lost.* This is called "*spontaneous symmetry breaking.*" Thus, as k goes from positive to negative values, the stable point at $q = 0$ becomes unstable and is replaced by an unstable position and two stable

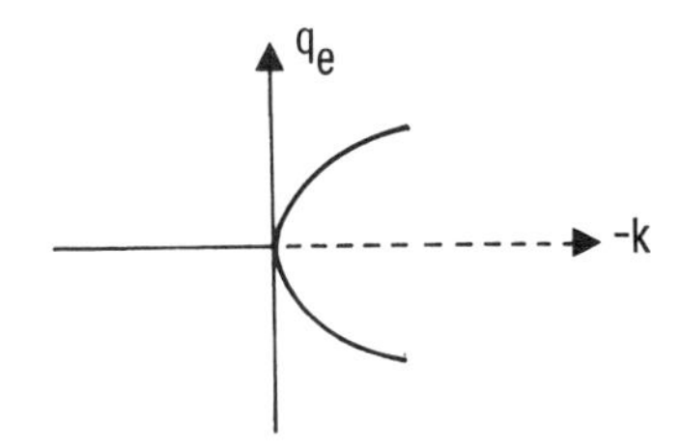

FIGURE 1-5. Equilibrium position of an anharmonic oscillator (see Fig. 1-4) as a function of the control parameter $-k$.

positions:

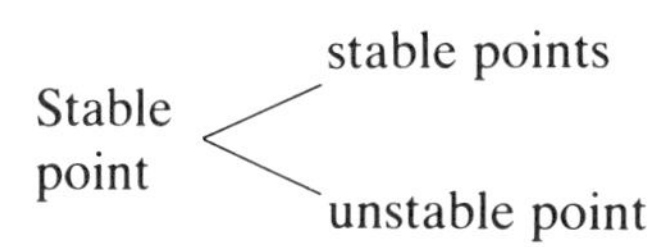

Equation (1-3) is readily generalized to n dimensions by writing

$$\frac{d\mathbf{q}}{dt} = \mathbf{F}(\mathbf{q}, \mathbf{k}) \tag{1-9}$$

where $\mathbf{q} \equiv (q_1, q_2, \ldots, q_n)$ is a formal n-dimensional vector of dependent variables, and $\mathbf{F} \equiv (F_1, F_2, \ldots, F_n)$ are n (generally nonlinear) functions of these n variables, which depend additionally upon m parameters, the so-called control parameters $(k_1, k_2, \ldots, k_m) \equiv \mathbf{k}$. Equations of this type are familiar in a variety of disciplines. Below we will use some examples from classical mechanics, laser physics, and electronics. The set of differential equations (1-9) is called a *dynamic system* (Andronov et al., 1973) because it determines the dynamic evolution of a system in time if the initial values of the dependent variables $\mathbf{q}(t_0)$ are specified. By varying the control parameters the system itself can be changed, and the nature of its solutions may change dramatically, as for example in the bifurcation discussed above. The larger the number of dependent variables, the richer is the possible dynamic behavior of the system. In going from one to two variables self-sustained oscillatory solutions become possible. These are called limit cycles if they possess certain stability properties. We shall elaborate on this later.

A special case of Eq. (1-9) occurs if the functions F_j do not depend explicitly on time. In this case the dynamic system (1-9) is called *autonomous,* and all the time dependence is through $\mathbf{q}(t)$. Examples are semiconductors under steady irradiation or dc bias.

Further special cases of Eq. (1-9) follow:

A *Hamiltonian* system is of the form

$$\frac{dx_i}{dt} = \frac{\partial H}{\partial p_i}, \qquad \frac{dp_i}{dt} = -\frac{\partial H}{\partial x_i}, \qquad i = 1, \ldots, 3N \tag{1-10}$$

where x_i and p_i are the generalized positions and the conjugate momenta, respectively, of N particles, and $H = H(x_i, p_i)$ is the Hamiltonian function. H is equal to the sum of the kinetic and potential energy expressed as functions of position and momenta. Equations (1-10) are the canonical equations of classical mechanics describing a conservative system of N mass points. For systems in which the energy E is conserved, $H(x_i, p_i) = E = \text{const}$. Thus $H(x_i, p_i)$ represents a first integral of Eq. (1-10), and the problem of finding a solution of Eq. (1-10) is automatically reduced by one integration step. As an example, consider a frictionless mass point in the potential $\Phi = \frac{1}{2}kx^2 + \frac{1}{4}k_1x^4$ [cf. Eq.

(1-8)] with the kinetic energy expressed as $p^2/2m$. The Hamiltonian is

$$E = \frac{p^2}{2m} + \tfrac{1}{2}kx^2 + \tfrac{1}{4}k_1x^4 \tag{1-11}$$

and the dynamic system is

$$\frac{dx}{dt} = \frac{p}{m}, \qquad \frac{dp}{dt} = -kx - k_1x^3 \tag{1-12}$$

[cf. (1-2) with $\gamma = 0$].

If Eq. (1-11) is then used to eliminate p from the first equation of (1-12), the general solution is an elliptic integral

$$t = \int_{x_0}^{x} [m/(2E - kx^2 - \tfrac{1}{2}k_1x^4)]^{1/2}\, dx \tag{1-13}$$

For a harmonic potential ($k_1 = 0$) the sinusoidal solution $x = (2E/k)^{1/2} \sin[(k/m)^{1/2}t + \phi_0]$ is, of course, obtained from Eq. (1-13). Hamiltonian systems are a subset of the more general class of *conservative* systems. These are general dynamic systems that possess a first integral.

In the following we shall generally be concerned with *dissipative systems.* These are dynamic systems in which energy is dissipated, e.g., via friction or Joule heating. A mathematical definition is given by the property that n-dimensional "volume elements" ΔV in the *phase space* of the variables $(q_1, \ldots, q_n)$ shrink as time increases: $\Delta V \to 0$ for $t \to \infty$. Therefore the dynamic "flow" $\mathbf{q}(t)$ is asymptotically attracted into bounded subsets of the phase space, into so-called *attractors.* They govern the longtime behavior of the dynamic systems. An example of an attractor is the steady state of the damped harmonic oscillator in Fig. 1-3.

A *gradient* system (or *potential* system) is described by the equation

$$\frac{dq_i}{dt} = -\frac{\partial}{\partial q_i} V(\mathbf{q}, \mathbf{k}), \qquad i = 1, \ldots, n \tag{1-14}$$

where $V(\mathbf{q}, \mathbf{k})$ is a potential function that combines all the information of the dynamic system. Hence, it is a straightforward generalization of a one-dimensional system which can always be written as $dq/dt = -dV(q, \mathbf{k})/dq$.

The gradient system should not be confused with the Hamiltonian system given in Eq. (1-10), where the derivatives are with respect to conjugate, rather than defined variables. In fact, as we shall see below, the structure of the phase portraits is completely different for the Hamiltonian system and the gradient system. In terms of the mechanical example of Eq. (1-2), the Hamiltonian (conservative) system and the gradient system represent two opposite approximations: the frictionless case corresponds to a conservative system [cf. Eq. (1-12)], whereas the overdamped case leads to a gradient system [cf. Eq. (1-3)].

Of more direct interest for us is the immediate discussion below of the two-dimensional autonomous system:

$$\frac{dq_1}{dt} = F_1(q_1, q_2, k)$$
$$\frac{dq_2}{dt} = F_2(q_1, q_2, k) \tag{1-15}$$

Such systems are conveniently discussed in the (q_1, q_2) *phase plane* of dependent variables. For every *phase point* (q_1, q_2), except for *singular* or *fixed points* (q_1^0, q_2^0) where

$$F_1(q_1^0, q_2^0, k) = F_2(q_1^0, q_2^0, k) = 0 \tag{1-16}$$

the differential equations (1-15) give a unique direction of motion (F_1, F_2) in the phase plane. (Note that the uniqueness property is lost for nonautonomous systems, since F_1 and F_2 may have different values for the same values of q_1 and q_2 at different times, owing to the explicit dependence on t.) By plotting the vector field (F_1, F_2) in the whole phase plane, one can construct a *field of directions* that is tangent to the curve or *trajectory* along which the system moves in the phase plane. For a given initial phase point $(q_1(0), q_2(0))$ a unique trajectory can be graphically constructed from this field of directions. By choosing other initial points other trajectories are obtained and a phase portrait emerges, similar to the streamlines of a fluid. Because of the uniqueness of the field of directions a trajectory can never cross itself or other trajectories. (This is not true for nonautonomous systems!) The graphical construction will be illustrated for the Van der Pol oscillator in Sec. 1.2.2.

A simple way of obtaining a qualitative phase portrait is by drawing the two curves $F_1(q_1, q_2) = 0$ and $F_2(q_1, q_2) = 0$, the so-called null-isoclines. On the $F_1 = 0$ curve dq_1/dt vanishes and $dq_1/dq_2 = 0$, hence the phase plane trajectories cross this curve parallel to the q_2 axis. Similarly, on the $F_2 = 0$ curve dq_2/dt and dq_2/dq_1 vanish, hence the trajectories cross parallel to q_1 axis. This, together with the determination of the sign of dq_1/dt on either side of the $F_1 = 0$ curve, and of dq_2/dt on either side of the $F_2 = 0$ curve, yields a qualitative idea of the phase portrait. An example, corresponding to a semiconductor laser, is shown in Fig.

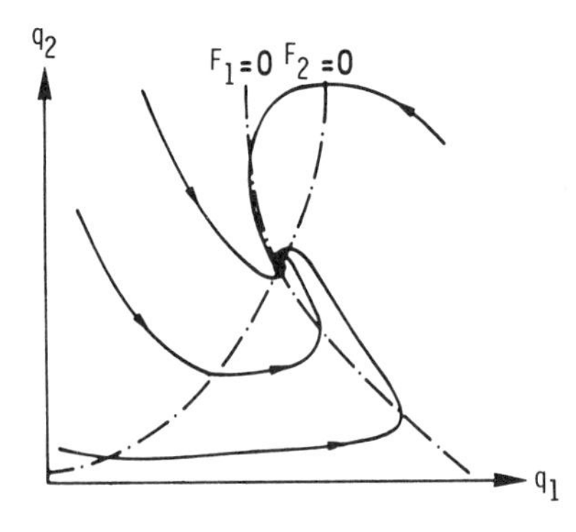

FIGURE 1-6. Phase portrait of a semiconductor laser. The coordinates q_1 and q_2 represent electron and photon concentrations, respectively. Different trajectories correspond to different initial conditions. The dash-dotted lines are the null isoclines. (After Schöll and Landsberg, 1983.)

1-6. Another important example of such a phase portrait analysis will be given in Chap. 2, when we discuss NNDC characteristics.

The intersections of the two null-isoclines are, by Eq. (1-16), the singular points, and represent time-independent *steady states* ($d\mathbf{q}/dt = 0$). They are sometimes also called equilibrium states, which is, however, misleading (since they are usually states far from thermal equilibrium!) and should be avoided.

An analytical expression for the geometrical form of the trajectories can be obtained by eliminating time from Eq. (1-15), assuming $F_1 \neq 0$:

$$\frac{dq_2}{dq_1} = \frac{F_2(q_1, q_2, k)}{F_1(q_1, q_2, k)} \quad \text{or} \quad \frac{dq_1}{dq_2} = \frac{F_1(q_1, q_2, k)}{F_2(q_1, q_2, k)} \quad \text{for } F_2 \neq 0 \tag{1-17}$$

The solution of the differential equation (1-17) is the phase plane trajectory. In order to obtain the actual time dependence $(q_1(t), q_2(t))$ as the system moves along a trajectory, we must perform one further integration. In the example of the undamped anharmonic oscillator [Eq. (1-12)] the differential equation (1-17) becomes

$$\frac{dp}{dx} = -\frac{m}{p}(kx + k_1x^3) \tag{1-18}$$

and the phase plane trajectories are closed curves of constant energy:

$$\frac{p^2}{2m} + \tfrac{1}{2}kx^2 + \tfrac{1}{4}k_1x^4 = E \tag{1-19}$$

which is Eq. (1-11). Here the energy E specifies the initial condition up to a phase angle (Fig. 1-7). Note that for $k < 0$ there are three singular points, and the oscillations around either of the two extreme points correspond to oscillations around the stable minima of the potential in Fig. 1-4b. In the special case of a harmonic oscillator ($k_1 = 0$, $k > 0$) the trajectories are concentric ellipses.

1.2.1.1. Stability

In the mechanical example discussed above the change of stability of the central steady state plays a crucial role for the nature of the phase portraits and

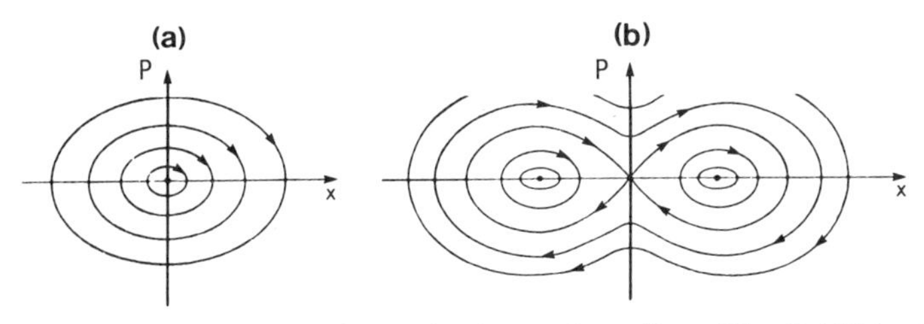

FIGURE 1-7. Phase portraits of an undamped anharmonic oscillator [Eq. (1-11)] for $k > 0$ (a) and $k < 0$ (b).

for the bifurcation behavior. A more general discussion of stability (Hahn, 1967) within the framework of the general dynamic system given by Eq. (1-9) follows.

Any solution $\mathbf{q} = \mathbf{u}_{\mathbf{q}_0}(t)$, which may be regarded as the path of a particle in $\mathbf{q}$ space, is uniquely determined by its initial value $\mathbf{q}_0$ at the initial time $\mathbf{t}_0$. Since the initial values are normally subject to perturbations, the question of interest becomes: What happens to the path of the particle if the initial condition is slightly different from the initial choice? A trajectory is taken to be *locally stable* if other trajectories, obtained from initial conditions, different from but close to $\mathbf{q}_0$, remain close to the original trajectory for all later times as both phase points move about in the phase plane. If a neighborhood of $\mathbf{q}_0$ cannot be found such that all initial departures within this neighborhood satisfy this criterion, the trajectory $\mathbf{u}_{\mathbf{q}_0}$ is unstable. It is *locally asymptotically stable* if for all trajectories $\mathbf{v}_{\mathbf{q}_0'}$, close to $\mathbf{u}_{\mathbf{q}_0}$ whose phase points fulfill the stability criterion, the condition

$$|\mathbf{u}_{\mathbf{q}_0}(t) - \mathbf{v}_{\mathbf{q}_0'}(t)| \rightarrow 0 \qquad \text{as } t \rightarrow \infty$$

also holds. These definitions include the special, but very important, case that $\mathbf{u}$ consists of a single (singular!) point, i.e., a steady state.

A weaker form of stability, which is important when limit cycles (defined below) are considered, is *orbital stability*. It involves only the geometric form of the trajectories in phase space, and not the distance of neighboring phase points on these trajectories at fixed times. A trajectory $\mathbf{u}$ is orbitally stable if all phase points sufficiently close to $\mathbf{u}$ at some time t_0 remain within a certain given distance from the curve $\mathbf{u}$ for all later times. Orbital stability allows for a phase lag between phase points on the original and neighboring trajectories. *Asymptotic orbital stability* is then defined by requiring additionally that the distance between $\mathbf{u}$ and an arbitrary phase point $\mathbf{q}(t)$ on a neighboring trajectory tends to zero as $t \rightarrow \infty$.

Global stability requires that not only neighboring phase points, but all points of the relevant phase plane meet the above conditions of local stability. Thus, there cannot exist more than one globally stable trajectory. Note, however, that a phase plane may contain several locally stable trajectories or singular points. In Hamiltonian and in gradient systems global stability can readily be determined by looking at the form of the potential curves. In many systems that do not have a potential a "Lyapunov function" (a generalization of a potential) can be found (Haken, 1983a).

Yet another notion is that of structural stability. It involves not the evolution of an individual phase point as time increases, but rather the change of the whole field of trajectories as the control parameters are varied. A dynamic system is *structurally stable* if the topological structure of the phase portrait is not affected by small changes in the differential equations. The anharmonic oscillator discussed at the beginning of this section is certainly not structurally stable at the bifurcation point $k = 0$, since the phase portraits change qualitatively from $k > 0$ to $k < 0$ (cf. Fig. 1-7). A more precise mathematical definition is given by Thom (1975) and Andronov et al. (1966).

Finally, as a special case, which will be needed in later chapters, the local asymptotic stability of a singular point (fixed point) of a two variable autonomous system is examined in some more detail. In the vicinity of a singular point (q_1^0, q_2^0) the dynamic system given by Eq. (1-15) may be linearized around (q_1^0, q_2^0). In this approximation the trajectories near a singular point can be computed explicitly. Setting

$$\xi_1 = q_1 - q_1^0, \qquad \xi_2 = q_2 - q_2^0 \tag{1-20}$$

we obtain from (1-15)

$$\frac{d\boldsymbol{\xi}}{dt} = A(q_1^0, q_2^0)\boldsymbol{\xi} \tag{1-21}$$

where the Jacobian matrix $A(q_1^0, q_2^0)$ is defined by

$$A_{ij} \equiv \frac{\partial F_i}{\partial q_j}, \qquad i, j = 1, 2 \tag{1-22}$$

(Note that this definition can easily be generalized to arbitrary dimensions.) The general solution of the system of linear differential equations (1-21) is

$$\boldsymbol{\xi}(t) = c_1\boldsymbol{\eta}_1 e^{\lambda_1 t} + c_2\boldsymbol{\eta}_2 e^{\lambda_2 t} \tag{1-23}$$

where c_1, c_2 are constants, $\boldsymbol{\eta}_1, \boldsymbol{\eta}_2$ are eigenvectors of A, and λ_1, λ_2 are the eigenvalues of A, given by

$$\lambda^2 - T\lambda + D = 0 \tag{1-24}$$

or

$$\lambda = \tfrac{1}{2}[T \pm (T^2 - 4D)^{1/2}] \tag{1-25}$$

Here

$$T \equiv A_{11} + A_{22} \quad \text{and} \quad D \equiv A_{11}A_{22} - A_{12}A_{21} \tag{1-26}$$

are the trace and the determinant of the matrix A, respectively. For an asymptotically stable fixed point ("sink" or *point attractor*) all trajectories (1-23) are required to approach it as $t \to \infty$, hence the real parts of both λ_1 and λ_2 must be negative.

The five qualitatively different regimes of solutions [Eq. (1-23)] are shown in Fig. 1-8.

(*a*) *Focal Stability*: $D > 0$, $T < 0$, $T^2 < 4D$. The eigenvalues are complex conjugate and have negative real parts. The solutions are damped and complex, causing $\boldsymbol{\xi}$ to execute damped oscillations in the phase plane with angular frequency $\omega = (D - \frac{1}{4}T^2)^{1/2}$ and damping constant $\frac{1}{2}T$. The general form of the trajectory is an elliptical spiral converging on the origin. A weakly damped harmonic oscillator is an example, where ξ_1 and ξ_2 correspond to position and momentum, respectively.

(*b*) *Focal Instability*: $D > 0$, $T > 0$, $T^2 < 4D$. The eigenvalues are complex and have positive real parts; here the elliptical spiral diverges. A harmonic

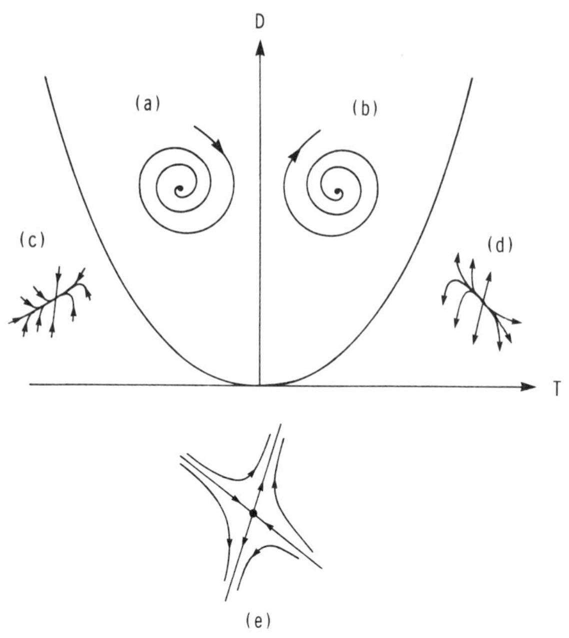

FIGURE 1-8. Regimes of solutions of a two-variable autonomous dynamic system in the neighborhood of a singular point. T and D denote the trace and the determinant, respectively, of the Jacobian matrix defined in (1-22). The insets show the qualitative nature of the plane portraits near the singular point for the five generic cases (a)–(e), corresponding to (a) stable focus; (b) unstable focus; (c) stable node; (d) unstable node; (e) saddle point. A *center* (Fig. 1-7a) is a nongeneric case and can occur on the line $T = 0$, $D > 0$ only.

oscillator with enough positive feedback to overcome frictional losses and provide gain is an example.

(*c*) *Nodal Stability*: $D > 0$, $T < 0$, $T^2 > 4D$. Both eigenvalues are negative. The solutions are real and the oscillations are replaced by an exponential decay. The overdamped harmonic oscillator is of this type.

(*d*) *Nodal Instability*: $D > 0$, $T > 0$, $T^2 > 4D$. Both eigenvalues are positive; the trajectories diverge from the origin. A harmonic oscillator with very strong positive feedback is of this type.

(*e*) *Saddle Point Instability*: $D < 0$. The eigenvalues are real, with one value positive and the other negative. The positive value causes an instability in the direction of the corresponding eigenvector in phase space. An example is a ball sitting on top of a hill, on the local maximum of the anharmonic potential in Fig. 1-4. It will roll off the hill once it is infinitesimally disturbed.

The interfaces between these five regimes obviously represent structurally unstable cases. They require more detailed investigations, involving higher orders in the Taylor expansion of the system around the singular point (Andronov et al., 1973). In particular, for $T = 0$ and $D > 0$ the trajectories near the singular point are either ellipses (as in the example of an undamped harmonic oscillator, cf. Fig. 1-7a), or slowly converging or diverging spirals (depending on the nonlinear terms of the expansion). In the first case the singular point is called a center, in the second a weakly stable/unstable focus.

The interfaces between regimes (a) and (b), and between (c) and (e) are associated with various bifurcations which we shall study below.

We note that by Eqs. (1-14), (1-22), and (1-26) a two-variable gradient system always satisfies

$$T^2 - 4D = \left(\frac{\partial^2 V}{\partial q_1^2} - \frac{\partial^2 V}{\partial q_2^2}\right)^2 + 4\left(\frac{\partial^2 V}{\partial q_1\, \partial q_2}\right)^2 > 0 \tag{1-27}$$

and hence the cases (a) and (b) (focus) cannot occur. In case (c) (node) the potential V has a minimum at the singular point, in case (d) a maximum, while in (e) it is saddle-shaped. On the other hand, a conservative system fulfills [by Eqs. (1-10), (1-22), (1-26)] $T = 0$, and only saddle points (case e) or centers can occur (cf. Fig. 1-7).

1.2.1.2. Bifurcation

Bifurcation is a phenomenon peculiar to nonlinear dynamic systems and is closely related to the loss of stability (Sattinger, 1973). It describes the branching of solutions (steady states, oscillatory or spatially nonuniform solutions) as a control parameter k of the system, described by Eq. (1-9), is varied. When different solution branches intersect or coalesce, they usually change their stability character. For steady states this means that at least one of the eigenvalues of the Jacobian A [Eq. (1-22)] has a vanishing real part. Also, at the bifurcation point k_0 the whole system is structurally unstable.

In the following a topological classification of the most important types of bifurcations is given. Although most of these bifurcations are more general, the restriction here is, except for case F, to autonomous two-variable systems (Andronov et al., 1971), whose behavior can readily be visualized in the corresponding phase portraits.

A. Zero Eigenvalue Bifurcations. Local bifurcations (Guckenheimer and Holmes, 1983) can be completely characterized by the phase portrait near the bifurcating singular point. All the information about the bifurcation is contained in the linearized system given by the matrix A in Eq. (1-22). The simplest bifurcation occurs if one simple real eigenvalue λ of A changes from negative to positive values upon variation of the control parameter k. At the bifurcation value, k_0, the eigenvalue is zero, and the singular point does not fall in any one of the simple categories (a)–(e) of Fig. 1-8. Rather, it is a *multiple singular point,* which is generated by the coalescence of several simple singular points. From Eq. (1-24) it follows that $\lambda = 0$ results in $D = 0$, while $T \neq 0$. Thus, the bifurcation is described by a simple analytical condition for the Jacobian matrix A of a two-variable system. In the physical literature this type of bifurcation is often called a *soft mode instability* because $\lambda \to 0$ implies that the corresponding linear mode softens, i.e., slows down critically in its temporal evolution. In the next chapter, in Sec. 2.2.2, we shall find an explicit example of such zero eigenvalue bifurcations when we study the circuit response for an NNDC element.

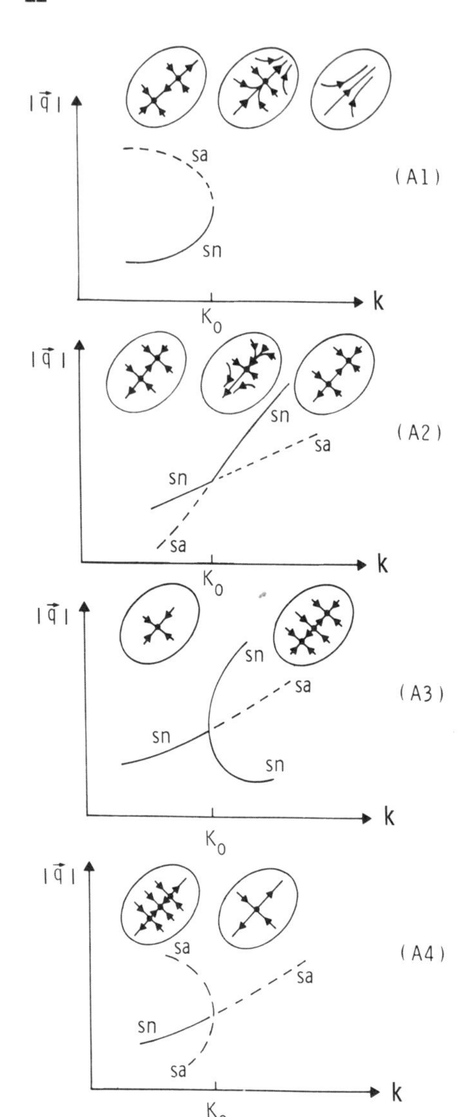

FIGURE 1-9. Bifurcation diagrams for a soft-mode instability (zero eigenvalue bifurcation). The singular points are plotted versus the control parameter k. Solid lines represent stable fixed points; broken lines are unstable. Saddle points and stable nodes are denoted by sa and sn, respectively. The insets represent schematic phase portraits corresponding to $k < k_0$, $k = k_0$, and $k > k_0$, where k_0 is the bifurcation point. (The four different cases, A1–A4, are described in the text.)

The following four subclasses may occur (Fig. 1-9):

1. Saddle-node bifurcation: A saddle point (sa) and a stable node (sn) coalesce and disappear. At the bifurcation point they form a so-called saddle node. This is typical of threshold switching and SNDC (Fig. 1-9, A1).

2. Transcritical bifurcation: A saddle point and a stable node coalesce and separate again. Often the unstable branch for $k < k_0$ turns back again in a saddle-node bifurcation of type (1), or it lies in a nonphysical regime of $\mathbf{q}$ space (Fig. 1-9, A2).

3. Supercritical pitchfork bifurcation: A saddle point and two stable nodes coalesce. This is typical of a symmetry-breaking bifurcation, as shown in Fig. 1-5 (Fig. 1-9, A3).

4. *Subcritical pitchfork bifurcation*: Two saddle points and a stable node coalesce. Often the two unstable branches for $k < k_0$ turn back again in a saddle-node bifurcation (Fig. 1-9, A4).

B. *Hopf Bifurcation of Limit Cycles*. If a pair of conjugate complex eigenvalues of A cross the imaginary axis, the singular point changes from a stable to an unstable focus, and a time-dependent periodic solution bifurcates from the steady state solution branch with zero amplitude (Hopf, 1942), as shown in Fig. 1-10. This periodic solution is asymptotically orbitally stable, i.e., all trajectories starting in a neighborhood (inside and outside) of this closed trajectory in the phase plane will asymptotically approach it. It is therefore called a *periodic attractor* or a stable *limit cycle*. (The inverted configuration with an unstable limit cycle and a stable focus for $k_0 < 0$, and an unstable focus for $k_0 > 0$, also exists.)

A system that possesses a limit cycle is structurally stable away from the bifurcation points. These two properties—orbital and structural stability—distinguish a limit cycle from conservative oscillations around a center in conservative systems. The latter are also self-sustained oscillations, but not asymptotically orbitally stable, since there exists an infinite continuum of neighboring oscillations in the phase plane (see, e.g., the undamped anharmonic oscillator in Fig. 1-7a). Also, conservative oscillations are not structurally stable since a damping term—however small it may be— destroys the closed trajectories and converts the center into a stable focus. For these two stability reasons, limit cycles are extremely important in practical devices. We shall encounter specific examples of this behavior in the next section and in Chap. 2.

The Hopf bifurcation may be supercritical with the bifurcation scheme

$$\text{Stable focus} < \begin{matrix} \text{stable limit cycle} \\ \\ \text{unstable focus} \end{matrix} \tag{1-28}$$

or subcritical with the bifurcation scheme

$$\text{Unstable focus} < \begin{matrix} \text{unstable limit cycle} \\ \\ \text{stable focus} \end{matrix} \tag{1-29}$$

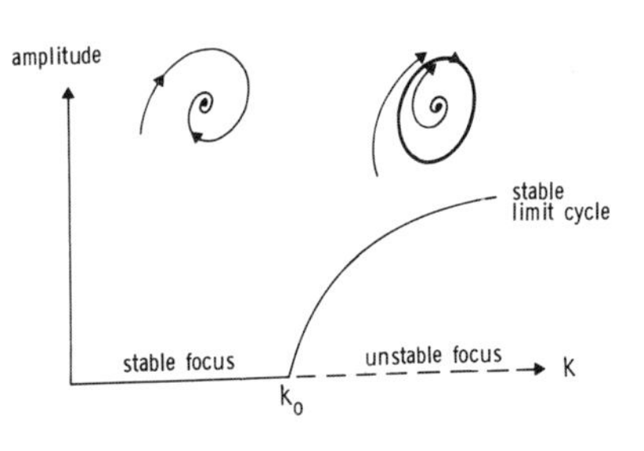

FIGURE 1-10. Bifurcation diagram for a Hopf bifurcation of a limit cycle. The amplitude of the limit cycle oscillation is plotted versus the control parameter k. The insets show schematic phase portraits for $k < k_0$ (stable focus) and $k > k_0$ (unstable focus and stable limit cycle).

The bifurcation point is given, cf. (1-24), by the condition $T = 0$, $D > 0$; it corresponds to a weakly stable or unstable focus, from which the limit cycle bifurcates at frequency $\omega = \sqrt{D} \neq 0$. Since the frequency does not tend to zero, i.e., does not "soften," the bifurcation is sometimes called a *hard mode instability*.

C. Limit Cycle Bifurcations by Condensation of Paths. If a stable and an unstable limit cycle are around the same singular point (e.g., a stable focus), they can coalesce and disappear (Fig. 1-11). At the bifurcation both the amplitude and the frequency are nonzero, as opposed to the Hopf bifurcation (b).

D. Bifurcation of Limit Cycles from a Separatrix. Such bifurcations are called *global* bifurcations, since they do not merely involve the neighborhood of a singular point, but a larger region of phase space. Two possible topologies are shown in Fig. 1-12. In Fig. 1-12a a saddle point and a stable node, which lie on a closed curve formed by two trajectories ("separatrices"), coalesce and disappear in a saddle-node bifurcation whereby a stable limit cycle bifurcates from the separatrix.

In Fig. 1-12b a saddle-to-saddle loop is formed from which the limit cycle bifurcates. Here the saddle point does not disappear. Since the limit cycle appears "out of the blue sky," this bifurcation has also been termed *blue-sky catastrophe* (Thompson and Stewart, 1986).

In both cases the limit cycle appears immediately with a nonzero amplitude, but with zero frequency. Thus the bifurcation is associated with a critical slowing down of the oscillation frequency according to a universal scaling law. Such bifurcations have indeed been observed experimentally (Peinke et al., 1989) and modeled theoretically (Hüpper et al., 1989) in semiconductors in the regime of low-temperature impurity breakdown.

E. Continuous Bifurcation. In dynamic systems allowing for diffusion and/or drift, i.e., containing spatial derivatives, there is yet another type of bifurcation: the bifurcation of nonuniform spatial structures ("dissipative structures") from a uniform steady state. The condition for bifurcations is obtained in a similar way as for uniform structures by linearizing the dynamic system around the uniform steady state and using the ansatz

$$\xi \sim \exp(\lambda t + i\mathbf{K}\mathbf{x}) \tag{1-30}$$

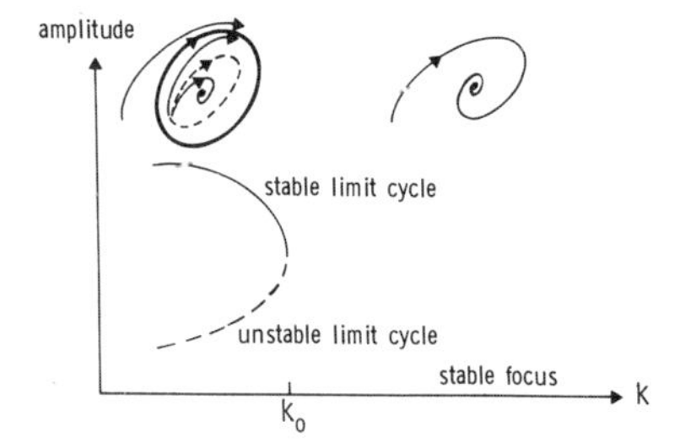

FIGURE 1-11. Bifurcation diagram for a limit-cycle bifurcation by condensation of paths. The insets show a stable (—) and an unstable limit cycle (- - -) around a stable focus for $k < k_0$, and a stable focus for $k > k_0$.

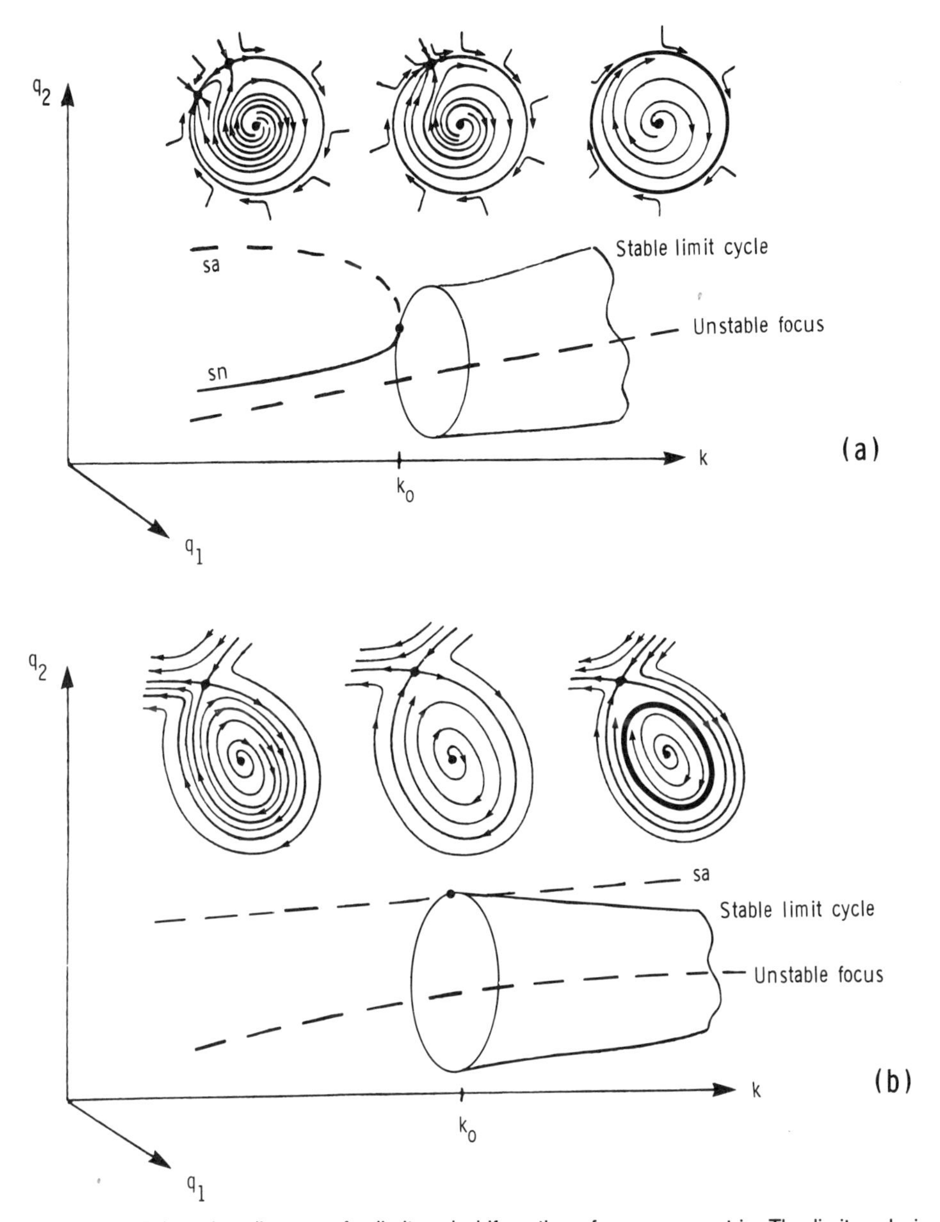

FIGURE 1-12. Bifurcation diagrams for limit-cycle bifurcations from a separatrix. The limit cycle is represented by a cylindrical surface in the three-dimensional space of dynamic variables (q_1, q_2) and control parameter (k). (The different cases, a and b, are described in the text.)

where **K** is a wave vector and **x** the spatial coordinate. From this a dispersion relation $\lambda(\mathbf{K})$ is obtained. It describes the damping out of small periodic spatial fluctuations if the real part of $\lambda(\mathbf{K})$ is negative. If Re $\lambda(\mathbf{K})$ changes sign at some value of **K**, a spatially periodic solution with the wave vector **K** bifurcates from the uniform steady state (Kirchgässner, 1977; Büttiker and Thomas, 1981). Upon variation of the control parameters the dispersion relation $\lambda(\mathbf{K})$ is changed, and a whole family of spatially nonuniform solutions bifurcates. The family is specified by appropriate boundary conditions and may include a "solitary solution" of infinite period as a limit case.

If at the bifurcation point $\lambda(\mathbf{K})$ is zero, the bifurcating solutions are time independent. An important example is the current filamentation associated with SNDC. We shall treat this case in Chap. 2.

If at the bifurcation point $\lambda(\mathbf{K})$ is purely imaginary, a traveling or standing wave bifurcates. Such nonlinear propagating waves may be periodic, or solitary, like the high electric field domains occurring in the Gunn–Hilsum effect. This can lead to current oscillations governed by the transit time of a domain between cathode and anode.

F. Transition to Chaos. First, we shall clarify some of the basic notions and ideas of *deterministic* chaos. In two-variable autonomous dynamic systems, such as we have encountered so far, the field of directions of the phase portrait is uniquely defined, and therefore trajectories cannot cross. All trajectories in a bounded dissipative system must tend asymptotically to a singular point, or a limit cycle, i.e., a zero- or one-dimensional attractor (Andronov et al., 1973). Now, we shall consider nonautonomous (that is, explicitly time-dependent) two-variable systems, or more generally, autonomous systems involving at least three variables. Note that nonautonomous systems

$$\dot{\mathbf{q}} = \mathbf{F}(\mathbf{q}, t)$$

can always be represented as autonomous systems with time as an additional dynamic variable:

$$\dot{\mathbf{q}} = \mathbf{F}(\mathbf{q}, \theta)$$
$$\dot{\theta} = 1$$

These systems may have completely irregular trajectories which do not tend asymptotically to a singular point or a limit cycle. Such irregular motion of a deterministic system with few degrees of freedom, which depends very sensitively upon the initial conditions, is called *chaotic*. It is distinct from the irregularities caused by stochastic fluctuations in systems with many microscopic degrees of freedom.

The third variable adds a whole wealth of qualitatively new phenomena of the dynamic system, in addition to the singular points and limit cycles familiar from two-variable autonomous systems, where the trajectories in the phase plane can never cross. In three dimensions, however, the trajectories can be wrapped or twisted around each other in quite intricate ways. Consider, for example, the combination of a two-dimensinal unstable focus with a "bistable slow manifold" folded up in the third dimension as shown in Fig. 1-13 (Rössler, 1977). The "slow manifold" is a surface in which the phase point moves slowly, whereas in the rest of the phase space the motion is fast. Thus, when a trajectory reaches the edge of the upper sheet of the slow manifold, it suddenly jumps down to the lower sheet, and when it reaches the edge of the lower sheet, it jumps up to the upper sheet.

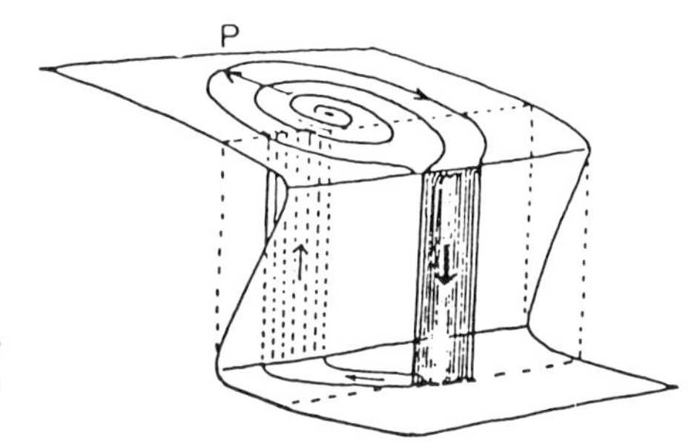

FIGURE 1-13. Chaotic motion produced by the combination of a slow bistable manifold with an unstable focus (after Rössler, 1977).

The reinjection into the upper sheet depends sensitively on its position before the jump, which explains (intuitively) the origin of the seemingly random jumps. This produces the complicated entangling of trajectories that is characteristic of chaotic motion.

The *spiral-type chaos* discussed above represents the simplest form of chaotic behavior in continuous dissipative dynamic systems. It is called a *strange attractor*. A strange attractor can be characterized as follows: It consists of a set of phase points embedded in a finite region of phase space. Trajectories outside that region but close enough are attracted to that region. Trajectories within that region will remain in it. The "strangeness" of the attractor consists in the fact that it is neither a singular point (dimension zero) nor a limit cycle (dimension one) nor a torus (dimension two), but has in general a noninteger "fractal" dimension. Trajectories that are initially very close to each other on the attractor separate exponentially fast as time goes on. Intuitively, this implies unstable motion *within* the attractor.

The onset of chaos is usually preceded by other bifurcations, like those discussed above. For example, with increasing value of the control parameter, the system may first exhibit a Hopf bifurcation of a limit cycle, and then, at still higher values of the control parameter, a transition to chaos. Different routes to chaos have been found in different physical systems. In the following we discuss some of the most important scenarios (Haken, 1983b; Guckenheimer and Holmes, 1983; Eckmann, 1981). They represent sequences of global bifurcations leading to chaos.

First, we consider the transition to chaos by *period doubling*, or *subharmonic generation*. When a limit cycle embedded in a three-dimensional phase space becomes unstable, it may bifurcate into another limit cycle on which the system takes twice the time to return to its original state. That is, the period has doubled, or the original frequency has been cut by half. The projection of the phase trajectory onto a plane is shown schematically in Fig. 1-14. Upon further increase of the control parameter another period doubling bifurcation occurs, so that the original period T is multiplied by 4, and so on. Thus, a whole sequence of period doubling bifurcations occurs at successive values of the control parameter k_m, $m = 1, 2, 3, \ldots$, with period $2^m T$. For $m \to \infty$ the k_m converge to a critical value k_∞, at which the motion ceases to be periodic, and becomes chaotic.

A convenient description of period doubling and other nonlinear dynamic effects is in terms of discrete iterated maps (Collet and Eckmann, 1980). These arise if the set of variables $\mathbf{q}(t)$ is not taken as a function of the continuous time t,

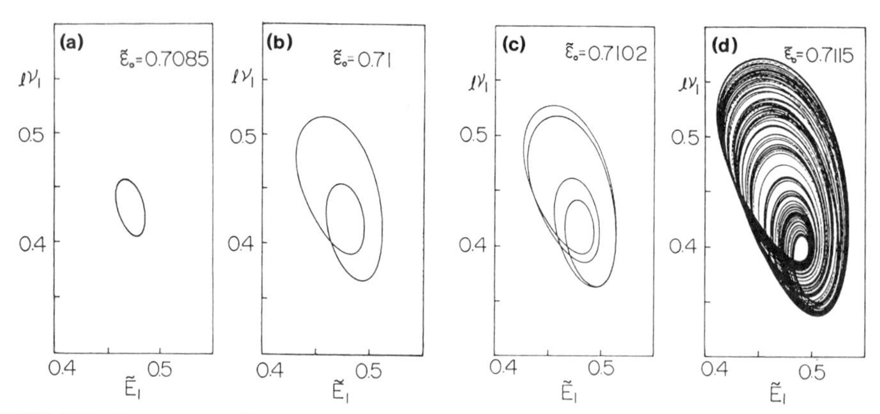

FIGURE 1-14. A sequence of period doublings of a limit cycle in a two-dimensional phase plane for different control parameters $\tilde{\epsilon}_0$. (a), (b), and (c) correspond to periods T, $2T$, and $4T$, respectively; (d) corresponds to chaos. (After Aoki et al., 1989.)

but instead is considered only at a certain discrete sequence of times, $t_0, t_1, t_2, t_3, \ldots$. The sequence $\mathbf{q}_m \equiv \mathbf{q}(t_m)$ can be constructed, for example, by taking the transverse intersections of the trajectories in n-dimensional phase space with an $(n - 1)$-dimensional hypersurface. This is called a *Poincaré section.** For example, as the Poincaré section of a two-dimensional system oscillating around the origin, we may choose the intersections with one of the coordinate half-axes, x_m. Each intersection point x_{m+1} is determined, once the previous point x_m is fixed. Therefore, a relation $x_{m+1} = f(x_m)$ exists: a one-dimensional discrete map (*Poincaré map,* or first return map). A limit cycle is then represented by a single point of intersection x_∞, to which the intersection points of all neighboring phase trajectories converge for $m \rightarrow \infty$. The period-doubling bifurcation sequence corresponds to a cascade of pitchfork bifurcations of the Poincaré map (Fig. 1-15). The cascade of bifurcation values k_m has been shown to follow a universal law

$$\lim_{m\to\infty} \frac{k_{m+1} - k_m}{k_{m+2} - k_{m+1}} = \delta \equiv 4.6692016 \cdots$$

for a large class of one-dimensional maps with quadratic maxima (Feigenbaum, 1978). Beyond k_∞, the chaotic regime can be interrupted by narrow bands of k-values in which the motion is again periodic. In such "periodic windows" further period-doubling cascades may occur. A variety of universal features and scaling laws have been established for these one-dimensional maps (Collet and Eckmann, 1983).

A second route to chaos is the *quasi periodic* breakdown. A limit cycle with a single frequency ω may undergo a second Hopf bifurcation leading to a doubly periodic motion which is characterized by two fundamental frequencies ω_1, ω_2. It

* The phase space flow through the Poincaré surface induces an $(n - 1)$-dimensional invertible map $\mathbf{q}_m \rightarrow \mathbf{q}_{m+1}$, the *Poincaré map,* whereby the differential equations are replaced by difference equations.

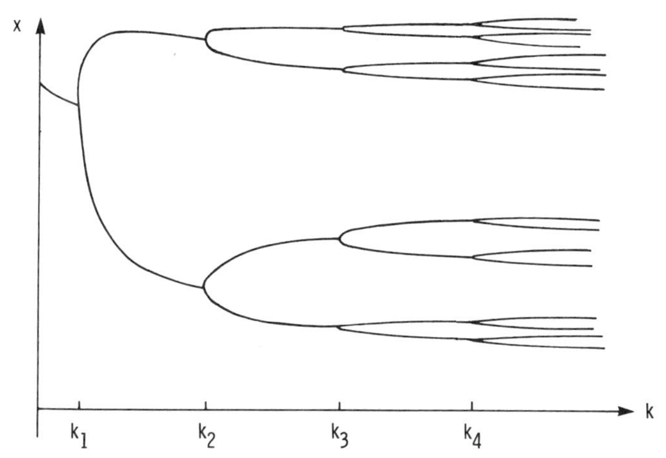

FIGURE 1-15. Bifurcation diagram of a one-dimensional map (Poincaré sections of limit cycles). The dynamic variable x is plotted versus the control parameter k (logarithmic scale). There is a sequence of period-doubling bifurcations at k_1, k_2, k_3, and k_4. (After Collet and Eckmann, 1980.)

is convenient to parametrize the phase point by the two angular coordinates $\phi_1 = \omega_1 t$ and $\phi_2 = \omega_2 t$. The doubly periodic motion can thus be visualized as the motion on a two-dimensional torus (Fig. 1-16). If the ratio ω_1/ω_2 is rational, the trajectory is closed, as shown in Fig. 1-17. If, on the other hand, ω_1 and ω_2 are incommensurate (i.e., of irrational ratio), the motion is called *quasiperiodic* and the trajectory fills up the entire torus. A subsequent bifurcation may either take the system back to a limit cycle with a single frequency (which is called *frequency-locking* or *mode-locking*), or add a third fundamental frequency ω_3 (which transforms the two-dimensional torus into a "three-dimensional torus"). A condition for the latter is that the frequencies are sufficiently irrational with respect to each other. A more rigorous mathematical definition is given by the so-called KAM condition (Haken, 1983b).

Another, more likely possibility of transition to chaos (the Ruelle–Takens scenario) is that the three-dimensional torus becomes unstable and decays to a *strange attractor*. The observed quasiperiodic behavior of many different dissipative systems obeys a universality which can be most easily studied in terms of an iterated map, viz., the *circle map* (Schuster, 1987).

$$\theta_{n+1} = \theta_n + \Omega - (K/2\pi)\sin(2\pi\theta_n)$$

where θ is the iterated angular variable on the circle (modulo 1), and the

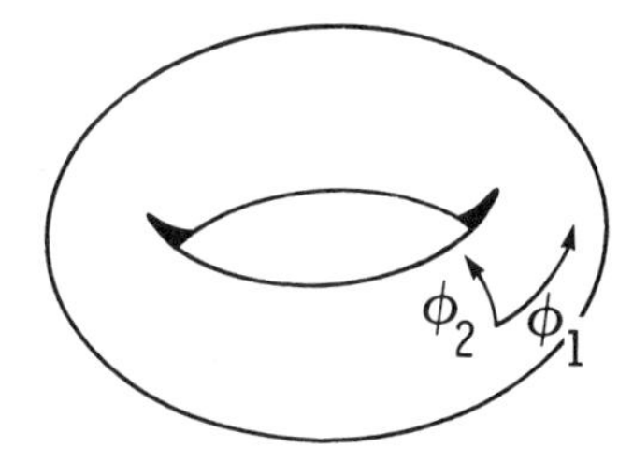

FIGURE 1-16. Two-dimensional torus parametrized by two angular coordinates ϕ_1, ϕ_2.

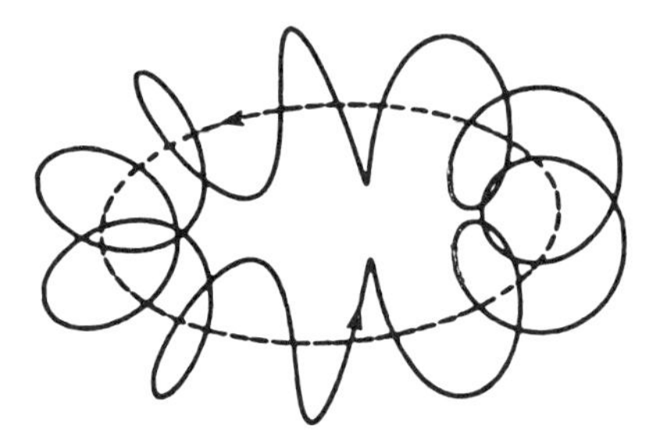

FIGURE 1-17. Limit cycle, representing mode-locked motion on a two-dimensional torus (after Haken, 1983b).

parameters Ω and K denote the frequency ratio and the coupling strength of the two competing oscillatory modes, respectively. Increasing the strength of the nonlinear coupling between the oscillators develops an increasing tendency to lock into commensurate motion where the ratio of the oscillation frequencies is rational. The resonant frequency-locked states form hornlike structures within the (Ω, K) control parameter plane known as *Arnold tongues.* They are ordered hierarchically according to the *Farey tree,* which orders all rationals in the interval [0, 1] with increasing denominator after the rule that the rational with the smallest denominator between p/q and p'/q' is $(p + p')/(q + q')$, where p, p', q, q' are integers. The Farey tree has been observed in Si p-i-n diodes (Chap. 7) at 4.2 K (Coon et al., 1987). Such behavior was also obtained in a model for spatially coupled impact-ionization induced oscillations at low temperatures (Naber and Schöll, 1990a). At the critical line $K = 1$, the mode-locking intervals form a characteristic self-similar structure (complete *devil's staircase*), i.e., it looks the same in any scale of magnification, and the Arnold tongues cover the entire Ω interval. For $K > 1$ the map ceases to be invertible, and chaos may set in.

Finally, a transition to chaos may occur via *intermittency.* This means that the phase trajectory appears approximately as a limit cycle during long periods of time; but these periods are interrupted by sudden outburst of chaotic motion. The associated Poincaré sections show saddle-node bifurcations, in which a stable node and a saddle point collide and annihilate.

Examples of chaotic behavior occur in a great variety of disciplines. Among these examples are turbulence in hydrodynamic systems, chemical reaction systems, lasers at high pumping or with feedback, optically bistable elements, nonlinear electronic circuits, and Josephson junctions (Chap. 6). Chaos in semiconductors induced by optical irradiation, electric and magnetic fields, or a combination of these, has also recently been found (Aoki et al., 1981, Teitsworth et al., 1983, Held et al., 1984, Seiler et al., 1985, Peinke et al., 1985, Maracas et al., 1985, Bumeliene et al., 1985, Brandl et al., 1987, Knap et al., 1988, Yamada et al., 1988, Spinnewyn et al., 1989, for reviews see Schöll, 1987; Abe, 1989); we shall review the situation in Section 1.2.4.

1.2.2. Self-Sustained Oscillations

The last section contained an introduction to concepts salient to the understanding of the general problem of the stability of an arbitrary system. The next chapter discusses the stability of a system specific to our concern, the NDC element. To provide further basis for that analysis, we discuss the general problem of self-sustained oscillations, culling our example from the area of electronics. Here the example given by Stoker (1950, 1980) is useful.

Consider the vacuum tube triode feedback circuit shown in Fig. 1-18 and neglect the small grid current. The potential drop in the inductor is $L(dI/dt)$, and this is also the drop across the resistor R and capacitor C. If I_R is the current through R, I_c the current through C, and q_c the charge on C, then

$$L\frac{dI}{dt} = RI_R \tag{1-31}$$

$$L\frac{dI}{dt} = \frac{q_c}{C} \quad \text{or} \quad L\frac{d^2I}{dt^2} = \frac{1}{C}\frac{dq_c}{dt} = \frac{1}{C}I_c \tag{1-32}$$

$$I_p = I + I_R + I_c \tag{1-33}$$

This leads to

$$LC\frac{d^2I}{dt^2} + \frac{L}{R}\frac{dI}{dt} + I = I_p \tag{1-34}$$

For an assumption of negligible current in the grid circuit,

$$\Phi_G = M\frac{dI}{dt} \tag{1-35}$$

where M is the mutual inductance between the anode and grid circuits, and

$$\Phi_p = \Phi_B - L\frac{dI}{dt} \tag{1-36}$$

At this point the properties of the vacuum tube itself are introduced. While the specific properties of the triodes are discussed only for illustrative purposes, the overall electrical similarities between it and field effect transistors necessitate some brief description.

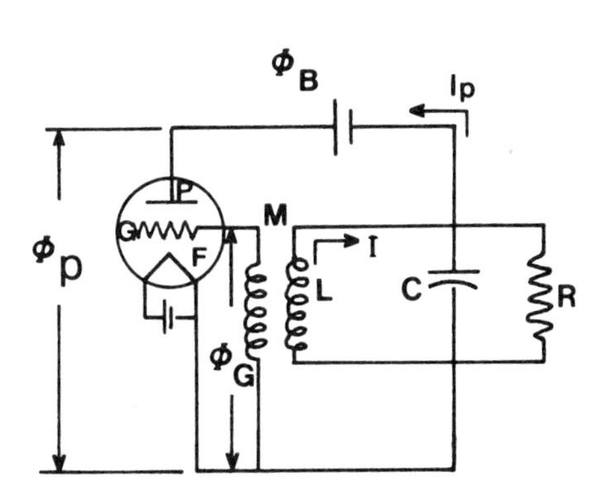

FIGURE 1-18. Feedback circuit of a vacuum tube triode (F, cathode; G, grid; P, plate). All parameters are defined in the text.

In a two-terminal diode the space-charge current is limited by the distribution of excess carriers at the cathode, which in turn alters the potential distribution in this region. (This particular phenomenon will again arise when transport in submicron devices is examined in Chap. 5) If a third electrode is placed between the cathode and anode the potential distribution is further altered, and the flow of space charge is controlled. The third electrode is the grid. As the tube is generally used, the grid is kept negative with respect to the cathode with most of the current flowing to the plate. The plate and grid currents depend on both the plate potential and the grid potential, and for small currents and potentials a Taylor series expansion up to linear order is possible. Retaining, for the moment, the possibility of a nonvanishing grid current I_G, clearly important during transients, the linear relation between grid and plate currents and potentials can be written as

$$I_G = k_G \Phi_G + S_G \Phi_p \tag{1-37}$$

$$I_p = S_p \Phi_G + k_p \Phi_p \tag{1-38}$$

Here k_G and k_p represent, respectively, the grid and plate conductance; and S_G and S_p, the reflex transconductance and transconductance, respectively. Similarly, the plate potential can be expressed as a linear combination of the grid potential and plate current (or any two quantities may be chosen as independent variables).

We now consider, more generally, a nonlinear triode current–voltage characteristic, but assume that the combined dependence of plate current on grid and plate potentials is still through a linear combination of Φ_G and Φ_p:

$$I_p = \lambda(\Phi) \tag{1-39}$$

where, by (1-35) and (1-36), we have

$$\Phi \equiv \Phi_G + D\Phi_p = D\Phi_B + (M - DL)\frac{dI}{dt} \tag{1-40}$$

Here $D = k_p/S_p$ is a positive constant, the reciprocal of which is sometimes called the plate amplification factor.

The function $\lambda(\Phi)$ is nonlinear; it is shown in Fig. 1-19. Equations (1-35),

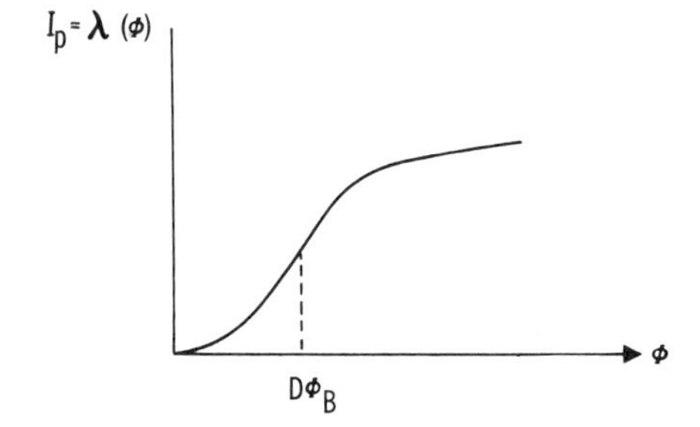

FIGURE 1-19. Nonlinear current (I_p)–voltage (Φ) characteristic of a triode. The inflection point is noted by the dashed line.

(1-36), (1-39), and (1-40) when substituted into Eq. (1-34) yield

$$LC\frac{d^2I}{dt^2} + \frac{L}{R}\frac{dI}{dt} + I = \lambda(\Phi) \tag{1-41}$$

and since the tube characteristic is nonlinear, Eq. (1-41) is a nonlinear differential equation for I in which the nonlinearity λ appears in the first derivative, or damping term dI/dt by (1-40). This type of differential equation is central to our theme—we will emphasize it now and several times throughout the text.

In order to discuss the differential equation (1-41) it is useful to introduce a new dependent variable x defined by

$$x = I - I_0, \tag{1-42}$$

where I_0 is a constant, whence Eq. (1-41) becomes, by (1-40)

$$LC\frac{d^2x}{dt^2} + \frac{L}{R}\frac{dx}{dt} + x + I_0 = \lambda\left[D\Phi_B + (M - DL)\frac{dx}{dt}\right] \tag{1-43}$$

Note that $dx/dt = dI/dt$. Defining

$$f\left(\frac{dx}{dt}\right) \equiv I_0 - \lambda\left[D\Phi_B + (M - DL)\frac{dx}{dt}\right] \tag{1-44}$$

allows us to write Eq. (1-43) as

$$LC\frac{d^2x}{dt^2} + \frac{L}{R}\frac{dx}{dt} + f\left(\frac{dx}{dt}\right) + x = 0 \tag{1-45}$$

Now let us choose I_0 so that $\lambda(\Phi) = I_0$ is near the inflection point of the characteristic curve (Fig. 1-19). Also assume $M - DL > 0$, an essential necessary condition for self-generated oscillations, as we shall soon see. The graph of $f(dx/dt)$ as a function of dx/dt has the general shape shown in Fig. 1-20a, and in Fig. 1-20b we show the graph of the function $\tilde{f}(dx/dt)$, called the *characteristic*, defined by

$$\tilde{f}\left(\frac{dx}{dt}\right) = \frac{L}{R}\frac{dx}{dt} + f\left(\frac{dx}{dt}\right) \tag{1-46}$$

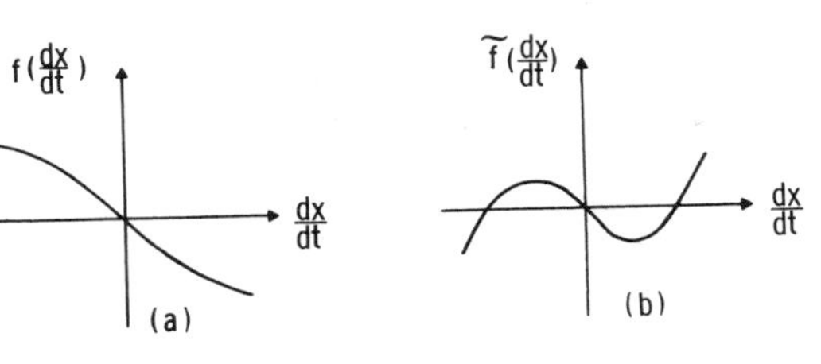

FIGURE 1-20. Nonlinear characteristics of the Van der Pol oscillator: (a) $f \equiv I_0 - I_p(\Phi)$ versus $dx/dt \equiv dI/dt$; (b) $\tilde{f} \equiv (L/R)(dx/dt) + f$ versus dx/dt.

assuming that the slope of $\tilde{f}(dx/dt)$ for $dx/dt = 0$ is negative, i.e., that

$$L/R + f'(0) < 0 \tag{1-47}$$

where the prime denotes differentiation with respect to dx/dt, or

$$L/R < (M - DL)\lambda'(D\Phi_B) \tag{1-48}$$

This can only occur if $f'(0) < 0$, i.e., $M - DL > 0$, as stated above. Equation (1-45) then becomes

$$LC\frac{d^2x}{dt^2} + \tilde{f}\left(\frac{dx}{dt}\right) + x = 0 \tag{1-49}$$

where

$$\frac{dx}{dt}\tilde{f}\left(\frac{dx}{dt}\right) < 0$$

for $|dx/dt|$ sufficiently small and

$$\frac{dx}{dt}\tilde{f}\left(\frac{dx}{dt}\right) > 0$$

for $|dx/dt|$ large. Thus, the "damping" is negative for small $|dx/dt|$; the system gains energy and x increases in amplitude. But for large $|dx/dt|$ the system dissipates energy and the amplitude of x becomes limited. A steady oscillation is expected after a transient response occurs. Further, we note again the requirement that the slope of $\tilde{f}(dx/dt)$ be negative for small $|dx/dt|$, the amplification factor being large enough, or else no oscillation will be expected.

[It is often preferred that a different differential equation than (1-49) is analyzed. Instead of x, the variable $v = dx/dt$ is used; differentiating Eq. (1-49) with respect to time yields

$$LC\frac{d^2v}{dt^2} + \chi(v)\frac{dv}{dt} + v = 0 \tag{1-50}$$

where $\chi(v) = \tilde{f}'(v)$. If $\tilde{f}(dx/dt)$ has the form shown in Fig. 1-20b, then $\chi(v)$ has the form shown in Fig. 1-21.]

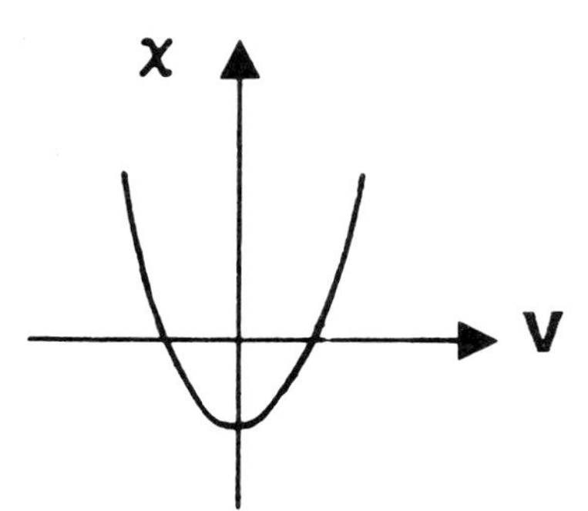

FIGURE 1-21. Nonlinear characteristic $\chi(v) \equiv d\tilde{f}/dv$ versus $v \equiv dx/dt$.

If the vacuum tube is operated near the inflection point of its $\tilde{f}(dx/dt)$ characteristic, a good approximation for $\tilde{f}(dx/dt)$ is

$$\tilde{f}\left(\frac{dx}{dt}\right) \cong -\alpha\frac{dx}{dt} + \frac{\beta}{3}\left(\frac{dx}{dt}\right)^3 \tag{1-51}$$

where α, $\beta > 0$ and dx/dt is not too large. From Eqs. (1-46) and (1-44) it follows that $\alpha = (M - DL)\lambda'(D\Phi_B) - L/R$. Equation (1-49) then becomes

$$LC\frac{d^2x}{dt^2} + \left[-\alpha\frac{dx}{dt} + \frac{\beta}{3}\left(\frac{dx}{dt}\right)^3\right] + x = 0 \tag{1-52}$$

a well-known special case of the "*van der Pol*" equation. If we introduce dimensionless variables $t_1 = \omega t$, $\omega^2 = 1/LC$, $x_1 = (\omega\sqrt{\beta/\alpha})x$ and $\epsilon \equiv \alpha/\sqrt{LC} = \alpha\omega$, Eq. (1-52) becomes

$$\frac{d^2x_1}{dt_1^2} - \epsilon\left[\frac{dx_1}{dt_1} - \frac{1}{3}\left(\frac{dx_1}{dt_1}\right)^3\right] + x_1 = 0 \tag{1-53}$$

We see that ϵ occurs as the coefficient of the damping term, a parameter we shall refer to liberally in the text, often in different notation. For convenience in analyzing Eq. (1-53) we shall drop the subscripts and write it as

$$\frac{d^2x}{dt^2} + \epsilon\tilde{f}\left(\frac{dx}{dt}\right) + x = 0, \qquad \epsilon > 0 \tag{1-54}$$

with

$$\tilde{f}\left(\frac{dx}{dt}\right) = -\frac{dx}{dt} + \frac{1}{3}\left(\frac{dx}{dt}\right)^3 \tag{1-55}$$

Equation (1-54) is the governing differential equation for the description of "free" oscillations (as opposed to "forced" oscillations, which involve an explicitly time-dependent force). Introducing the "velocity" $v = dx/dt$, we can write Eq. (1-54) as an autonomous two-variable dynamic system

$$\begin{aligned} \frac{dx}{dt} &= v \\ \frac{dv}{dt} &= -\epsilon\tilde{f}(v) - x = \epsilon(v - v^3/3) - x \end{aligned} \tag{1-56}$$

The only singular point is $x_0 = v_0 = 0$. The Jacobian matrix [Eq. (1-22)] at this point is

$$A = \begin{pmatrix} 0 & 1 \\ -1 & \epsilon \end{pmatrix}$$

Thus $D = 1$ and $T = \epsilon$. For $0 < \epsilon < 2$ the origin is an unstable focus, for $\epsilon > 2$ an unstable node. For $\epsilon = 0$ we have a center surrounded by concentric circles: the system represents an undamped harmonic oscillator.

A complete phase portrait for $\epsilon > 0$ can be obtained graphically. First, we find the field of directions

$$\frac{dv}{dx} = -\frac{\epsilon\tilde{f}(v) + x}{v} \tag{1-57}$$

at any point P in the (x, v) plane. This method is called the Liénard construction.

We first plot the null-isocline $x = K(v) = -\epsilon\tilde{f}(v)$, where $dv/dt = 0$. This is shown in Fig. 1-22. Then, for a point $P(x, v)$, a horizontal line is drawn until it intersects the curve $K(v)$ at the point $R(x_1, v)$. The distance $x - x_1 = x + \epsilon\tilde{f}(v)$. Then a perpendicular is dropped from R to S and the diagonal PS is constructed. The slope of PS is

$$m(\overline{PS}) = \frac{v}{x + \epsilon\tilde{f}(v)} \tag{1-58}$$

The normal to $\overline{PS}$ at P then has the slope

$$\hat{m} = -\frac{x + \epsilon\tilde{f}(v)}{v} \tag{1-59}$$

Thus a technique is available for obtaining the slope of the trajectory at any point (x, v). In particular, on the null-isoclines $x = K(v)$ and $v = 0$ we have $dv/dx = 0$ and $dx/dv = 0$, respectively.

Based on this discussion, the phase portrait of the Van der Pol oscillator can be drawn. For $\epsilon = 0$ Eq. (1-56) is a conservative linear system whose integral curves (solution trajectories) are concentric circles having a common center at the origin; we have simple harmonic motion of frequency $\omega = 1$ and arbitrary amplitude. For $\epsilon > 0$ the singularity becomes an unstable focus and the integral curves depart from the origin; the origin is now an unstable singular point. The outward-directed spirals near the origin do not spread indefinitely, however, since the damping becomes positive for large values of v. Far away from the origin the

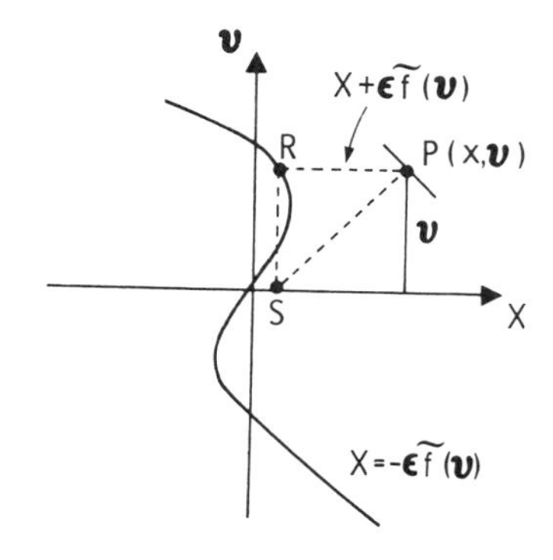

FIGURE 1-22. Liénard construction of the field of directions for the Van der Pol oscillator in the (v, x) phase plane. The solid line is the null-isocline $dv/dt = 0$. For an explanation see the text.

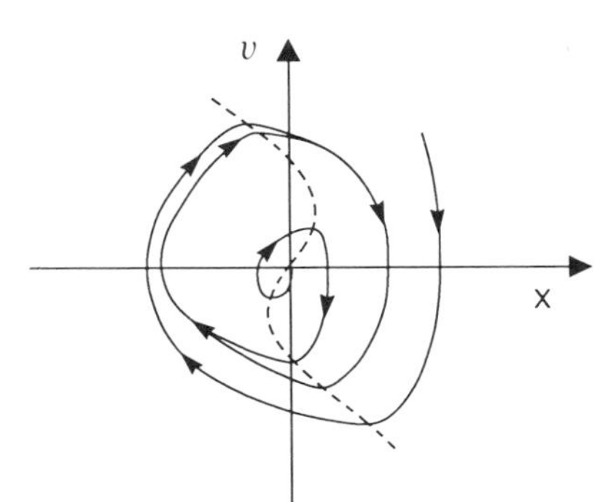

FIGURE 1-23. Limit cycle of the Van der Pol oscillator. Two transient trajectories approaching the limit cycle from the inside and from the outside are also shown.

trajectories are spirals which spin about the origin as they move towards it with increasing t. Indeed, the trajectories near the origin as well as those far away from it tend to a single closed curve, a periodic solution, as shown in Fig. 1-23, sketched for the case $\epsilon = 1$. Hence, every solution tends to a periodic solution as $t \to \infty$; the system will always approach a limit cycle. The limit cycle in this case encloses one singular point.

For small ϵ, which corresponds to large LC, the limit cycle is approximately a circle of radius 2 and the motion is that of a near sinusoid of frequency $\omega = 1$ (Stoker, 1950), i.e., near simple harmonic motion. Thus, as ϵ increases from zero, the limit cycle bifurcates from the conservative harmonic solution which has the particular amplitude 2. This is not a Hopf bifurcation since the limit cycle appears immediately with a nonzero amplitude. This singular behavior is, however, artificially introduced by the scaling of variables in Eq. (1-53). In the original dimensioned variables, Eq. (1-52), we do find a Hopf bifurcation of a limit cycle with zero amplitude (Minorsky, 1962), as α is changed from negative to positive values, e.g., by increasing the mutual inductance M, and $\beta > 0$. For $\alpha < 0$ the origin is a stable focus, for $\alpha = 0$ a weakly stable focus (stabilized by the nonlinear β-term), and for $\alpha > 0$ an unstable focus. In the case $\alpha < 0$ the inequality of Eq. (1-48) is violated, i.e., the circuit is dominated by the linear resistor (R small), and no limit cycle oscillations occur.

As ϵ increases the limit cycle deviates more and more from simple harmonic motion, developing an appreciable anharmonic content. For large $\epsilon(\geq 10)$, corresponding to small LC, the oscillations become quite "jerky" and are referred to as "relaxation oscillations" in the electrotechnical and mathematical literature. [Unfortunately, in laser physics the term "relaxation oscillation" is used in a completely different sense: here it denotes a damped, i.e., not self-sustained oscillation (Haken, 1970; Thompson, 1980). These definitions should not be confused!] For the case where $\epsilon \to \infty$, the relaxation oscillations appear as shown in Fig. 1-24, where we introduce the new variable $\xi = x/\epsilon$.

Relaxation oscillations exhibit two different phases. During one phase, energy is slowly stored; during the other it is discharged rapidly when a critical threshold is reached. In terms of ξ Eq. (1-57) can be written

$$\frac{dv}{d\xi} = \epsilon^2 \frac{(v - v^3/3) - \xi}{v} = \epsilon^2 \frac{-\tilde{f}(v) - \xi}{v} \tag{1-60}$$

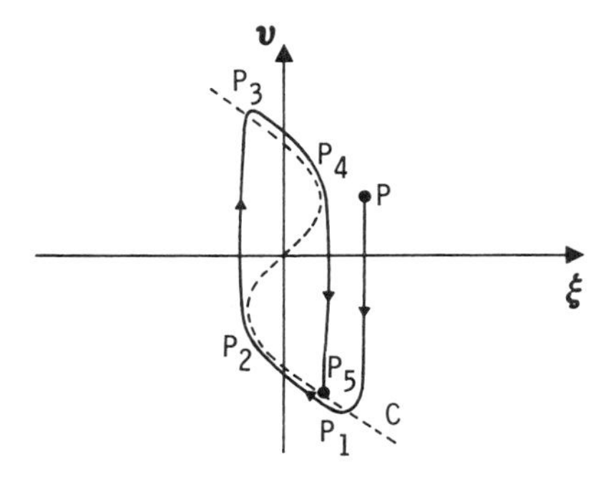

FIGURE 1-24. Relaxation oscillations of the Van der Pol oscillator for small *LC*. The limit cycle P_5-P_2-P_3-P_4-P_5 is shown in the phase plane.

The curve $\xi = -\tilde{f}(v)$ is plotted as the broken line C in Fig. 1-24. On this characteristic curve $(dv/d\xi = 0)$ the field of directions is horizontal for all values of ϵ. However, for all other points the slope tends to ∞ as $\epsilon \to \infty$. The field directions are nearly vertical when we are off the characteristic curve. We can now imagine what the limit cycle will look like. Starting at P, the solution curve is nearly vertical down to P_1, where it hits C. The slope goes to zero here, but since the field of direction off C is nearly vertical, the solution curve will tend to follow C, staying a little below it, until P_2 is reached. It then shoots up vertically to P_3, stays slightly above C until P_4, then drops sharply to P_5. The limit cycle will then be similar to the path $P_5 \to P_2 \to P_3 \to P_4 \to P_5$. Indeed, for $\epsilon \to \infty$ the limit cycle will simply consist of portions of the characteristic curve plus vertical straight lines. This approximation is a manifestation of the general "slaving principle" (Haken, 1983a), which states that the fast relaxing variables (here: v) are enslaved by the slow variable (here: ξ)—i.e., v follows adiabatically along the characteristic as ξ changes.

To compute the period of the relaxation oscillations with $\epsilon \to \infty$ we can calculate a line integral over the limit cycle. In terms of the original variable $\xi = x_1/\varepsilon$ and $v = dx_1/dt_1$, we have

$$T_1 = \epsilon \oint \frac{d\xi}{v} \tag{1-61}$$

Since the straight vertical line portions make no contribution to T_1 $(d\xi = 0)$, we have, with $\xi = v - v^3/3$ along the curved paths on C,

$$T_1 = 2\epsilon \int_2^1 \frac{d(v - v^3/3)}{v} = 2\epsilon[\ln v - (v^2/2)]\Big|_2^1 = 1.614\epsilon \tag{1-62}$$

and since $T = T_1\sqrt{LC}$, then

$$T = 1.614\sqrt{LC}\epsilon = 1.614\alpha \tag{1-63}$$

which is good for large ϵ only. For small ϵ we have near sinusoids whose period is given by

$$T = 2\pi\sqrt{LC} \tag{1-64}$$

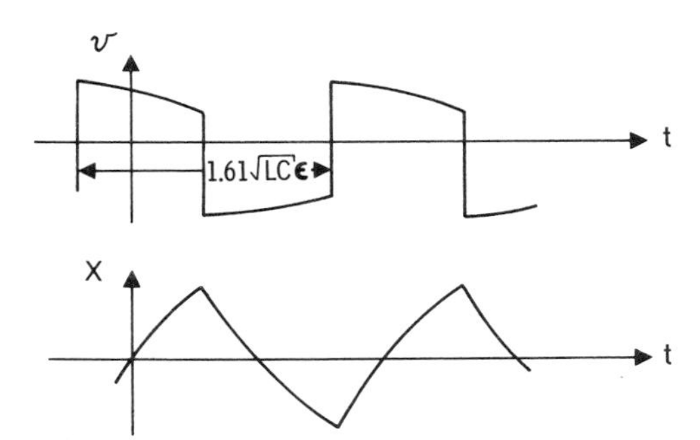

FIGURE 1-25. Relaxation oscillations of the Van der Pol oscillator. The time series $v(t)$ and $x(t) = I(t) - I_0$ are shown schematically.

The form of the relaxation oscillations for large ϵ as a function of time is shown in Fig. 1-25. These waveforms are of fundamental importance with regard to a wide variety of solid state devices and will be emphasized throughout the text.

We have so far assumed a dc bias, and we have seen that the free Van der Pol oscillator can exhibit strongly nonlinear limit cycle oscillations. Now, if an ac bias is applied, we obtain a periodically forced Van der Pol oscillator, which can be described by the basic equation

$$\frac{d^2x}{dt^2} + \epsilon\tilde{f}\left(\frac{dx}{dt}\right) + x = \beta\cos(\omega_0 t) \qquad (1\text{-}65)$$

where β and ω_0 are the driving amplitude and frequency, respectively. Equation (1-65) can be represented as a nonautonomous two-variable dynamic system, or an autonomous three-variable system, as discussed in Sec. 1.2.1F:

$$\frac{dx}{dt} = v \qquad (1\text{-}66)$$

$$\frac{dv}{dt} = \epsilon(v - v^3/3) - x + \beta\cos\theta \qquad (1\text{-}67)$$

$$\frac{d\theta}{dt} = \omega_0 \qquad (1\text{-}68)$$

This system exhibits a number of interesting bifurcations (Guckenheimer and Holmes, 1983). Besides periodic limit cycle oscillations and quasiperiodic motion on a two-dimensional torus ("drifting" solutions in the engineering terminology), chaotic motion is also possible. In the (β, ω_0) control parameter plane there occur saddle-node bifurcations of two singular points, Hopf bifurcations of a limit cycle and global bifurcations of a limit cycle from a separatrix (cf. Sec. 1.2.1D).

1.2.3. Solitons

The nonlinear differential equations associated with NDC systems often admit solutions that propagate through a sustaining medium without change of shape. These solutions are stable against fluctuations and have local field regions different from the uniform fields at extended points. Indeed, they might be described as solitary solutions of the governing differential equations; a freely propagating Gunn domain (Chap. 5) is an example of this situation (Shaw et al., 1979). Although such solitary wavelike disturbances may not be true *solitons* (Bishop and Schneider, 1978, Novikov et al., 1984) they are sufficiently similar so that no discussion of the physics of instabilities in solid state electron devices would be complete without the inclusion of this subject.

Most systems with soliton solutions are integrable models with a Hamiltonian structure described by nonlinear, dispersive wave equations. A solitary wave is a soliton if its amplitude, velocity, shape, and internal frequency are preserved after a collision with other solution components. In solid state physics solitons are manifest in a variety of areas, including: quantum calculations; statistical mechanics; structural phase transitions; low-dimensionality solids; Josephson junction transmission lines (Chap. 6). They were first reported on by Scott-Russell (1845), who observed the unusual stability of certain water waves propagating in shallow water. The first theoretical treatment of soliton motion was given in connection with the propagation of dislocations in solids (Seeger et al., 1950; 1953). In this context a nonlinear, hyperbolic differential equation arose, which is now called the *sine-Gordon equation*:

$$\frac{\partial^2 u(x, t)}{\partial x^2} - \frac{\partial^2 u(x, t)}{\partial t^2} = \sin u(x, t) \tag{1-69}$$

The sine-Gordon equation possesses solitonic solutions, which are kink shaped (i.e., a propagating pulse with different asymptotic values at $x \rightarrow \pm\infty$) and spatially localized, pulsating solutions, frequently called *breathers* (S. Novikov et al. 1984). Another famous equation with solitonic solutions is the Korteweg–De Vries equation, which describes wave propagation in shallow water (Novikov et al., 1984):

$$\frac{\partial u}{\partial t} + \frac{\partial u}{\partial x} + u\frac{\partial u}{\partial x} + \frac{\partial^3 u}{\partial x^3} = 0 \tag{1-70}$$

An alternative equation to model this phenomenon has been proposed by Peregrine (1966), the "regularized long-wave equation":

$$\frac{\partial u}{\partial t} + \frac{\partial u}{\partial x} + u\frac{\partial u}{\partial x} - \frac{\partial^3 u}{\partial x^2\,\partial t} = 0 \tag{1-71}$$

Figure 1-26 shows computer solutions of this equation; three solitons collide and re-emerge.

1.2.4. Deterministic Chaos

In this section we elaborate on the phenomenon of chaos, which we discussed briefly in Sec. 1.2.1F. Nonlinear electrical oscillations in semiconductor

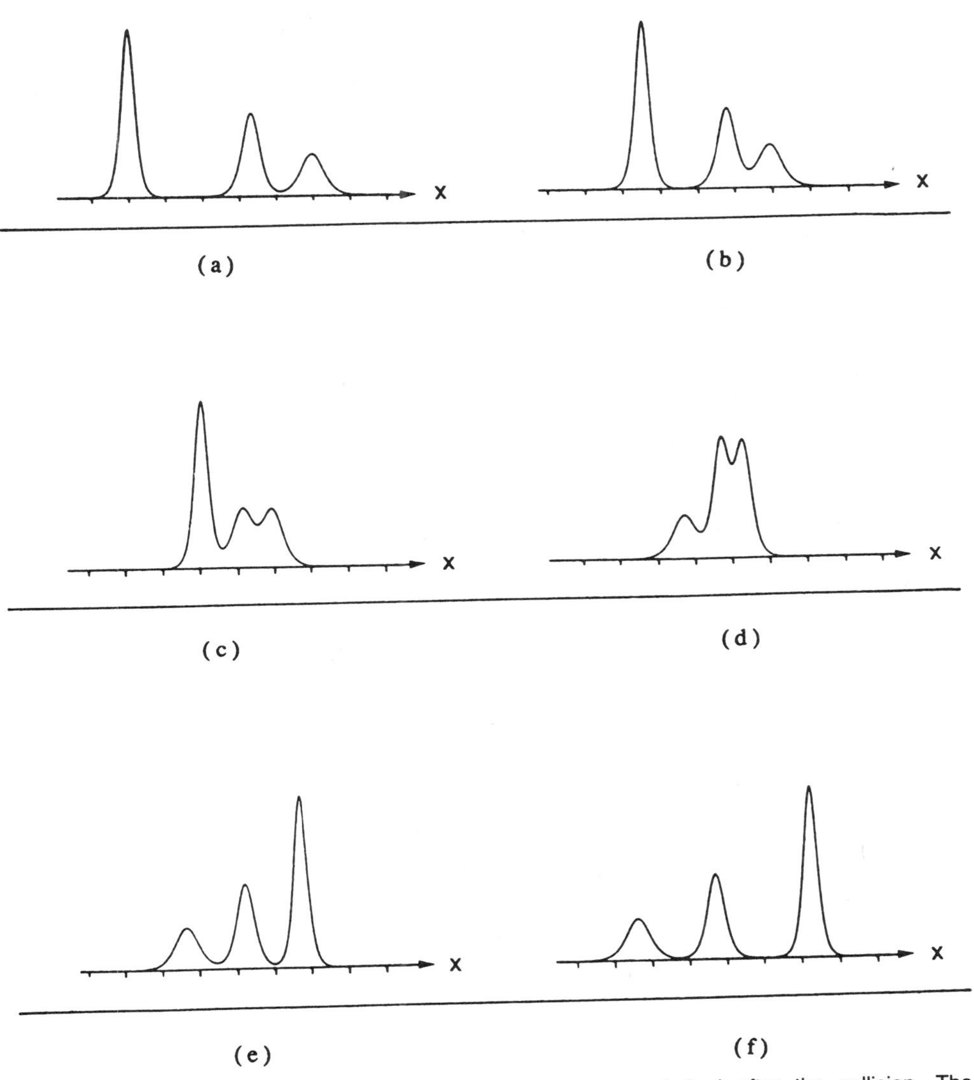

FIGURE 1-26. Three solitons that collide and reemerge unperturbed after the collision. The amplitude u is plotted versus the spatial coordinate x at six successive times (a)–(f).

devices have been studied extensively for a long time, but the analysis of chaotic behavior in semiconductors is a fairly new development. It has only become possible since sophisticated mathematical tools to deal with this have been developed in recent years. Semiconductors have been singled out by the ready availability of high-resolution measurement techniques and reproducable materials engineering technologies, which make them often superior to the "classical" objects of study like hydrodynamic or chemical reaction systems. In the field of solid-state devices, there exist at least two reasons that make the investigation of chaotic behavior worthwhile: Firstly, a profound knowledge of the nature and the conditions of chaos are necessary in order to avoid undesirable destructive chaos in device operation. Secondly, intentional use of deterministic chaos might serve as a simple source of electron broadband noise. Deterministic chaos in semiconductors has become a new, rapidly expanding field of research, and we can only discuss some exemplary results here. More details can be found elsewhere (Schöll, 1987, 1989a, b, c; Abe, 1989).

We shall first outline some criteria that allow us to distinguish chaotic motion from complicated multiply periodic motion, as well as from irregular behavior caused by stochastic fluctuations such as thermal noise. The latter are true random processes, and are thus in principle unpredictable, while chaotic motion is in principle fully determined by a few nonlinear differential equations supplemented by initial conditions, and is hence deterministic, although the actual time dependence of the trajectories looks extremely irregular and most of the analysis has to be carried out numerically.

One characteristic feature of chaos is its *low-dimensional phase space.* Consider the dynamic variable q at discrete times t_n, $n = 1, 2, 3, \ldots$: this yields a sequence of values $q_1, q_2, q_3, \ldots$ Each value q_{n+1} is then uniquely determined by its predecessor q_n, and we may present the pairs (q_n, q_{n+1}) for all possible values of n as points in a (q_n, q_{n+1}) plane. This is a graph of the one-dimensional Poincaré map introduced in Sec. 1.2.1. Similarly, we might plot the triples (q_n, q_{n+1}, q_{n+2}) as points in a three-dimensional space, or more generally, conceive the N-tuples $(q_n, q_{n+1}, \ldots, q_{n+N-1})$ as points in an N-dimensional space. If the irregular motion were stochastic, we would expect all these N-dimensional spaces to be filled uniformly and randomly by points, since the sequences $\{q_n, \ldots, q_{n+N-1}\}$ would be random. Chaotic motion resulting from deterministic equations, however, will fill only certain low-dimensional regions (often of fractal dimension d) of the n-dimensional spaces, even if N is chosen very large. We shall come across examples of this behavior below, where the dynamic variable q will be the electric current or field.

A quantitative measure of the low dimensionality of chaotic motion is provided by the concept of the *fractal dimension* of a chaotic attractor (Schuster, 1987). To this purpose we partition the n-dimensional phase space into boxes of length l. The probability p_i of finding a point on the attractor in cell number $i[i = 1, \ldots, M(l)]$ is then given by

$$p_i = \lim_{N_{\text{tot}}\to\infty} \frac{N_i}{N_{\text{tot}}}$$

where N_i is the number of points $\{\mathbf{q}(t_j)\}$ in this cell, and $N_{\text{tot}} = \sum N_i$. The generalized dimensions D_q ($q = 0, 1, 2, \ldots$) are defined by

$$D_q = \lim_{l\to 0} \frac{1}{q-1} \log\Bigl(\sum_{i=0}^{M(l)} p_i^q\Bigr) \Big/ \log l \tag{1-72}$$

For $q \to 0$ we obtain

$$D_0 = \lim_{l\to 0}\left(\log \sum_{l=0}^{M(l)} 1\right)\Big/\log l = \lim_{l\to 0}\frac{\log M(l)}{\log(1/l)} \tag{1-73}$$

which means that for $l \to 0$ the minimum number $M(l)$ of N-dimensional cubes of length l needed to cover the attractor scales as

$$M(l) \sim l^{-D_0}$$

For a point, a line, a surface, and a volume this gives just the usual integer value of the dimension $D_0 = 0, 1, 2, 3$, respectively. For more complicated, "strange" attractors D_0 is called the *Hausdorff dimension* and may take on noninteger values. For sufficiently large $N > 2D_0 + 1$, D_0 is independent of the embedding dimension N. In contrast, for random stochastic motion $D_0 = N$ would be expected.

In practical experiments or numerical simulations, the computation of D_0 is often made difficult because a very large number of points on the attractor are needed to ensure the validity of the asymptotic scaling law. A more rapidly converging algorithm has been proposed by Grassberger and Procaccia (1983), which generally gives a very good lower bound $d \approx D_2 \leq D_0$. Here the *correlation dimension* d is defined via the scaling law for $l \to 0$

$$C(l) \sim l^d \quad \text{or} \quad d := \lim_{l\to 0}\frac{\log C(l)}{\log l}$$

where the correlation integral

$$C(l) = \lim_{N\to\infty}\frac{1}{N^2}\sum_{i,j=1}^{N}\theta(l - |\mathbf{q}_i - \mathbf{q}_j|)$$

measures the numbers of pairs of points $\mathbf{q}_n \equiv \{q_n, \ldots, q_{n+N-1}\}$ whose distance is less than l in an *N-Sdimensional embedding space with sufficiently large N.* θ is the Heaviside function. In order to characterize the global behavior of a strange attractor, the whole spectrum of dimensions D_q, $q = 0, 1, 2, \ldots$ is necessary. Its Legendre transform $f(\alpha)$, defined by

$$f(\alpha) = -(q - 1)D_q + q\alpha$$

with

$$\alpha(q) \equiv \frac{\partial}{\partial q}[(q - 1)D_q]$$

has been shown to be a universal function for very different systems. It measures the Hausdorff dimension of the set of points $\{\mathbf{q}_i\}$ on the attractor that have the same strength of singularity α, i.e., $p_i(l \to 0) \sim l^\alpha$. Another useful quantity is the bit-number variance $C_2 = \langle (b - \langle b \rangle)^2 \rangle$, $b \equiv -\log p_i$, which has been shown to represent a characteristic measure of correlations (Schlögl and Schöll, 1988; Naber and Schöll, 1990a).

A second characteristic feature of chaotic motion shows up in the correlations of the variables at different times: The *autocorrelation function*

$$\langle q(t)q(t + \tau) \rangle \equiv \lim_{T \to \infty} \frac{1}{2T} \int_{-T}^{T} q(t)q(t + \tau)\, dt \tag{1-74}$$

tends to zero for $\tau \to \infty$.

Such behavior is different from simply and even multiply periodic motion. For periodic motion we obtain always an undamped oscillatory correlation function, e.g.,

$$\langle q(t)q(t + \tau) \rangle = \tfrac{1}{2}\cos \omega\tau \tag{1-75}$$

for

$$q(t) = \sin \omega\tau \tag{1-76}$$

In case of thermal fluctuations the autocorrelation function is often zero or negligible for τ greater than a finite autocorrelation time τ_c.

A third feature of chaos is related to the Fourier spectrum of the temporal evolution. Defining the spectral power density or *power spectrum* by

$$S(\omega) \equiv \lim_{T \to \infty} \frac{\pi}{T} |q(\omega; T)|^2 \tag{1-77}$$

where

$$\hat{q}(\omega; T) \equiv \frac{1}{2\pi} \int_{-T}^{T} q(t) e^{i\omega t}\, dt \tag{1-78}$$

is the Fourier transform of $q(t)$ in a finite, but sufficiently long observation interval T, we find, using Eq. (1-73),

$$S(\omega) = \frac{1}{2\pi} \int_{-\infty}^{\infty} \langle q(t)q(t + \tau) \rangle e^{i\omega t}\, dt \tag{1-79}$$

This means that the power spectrum is the Fourier transform of the autocorrelation function (Wiener-Khinchin theorem). For a simply or multiply periodic motion the power spectrum has discrete lines at the fundamental frequencies. Chaotic motion, however, is characterized by a continuous broad band of frequencies.

A fourth indication of chaos is given by the so-called *Lyapunov exponents*. These are generalizations of the eigenvalues of the linear stability matrix of a singular point, Eq. (1-22)

$$\lambda = \lim_{t \to \infty} \sup \left[\frac{1}{t} \ln |\mathbf{q}(t) - \mathbf{q}^*(t)| \right]$$

where $\mathbf{q}(t)$ are trajectories in the neighborhood of a reference trajectory $\mathbf{q}^*(t)$. The sign of the Lyapunov exponents indicates whether or not $\mathbf{q}^*(t)$ is asymptotically approached by other trajectories $\mathbf{q}(t)$. If, in case of a three-variable dynamic system, one of the Lyapunov exponents is positive, one is zero, and one is negative, the motion is likely to be chaotic, although some more detailed analysis may be necessary (Haken, 1983b).

Finally, we recall that the onset of chaos is often preceded by one of the characteristic scenarios discussed in Section 1.2.1 F: a period-doubling cascade, quasiperiodic breakdown, or intermittency. The key experiments and some theoretical models for driven and self-generated chaos have been reviewed in the book by Schöll (1987). We also note that chaos has recently also been observed in silicon *p-i-n* diodes in the regime of SNDC induced by double injection of electrons and holes (Jäger, 1991). It is linked to the self-organized formation of current density filaments (Jäger et al., 1986; Symanczyk et al., 1990). The mechanism for producing SNDC in *p-i-n* diodes will be briefly discussed in Sec. 1.3.4, and elaborated on in Chap. 7.

In *p-n* diodes incorporated into electronic circuits, chaos can also be induced under specific conditions (Moon, 1987).

1.3. MECHANISMS FOR PRODUCING NDC—EXAMPLES

1.3.1. NNDC via Intervalley Transfer of Electrons—The McCumber–Chynoweth Model

In the last two sections we defined NDC and discussed stability in general terms. In Chap. 2 we show that NDC elements are often intrinsically unstable both against circuit-controlled oscillations and the formation of inhomogeneous field and/or current density distributions. Now, it is useful to show just how NDC can occur; the first case we consider is the mechanism of intervalley transfer of electrons which leads to *negative differential mobility* (NDM) in a variety of semiconductors, the Gunn–Hilsum effect (Shaw et al., 1979).

Ridley and Watkins (1961) first pointed out this possibility and Hilsum (1962) made the first calculation of the velocity versus electric field, $v(F)$, curve for the semiconductor *n*-GaAs two years before Gunn (1964) made his observations of bias induced instabilities of current in thin samples of this substance. It is remarkable that Hilsum's rather simple model predicted a field at peak velocity of 3 kV/cm, a value that is well into the expected range for typical low-field mobilities in *n*-GaAs (see Fig. 1-27). In his calculation Hilsum assumed that the electric field caused the electrons to heat up above ambient temperature and that electron collisions produced a common electron temperature, T_e, for carriers in both the lower and upper conduction band valleys in the material (see Fig. 1-28).

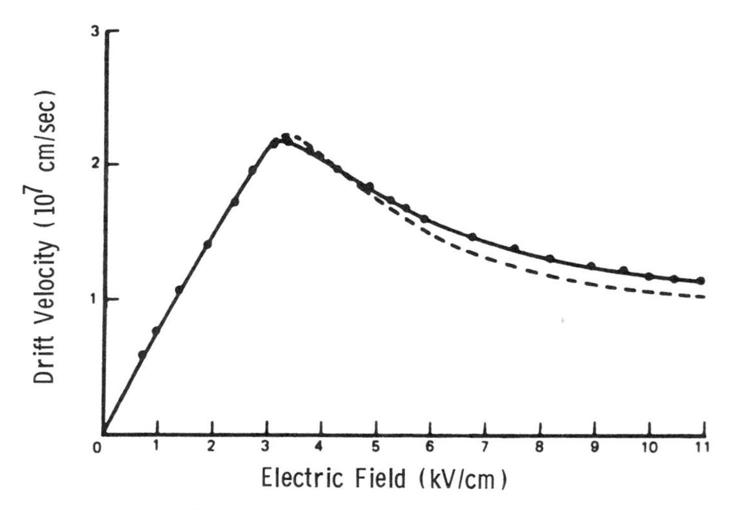

FIGURE 1-27. Experimental points (—) from Ruch and Kino (1967) and theoretical (- - -) values from Butcher and Fawcett (1966) for the electric field-dependent average drift velocity of electrons in GaAs (after Shaw et al., 1979).

He also assumed that the electrons distributed themselves between these valleys as if the system were at thermal equilibrium at a temperature T_e. A similar simple and insightful calculation was made later by McCumber and Chynoweth (1966); we review it now. But, before we begin, it is important to point out that the McCumber–Chynoweth calculation does not yield a quantitatively accurate model of the $v(F)$ curve for n-GaAs. Rather, it just provides a qualitative

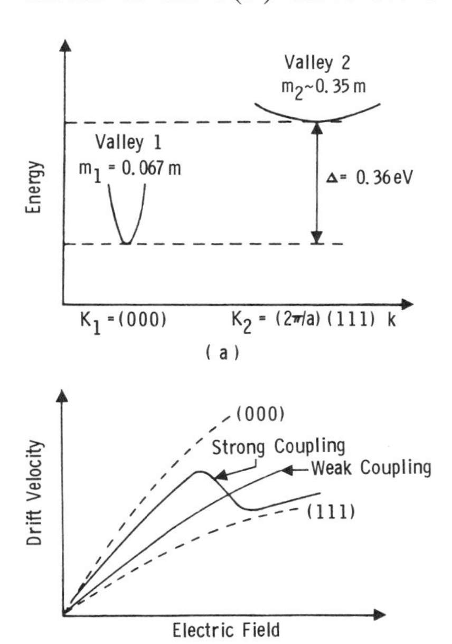

FIGURE 1-28. (a) Schematic conduction band structure for GaAs (from Butcher, 1967). The intervalley separation Δ and effective masses are given. (b) Schematic average drift velocity versus electric field curves for conduction in valley 1 (000), valley 2 (111), and when either strong or weak intervalley coupling is present. (After Shaw et al., 1979.)

description for the problem. More detailed calculations were performed by, e.g., Butcher and Fawcett (1966; Butcher, 1967) and Conwell and Vassell (1966). Although Fig. 1-27 shows a good fit between experiment and the curve derived by Butcher and Fawcett, there has been continued refinement of the $v(F)$ curve in more recent times (Fawcett and Herbert, 1974; Littlejohn et al., 1977).

McCumber and Chynoweth treated the system shown in Fig. 1-28 by assuming that carriers in the high-mobility lower valley having energies greater than Δ were instantaneously transferred to the low-mobility upper valley. Both bands were taken as parabolic. The basic assumption, however, was that carrier–carrier interactions were strong enough to keep electrons in thermodynamic equilibrium with each other with a common temperature T_e, and that the application of a field had the primary effect of creating a new population distribution, one that was consistent with Boltzmann statistics. The problem is to determine the distribution of carriers in each section of the conduction band. This distribution is a function of temperature, which in turn is a function of field.

For electrons of mass m_c in a simple parabolic conduction band, the density of electrons with energies between E_c and E, where E_c is the bottom of the conduction band is given by

$$n(E) = 2\int_{E_c}^{E} f(E)N(E)\,dE \tag{1-80}$$

where $N(E)$ is the density of states,

$$N(E) = \frac{\pi}{4h^3}(8m_c)^{3/2}(E - E_c)^{1/2} \tag{1-81}$$

and

$$f(E) \cong e^{(E_F - E)/kT_e} \tag{1-82}$$

E_F being the Fermi level energy. h and k are the Planck and Boltzmann constants, respectively. For transport carried by electrons in both lower, l, and upper, u, energy portions of the conduction band, the total density of carriers is approximated by

$$\begin{aligned} n &= \frac{\pi}{4h^3}(8m_l)^{3/2}e^{E_F/kT_e} \\ &\quad \times \left[\int_0^\infty E^{1/2}e^{-E/kT_e}\,dE + \alpha\int_\Delta^\infty (E-\Delta)^{1/2}e^{-E/kT_e}\,dE\right] \\ &= \frac{\pi}{4h^3}(8m_l)^{3/2}e^{E_F/kT_e}\tfrac{1}{2}\sqrt{\pi}\,(kT_e)^{3/2}(1 + \alpha e^{-\Delta/kT_e}) \end{aligned} \tag{1.83}$$

where

$$\alpha = \frac{A_u}{A_l}\left(\frac{m_u}{m_l}\right)^{3/2} \tag{1-84}$$

is the dimensionless ratio of the upper to lower valley density of states. The A's represent the number of equivalent valleys. For gallium arsenide with a $\Gamma - L$ model, when the upper X valleys are not taken into consideration, $A_l = 1$ and $A_u = 4$. Note that in Eq. (1-83) the bottom of the conduction band has been taken as zero energy, and the integrals have been extended to $E = \infty$ since the exponential factors fall off rapidly.

In the analysis of McCumber and Chynoweth it was assumed that all Γ valley carriers with energy greater than the energy separation between Γ and L would transfer to the L valley. Thus, the fractional population of the lower valley is

$$\frac{n_l}{n} = \left(\int_0^{\Delta} E^{1/2} e^{-E/kT_e}\, dE\right) / [\tfrac{1}{2}\sqrt{\pi}(kT_e)^{3/2}(1 + \alpha e^{-\Delta/kT_e})] \tag{1-85}$$

which is more conveniently written as

$$\frac{n_l}{n} = \frac{1 - \gamma}{1 + \alpha e^{-\Delta/kT_e}} \tag{1-86}$$

where

$$\gamma = [\tfrac{1}{2}\sqrt{\pi}(kT_e)^{3/2}]^{-1} \int_{\Delta}^{\infty} E^{1/2} e^{-E/kT_e}\, dE \tag{1-87}$$

(See, e.g., Hasty et al., 1968.) The fractional population of the upper valley is then

$$\frac{n_u}{n} = 1 - \frac{n_l}{n} = \frac{\alpha e^{-\Delta/kT_e} + \gamma}{1 + \alpha e^{-\Delta/kT_e}} \tag{1-88}$$

and the ratio of upper to lower valley carriers is

$$\frac{n_u}{n_l} = \frac{\alpha e^{-\Delta/kT_e} + \gamma}{1 - \gamma} \tag{1-89}$$

Hasty et al. obtained asymptotic expressions for γ in two limits:

for $kT_e \gg \Delta$

$$\gamma \approx 1 - \left(\frac{\Delta}{kT_e}\right)^{3/2} \frac{\sqrt{\pi}}{2} \tag{1-90}$$

for $kT_e \ll \Delta$

$$\gamma \approx \left(\frac{\Delta}{kT_e}\right)^{1/2} e^{-\Delta/kT_e} / \sqrt{\pi} \tag{1-91}$$

and at elevated electron temperatures a substantial number of carriers are in the upper valleys.

To determine how the carrier temperature increases with increasing field, a consistent set of transport equations are needed. Under uniform field conditions the first relevant transport equation is that for drift current

$$J = en\mu F \tag{1-92}$$

where in the McCumber and Chynoweth model,

$$\mu \equiv (\mu_l n_l + \mu_u n_u)/n \tag{1-93}$$

and both μ_l and μ_u are taken to be constant:

$$J = e(n_l\mu_l + n_u\mu_u)F \tag{1-94}$$

The second transport equation under homogeneous field conditions is associated with energy transport. If $\langle E \rangle$ denotes the mean energy per electron, then the time rate of change of energy due to both an applied field and a relaxation process is given by

$$\frac{d}{dt}(n\langle E \rangle) = \mathbf{J} \cdot \mathbf{F} - \frac{3}{2}\frac{nk(T_e - T_0)}{\tau_E} \tag{1-95}$$

where τ_E is an effective energy relaxation time and T_0 is the lattice temperature. $\langle E \rangle$ is related to the electron temperature T_e through the relation

$$\langle E \rangle = \frac{\langle nE \rangle}{n} = \frac{\int_0^\infty E^{3/2}e^{-E/kT_e}dE + \alpha\int_\Delta^\infty E(E-\Delta)^{1/2}e^{-E/kT_e}dE}{\int_0^\infty E^{1/2}e^{-E/kT_e}dE + \int_\Delta^\infty (E-\Delta)^{1/2}e^{-E/kT_e}dE} \tag{1-96}$$

or

$$\langle E \rangle = \frac{3}{2}kT_e + \frac{\alpha\Delta e^{-\Delta/kT_e}}{1 + \alpha e^{-\Delta/kT_e}} \tag{1-97}$$

Under steady state conditions we have

$$JF = \frac{3}{2}n\frac{k}{\tau_E}(T_e - T_0) \tag{1-98}$$

and the field and temperature relations are obtained as consistent solutions to the following equation:

$$T_e = T_0 + \frac{2}{3}\frac{e\tau_E}{k}\left[\left(\mu_l + \frac{n_u}{n_l}\mu_u\right)\Big/\left(1 + \frac{n_u}{n_l}\right)\right]F^2 \tag{1-99}$$

which is obtained from Eqs. (1-94) and (1-98) with n_u/n_l given as functions of T_e by Eq. (1-89). From this, the magnitude of the mean velocity, the drift velocity, v, may be obtained: $v(F) = \mu(T_e(F))F = [\mu_l - (\mu_l - \mu_u)n_u/n]F$. These results are displayed in Fig. 1-29 for different values of T_0.

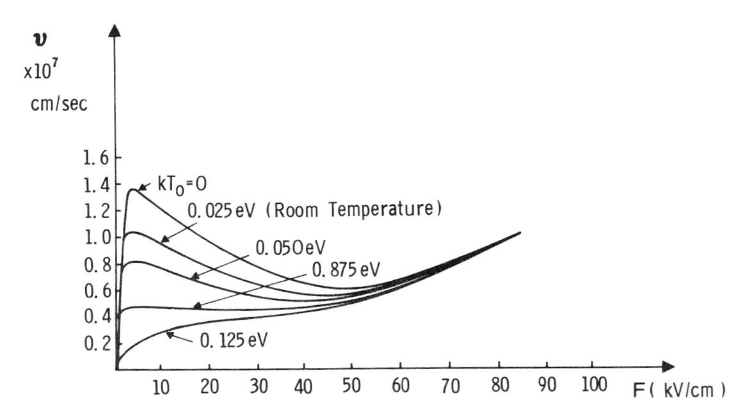

FIGURE 1-29. Calculated average drift velocity versus electric field curves for different ambient (lattice) temperatures T_0 in GaAs (after McCumber and Chynoweth, 1966).

We see from the figure that a finite range of temperatures exist where the $v(F)$ curve exhibits a region of NDM over a range of electric field values. The reason that the NDM region disappears at high temperatures is that the thermal population of the upper valley is already sufficiently high at zero electric field to inhibit field-induced NDM from appearing. In fact, as Δ is made smaller, the temperature at which NDM disappears is reduced. These features have all been discussed in detail in a variety of books on transferred electron devices (Bulman et al., 1972; Bosch and Engelmann, 1975, Shaw et al., 1979).

The essential physics of electron transfer induced NDC is contained in the above discussion. The presence of NDC, however, results in electrical instabilities, and our interest will be in determining how the instability manifests itself. For this discussion a broad device physics is necessary, one that includes spatial inhomogeneities and temporal constraints. This will be considered in detail in Chap. 5.

1.3.2. NNDC via Intervalley Transfer of Electrons—Semiconductors with Equivalent Valleys

In the previous section we provided an example of NNDC in a many-valley semiconductor with at least two valleys. Electrons in the lower valley have a small effective mass ("light states"), and a high mobility, but a low density of states. The upper valley is the opposite: it has a large effective mass ("heavy states") and low mobility, but the density of states is high. NNDC arises due to field-induced electron transfer from light states in the central valley to heavy states in the satellite valleys.

An analogous situation arises in semiconductors with equivalent valleys for some directions of current. As examples of indirect semiconductors with equivalent valleys we consider n-Ge or n-Si. In Ge there are four valleys located at the Brillouin zone boundary along the $\langle 111 \rangle$ axes. In Si there are six valleys or three pairs of valleys along the $\langle 100 \rangle$ axes located 85% of the distance toward the zone boundary. Electrons in these valleys have the same density of states but a different effective mass along the electric field because of the anisotropy in each valley (Asche et al., 1979; Conwell, 1967). In Ge the NNDC is most pronounced for $J \parallel \langle 111 \rangle$ axes, but in Si, for $J \parallel \langle 100 \rangle$. In each of these two cases there are two groups of valleys: (1) one valley in Ge and one pair of valleys in Si, oriented with their axes of symmetry along the current, so that the mobility in these valleys is small, since it is determined by a large effective mass m_l (heavy states); (2) all other valleys with essentially high mobility (light states). As the electric field increases, electrons in these other valleys, because of their high mobility, receive more energy from the electric field than electrons in the heavy valleys aligned along the current; so electrons in the light valleys will be hotter than in the heavy valleys. As a matter of fact, the time for intervalley scattering, τ_α, decreases when the mean energy of the electrons increases with increasing field. If the rate of decreasing τ_α with increasing F is large, the electron transfer from light to heavy valleys occurs in a narrow field region; as a result the current can decrease. Electron transfer to the heavy valleys here is not due to transitions to the valleys with higher energy and larger density of states as in GaAs, but rather to the rigid decrease of the intervalley scattering time τ_α with an increase in the electron heating with electric field. The heating in different valleys is different because of the different orientation of the electric field. Hence, two qualitative differences between NNDC in Ge and Si, and that in GaAs, appear: (1) a pronounced anisotropy of the effect due to the fact that the heating must be different in different groups of valleys; (2) the effect can exist only in relatively pure semiconductors when the electron concentration is not too high and the intervalley scattering must be via intervalley phonons with their energy, $\hbar\omega_0$, essentially higher than kT_0. In the previous example the inequality required was that $\Delta \gg kT_0$. Now we have $\hbar\omega_0 \gg kT_0$. The effect occurs at lower temperatures than as shown in Fig. 1-29 for GaAs because $\Delta \gg \hbar\omega_0$.

Let us now consider a quantitative description of the effect. For simplicity we employ the two-valley semiconductor shown in Fig. 1-30. For current along the x axis, valley 1 is heavy and valley 2 is light. We can use Eq. (1-94) for the current; it is useful to replace the indices u and l by 1 and 2:

$$J = e(n_1\mu_1 + n_2\mu_2)F \tag{1-100}$$

Here it is necessary to find the concentrations n_1 and n_2 in valleys 1 and 2 in the same manner as before [Eqs. (1-85)–(1-94)], i.e., using the condition of quasi neutrality:

$$n_1 + n_2 = n \tag{1-101}$$

and the balance of intervalley transitions

$$\frac{n_1}{\tau_1} = \frac{n_2}{\tau_2} \tag{1-102}$$

We find that

$$n_1 = n\frac{\tau_1}{\tau_1 + \tau_2}, \qquad n_2 = n\frac{\tau_2}{\tau_1 + \tau_2} \tag{1-103}$$

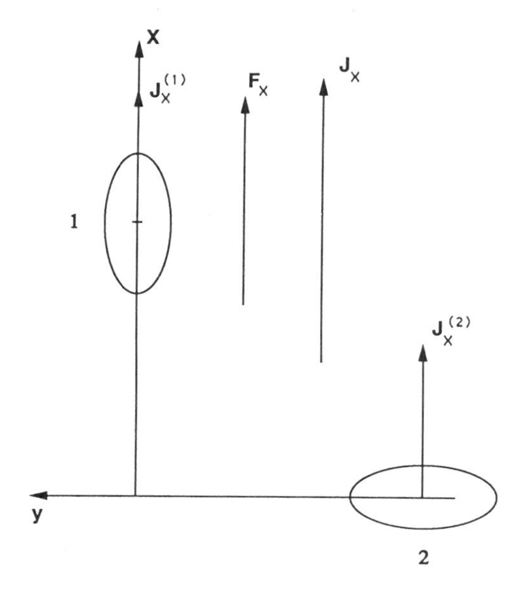

FIGURE 1-30. Two-valley semiconductor. Electrons in valley 2 have a larger mobility in the direction of the electric field than electrons in valley 1.

and using (1-103) in Eq. (1-100) we have

$$J = enF\frac{(\tau_1\mu_1 + \tau_2\mu_2)}{\tau_1 + \tau_2} \tag{1-104}$$

NNDC can arise only if τ_α decreases sufficiently strongly with increasing F. Let us simplify our consideration by assuming that in (1-104) only τ_α depends on F. (We neglect the weak dependence of μ_α on F.) In this case μ_1 and μ_2 are field independent and they are equal to the longitudinal and transverse components of the mobility tensor;

$$\mu_1 = \mu_l = \mu(1 - a), \qquad \mu_2 = \mu_t = \mu(1 + a), \qquad a = \frac{\mu_t - \mu_l}{\mu_t + \mu_l} \tag{1-105}$$

Here the parameter a is a characteristic of the anisotropy of the conductivity in each of these valleys, with μ_t and μ_l denoting the transverse and longitudinal component of the mobility tensor. We put Eq. (1-105) in Eq. (1-104)

$$J = en\mu F[1 - a(\tau_1 - \tau_2)/(\tau_1 + \tau_2)] \tag{1-106}$$

The expression in the brackets will decrease from unity at low fields, when $\tau_1 = \tau_2$, to $1 - a$ if τ_α decreases fast enough with field F and the condition $\tau_2 \ll \tau_1$ can be reached.

Differentiation of Eq. (1-106) yields

$$dJ/dF < 0$$

if

$$\left(1 - a\frac{\tau_1 - \tau_2}{\tau_1 + \tau_2}\right) + F\frac{d}{dF}\left(1 - a\frac{\tau_1 - \tau_2}{\tau_1 + \tau_2}\right) < 0 \tag{1-107}$$

or

$$d \ln[1 - a(\tau_1 - \tau_2)/(\tau_1 + \tau_2)]/d \ln F < -1$$

or

$$2a\frac{\tau_1\tau_2}{\tau_1 + \tau_2}\left(\frac{d \ln \tau_1}{d \ln F} - \frac{d \ln \tau_2}{d \ln F}\right) > \tau_1(1 - a) + \tau_2(1 + a)$$

To find the interval of F where inequality (1-107) is true we must introduce the dependence of τ_α on F. In Asche et al., (1979) different dependencies of $\tau_\alpha(F)$ are analyzed; it is shown that NNDC exists if $a > 0.6$.

If instead of a two-valley semiconductor we consider Si, with current along $\langle 100 \rangle$, we have three pairs of valleys. Two of them have mobility μ_t and one μ_l (1-105), so instead of (1-106) for the current, we can write

$$J = en\mu F\left(1 - a\frac{\tau_1 - 2\tau_2}{\tau_1 + 2\tau_2}\right) \tag{1-108}$$

That is Eqs. (1-108) and (1-106) differ only in the fact that instead of one light valley we now have two pairs of light valleys, and under electron heating they change from two pairs with high mobility μ_t to one pair with low mobility μ_l. The term in parentheses in (1-108) can decrease from $(1 + a/3)$ at $F = 0$, when $\tau_2 = \tau_1$, to $1 - a$ when $\tau_2 \ll \tau_1$. The ratio $(1 + a/3)/(1 - a)$ is larger than it was in a two-valley semiconductor $[1/(1 - a)]$, owing to the fact that electrons from two pairs of hot (light) valleys transfer to one heavy pair. Differentiating (1-108) we obtain the following condition for NNDC instead of (1-107):

$$4a\frac{\tau_1\tau_2}{\tau_1 + 2\tau_2}\left(\frac{d \ln \tau_1}{d \ln F} - \frac{d \ln \tau_2}{d \ln F}\right) > \tau_1(1 - a) + 2\tau_2(1 + a) \tag{1-109}$$

It is easy to see that Eq. (1-109) will be true for smaller a if the dependence τ_α on F is the same as in (1-107). Calculations give $a \geq 0.5$.

We have demonstrated here the possibility of NNDC, neglecting the dependence of mobility on electric field. (It is necessary to stress that in spite of the fact that μ was taken as field independent, this NNDC is really NDM, because the total carrier concentration remains the same and the current changes due to intervalley transfer, i.e., due to changes of contributions of separate valleys to the current.) We can find more rigorous calculations. For example, Asche et al. (1982) calculated the dependence of the average drift velocity $v = J/en$ on electric field in n-Si at different temperatures, as shown in Fig. 1-31. In addition to optical phonon scattering, additional intervalley scattering characterized by a field-independent time τ_0 was considered. When τ_0 is introduced for temperatures below 50K, the NNDC is well pronounced. Figure 1-32 shows the

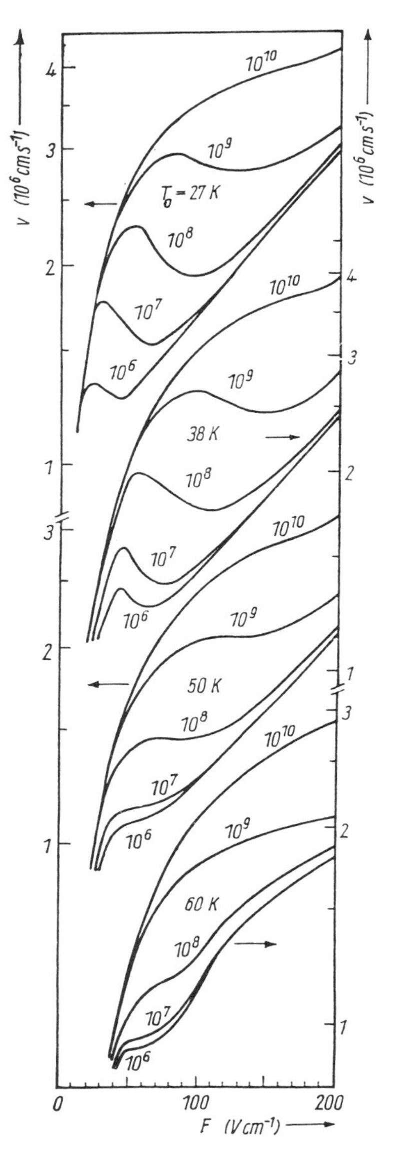

FIGURE 1-31. Calculated average drift velocity **v** as a function of field strength **F** for current **J** || ⟨100⟩ in *n*-Si at different lattice temperatures T_0 with field-independent intervalley scattering rate τ_0^{-1} as a parameter. (τ_0^{-1} is given in s^{-1}, and is defined in the text.) (After Asche et al., 1982.)

experimental current–voltage characteristics measured in *n*-Si with doping concentration $N_D = 9.9 \times 10^{13}\,\text{cm}^{-3}$ and $N_A = 2.6 \times 10^{13}\,\text{cm}^{-3}$ (Gram, 1972). NNDC was observed at temperatures below 52K.

To conclude this section we note that this type of NNDC was first observed in *n*-Ge by Kastalskii (1968) and Kastalskii and Ryvkin (1968). In the paper of Asche et al. (1982), it was shown that NNDC in Si as well as in Ge and other semiconductors with equivalent valleys can be observed for any direction of current with respect to the crystallographic axes. An additional decrease of current is due to the transverse electric field, which arises for directions other

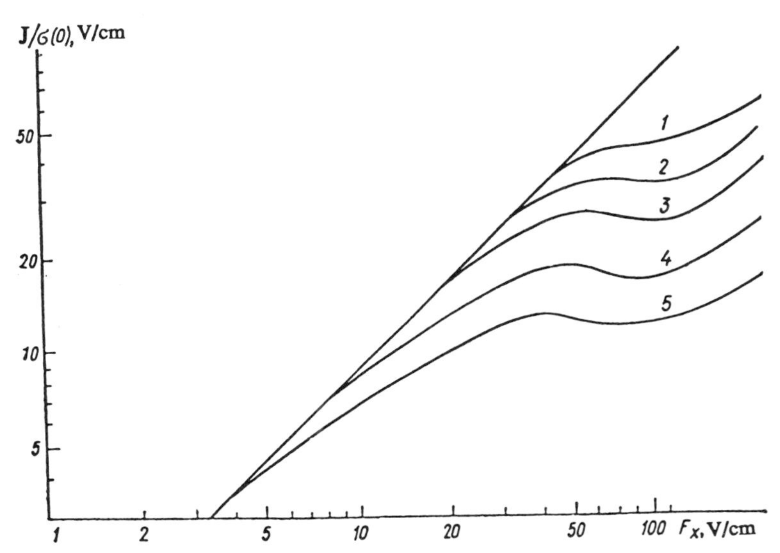

FIGURE 1-32. Measured current density–field characteristics for n-Si with **J** ∥ ⟨100⟩ at different lattice temperatures T_0, (1) T_0 = 62K, (2) 52K, (3) 43K, (4) 38K, (5) 34K. (After Gram, 1972.)

than those discussed above. Hence, NNDC in semiconductors with equivalent valleys can occur for any direction of current and is analogous to NNDC in GaAs, but a lower temperature is required for the effect to occur.

1.3.3. NNDC via Real-Space Transfer of Electrons

Intervalley transfer of electrons in momentum space was the basic NDM mechanism in the previous two sections. In this section we shall discuss a mechanism for NDC which is the real space analogue of the Gunn–Hilsum effect.

The theory of the effect of NDC due to electron transfer in real space in a semiconductor heterostructure has been developed by Gribnikov (1973), by Pacha and Paschke (1978), and by Hess et al. (1979). The effect was experimentally demonstrated by Keever et al. (1981) for modulation-doped GaAs/Al_xGa_{1-x}As heterojunction layers. New transistor concepts were also based upon real-space transfer (Kastalsky and Luryi, 1983): the charge injection effect transistor (CHINT) and the negative resistance field effect transistor (NERFET). The high-electron mobility transistor (HEMT) also uses modulation-doped GaAs/Al_xGa_{1-x}As layers as the conducting channel. A review of these novel heterostructure device concepts has been given by Hess (1988, 1990).

Figure 1-33 shows the energy-band diagram of a modulation doped GaAs/Al_xGa_{1-x}As heterostructure. The Al_xGa_{1-x}As layer is heavily n-doped, while the GaAs is undoped. There is a band edge discontinuity ΔE_c between the conduction bands in the two layers because of the smaller band gap and the larger electron affinity of GaAs.

In thermodynamic equilibrium the electrons fall into the lower GaAs well, where they experience strongly reduced impurity scattering if they are separ-

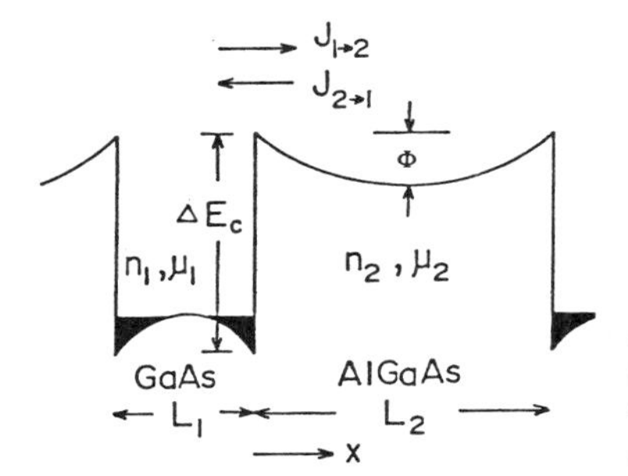

FIGURE 1-33. Schematic sketch of the conduction band edge energy for a $GaAs/Al_xGa_{1-x}As$ heterostructure. The band bending is due to the space charge that results from the positive ionized donors in the $Al_xGa_{1-x}As$ and the negative charge carriers in the GaAs.

ated from their parent donors by more than 200Å. Thus, for a layer thickness of typically 400Å, the mobility μ_1 in the GaAs layer will be high ($\geq 5000\,\mathrm{cm^2\,V^{-1}\,s^{-1}}$ at room temperature) and the mobility μ_2 of the electrons in the $Al_xGa_{1-x}As$ will be low ($\approx 500\,\mathrm{cm^2\,V^{-1}\,s^{-1}}$ or less). Application of a low electric field F *parallel* to the layer interface will result in current due primarily to electrons in the GaAs, since both the carrier density and mobility are much larger there than in the $Al_xGa_{1-x}As$. A high electric field, however, will result in heating of the high-mobility electrons to energies far above their thermal equilibrium values. The low-mobility carriers in the $Al_xGa_{1-x}As$ will not be heated significantly since the power input per electron is equal to $e\mu_2F^2$, which is small. As a result, the electrons in the GaAs acquire a high kinetic energy and are thermionically emitted into the $Al_xGa_{1-x}As$, where their mobility is much lower due to strongly enhanced impurity scattering. Thus, during the transfer from the high-mobility GaAs layer to the low-mobility $Al_xGa_{1-x}As$ layer, the sample exhibits NNDC, as occurs in the Gunn–Hilsum effect. The physical mechanism differs, however, in that the electrons are transferred in real space, rather than in momentum space, and the mobility in the two states is different because of different scattering times, rather than because of different effective masses. Also, the switching speed is higher owing to the fast thermionic emission time ($\approx 10^{-11}$ s), and the static and dynamic characteristics can easily be controlled over a wide range by varying the material parameters such as the layer widths L_1 and L_2, the doping concentration N_D, and the aluminum fraction x (which determines the band-edge discontinuity ΔE_c). Thus, a much larger peak-to-valley current ratio can be achieved than in the Gunn–Hilsum effect by choosing small ratios μ_2/μ_1 and L_1/L_2.

A quantitative understanding of the effect can be obtained by analytic two-electron-temperature models (e.g., Shichijo et al., 1980) or by Monte Carlo simulations (e.g., Sakamoto et al., 1989). Here we shall describe a simple analytic approach that uses Bethe's theory of thermionic emission for the current densities $J_{1\to2}$ and $J_{2\to1}$ between the two layers, denoted by subscripts 1 and 2, respectively (see Fig. 1-33):

$$J_{1\to2} = -en_1[kT_1/(2\pi m_1^*)]^{1/2}\exp[-\Delta E_c/(kT_1)] \qquad \text{(1-110a)}$$

$$J_{2\to1} = -en_2[kT_2/(2\pi m_2^*)]^{1/2}\exp[-\Phi/(kT_2)] \qquad \text{(1-110b)}$$

Here n_i, T_i, m_i^* are the spatially averaged electron density, electron temperature, and effective mass in the two layers, $i = 1, 2$, respectively, and ΔE_c and Φ are the potential barriers as shown in Fig. 1-33. The continuity equation for the carriers is given by

$$\frac{d}{dt} n_1 = (J_{1\to 2} - J_{2\to 1})/(eL_1) \tag{1-111}$$

where the conservation of the total number of carriers provided by the donor density N_D in the $Al_xGa_{1-x}As$ requires

$$L_1 n_1 + L_2 n_2 = L_2 N_D \tag{1-112}$$

The energy balance in the two layers $i = 1, 2$ gives

$$\frac{d}{dt}(n_i\langle E_i\rangle) = n_i\left\langle\frac{\partial E_i}{\partial t}\right\rangle + n_i e\mu_i F^2 \mp (J_{E_i}^{1\to 2} - J_{E_i}^{2\to 1})/L_i \tag{1-113}$$

where $\langle E_i\rangle = \frac{3}{2}kT_i$ is the mean energy per carrier, $\langle \partial E_i/\partial t\rangle$ is the energy loss rate per carrier due to optical phonon scattering, and

$$J_{E_1}^{1\to 2} = (-J_{1\to 2}/e)(\Delta E_c + 2kT_1), \qquad J_{E_2}^{2\to 1} = (-J_{2\to 1}/e)(\Phi + 2kT_2)$$

$$J_{E_1}^{2\to 1} = (-J_{2\to 1}/e)(\Delta E_c + 2kT_2), \qquad J_{E_2}^{1\to 2} = (-J_{1\to 2}/e)(\Phi + 2kT_1)$$

are the energy flow rates out of layer 1, out of layer 2, into layer 1 and into layer 2, composed of a convective, a pressure-induced, and a heat flow contribution (Aoki et al., 1989). The static $J(F)$ characteristic is then given by

$$J = (en_1\mu_1 L_1 + en_2\mu_2 L_2)F/(L_1 + L_2) \tag{1-114}$$

where n_1 and n_2 are determined by the time-independent simultaneous solution of Eqs. (1-111)–(1-113). The result is plotted in Fig. 1-34 for different values of the band-edge discontinuity ΔE_c. The threshold electric field for the onset of NDC is about 1.6 kV cm^{-1}, and the NDC is most pronounced for $\Delta E_c = 250$ meV, which corresponds to an aluminum fraction $x = 0.3$. The associated heating of the electron gases in the two layers is shown in Fig. 1-35. Whereas the electron temperature T_1 in the GaAs increases strongly, T_2 remains almost constant.

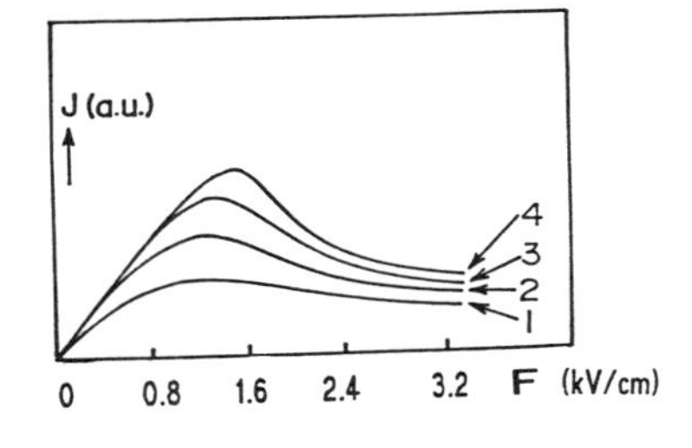

FIGURE 1-34. Calculated current density versus field characteristics for real space transfer across various potential barriers ΔE_c: 1–100 meV, 2–150 meV, 3–200 meV, 4–250 meV. The numerical parameters $\mu_1 = 5000\ \text{cm}^2\ \text{V}^{-1}\ \text{s}^{-1}$, $\mu_2 = 50\ \text{cm}^2\ \text{V}^{-1}\ \text{s}^{-1}$, $N_D = 10^{17}\ \text{cm}^{-3}$, $L_1 = 0.2\ \mu\text{m}$, $L_2 = 2\ \mu\text{m}$ have been used. (After Aoki et al., 1989.)

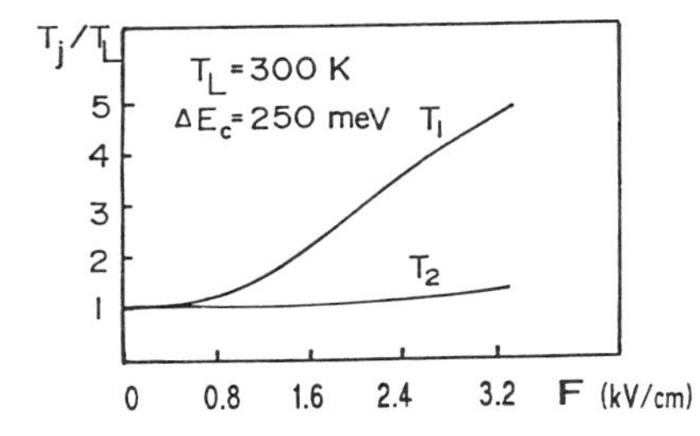

FIGURE 1-35. Calculated electron temperatures T_1 and T_2 in the GaAs and $Al_xGa_{1-x}As$ layers, respectively, as a function of field for $\Delta E_c = 250$ meV and the parameters of Fig. 1-34 (after Aoki et al., 1989.)

In the NDC regime the real-space-transfer device can be operated as an oscillator by applying an ac oscillating field in the 2–25-MHz frequency range, in addition to the dc bias field, parallel to the layers, thus inducing periodic transfer between the GaAs and the AlGaAs (Coleman et al., 1982).

The dynamic behavior has been analyzed theoretically by integrating the time-dependent equations (1-111)–(1-113) supplemented by dynamic equations for the parallel field F and the potential barrier Φ, thereby accounting for the inertia of dielectric relaxation of the applied field, and the space charge, respectively (Aoki et al., 1989; Schöll and Aoki, 1991). As a result, self-generated oscillations in the 10–80-GHz regime have been predicted to occur under dc bias. They are generated by a Hopf bifurcation in the NDC regime of the static $J(F)$ characteristic, and are followed by a period-doubling route to chaos, as the bias is increased. Phase portraits of the limit cycles and of the chaotic attractor are shown in Fig. 1-14, where the carrier density in the GaAs layer (n_1 in units of $N_D L_2/L_1$) is plotted versus the mean energy per carrier ($\langle E_1 \rangle$ in units of ΔE_c) for four values of the applied field $\tilde{F}_0$ (in units of 4 kV cm^{-1}). The motion of the trajectories is anticlockwise. Although the oscillation involves a complicated nonlinear interaction of the carrier densities, the carrier energies, the parallel field, and the interface potential barrier, the basic physical oscillation mechanism can be visualized as follows: real-space transfer of the electrons in the GaAs layer leads to a decrease of n_1 and an increase of n_2, which diminishes the positive space charge in the $Al_xGa_{1-x}As$ layer. Subsequently, the potential barrier Φ drops after some delay owing to the finite dielectric relaxation time. This leads to an increased backward thermionic emission current from 2 to 1, which increases n_1, and thereafter increases Φ. This, in turn, decreases the thermionic emission current from 2 to 1, which completes the cycle.

1.3.4. SNDC via Field-Induced Trapping and Double Injection

We now consider the simple idealized system shown in Fig. 1-36, where there are N_D ionized shallow donors and N_R deep acceptorlike traps (recombination centers) which are negatively charged when filled by electrons. The kinetics of these donors and acceptors is such that the hole lifetime increases dramatically with injection current, and from this a current-controlled NDC (SNDC) results. The recombination centers lie sufficiently below the Fermi level so that they are completely filled with electrons in thermal equilibrium. It is also assumed that the average cross section for capturing a free hole by a filled center greatly exceeds the average cross section for capturing a free electron by an empty center. (We

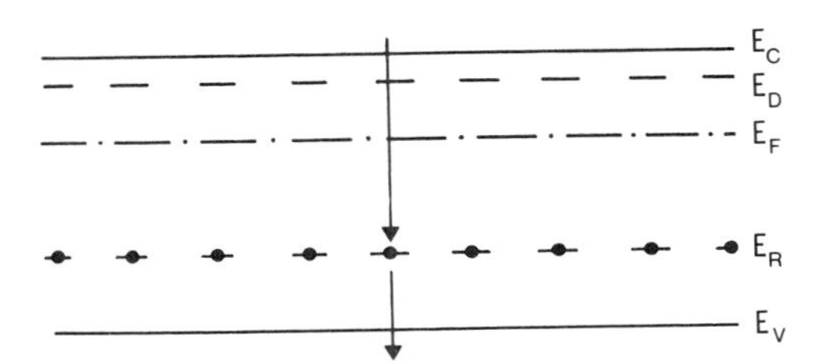

FIGURE 1-36. Schematic thermal equilibrium energy band diagram. E_F is the Fermi level, E_D and E_R are the energies of the shallow donors and the deep recombination centers, respectively. E_c is the conduction band minimum and E_v is the valence band maximum.

are referring to acceptorlike centers that are negatively charged when occupied by electrons.) Neutrality of the material in the absence of free electrons and holes is achieved by the presence of an equal number of shallow donors, which play no further role in the electrical behavior of the device other than to provide the electrons for the acceptors. In the absence of any trap kinetics the net charge density in the insulator through space charge injection is $\rho = -e(n - p)$. On the other hand, when injected holes transit through the semiconductor and are captured by a trap, which in turn captures an electron from the *conduction band,* the net charge density becomes

$$\rho = -e(n - [p + p_R]) \tag{1-115}$$

where p_R denotes the concentration of trapping sites occupied by a hole. Note: (1) We are neglecting recombination with a donor site; (2) we are referring to traps that were originally neutral. If N_R denotes the total concentration of trap sites, and n_R the concentration of traps occupied by an electron, then

$$p_R + n_R = N_R \tag{1-116}$$

The transport equations describing double injection space charge limited current flow under steady state conditions (Lampert and Mark, 1970; Weber and Ford, 1972; Kao, 1983) are: the current density equation

$$J = -nev_e + pev_h = e(n\mu_e + p\mu_h)F \tag{1-117}$$

where v_e, v_h are the drift velocities, and μ_e, μ_h the mobility of electrons and holes; Poisson's equation

$$\frac{\partial F}{\partial x} = \frac{\rho}{\epsilon} = \frac{-e}{\epsilon}[n - (p + p_R)] \tag{1-118}$$

and the continuity equations for electrons and holes under time-independent conditions:

$$\frac{\partial}{\partial x} nv_e = -r_e, \qquad \frac{\partial}{\partial x} pv_h = -r_h \tag{1-119}$$

where the net rate of electron and hole recombination r_e and r_h is regarded as occurring only through the recombination centers. Thus we have

$$r_e \propto np_R \tag{1-120}$$

$$r_h \propto pn_R \tag{1-121}$$

These results state that the recombination is proportional to the product of the number of carriers available for recombination and the number of unoccupied centers. In the process [Eq. (1-120)] an electron is captured by an empty center. For holes [Eq. (1-121)] the process is reversed. A hole is captured by a negatively charged center (density of charged centers n_R). Under time-independent conditions we have

$$r_e = r_h \equiv r \tag{1-122}$$

and thus

$$r \equiv v_{\text{th}}\sigma_n p_R n \equiv \frac{1}{\tau_e} \cdot n \tag{1-123}$$

$$= v_{\text{th}}\sigma_p n_R p \equiv \frac{1}{\tau_h} \cdot p \tag{1-124}$$

where σ_n and σ_p are the capture cross sections, τ_e and τ_h the effective carrier concentration-dependent lifetimes, and v_{th} the thermal velocities. The units of σ are those of cross-sectional area.

Some manipulation of the above equations is called for. Writing the continuity equations as

$$\frac{\partial}{\partial x} nF = \frac{n}{\mu_e \tau_e} \tag{1-125}$$

$$\frac{\partial}{\partial x} pF = -\frac{p}{\mu_h \tau_h} \tag{1-126}$$

The following equation results:

$$\frac{\partial}{\partial x}[(n - p)F] = \frac{(a + 1)}{\mu_h \tau_e} n = \frac{(a + 1)}{\mu_h \tau_h} p \tag{1-127}$$

where $a = \mu_h/\mu_e$.

The boundary conditions for the problem will be assumed to be $F = 0$ at $x = 0$ and $x = L$, which corresponds to ideal injecting contacts. The hole injecting contact is at $x = 0$, and the electron injecting contact is at $x = L$.

The physics of the problem, and the way it is solved, is as follows. At low current or voltage levels holes penetrate the material a short way where they are captured by the recombination centers. Recombination results in the ejection of an electron into the conduction band from the cathode. With the exception of a region near the cathode, where the space charge accumulates, electrons freely travel across the device from the cathode to the anode. They provide the entire mechanism by which current flows, and their characteristics are determined by the conditions for space charge limited current flow. At a given value of bias the holes penetrate a specific distance, which increases with increased bias. Until the holes penetrate the entire length of the insulator, the entire current is, to a good approximation, given by the electronic contribution

$$J \cong ne\mu_e F \tag{1-128}$$

At a critical value of bias the holes penetrate to the end of the material, and the occupancy of the centers is inverted, going from electron dominancy to hole dominancy. If we assume, as has been implied above, that the average cross section for capture of a free hole by a filled center greatly exceeds the average cross section for capture of a free electron by an empty center, then at a critical value of voltage the current is suddenly carried by both electrons and holes. In the absence of a constraining load line of the type discussed earlier, the device is expected to switch to the higher current state.

Reducing the voltage under high current conditions does not initially influence the hole occupancy of the centers until a sufficiently low bias is reached when conduction band electrons are captured by the centers and a low current state is reached. In the presence of a dc load, transition to the high current state is expected to follow the path of the load line and a dynamic "S"-shaped characteristic unfolds. The mathematics of the system develops as follows. It is convenient to divide the whole length of the semiconductor into four regions, and consider these individually. This is known as the regional approximation (Lampert and Mark, 1970).

(1) There is a region I, $0 < x < x_1$, in which all holes are trapped by the centers. We are interested in the potential drop across this region. Here, as in the other regions, almost all the current is carried by electrons, whose density n is assumed to be greater than the density of the recombination centers, i.e., $n > N_R$. Within this region it is also assumed that all centers are filled with holes, i.e., $p_R \approx N_R$. Hence the electron lifetime becomes $\tau_e = (v_{\text{th}}\sigma_n N_R)^{-1}$ by Eq. (1-123). It is also assumed that a condition of approximate charge neutrality exists, i.e., $n = p + p_R \approx p + N_R$. Under these conditions Eq. (1-127) becomes

$$N_R \frac{dF}{dx} = \frac{(a + 1)}{\mu_h \tau_e} n \tag{1-129}$$

which, when combined with the equation for total current, including holes,

$$J = [e\mu_e(1 + a)n - e\mu_h N_R]F \tag{1-130}$$

yields the differential equation

$$J = e\mu_h N_R\left(\mu_e \tau_e F \frac{dF}{dx} - F\right) \tag{1-131}$$

It may be verified that the solution to this differential equation, subject to the condition $F = 0$ at $x = 0$, is

$$\frac{F}{F_0} - \ln\left(\frac{F}{F_0} + 1\right) = \frac{x}{x_0} \qquad (1\text{-}132)$$

where

$$F_0 = \frac{J}{e\mu_h N_R} \quad \text{and} \quad x_0 = \mu_e \tau_e F_0 \qquad (1\text{-}133)$$

The potential drop across this region is

$$\Phi_I = \int_0^{x_1} F\,dx \qquad (1\text{-}134)$$

which may be written as

$$\Phi_I = \int_0^{x_1} \frac{dF}{dx}\frac{dx}{dF} F\,dx = \int_0^{F(x_1)} \frac{FdF}{dF/dx} = \frac{x_0}{F_0}\int_0^{F(x_1)} \frac{F^2 dF}{F_0 + F} \qquad (1\text{-}135)$$

or

$$\Phi_I = \frac{x_0}{F_0}\left\{\frac{F(x_1)^2}{2} - F_0 F(x_1) + F_0^2 \ln\left[\frac{F_0 + F(x_1)}{F_0}\right]\right\} \qquad (1\text{-}136)$$

To complete the evaluation of Φ_I, information is needed about $F(x_1)$. In the region (II) adjacent to I, where $0 < x < x_1$, we assume that all centers are largely occupied by electrons and that J is given by Eq. (1-128). But, in region I it was assumed that $n > N_R$. In region II we may expect that at low current levels $n < N_R$. The transition when $n = N_R$ is taken as the boundary between regions I and II. Thus, at $x = x_1$ $J \approx e\mu_e N_R F$ so

$$F(x_1) = aF_0 \qquad (1\text{-}137)$$

and

$$\Phi_I = \frac{\tau_e}{(eN_R)^2 a\mu_h} J^2\left[\frac{a^2}{2} - a + \ln(1 + a)\right] \qquad (1\text{-}138)$$

Also note that

$$x_1 = \frac{\tau_e}{eN_R a} J[a - \ln(1 + a)] \qquad (1\text{-}139)$$

thus both Φ_I and x_1 increase with increasing bias.

(2) In the adjacent region II, $x_1 < x < x_2$, the centers are occupied by electrons, $n < N_R$, $p \ll n$. We next manipulate various terms to obtain the governing equations for this region. First, equating the recombination rates

$$n v_{\text{th}} \sigma_n p_R = p v_{\text{th}} \sigma_p n_R \qquad (1\text{-}140)$$

and again assuming charge neutrality

$$n = p + p_R \tag{1-141}$$

with $p \ll n$, $n \approx p_R$ and thus $n_R \approx N_R$; i.e., almost all centers are occupied by electrons. Then from Eq. (1-140),

$$p \approx \frac{\sigma_n}{\sigma_p}\frac{n^2}{N_R} \cong \frac{\sigma_n}{\sigma_p}\frac{nJ}{Fe\mu_e}\frac{1}{N_R} = \frac{n}{F}F_0\frac{\sigma_n}{\sigma_p}a \tag{1-142}$$

We use this term along with the continuity equation for holes to obtain

$$\frac{d}{dx}pF = F_0\frac{\sigma_n}{\sigma_p}a\frac{dn}{dx}$$

$$= -\frac{r}{\mu_h} \tag{1-143}$$

But $r = nv_{\text{th}}\sigma_n p_R \approx n^2 v_{\text{th}}\sigma_n$, so

$$F_0\frac{\sigma_n}{\sigma_p}a\frac{dn}{dx} = -\frac{n^2 v_{\text{th}}\sigma_n}{\mu_h} \tag{1-144}$$

or

$$\frac{dn}{dx} = \frac{-n^2 v\sigma_P}{aF_0\mu_h} \cong \frac{-n^2}{N_R}\frac{1}{a\mu_h}\frac{1}{\tau_h F_0} \tag{1-145}$$

where

$$\tau_h = (v_{\text{th}}\sigma_p N_R)^{-1} = \tau_e\sigma_n/\sigma_p \tag{1-146}$$

is the hole lifetime in region II. Thus

$$\frac{dn}{dx} = -\frac{e}{aJ\tau_h}n^2 \tag{1-147}$$

is the governing equation for n. The solution is

$$\frac{1}{n} = \frac{e}{aJ\tau_h}(x - x_1) + C \tag{1-148}$$

where at $x = x_1$, $n = N_R$ so that

$$\frac{1}{n} = \frac{1}{N_R} + \frac{e}{aJ\tau_h}(x - x_1) \tag{1-149}$$

We next use the above expression to find the field

$$F = \frac{J}{ne\mu_e} \cong \frac{J}{N_R e\mu_e} + \frac{1}{\mu_h \tau_h}(x - x_1) \tag{1-150}$$

or

$$F = aF_0\left(1 + \frac{eN_R}{aJ\tau_h}[x - x_1]\right) \tag{1-151}$$

The potential drop across this region is

$$\Phi_{\text{II}} = aF_0\left([x_2 - x_1] + \frac{eN_R}{aJ\tau_h}[x_2 - x_1]^2\right) \tag{1-152}$$

We next require a determination of x_2. This is somewhat uncertain. It is assumed that region II ends when the approximation of local charge neutrality loses its self-consistency. Thus, at $x = x_2$ it is expected that

$$\left.\frac{dF}{dx}\right|_{x_2} \cong \frac{e}{\epsilon} p_R(x_2) \approx \frac{e}{\epsilon} n(x_2) \tag{1-153}$$

For mathematical consistency, from Eq. (1-150) we have

$$\left.\frac{\partial F}{\partial x}\right|_{x_2} = \frac{1}{\mu_h \tau_h} = \frac{e}{\epsilon} n(x_2) \tag{1-154}$$

and from the expression for carrier density (1-149) we have

$$\frac{1}{n(x_2)} = \frac{e}{\epsilon}\mu_h \tau_h = \frac{1}{N_R}\left[1 + \frac{eN_R}{aJ\tau_h}(x_2 - x_1)\right] \tag{1-155}$$

or

$$x_2 - x_1 = \left(\frac{e}{\epsilon}\frac{\mu_h}{v_{\text{th}}\sigma_p} - 1\right)\frac{aJ\tau_h}{eN_R} \equiv \left(\frac{\tau_h}{\tau_R} - 1\right)\frac{aJ\tau_h}{eN_R} \tag{1-156}$$

where $\tau_R = \epsilon/(N_R e\mu_h)$ is an effective dielectric relaxation time. Thus, by (1-152) we have

$$\Phi_{\text{II}} = aF_0\left(\frac{\tau_h}{\tau_R} - 1\right)\frac{aJ\tau_h}{eN_R}\left[1 + \frac{1}{2}\left(\frac{\tau_h}{\tau_R} - 1\right)\right] \tag{1-157}$$

or

$$\Phi_{\text{II}} = \frac{F_0 a^2 J\tau_h}{2\, eN_R}\left(\left[\frac{\tau_h}{\tau_R}\right]^2 - 1\right) \tag{1-158}$$

or

$$\Phi_{II} = \tfrac{1}{2}J^2 \frac{a^2 \mu_h \tau_h^3}{\epsilon^2}\left[1 - \left(\frac{\tau_R}{\tau_h}\right)^2\right] \tag{1-159}$$

Note that

$$x_2 - x_1 = a^2 \frac{\tau_h}{\tau_R}\left(1 - \frac{\tau_R}{\tau_h}\right)\frac{\mu_e \tau_h J}{N_R e \mu_h} \tag{1-160}$$

$$= J\frac{a^2 \mu_e}{\epsilon}\tau_h^2\left(1 - \frac{\tau_R}{\tau_h}\right) \tag{1-161}$$

so both $(x_2 - x_1)$ and ϕ_{II} increase with increasing current.

(3) Region III is $x_2 < x < x_3$. Within regions I and II the approximations to make are relatively clear. Region III is more difficult. It is here that the electric field begins to turn around and space charge effects are significant. Within region III the hole density is negligible compared to n and Poisson's equation is given approximately by

$$\frac{dF}{dx} = \frac{e}{\epsilon}[p_R - n] \tag{1-162}$$

As seen in the preceding discussion, the important quantities are the widths of the individual regions. More often than not these values are determined by consistency of the physical assumptions rather than on hard knowledge of its value. The procedure is interpolative and takes advantage of the fact that in regions where analytical procedures offer difficulty, the needed quantity emphasizes the order of the approach. It is the potential that is offered here and to obtain it an interpolation scheme is presented. It is assumed that the slope of the field is known at the end points of the region. If $F_2(x)$ represents the equation of the field near $x = x_2$, then a first-order Taylor expansion provides

$$F_2(x) \approx F(x = x_2) + \left(\frac{dF}{dx}\right)_{x=x_2}(x - x_2) \tag{1-163}$$

Similarly

$$F_3(x) = F(x = x_3) + \left(\frac{dF}{dx}\right)_{x=x_3}(x - x_3) \tag{1-164}$$

The field everywhere between x_2 and x_3 is obtained from the scheme

$$F(x) \doteq \frac{F_2(x)(x_3 - x) + F_3(x)(x - x_2)}{x_3 - x_2} \tag{1-165}$$

and from this expression the potential drop is computed. We need the equilibrium quantities. For this we concentrate on the left-hand region, namely, $x = x_2$. The field at $x = x_2$ is obtained from Eqs. (1-151) and (1-156):

$$F(x_2) = aF_0\tau_h/\tau_R = aJ\tau_h/\epsilon \tag{1-166}$$

The slope of the field at $x = x_2$ is obtained from Eq. (1-150);

$$\left.\frac{dF}{dx}\right|_{x_2} = \frac{1}{\mu_h \tau_h} \tag{1-167}$$

and from Eqs. (1-150) and (1-139),

$$x_2 = x_0\left\{[a - \ln(1 + a)] + \left(\frac{\tau_h}{\tau_R} - 1\right)a^2 \frac{\sigma_n}{\sigma_p}\right\} \tag{1-168}$$

To obtain expressions for these quantities at the $x = x_3$ boundary it is necessary to return to the governing equations. In region II it was assumed that $p \ll n \approx p_R \ll N_R$, $n_R \approx N_R$. The difference in region III is that *charge neutrality (Eq. 1-141) no longer holds*. Thus, $n = p_R$ no longer holds, and there will be some modifications of the results for region II. To develop these equations we start from the continuity equation as expressed by Eq. (1-143), and replace n by p_R. Then we have

$$\frac{d}{dx} pF \approx F_0 \frac{\sigma_n}{\sigma_p} a \frac{dp_R}{dx} \tag{1-169}$$

$$= \frac{-r}{\mu_h} \tag{1-170}$$

with $r = nv_{\text{th}}\sigma_n p_R \cong v_{\text{th}}\sigma_n p_R^2$. The relevant equation is

$$\frac{dp_R}{dx} = -\frac{e}{aJ\tau_h} p_R^2 \tag{1-171}$$

which is the same as Eq. (1-147) with n replaced by p_R. Thus, we have

$$\frac{1}{p_R} = \frac{1}{p_{R_2}} + \frac{e}{aJ\tau_h}(x - x_2) \tag{1-172}$$

To obtain an expression for p_R at the boundary $x = x_3$ some provisional information is needed. The highest concentration of mobile electrons is in region IV, and this concentration decreases as the carriers move toward the anode. Fairly arbitrarily, we will assume that at $x = x_3$, $p_R \approx n/2$. Thus, Poisson's equation, which in region IV is given approximately by

$$\frac{dF}{dx} = -\frac{en}{\epsilon} \tag{1-173}$$

becomes at $x = x_3$

$$\left.\frac{dF}{dx}\right|_{x_3} = -\frac{e}{\epsilon} 2p_R\bigg|_{x_3} \tag{1-174}$$

Further evaluation of this requires knowledge of the field within region IV. Borrowing a result to be obtained later, we have

$$F = \left(\frac{2J}{\epsilon\mu_e}\right)^{1/2}(L - x)^{1/2} \tag{1-175}$$

and

$$F(x_3) = \left(\frac{2J}{\epsilon\mu_e}\right)^{1/2}(L - x_3)^{1/2} \tag{1-176}$$

Also,

$$\frac{dF}{dx} = -\frac{1}{2}\left(\frac{2J}{\epsilon\mu_e}\right)^{1/2}(L - x)^{-1/2} \tag{1-177}$$

which at x_3 is

$$\left.\frac{dF}{dx}\right|_{x_3} = -\frac{1}{2}\left(\frac{2J}{\epsilon\mu_e}\right)^{1/2}(L - x_3)^{-1/2} \tag{1-178}$$

Combining Eqs. (1-172), (1-174), and (1-178), we obtain

$$\begin{aligned}\left.\frac{1}{p_R}\right|_{x_3} &= \frac{2e}{\epsilon}\left(\frac{\epsilon\mu_e}{2J}\right)^{1/2}(L - x_3)^{1/2} \\ &= \frac{1}{p_R(x_2)} + \frac{e}{aJ\tau_h}(x_3 - x_2)\end{aligned} \tag{1-179}$$

Using Eqs. (1-153) and (1-154) to eliminate $p_R(x_2)$ we find

$$\frac{2e}{\epsilon}\left(\frac{\epsilon\mu_e}{2J}\right)^{1/2}(L - x_3)^{1/2} = \frac{e\mu_h\tau_h}{\epsilon} + \frac{e}{aJ\tau_h}(x_3 - x_2) \tag{1-180}$$

which is a defining equation for x_3. Simplifying, we obtain

$$\left(\frac{2J\mu_e}{\epsilon}\right)^{1/2}(L - x_3)^{1/2} = \frac{J\tau_h\mu_h}{\epsilon} + \frac{x_3 - x_2}{a\tau_h} \tag{1-181}$$

Thus, with Eqs. (1-176) and (1-181) we now have a consistent specification of $F(x_3)$, and are now in a position to perform the necessary integration for the potential drop:

$$\Phi_{\text{III}} = \int_{x_2}^{x_3} \frac{F_2(x)(x_3 - x) + F_3(x)(x - x_2)}{x_3 - x_2}\,dx \tag{1-182}$$

$$\begin{aligned} &= \frac{1}{x_3 - x_2}\int_{x_2}^{x_3} [F(x_2)(x_3 - x) + F'(x_2)(x - x_2)(x_3 - x) \\ &\quad + F(x_3)(x - x_2) + F'(x_3)(x - x_3)(x - x_2)]dx \end{aligned} \tag{1-183}$$

$$= \tfrac{1}{2}[F(x_2) + F(x_3)](x_3 - x_2) + \frac{1}{6}\left(\left.\frac{dF}{dx}\right|_{x_2} - \left.\frac{dF}{dx}\right|_{x_3}\right)(x_3^2 - x_2^2 - 6x_2x_3) \tag{1-184}$$

or

$$\Phi_{\mathrm{III}} = \frac{1}{2}\left[\frac{aJ\tau_h}{\epsilon} + \left(\frac{2J}{\epsilon\mu_e}\right)^{1/2}(L - x_3)^{1/2}\right](x_3 - x_2)$$
$$+ \frac{1}{6}\left[\frac{1}{\mu_h\tau_h} + \frac{1}{2}\left(\frac{2J}{\epsilon\mu_e}\right)^{1/2}(L - x_3)^{-1/2}\right](x_3^2 - x_2^2 - 6x_2x_3) \quad (1\text{-}185)$$

where Eqs. (1-166), (1-167), (1-176), (1-178) have been used, and x_2 and x_3 are given by Eqs. (1-168) and (1-181), respectively.

(4) In region IV, $x_3 < x < L$, $p \ll p_R \ll n$, $n_R \approx N_R$, and only injected free electrons are of significance. Poisson's equation is given by Eq. (1-173), and the current by Eq. (1-128). The relevant equation for this region is

$$\frac{F\,dF}{dx} = -\frac{J}{\mu_e\epsilon} \quad (1\text{-}186)$$

with the solution given by Eq. (1-175). The carrier density away from the cathode is

$$n = -\frac{\epsilon}{e}\frac{dF}{dx} = \left(\frac{\epsilon}{2e^2}\frac{J}{\mu}\right)^{1/2}(L - x)^{-1/2} \quad (1\text{-}187)$$

The potential drop across region IV is

$$\Phi_{\mathrm{IV}} = \frac{2}{3}\left(\frac{2J}{\epsilon\mu_e}\right)^{1/2}(L - x_3)^{3/2} \quad (1\text{-}188)$$
$$= \tfrac{2}{3}F_3(L - x_3) \quad (1\text{-}189)$$

where F_3 is the field at $x = x_3$.

The net voltage drop across the element is the sum across all regions; using Eqs. (1-138), (1-159), (1-185), and (1-188), provided that all these regions can actually be accommodated within the length L of the semiconductor, we obtain

$$\Phi = J^2\tau_e\tau_R^2\mu_e\epsilon^{-2}[\tfrac{1}{2}a^2 - a + \ln(1 + a)] + \tfrac{1}{2}J^2a^2\mu_h\tau_h^3\epsilon^{-2}\left[1 - \left(\frac{\tau_R}{\tau_h}\right)^2\right]$$
$$+ \frac{1}{2}\left[\frac{aJ\tau_h}{\epsilon} + \left(\frac{2J}{\epsilon\mu_e}\right)^{1/2}(L - x_3)^{1/2}\right](x_3 - x_2)$$
$$+ \frac{1}{6}\left[\frac{1}{\mu_h\tau_h} + \frac{1}{2}\left(\frac{2J}{\epsilon\mu_e}\right)^{1/2}(L - x_3)^{-1/2}\right](x_3^2 - x_2^2 - 6x_2x_3) + \frac{2}{3}\left(\frac{2J}{\epsilon\mu_e}\right)^{1/2}(L - x_3)^{3/2} \quad (1\text{-}190)$$

where x_3 is obtained from the defining equation (1-181).

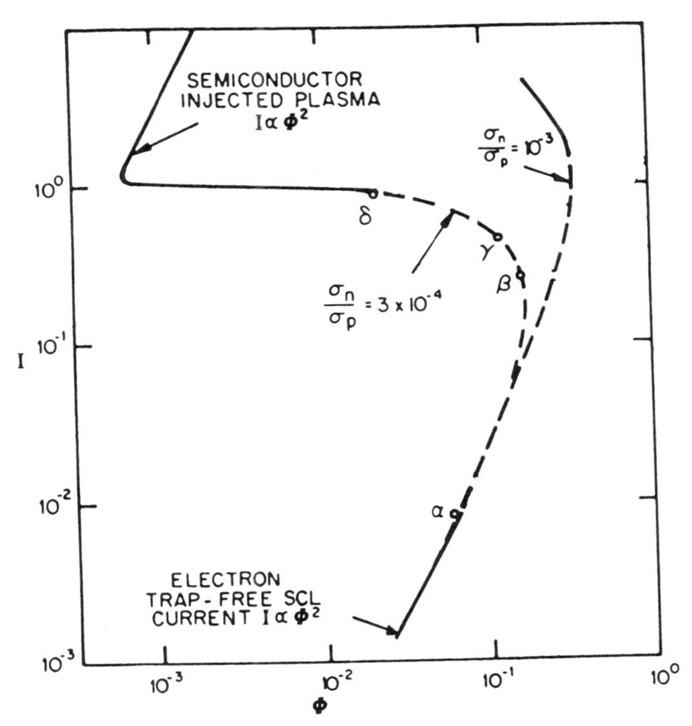

FIGURE 1-37. Universal current (I)-voltage (Φ) characteristics in arbitrary units for double injection and field-induced trapping in a pin diode. The two calculated cases are $\sigma_n/\sigma_p = 10^{-3}$ and $\sigma_n/\sigma_p = 3 \times 10^{-4}$. (After Lampert and Mark, 1970.)

$\Phi(I)$ curves for two cases are shown in Fig. 1-37. For both curves $\epsilon/\epsilon_0 = 12$, $\mu_e = \mu_h = 10^4\,\text{cm}^2/\text{V s}$, and $N_R = 10^{15}/\text{cm}^3$. For the lower curve $\sigma_n v_{\text{th}} = 9 \times 10^{-10}\,\text{cm}^3/\text{s}$ and $\sigma_p v_{\text{th}} = 3 \times 10^{-6}\,\text{cm}^3/\text{s}$. For the upper curve $\sigma_n v_{\text{th}} = 10^{-9}\,\text{cm}^3/\text{s}$ and $\sigma_p v_{\text{th}} = 10^{-6}\,\text{cm}^3/\text{s}$. With regard to the relative occupation of each region, on the lower curve at point α region IV occupied 83% of the device length and 81% of its voltage. The transition region III virtually occupies the rest. At point β, the neutrality regions I and II are important; they take up 32% of the semiconductor width but only a negligible fraction of voltage. Region III takes 43% of the width and 72% of the voltage; region IV 25% of the width and 26% of the voltage. At point δ regions I and II occupy 99% of the semiconductor width and 98% of the voltage. Note that in the high-current region the $\Phi(I)$ relation is that of an injected plasma. We will confront this problem again in Chap. 7.

1.4. SUMMARY

In our initial chapter we have introduced the subject of NDC in solids and pointed out the two major classes: SNDC and NNDC. After describing a variety of device structures, we developed some of the mathematics pertinent to our projected needs, with emphasis on stability and bifurcation, self-sustained oscillations, solitons, and chaos. We then provided four examples by which NDC occurs: three mechanisms for NNDC: the McCumber–Chynoweth model for intervalley transfer; transfer in equivalent valley semiconductors; real space transfer in modulation doped heterostructures, and a mechanism for SNDC: field-induced trapping in semiconductors. We are now fortified with some of the tools required to understand a wide variety of electron devices that depend on desirable NDC-induced instabilities for their useful properties. Our next task is to understand how these instabilities lead to inhomogeneous field and current density distributions, and how the local circuit environment both responds to the instability and controls its subsequent behavior.

2

Stability

2.1. SEMICONDUCTORS AS NONLINEAR DYNAMIC SYSTEMS—OSCILLATORY INSTABILITIES

Semiconductor devices may be considered as nonlinear dynamic systems that can exhibit electrical instabilities such as: current run-away; threshold switching between low- and high-conductivity states; self-organized formation of spatio-temporal patterns; and spontaneous periodic or chaotic current oscillations, if an external control parameter, e.g., an applied bias voltage, magnetic field, or optical irradiation intensity is varied (Schöll, 1987).

In this section we present a general theoretical approach to nonlinear charge transport and electrical instabilities in semiconductors in the language of dynamic systems, and apply it to obtain some general results on oscillatory instabilities (Schöll, 1989b).

We consider charge transport in a semiconductor described by the electric field F and a set of additional macroscopic dynamic transport variables $\mathbf{q} \equiv (q_1, \ldots, q_N)$. These may represent, for example, the mean carrier densities in the conduction band and in impurity levels, the mean carrier energy, or the radius of a current filament. In Sec. 2.1, however, we shall restrict ourselves to spatially uniform states only, although the results may be generalized to nonuniform current flow.

The dynamics are given by the equation for the total current density composed of the displacement current and the drift current:

$$J = \epsilon \frac{d}{dt} F + \sigma(\mathbf{q}, F)F \tag{2-1}$$

where ϵ is the permittivity, and σ is the (non-Ohmic) conductivity, and by a set of additional nonlinear transport equations

$$\frac{d}{dt}\mathbf{q} = \mathbf{f}(\mathbf{q}, F) \tag{2-2}$$

If the semiconductor is connected to a resistive external circuit with bias voltage Φ_B and load resistance R, (2-1) can be cast into the form

$$\frac{d}{dt}F = g(\mathbf{q}, F; J_0) \equiv [J_0 - (\sigma_L + \sigma(\mathbf{q}, F))F]/\epsilon \tag{2-3}$$

with control parameter $J_0 \equiv \Phi_B/(RS)$ and $\sigma_L \equiv l/(RS)$ where l and S are the length and the cross section of the semiconductor, respectively. Here the field F has been assumed spatially homogeneous.

The steady state solutions of the nonlinear dynamic system (2-2), (2-3) will be denoted by zero subscript. Their stability with respect to small fluctuations $(\delta\mathbf{q}(t), \delta F(t))$ follows from the linearization of (2-2) and (2-3) around $(\mathbf{q}_0, F_0)$:

$$\frac{d}{dt}\begin{pmatrix}\delta\mathbf{q}\\ \delta F\end{pmatrix} = A\begin{pmatrix}\delta\mathbf{q}\\ \delta F\end{pmatrix} \tag{2-4}$$

with

$$A_{ij} = \tilde{A}_{ij} \equiv \left.\frac{\partial f_i}{\partial q_i}\right|_0, \qquad \text{for } i, j \leq N$$

$$A_{i,N+1} = \left.\frac{\partial f_i}{\partial F}\right|_0$$

$$A_{N+1,i} = \left.\frac{\partial g}{\partial q_i}\right|_0$$

$$A_{N+1,N+1} = \left.\frac{\partial g}{\partial F}\right|_0$$

The static current density–field characteristic (including the load resistor) is determined by $g(\mathbf{q}(F_0), F_0; J_0) = 0$, where $\mathbf{q}(F_0)$ follows from $d\mathbf{q}/dt = 0$, and the static differential conductivity is given by

$$\sigma_{\text{diff}} \equiv \frac{d}{dF_0}[\sigma(\mathbf{q}(F_0), F_0)F_0] \tag{2-5}$$

Using

$$\frac{dg(\mathbf{q}(F_0), F_0; J_0)}{dF_0} = -(\sigma_L + \sigma_{\text{diff}})/\epsilon$$

and an expansion of $\det A$ in terms of $\det \tilde{A}$, we obtain the important general expression

$$\det A = \prod_{i=1}^{N+1} \lambda_i = -\det \tilde{A}(\sigma_L + \sigma_{\text{diff}})/\epsilon \tag{2-6}$$

This relation connects the differential conductivity σ_{diff} with the stability of the steady states, which is determined by the sign of the real parts of the eigenvalues λ_i of the Jacobian matrix A; see Sec. 1.2.1.

We note that λ_i factors into a part $\det \tilde{A}$, which is determined by the transport Eqs. (2-2), and the inverse differential dielectric relaxation time

$$\lambda_M^{\text{diff}} \equiv (\sigma_L + \sigma_{\text{diff}})/\epsilon$$

The eigenvalues λ_i are given by the roots of the characteristic polynomial of A

$$\sum_{n=0}^{N+1} g_n \lambda^n = 0 \tag{2-7}$$

where $g_{N+1} = 1$, $g_N = -\operatorname{tr} A = -\sum_i \lambda_i$, $g_0 = (-1)^{N+1} \det A$.

We shall now exploit the relation (2-6) to obtain general conditions for the onset of electrical instabilities, which are independent of the particular transport equations used. To this purpose we shall study different bifurcations of the dynamic system (2-2) and (2-3).

First, we shall consider saddle-node bifurcations of fixed points (steady states; see Fig. 1-9). They are given by a vanishing eigenvalue λ_i, and hence, by (2-6), (2-7), $g_0 = 0$, i.e., $\det \tilde{A} = 0$ or $\sigma_L + \sigma_{\text{diff}} = 0$.

These bifurcations mark the onset of the range of bistability of the steady states as a function of the control parameter J_0. For specific transport models describing impact ionization breakdown, $\det \tilde{A} = 0$ has been shown to be associated with the onset of SNDC in the voltage-controlled case ($R = 0$), whereas $\sigma_{\text{diff}} = 0$ is associated with the onset of NNDC in the current-controlled case ($R \to \infty$ or, equivalently, $\sigma_L = 0$). For a general load resistance R, $\sigma_L + \sigma_{\text{diff}} = 0$ marks the point where the load line becomes tangential to the current–voltage characteristic. This will be discussed in detail in Sec. 2.2 (see Fig. 2-1e).

Next, we consider Hopf bifurcations of limit cycles (i.e., self-generated oscillations). They mark one possible onset of oscillatory instabilities. There are others, associated, e.g., with global bifurcations (cf. Sec. 1.2.1 C, D and Figs. 1-11, 1-12), or with moving field domains (see Sec. 2.2.2, or circuit-induced, see Sec. 2.3). A Hopf bifurcation is singled out by a pair of complex conjugate eigenvalues crossing the imaginary axis. For $N = 1$, i.e., a single additional transport variable, the condition is $g_1 = -\operatorname{tr} A = 0$, $g_0 = \det A > 0$, and hence, by (2-4), (2-6) with $\tilde{\lambda} \equiv \det \tilde{A}$, $\tilde{\nu} \equiv \{\sigma_L + (\partial/\partial F)[\sigma(q, F)F]\}/\epsilon = -A_{22}$,

$$\begin{gathered} \tilde{\lambda} - \tilde{\nu} = 0 \\ -\tilde{\lambda}(\sigma_L + \sigma_{\text{diff}}) > 0 \end{gathered} \tag{2-8}$$

Physically, $\tilde{\lambda}$ is the eigenvalue determining the stability of the transport Eq. (2-2) for fixed field F, while $\tilde{\nu}$ is (for $\sigma_L = 0$) proportional to the differential mobility dv/dF, if the drift current density $J = \sigma F = env(F)$ is defined in the usual way by the carrier density n and the drift velocity $v(F)$.

Equations (2-8) can be satisfied by either

$$\text{(i)}\ \sigma_{\text{diff}} + \sigma_L < 0 \quad \text{and} \quad \tilde{\lambda} = \tilde{\nu} > 0 \tag{2-9}$$

or

$$\text{(ii)}\ \sigma_{\text{diff}} + \sigma_L > 0 \quad \text{and} \quad \tilde{\lambda} = \tilde{\nu} < 0 \tag{2-10}$$

For $\sigma_L = 0$, Eq. (2-9) corresponds to negative differential conductivity, positive differential mobility ($\tilde{\nu} > 0$) and instability of the q variable ($\tilde{\lambda} > 0$), and (2-10)

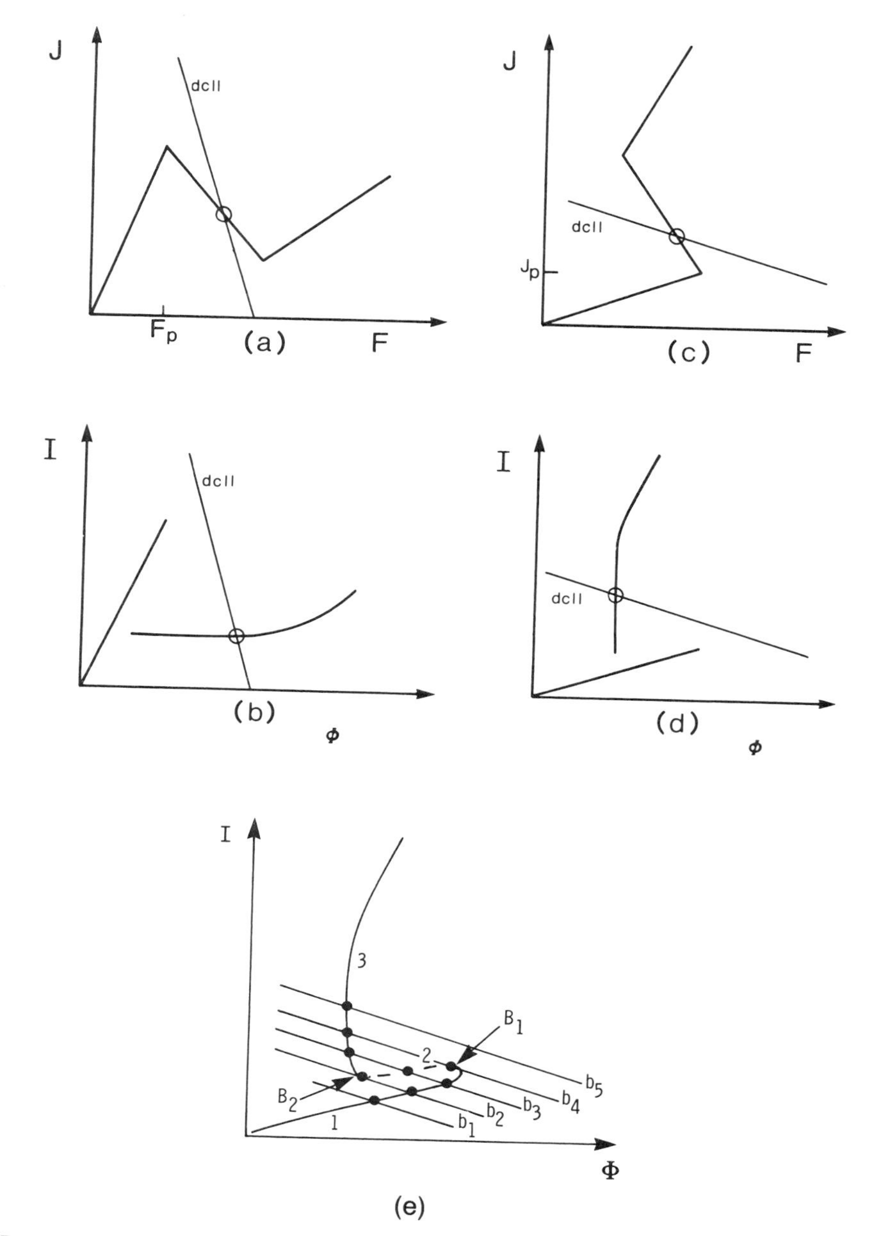

FIGURE 2-1. Schematic $J(F)$ (current density–field) and $I(\Phi)$ (current–voltage) characteristics. The dc load line is marked by dcll, and the circle marks the steady state operating point, which is unstable for (a) NNDC, and (c) SNDC. This leads to the formation of inhomogeneous field and current distributions corresponding to stable operating points with positive differential conductance states in (b) (field domain) and (d) (current filament), respectively. Part (e) shows the steady state operating points (solid circles) for a sequence of increasing applied bias values, corresponding to dc load lines $b_1, \ldots, b_5$, in an SNDC device. Branches 1, 2, 3 denote stable homogeneous, unstable filamentary, and stable filamentary $I(\Phi)$ branches, respectively. Upon increase and decrease of the bias, hysteretic jumps occur at B_1 and B_2, respectively.

corresponds to positive differential conductivity, negative differential mobility and stability of the q variable.

For $N > 1$, i.e., more than one additional transport variable besides the field F, more complex situations are possible, but the essential result is that oscillatory instabilities are still possible with either (i) negative differential conductivity or (ii) positive differential conductivity. These conditions are generalizations of the conditions that were first derived for oscillatory instabilities in the case where the additional dynamic variables $\mathbf{q}$ are given by carrier densities, and $\sigma_L = 0$ (Schöll, 1987).

The existence of limit cycle oscillations is a general precondition for the occurrence of further bifurcations, which can lead to more complicated, quasi-periodic or chaotic oscillations for $N > 1$, as discussed in Secs. 1.2.1 and 1.2.4. While chaotic solutions can only be found by numerical simulation of the nonlinear dynamic equations (2-2), (2-3), the simple analytical conditions (2-8), (2-9), (2-10) for the onset of periodic oscillations are still useful, since they describe typical situations where chaotic behavior may be expected upon further increase of a control parameter.

Examples of oscillatory instabilities occurring on the *negative* differential conductivity branch of the $J(F)$ characteristic are provided by impact-ionization induced self-generated dielectric relaxation oscillations in the regime of low-temperature impurity breakdown. Indeed, a Hopf bifurcation of a limit cycle and a subsequent period-doubling route to chaos were obtained numerically. The model has been extended to describe oscillations of breathing current filaments (Schöll and Drasdo, 1990) within the same approach as given by Eq. (2-9) where σ_{diff} now denotes the radially averaged differential conductance. Such oscillations were observed in p-Ge and n-GaAs (Rau et al., 1989).

Examples of oscillatory instabilities occurring on the *positive* differential conductivity branch of the $J(F)$ characteristic are given by models for an optical phonon induced subcritical Hopf bifurcation in p-Ge (Hüpper et al., 1989), for a Poole–Frenkel emission induced supercritical Hopf bifurcation describing low-frequency oscillations in semiinsulating GaAs (Schöll, 1989b), and for the dynamic Hall effect (Hüpper and Schöll, 1991).

An important conclusion of this section is that oscillatory electrical instabilities may occur not only in regimes of negative differential conductivity, but also *in regimes of positive differential conductivity* of the static current–voltage characteristic. Note that we have considered *spatially homogeneous* fluctuations only, and thus obtained bulk-dominated oscillations. Stability with respect to spatially inhomogeneous oscillations will be considered in Sec. 2.2.

2.2. STABILITY OF HOMOGENEOUS NDC POINTS—DOMAINS AND FILAMENTS

2.2.1. Introduction

In Chap. 1 we discussed the concept of a load line and defined the steady state operating point as the intersection of the dcll with the $I(\Phi)$ characteristics of the sample. For the usual case where the current density–electric field, $J(F)$, relation of the sample has $dJ/dF > 0$ for all values of $J(F)$, we have shown in Sec. 2.1 that the static operating points in the $I(\Phi)$ plane may or may not be stable (they are generally stable). On the other hand, NDC samples contain a range of fields where $dJ/dF < 0$, and this commonly leads to situations where instabilities are produced and no steady state operating point exists (Shaw et al., 1979). Because of these features, it becomes useful to define operating points and load lines in both the $J(F)$ (constitutive) and $I(\Phi)$ (laboratory) planes, where, as defined in Chap. 1,

$$I = \int_S \mathbf{J} \cdot \hat{\mathbf{n}}\, dS \tag{2-11}$$

$$\Phi = -\int_0^l \mathbf{F} \cdot d\mathbf{x} \tag{2-12}$$

where l is the length of the sample, S the cross-sectional area, and $\hat{\mathbf{n}}$ a vector normal to S. In what follows we again drop the vector notation on J and F for convenience, assuming isotropy.

We shall first ask about the stability of operating points in the $J(F)$ plane, where NDC regions can exist. For example, Fig. 2-1a shows a schematic NNDC $J(F)$ curve with the dcll intersecting it in the NDC region. As we shall show, this case is unstable against the formation of either moving or stationary inhomogeneous field distributions (*high field domains*). A homogenous-field NNDC sample is shown to be unstable against small fluctuations in field; the sample becomes electrically nonuniform and this nonuniformity displays itself in the $I(\Phi)$ characteristics. One possibility is shown in Fig. 2-1b, where we show the $I(\Phi)$ characteristics resulting from a moving high field domain. Note that here the dcll intersects the $I(\Phi)$ characteristics in a P (positive) DC (conductance) region ($dI/d\Phi > 0$). In fact, the $I(\Phi)$ curve essentially *does not display a region of NDC (conductance)*. The intersection now lies on the "domain characteristic" part of the $I(\Phi)$ curve. This is a common phenomenon in instability electronics: a specific mechanism produces a region of NDC in the constitutive curve $J(F)$; this manifests itself as an instability that causes the $I(\Phi)$ curve to become nonlinear, show discontinuities, switch along the load line, or develop oscillatory modes. In most of these cases, however, $dI/d\Phi < 0$ *is not observed* in the $I(\Phi)$ plane. We do not mean to imply that the situations where $dI/d\Phi < 0$ *never* occur. On the contrary, they are sometimes observed in special situations. For example, we have been considering the case of *isothermal, bulk* NDC effects. Here cases where $dI/d\Phi < 0$ can only be observed during certain times when inhomogeneous (domain or filamentary) situations are near the limits of their maintainability. Indeed, a controversial "positive conductance theorem" exists (Shockley, 1954; Kroemer, 1970, 1971) which states that bulk-induced $dI/d\Phi < 0$ situtations can *never* be observed. However, for samples where the NDC is produced by a junction effect (Lampert and Mark, 1970; Weber and Ford, 1970) or a filamentary generation–recombination mechanism (Schöll, 1983) or in nonisothermal situations where $dI/d\Phi < 0$ results from heating (Jackson and Shaw, 1974), situations where $dI/d\Phi < 0$ can be stabilized and observed.

As discussed in a general context in Chap. 1, the loss of stability is intimately connected with the intersection or coalescence of different solution branches, i.e., a bifurcation. Now, choosing J as the control parameter in Fig. 2-1a, or F in Fig. 2-1c, we find a close similarity between the $F(J)$ or $J(F)$ curves and the bifurcation shown in Fig. 1-9 (A1). If the PDC states are stable, the NDC states must necessarily be unstable. The instability of certain branches of the $J(F)$ or $I(\Phi)$ characteristics thus appear to be a generic property of the topology and the bifurcation behavior, rather than being simply determined by the sign of the differential conductivity. This is of particular relevance for $I(\Phi)$ curves that have a more complicated shape than a simple N or S. Such curves can occur if the external circuit, or filamentary states are included (Schöll, 1987), and they may contain stable portions with $dI/d\phi < 0$. For general circuit conditions the applied bias can be taken as a control parameter, rather than I or Φ, and the bifurcation point then marks the separation between stable and unstable states. This is illustrated in Fig. 2-1e.

Operating points in the $J(F)$ plane are generally analyzed for stability against variations in electric field profiles, current density distributions, additional internal transport variables such as carrier densities, and external circuit parameters. The stability of field and current density distributions are analyzed via Maxwell's equations, whereas circuit stability is analyzed via the appropriate circuit equations. A complete analysis requires the simultaneous solution of both stability problems. In the next sections we will first discuss the stability of operating points in an ideal circuit where the resistive components dominate completely over the reactive components. Afterward, we will attack the more realistic case where circuit stability becomes important. Here we will find that situations can occur where no stable operating point can be found in the $I(\Phi)$ plane and only oscillatory solutions occur.

In order to investigate the stability of spatially uniform operating points in an ideal nonreactive circuit, we will study the evolution of small space-and time-dependent fluctuations of the electric field and the charge density. We will see that for NDC points certain fluctuations can grow and finally lead to steady, nonuniform spatial field or current distributions. In particular, NDC points can be unstable against fluctuations with a spatial dependence in the direction of the current flow (Fig. 2-2a), which leads to the formation of moving or static layerlike field inhomogeneities (*domains*), with $I(\Phi)$ characteristics as in Fig. 2-1b, or unstable against fluctuations with a spatial modulation perpendicular to the

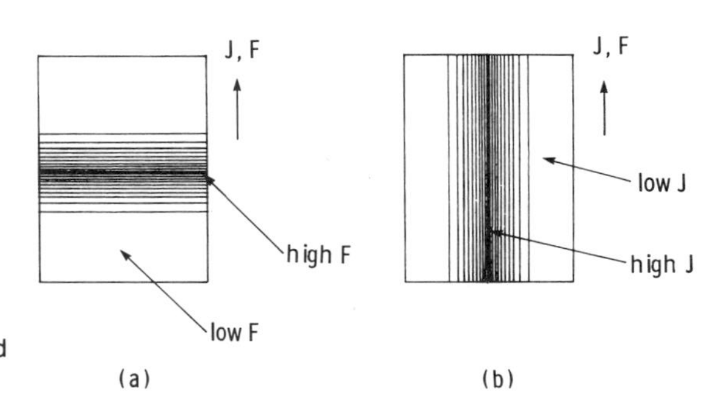

FIGURE 2-2. Sketch of (a) a high-field domain, (b) a high-current filament.

current flow (Fig. 2-2b), which leads to the formation of inhomogeneous current density distributions over its cross section (*filaments*) with $I(\Phi)$ characteristics as in Fig. 2-1d. Whether the final steady state of the system contains a domain, a filament, or both, will be shown to depend upon the detailed shape of the entire constitutive $J(F)$ curve. A general thermodynamic argument (Ridley, 1963) which predicted the formation of domains in NNDC systems and the formation of filaments in SNDC systems has turned out to be invalid since the invoked principle of minimum entropy production holds only in the linear, near equilibrium regime of thermodynamics for systems with Onsager symmetry (e.g., Schlögl, 1980), whereas instabilities must occur far from thermal equilibrium.

We will also show that the final steady state inhomogeneous solutions are governed by "equal areas" rules (Butcher, 1967; Adler et al., 1980; Schöll, 1981, 1986a, 1987; Schöll and Landsberg, 1988, Wu et al., 1990). The steady state $I(\Phi)$ curve will then reflect the internal field and/or current density distribution. We outlined one example of this in a qualitative manner in Figs. 2-1a and 2-1b; Figs. 2-1c and 2-1d show another—the SNDC case.

Consider slowly raising the bias voltage for the case shown in Fig. 2-1c. For low values of bias we are in the PDC region. Eventually the point of intersection of the dcll and the $J(F)$ curve will reach the NDC region and a current filament will form. If the boundary conditions are such that the current density at the lateral surfaces of the sample are fixed at a low value, and the lateral dimension is large, then the $I(\Phi)$ characteristic seen in the laboratory will be as shown schematically in Fig. 2-1d; the $dI/d\Phi < 0$ region associated with the homogeneous NDC state of Fig. 2-1c will not be observed. Instead, the system will jump from the homogeneous low-conductivity state to the almost vertical filamentary branch along the load line. When the bias is decreased, the sample will switch back at a lower value of bias, thus exhibiting hysteresis. If the boundary conditions or the sample width are changed, the $I(\Phi)$ curve also changes in a characteristic way (Schöll, 1985a). The observed filamentary branch will, in general, have a small region of negative conductance at its lower end. In this regime current oscillations generated by "breathing current filaments" may occur (Rau et al., 1989; Schöll, 1989c; 1990).

The almost vertical branch of the $I(\Phi)$ characteristic in Fig. 2-1d corresponds to a *filament,* rather than a *domain,* since I must rise and Φ must drop in order to reach this state from the low-conductivity homogeneous state, which can be achieved by the formation of a high current density filament. On the other hand, the situation of Fig. 2-1b involves the formation of a high-field domain, since here the current drops and the voltage rises.

In summary, we have pointed out that homogeneous NDC states in the $J(F)$ plane are unstable and are not observed in the $I(\Phi)$ (laboratory) plane. Instead, the system switches to domain or filamentary states along a load line if a purely resistive (ideal) circuit is present.

As we shall see, sometimes the circuit conditions are such that these stable inhomogeneous points can never be reached and only oscillatory solutions occur. But, first, we begin by considering the spatial and temporal stability of NNDC and SNDC curves in a simple manner, neglecting the reactive components that are always present in a real circuit.

2.2.2. NNDC and Domains

2.2.2.1. Stability Against Small Charge Fluctuations

To analyze the stability of an NDC element it is useful to inquire into its response to a charge fluctuation. By way of example, we consider the case where the NDC results from NDM (negative differential mobility), so that a range of fields where $dv/dF < 0$ exists; v is the carrier drift velocity. An example of this would be an n-type semiconductor where the carrier mobility is affected by intervalley transfer (Shaw et al., 1979).

To illustrate the response of an NNDC element to a charge fluctuation, consider a uniform field with a domain of increased field in the center of the element as shown in Fig. 2-3a. The charge distribution that produces this field fluctuation is shown in Fig. 2-3b. There is a net accumulation of charge on the left side of the domain and a depletion layer on the right. If we consider *positively* charged carriers, the carriers and hence the domain will be moving to the right. Assuming that the field within the domain is within the NDC range and the field outside the domain is within the ohmic range, but close to the field of peak velocity, then it is clear that the field fluctuation will initially grow with time. This happens because the higher upstream field in the center of the domain results in carriers moving more slowly than those at the edges, where the field is lower. Charge will therefore deplete on the right (leading) edge of the domain and accumulate at the left (trailing) edge. This charge will add to what is already there, increasing the field in the domain. If the element is in a resistive circuit the

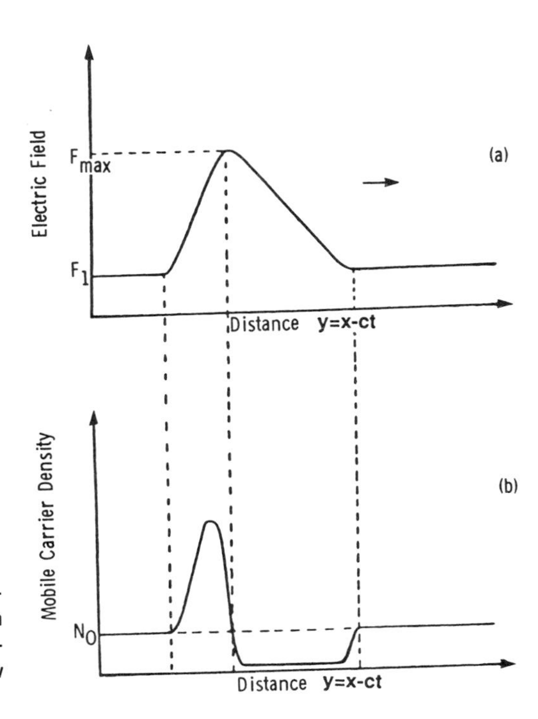

FIGURE 2-3. (a) Electric field profile; (b) carrier density profile of a moving Gunn domain (schematic). The domain is moving with velocity c in the positive x direction. (After Shaw et al., 1979.)

increasing voltage across the domain will decrease the current in the circuit and lower the field outside the domain.

The preceding situation constitutes a runaway process. The field will continue to grow in the interior of the domain and drop outside. Given enough time, the domain field will grow until the domain velocity is equal to the velocity of the carriers outside the domain. (This is possible with the NNDC curve we are considering.) The final velocity of the domain depends on the circuit load line. If the original field fluctuation were produced by a doping nonuniformity, the decreased field around the nucleation site which would result after the domain moved from the region would prevent nucleation of subsequent domains. This situation would prevail until the domain reached the right-hand edge of the element, where it could disappear. The field would then rise to nucleate another domain, resulting in a periodic current oscillation whose period is determined by the transit time of the domain (Gunn, 1964).

We next consider the response of the NNDC element to a charge fluctuation in a more rigorous manner by asking: given a homogeneous bulk NNDC characteristic produced by an arbitrary isothermal nonmagnetic NDM mechanism, what response occurs when a small fluctuation of charge is induced when this medium is biased into its NDC region? This question was answered analytically by Butcher (1967), Butcher and Fawcett (1966), and Knight and Peterson (1966, 1967), and numerically by Kroemer (1966) and McCumber and Chynoweth (1966). Knight and Peterson used the method of characteristics to analyze a model for the growth and propagation of charge disturbances in a medium with an arbitrary electric-field-dependent mobility. They developed possible forms of propagating field nonuniformities in a material exhibiting NDC, obtaining expressions for the velocity of both charge layers and dipole domains. Bonch-Bruevich and his colleagues have also contributed substantially to the understanding of the problem (1965, 1966). [See also the monographs by Bonch-Bruevich et al. (1975), and by Asche et al. (1979).]

To determine the response to a charge fluctuation, we use the continuity equation in one dimension,

$$\frac{\partial}{\partial x} n(x, t) e v(F) + \frac{\partial}{\partial t} n(x, t) e = 0 \tag{2-13}$$

where $n(x, t)$ is the space- and time-dependent *positive* mobile charge density and e the electronic charge, and the Poisson equation,

$$\frac{\epsilon}{e} \frac{\partial F}{\partial x} = n(x, t) - N_0 \tag{2-14}$$

where eN_0 is the fixed negative uniform background charge density and ϵ the permittivity of the semiconductor material. Diffusion is neglected in the argument; its inclusion will be considered later.

Equations (2-13) and (2-14) can be combined into

$$\frac{\partial}{\partial x}[env(F)] + \epsilon \frac{\partial^2 F}{\partial x \, \partial t} = 0 \tag{2-15}$$

A spatial integration of Eq. (2-15) yields

$$env(F) + \epsilon \frac{\partial F}{\partial t} = J(t) \tag{2-16}$$

which states that the sum of the conduction and displacement current densities is only a function of time. $J(t)$ is identified as the current density in the external circuit. Putting Eq. (2-14) into Eq. (2-16) yields

$$\epsilon\left[\frac{\partial F}{\partial t} + v(F)\frac{\partial F}{\partial x}\right] = J(t) - eN_0 v(F) \tag{2-17}$$

Equation (2-17) is a nonlinear partial differential equation whose solutions describe the influence of an electric field on the motion of a charge carrier. As an illustration, we consider the motion of a carrier along a path (trajectory) described by the equation

$$v(F) = d\bar{x}/dt \tag{2-18}$$

Using this, we obtain for Eq. (2-17)

$$\epsilon\left(\frac{\partial F}{\partial t} + \frac{d\bar{x}}{dt}\frac{\partial F}{\partial x}\right) = J(t) - eN_0 v(F) \tag{2-19}$$

Since the expression in the parentheses in Eq. (2-19) is the total time derivative along a trajectory, Eq. (2-17) can be replaced by two equations, (2-18) and

$$\epsilon \frac{dF(\bar{x}, t)}{dt} = J(t) - eN_0 v(F) \tag{2-20}$$

where $J(F)$ is proportional to $v(F)$. We are now moving along with each carrier in the $\bar{x}$–t plane. We can use Eq. (2-20) to examine the stability of the PDC and NDC operating points shown in Fig. 2.4. For simplicity we take an infinite load

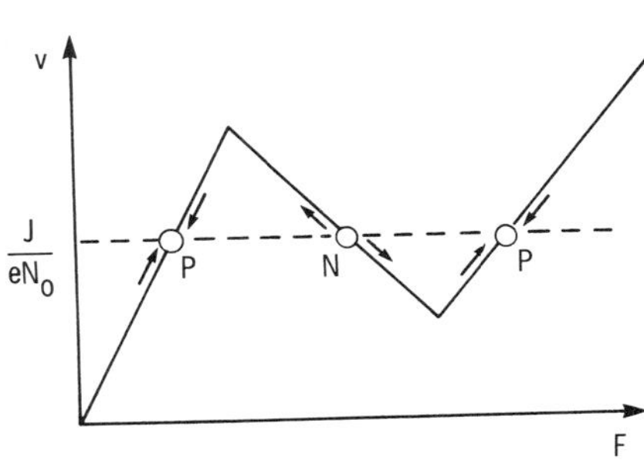

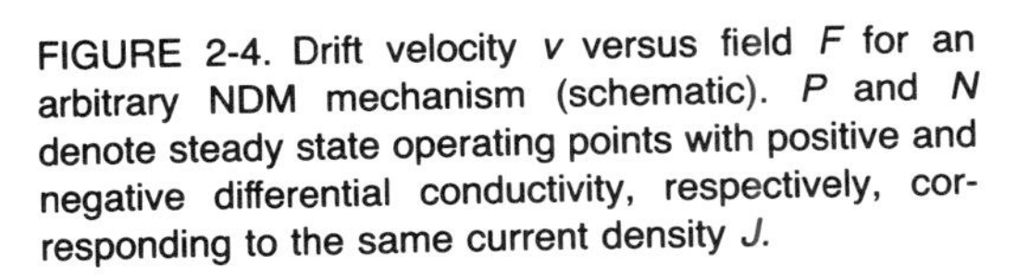
FIGURE 2-4. Drift velocity v versus field F for an arbitrary NDM mechanism (schematic). P and N denote steady state operating points with positive and negative differential conductivity, respectively, corresponding to the same current density J.

resistor so that the load line is horizontal (constant current source; $dJ/dt = 0$). Consider the P points. A small charge fluctuation that *increases* F locally will take us along the $v(F)$ curve to a nearby point where $v(F) > J/eN_0$. Hence, from Eq. (2-20), $\epsilon\, dF/dt$ will be negative; thus the displacement current will drive us back to point P. A local charge fluctuation about point P that *decreases* F will similarly be returned back to point P. Point P is stable. The same argument applied to point N shows us that it is unstable, and, in fact, were we to start at point N, a fluctuation about N at fixed J would split the sample up into two regions of field, the high and low fields at the P points. The sample becomes electrically heterogeneous. This is similar to the phenomenon of spinodal decomposition in thermodynamic equilibrium phase transitions. Here, however, we are dealing with a nonequilibrium phase transition.

For simple cases the temporal growth of such fluctuations can be studied analytically. For example, along a characteristic path a small charge fluctuation may be expected to produce a small change in electric field, δF, whose time dependence is governed by Eq. (2-20). Assuming a time-independent solution represented by F_0, we have that

$$\epsilon \frac{d}{dt}(F_0 + \delta F) = J - eN_0 v(F_0) - eN_0\, \delta v \tag{2-21}$$

Since F_0 is a solution to Eq. (2-20), δF must satisfy

$$\frac{\epsilon}{eN_0} \frac{d}{dt} \delta F = -\delta v = -\frac{dv}{dF}(F_0)\, \delta F \tag{2-22}$$

where we have linearized the $v(F)$ characteristic around F_0. Solutions of Eq. (2-22) have the form

$$\delta F(\bar{x}, t) = \delta F(\bar{x}, 0) e^{-t/\tau} \tag{2-23}$$

where $\tau = \epsilon/(N_0 e\, dv/dF)$, and describe a propagating disturbance that either grows or decays while propagating. Growth occurs when $dv/dF < 0$; decay occurs when $dv/dF > 0$. Thus, if we had an NNDC curve biased in the positive mobility, μ, region, where, e.g., $v = \mu F$, a small local increase in electric field would decay in time. If the bias were sufficient to take the electric field into the NDC region, a small local increase in electric field would grow.

Kroemer (1968) discussed the criterion for NDC element stability by considering the growth of charge fluctuations as described by Eq. (2-23). He assumed that the linearization invoked in Eq. (2-23) remained valid for the growth of the charge fluctuation throughout its transit across the element. The growth

$$G = \frac{\delta F(\bar{x}, t)}{\delta F(\bar{x}, 0)}$$

at the end of the transit is then given by

$$G = \exp[L/(v\,|\tau|)] = \exp[(LN_0e\,|dv/dF|)/(\epsilon v_0)] \tag{2-24}$$

where v_0 is an average domain velocity and L is the length of the device. Substantial growth, i.e., a space charge instability, occurs if the exponent in Eq. (2-24) is larger than unity, or

$$N_0L > \epsilon v_0/(e\,|dv/dF|) \tag{2-25}$$

The above N_0L product stability criterion was first described by McCumber and Chynoweth (1966).

If an NNDC element is unstable to charge fluctuations, how do the fields rearrange? Are there stable nonuniform field configurations? What happens to the current during the transient? While these questions require the use of numerical calculations, there are some very general features that emerge from the principles of Chap. 1. Stability of the nonuniform configurations requires a more detailed discussion, which is deferred to Chap. 5. Some qualitative features emerge from the following discussion.

The one-dimensional equation for total current flow, including diffusion, is

$$J(t) = e\left[nv(F) - D\frac{\partial n}{\partial x}\right] + \epsilon\frac{\partial F}{\partial t} \tag{2-26}$$

which, when combined with Poisson's equation yields

$$J(t) = N_0ev + \epsilon\frac{v\,\partial F}{\partial x} - \epsilon D\frac{\partial^2 F}{\partial x^2} + \epsilon\frac{\partial F}{\partial t} \tag{2-27}$$

The generalizations of Eq. (2-27) over that of Eq. (2-17) lies in the contribution of the diffusion current. A principal consequence of diffusion is that it is dissipative. To see this, assume that F_0 represents a spatially and temporally uniform solution, and track a small departure $\delta F(x, t) = F(x, t) - F_0$, from equilibrium. $\delta F(x, t)$ satisfies the linearized differential equation

$$\delta J = N_0e\frac{dv}{dF}(F_0)\,\delta F + \epsilon v(F_0)\frac{\partial\,\delta F}{\partial x} - \epsilon D\frac{\partial^2\,\delta F}{\partial x^2} + \epsilon\frac{\partial\,\delta F}{\partial t} \tag{2-28}$$

Assume a solution to the homogeneous equation of the form

$$\delta F(x, t) \sim \exp\lambda t\,\exp\left[i\frac{2\pi}{L}(x - v_0t)\right] \tag{2-29}$$

which is one Fourier component of a growing or decaying wave of wavelength L and phase velocity $v_0 = v(F_0)$. To find λ, insert Eq. (2-29) into the homogeneous part of Eq. (2-28), which corresponds to a nonfluctuating external circuit. λ

satisfies the equation

$$\lambda = -\left[\frac{N_0 e\, dv/dF}{\epsilon} + D\left(\frac{2\pi}{L}\right)^2\right] \tag{2-30}$$

The first part represents growth as before, when F_0 falls within the NNDC region; the second part is dissipative, and is pronounced at short wavelengths. It always adds a negative, and hence stabilizing, contribution to λ. We note that the first term in Eq. (2-28) leads to the same time dependence as that of Eq. (2-23). Modifications due to the consequences of Eq. (2-24) follow by including diffusion. [Note that Eq. (2-29) only holds in general during the initial stages of the instability as long as δF is still small enough to allow for linearization.]

2.2.2.2. The Method of Phase Portraits Applied to NNDC Elements

To provide a more general discussion of possible nonuniform field arrangements within an NDC element, we proceed to a more detailed study of the nonlinear Eq. (2-26) as it is coupled to Poisson's equation. In doing this it is convenient to introduce dimensionless variables as follows:

$$\begin{aligned} t &= T_0 T \\ v &= V_0 V \\ x &= V_0 T_0 X \\ F &= F_0 F^* \\ n &= N_0 N \\ J &= N_0 e V_0 J^* \\ D &= D_0 D^* \end{aligned} \tag{2-31}$$

Then, with $T_0 = \epsilon F_0/(N_0 e V_0)$, $V_0 = v(F_0)$ and $D_0 = V_0^2 T_0$, Eq. (2-26) becomes

$$J^* = NV - D^* \frac{\partial N}{\partial X} + \frac{\partial F^*}{\partial T} \tag{2-32}$$

Poisson's equation reads

$$\frac{\partial F^*}{\partial X} = N - 1 \tag{2-33}$$

In the following we drop the asterisks, and first consider time-independent solutions.

Under time-independent conditions, i.e., $\partial/\partial T = 0$, the two relevant equations are rearranged as

$$\frac{dF}{dX} = N - 1 \tag{2-34}$$

and

$$\frac{dN}{dX} = \frac{1}{D}[NV(F) - J] \tag{2-35}$$

For semiconductors with a region of negative differential conductivity there is a rich family of solutions to these coupled equations. To develop insight into these solutions we concentrate on phase plane solutions of the equation

$$\frac{dN}{dF} = \frac{1}{D}\frac{(NV - J)}{N - 1} \tag{2-36}$$

We shall use the general methods for phase portrait analysis of dynamic systems outlined in Chap. 1. However, it should be noted that now the parameter of the trajectories is the spatial coordinate X rather than the time, and hence the terminology "unstable node," "unstable focus," etc. does not imply a *runaway process in time,* but rather indicates the spatial variation. The qualitative features of these solutions are developed graphically beginning with Fig. 2-5. For assistance in sketching these curves we have also sketched the auxiliary curves (null-isoclines)

$$(N_\infty) \qquad N - 1 = 0 \tag{2-37}$$

and

$$(N_0) \qquad J - NV(F) = 0 \tag{2-38}$$

Each time a phase plane trajectory crosses the N_∞ curve, its slope dN/dF is infinite. When it crosses the N_0 curve the slope is zero. For a material in which negative differential conductivity is N-shaped and includes two positive mobility regions, as in the inset to Fig. 2.5, the N_0 curve has the general feature of the solid curve. If the N-shaped curve saturated at high fields, as represented by the dashed section of the inset, the N_0 curve would also be modified. The N_0 curve in Fig. 2-5 is for low current levels J; at elevated values of current the curve moves vertically upward as in Fig. 2-6. For low currents the intersection of the N_0 and N_∞ curves occurs at one singular point. At higher current levels there are three intersection points, each being singular.

In the first quadrant of the N–F plane, the N_0 and N_∞ curves divide the region into several sections, each with a characteristic path direction. The path is taken in the sense of a solution originating at the cathode, and thus *describes the influence of the cathode boundary* on the resulting solution. In Fig. 2-5 and region 1, $dF/dX > 0$ and $dN/dX > 0$ (or $d^2F/dX^2 > 0$). Thus, for increasing X, the slope $dN/dF > 0$, and the general direction of any trajectory in this region has a positive slope, as indicated in the figure. In region 2, $dF/dX > 0$, $dN/dX < 0$, (or $d^2F/dX^2 < 0$); an increasing F is accompanied by a decreasing carrier

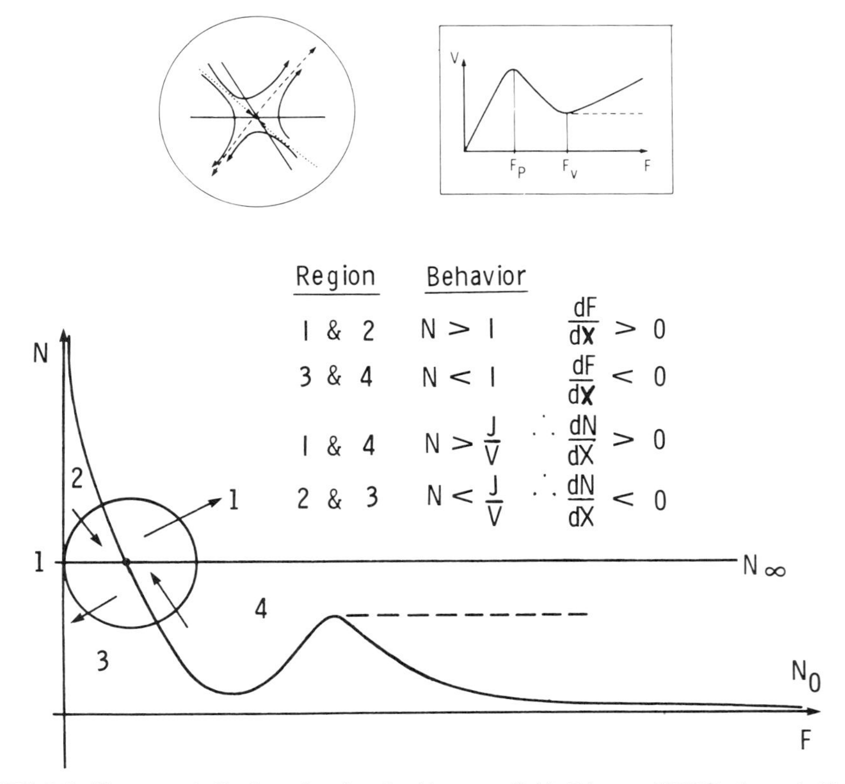

FIGURE 2-5. Phase portrait of carrier density N versus field F for an NNDC element. The solid curves marked N_∞ and N_0 represent the null isoclines $dF/dX = 0$ and $dN/dX = 0$, respectively. The inset on the top right shows the underlying velocity–field characteristic; the dashed lines in the inset and in the main figure correspond to a simplified $V(F)$ relation. The inset on the top left shows the detailed structure of the phase portrait near the singular or fixed point (intersection point of N_∞ and N_0 curves). (Schematic.)

density, and $dN/dF < 0$ as shown by the region 2 arrow. In region 3, dF/dX and $dN/dX < 0$. Here a decreasing F is accompanied by a decreasing N and dN/dF is as shown by the region 3 arrow. In region 4, $dN/dX > 0$ while $dF/dX < 0$. Thus, in going to increasing values of X, a decreasing F is accompanied by an increasing N and dN/dF as is shown.

On the basis of the phase path direction, the singular point in Fig. 2-5 has the characteristics of a saddle point. This is verified directly using the methods discussed in Chap. 1. Letting N_1 and F_1 represent the singular point given by $N_1 = 1$ and $V(F_1) = J$, and

$$\xi_1 = N - N_1 \tag{2-39}$$

$$\xi_2 = F - F_1 \tag{2-40}$$

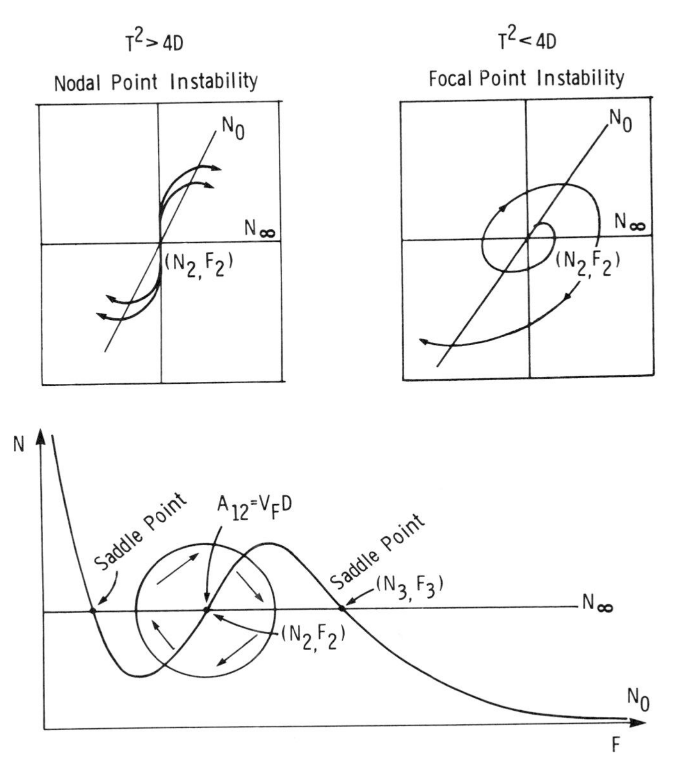

FIGURE 2-6. Phase portrait as in Fig. 2-5, but for a larger value of J, such that three singular points exist. The two insets show possible structures of the phase portrait near the middle singular point (N_2, F_2). (Schematic.)

Eqs. (2-34) and (2-35) become (in the small signal sense)

$$\frac{d\xi_1}{dX} = \frac{J}{D}\xi_1 + \frac{1}{D}\frac{dV}{dF}\xi_2 \tag{2-41}$$

$$\frac{d\xi_2}{dX} = \xi_1 \tag{2-42}$$

where we have used the fact that at the singular point $N_1 = 1$ and $V = J$. In the notation of Eqs. (1-20), (1-21), (1-22), and (1-26) $A_{11} = J/D$, $A_{12} = D^{-1}\,dV/dF$, $A_{21} = 1$, $A_{22} = 0$, and $T = J/D > 0$, and $D = -D^{-1}\,dV/dF < 0$. Thus, the intersection point is a saddle point. The phase plane curves in the vicinity of the saddle are shown in the inset, where it is noted that the $N = 1$ line is intersected by trajectories at 90° with the exception of the separatrices (dotted and dashed lines). The slope of the separatrices is tangent to the eigenvectors of A at the singular point, and is given by

$$dN/dF = \lambda \tag{2-43}$$

where

$$\lambda = \frac{1}{2D}\left[J \pm \left(J^2 + 4D\frac{dV}{dF}\right)^{1/2}\right]$$

are the two eigenvalues of A. For small values of D the slope of the eigenvector with the positive eigenvalue (dashed line) tends to infinity, while the eigenvector with negative eigenvalue (dotted line) approaches the slope $-J^{-1}\,dV/dF$, which means that it becomes tangent to the N_0 curve. In this case practically all trajectories starting in region 2 will remain below the N_0 curve and cross into region 3. For large values of D, on the other hand, the slopes of both the dotted and the dashed separatrices becomes very small, and trajectories starting in region 2 are likely to cross over into region 1. The significance of the saddle point is that only a point whose initial condition places it on the dotted line will move toward the singular point. All other points diverge away. Any physical situation is likely to have some fluctuations, which as a result will move it off the dotted line. The dotted line is not a likely physical curve.

Before proceeding to sketch possible field distributions it is necessary to point out that some solution curves are likely to cross the $N = 0$ axis and go into the fourth quadrant of the N–F plane. While this is certainly a possible mathematical parallel, physical constraints exclude this solution. There cannot be a negative number of particles. We first consider the variety of solutions associated with the one singular point problem, and then discuss the three singular point problem (Fig. 2-6).

Consider Fig. 2-7 and case A where as an initial condition the cathode field is near zero (region 2) and the cathode boundary sustains an accumulation of carriers. The direction of the trajectory depends on the initial slope at N_c, F_c, where the subscript c indicates that the initial point represents the cathode

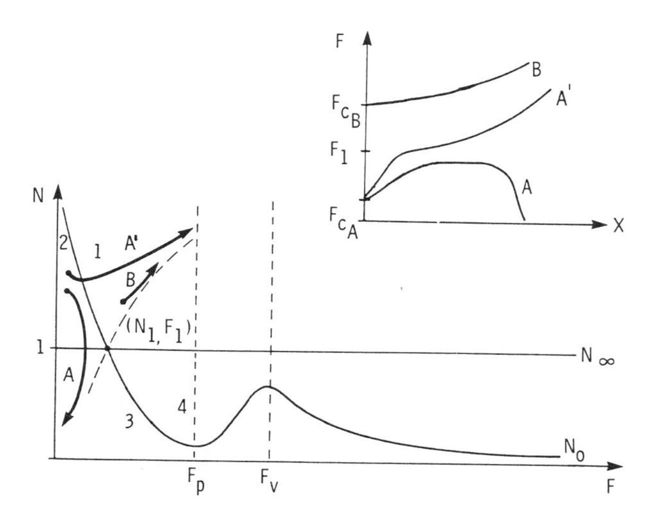

FIGURE 2-7. Phase portrait with three individual trajectories A, A′, B corresponding to accumulation layers at the cathode boundary $X = 0$. The inset shows the corresponding spatial field profiles $F(X)$. (N_1, F_1) marks the singular point. (Schematic.)

boundary. The initial slope is

$$\left.\frac{dN}{dF}\right|_c = \frac{1}{D}\frac{(N_cV_c - J)}{N_c - 1}$$

and depends critically on D. For small values of D (e.g., low carrier density), the slope is large and negative, and the trajectory will have the qualitative features of curve A. For large values of D, the trajectory is likely to intersect the N_0 curve, and display features of curve A'. For curve A the trajectory follows the N_0 curve closely until it crosses the N_∞ line, where the carrier density experiences depletion. Below the N_∞ curve strong diffusion currents contribute. A sketch of $F(X)$ is shown in the inset. Several points should be noted about the sketch. First, it is qualitative; the extent of each region is determined to a large extent by the constraint imposed by the potential drop,

$$\Phi = \int_0^L F\,dX \tag{2-44}$$

across the device. Secondly, solutions that "hug" the N_0 curve are adequately represented by Poisson's equation and the approximate equation

$$J \cong NV \tag{2-45}$$

which neglects the diffusion current. Departures from the N_0 curve indicate significant diffusive contributions, as at the terminus of trajectory A, and the full trajectory A'. Thus A and A' are possible solutions, the "correct" one being determined by the constraint of Eq. (2-44) and the value of current flowing through the device. J and Φ are in turn determined by the constraint of the external circuit. (Extension of the solution into the NDM region will be considered subsequently.)

Figure 2-7 also shows a sketch of a solution B beginning with a field value above the singular point field F_1. The phase plane sketch indicates that the slope away from the initial or cathode condition is increasing everywhere. The $F(X)$ plot is also shown qualitatively in Fig. 2-7.

Curves A, A', and B show accumulation layers at the cathode ($N_c > N_1$). Curves originating in regions 3 and 4 show cathode depletion ($N_c < N_1$) followed by either further depletion or accumulation. The results are displayed in Fig. 2-8. For curves C and C', the initial slope dN/dF is large enough so that the trajectory does not cross the N_0 curve. Eventually charge neutrality occurs and the solution represents charge accumulation near the anode boundary. The trajectory C″ is qualitatively different. Here, because of a relatively flat slope, the trajectory crosses the N_0 curve and enters the third region. The trajectory is as shown in Fig. 2–8. For a trajectory beginning in the third region, case D, the curve is similar to the second half of curve C″.

The field profiles in Figs. 2-7 and 2-8 are for fields beginning within the ohmic or lossy portion of the velocity–field curve. The curves were all sketched for the same value of current. The variability of solutions indicates, or rather

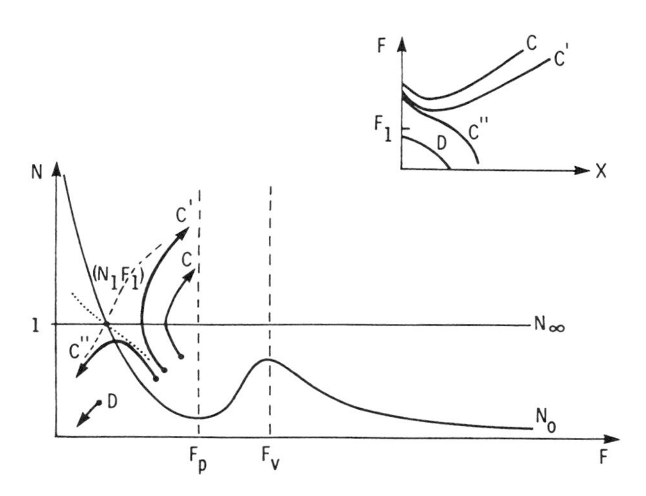

FIGURE 2-8. Phase portrait with four trajectories C, C′, C″, D corresponding to depletion layers at the cathode boundary $X = 0$. (Schematic.)

suggests, that simple pictures of transport in solids, i.e., those represented by Eq. (2-45), may be ignoring a larger class of possible solutions and, hence, physics. It would appear that more reliance on numerical solutions may be necessary.

We now consider the trajectory when there are three singular points, and first concentrate on the singular point in the NDM region (see Fig. 2-6). Proceeding as before, the directions of the phase plane trajectories are indicated by the arrows in the enclosed circles. The nature of the singular point which we denote as (N_2, F_2) is determined by the relative values of $T = J/D$ and $\underset{\sim}{D} = -D^{-1}\, dV/dF$. For the case of the one singular point discussed earlier, $\underset{\sim}{D} < 0$ and the singular point was a saddle point. For the singular point within the NDM region $\underset{\sim}{D} > 0$. However, the relative value of T^2 and $4\underset{\sim}{D}$ is dependent on both current level and diffusion. For example, at small values of D, $T^2 = J^2/D^2$ may exceed $4D^{-1}\,|dV/dF|$. For large values of D, $T^2 < 4\underset{\sim}{D}$. For $\underset{\sim}{D} > 0$, $T^2 > 4\underset{\sim}{D}$, i.e., $J^2 > 4D\,|dV/dF|$, a nodal point instability occurs. For $\underset{\sim}{D} > 0$, $T^2 < 4\underset{\sim}{D}$, a focal point instability occurs. Both are sketched in the insets of Fig. 2-6. The distinguishing feature of the node is that all trajectories except one pair approach the critical point *tangent to the same line*; the exceptional pair of trajectories approach tangent to a different line. These two lines are tangent to the two eigenvectors of A and have the slopes $dN/dF = \lambda$, as given by Eq. (2-43). Since both eigenvalues λ of an unstable node are positive, the slopes of both lines are also positive. The critical point (N_3, F_3) has the same properties as that associated with (N_1, F_1). It will not be discussed further, except as it relates to solution curves either originating or terminating on it.

The first solution curves we consider are shown in Fig. 2-9. Here we are fixing the cathode field at a value within the region of negative differential mobility. For initial conditions just below the N_0 curve there is a region of charge depletion adjacent to the cathode followed further downstream by the field moving into the lossy (PDM) region. The space charge trajectory closely follows

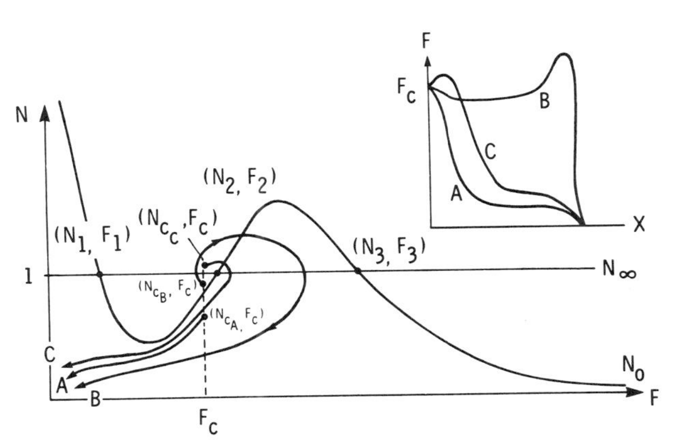

FIGURE 2-9. Phase portrait for the regime of three singular points, with three trajectories A, B, C orginating at the same cathode field F_c, but with different carrier densities N_{cA}, N_{cB}, N_{cC}. The singular point (N_2, F_2) is an unstable focus. (Schematic.)

the N_0 curve until the PDM region is reached. For a fixed cathode field an increase in current results in an increase in the net charge density adjacent to the cathode as long as (N_c, F_c) stays close to, but below, the N_0 curve. (See trajectory A of Fig. 2-9). For the point (N_c, F_c) above N_0 and below N_∞ a qualitatively different field profile may occur in which a high field region adjacent to the anode forms (curve B). For (N_c, F_c) above both N_0 and N_∞ curves similar to A result but with a region of charge accumulation at the cathode (see curves C, Fig. 2-8). Since the origin of curves A, B, and C all fall near each other it may be expected that fluctuations or changes in circuit conditions could cause switching from one state to the next. This indeed does occur and will be discussed in Chap. 5.

A feature of Figs. 2-7, 2-8, and 2-9 is that none of the solutions follow a closed path in phase space, as shown in Fig. 2-10. It is a direct matter to show

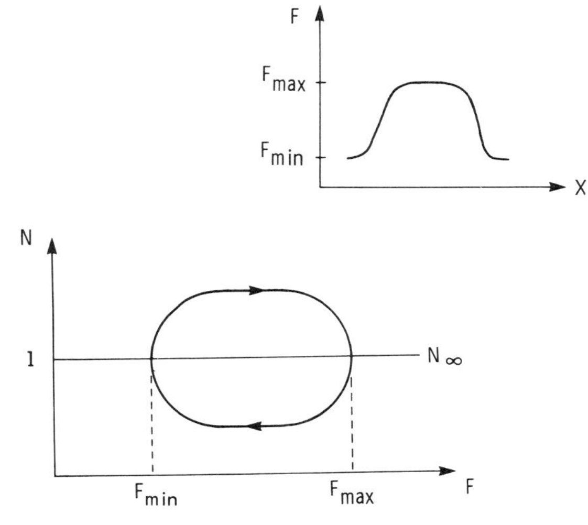

FIGURE 2-10. Phase portrait showing a hypothetical closed trajectory. (Schematic.)

that such a solution cannot occur. To do this, Eq. (2-36) is rearranged as

$$\frac{(N-1)\,dN}{N} = \frac{1}{D}\frac{(NV-J)}{N}\,dF \tag{2-46}$$

Then, defining

$$\psi_0(F, N) = N - \ln N - \frac{1}{D}\int V\,dF \tag{2-47a}$$

ψ_0 satisfies the differential equation

$$d\psi_0 = -\frac{J}{DN}\,dF \tag{2-47b}$$

and any two points (F', N') and (F'', N'') in the phase space that are connected by a solution to Eq. (2-46) satisfy the relation

$$\psi_0(F'', N'') = \psi_0(F', N') - \frac{J}{D}\int_{F'}^{F''}\frac{dF}{N} \tag{2-48}$$

If a closed solution were possible, where $\psi_0(F'', N'') = \psi_0(F', N')$, then either

$$\oint \frac{dF}{N} = 0 \tag{2-49}$$

or $J = 0$. The latter is not the case of interest. The former is not manifestly possible since $N > 0$. Thus, closed phase space solutions are not possible unless $J = 0$.

We next consider modifications when time-dependent considerations enter. Ignoring true spatial constraints associated with finite device size, we assume $F(X, T) = F(X - CT)$. In words, a disturbance propagates without change in shape. With $\xi = X - CT$, $\partial F/\partial T = C\,dF/d\xi$, $\partial F/dX = dF/d\xi$. In terms of X, the equation for total current [Eq. (2-32)] becomes

$$J = NV - D\frac{dN}{dX} - C(N-1) \tag{2-50}$$

where Poisson's equation, Eq. (2-34), has been used. For correspondence with previous phase plane analyses, Eq. (2-50) is rewritten as

$$\frac{dN}{dX} = \frac{1}{D}[N(V-C) - (J-C)] \tag{2-51}$$

and the trajectory within the plane has a slope

$$\frac{dN}{dF} = \frac{1}{D}\left[\frac{N(V-C) - (J-C)}{N-1}\right] \tag{2-52}$$

The singular points are given by $N_1 = 1$, $V(F_1) = J$ and do not depend upon C. Near the singular points (F_1, N_1), $\xi_1 = N - N_1$ and $\xi_2 = F - F_1$ satisfy the equations

$$\frac{d\xi_1}{dX} = \frac{J - C}{D}\xi_1 + \frac{1}{D}\frac{dV}{dF}\xi_2 \tag{2-53}$$

and

$$\frac{d\xi_2}{dX} = \xi_1 \tag{2-54}$$

where $J = V(F_1)$. In the notations of Eqs. (1-20), (1-21), (1-22), and (1-26), $A_{11} = (J - C)/D$, $A_{12} = D^{-1}\,dV/dF$, $A_{21} = 1$, $A_{22} = 0$; $T = (J - C)/D$, which is greater or less than zero, and depends on the relative values of J and C; and $\underset{\sim}{D} = -D^{-1}\,dV/dF$.

The qualitative features of the solutions for $C \neq 0$ are richer than those associated with the $C = 0$ case. There are two groups of solutions; those for $J > C$ and those for $J < C$. For either case the N_0 auxiliary curve (null isocline)

$$N(F) = \frac{J - C}{V(F) - C} \tag{2-55}$$

has a different shape. For $J > C$, the fields of direction near the singular point within the (PDM) region are as shown in Fig. 2-11. For $J < C$ the N_0 curve is different, as indicated in Fig. 2-12, but the characteristic direction of the trajectories in the four sections divided by the N_0 and N_∞ curves is not changed, and is the same as in Fig. 2-5. The singular point is for both $J > C$ and $J < C$ a saddle point, since $\underset{\sim}{D} < 0$. The slope of the eigenvectors or separatrices near the

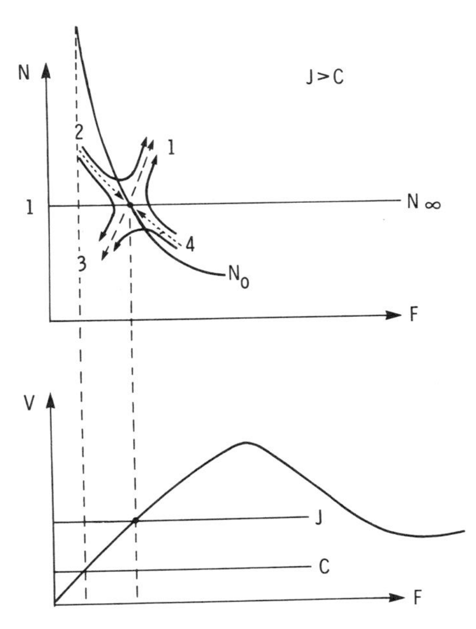

FIGURE 2-11. Phase portrait for field and carrier density profiles propagating with velocity $C < J$. The lower plot shows the drift velocity versus field characteristic displaying one static operating point (singular point in the phase portrait) in the PDM region. (Schematic.)

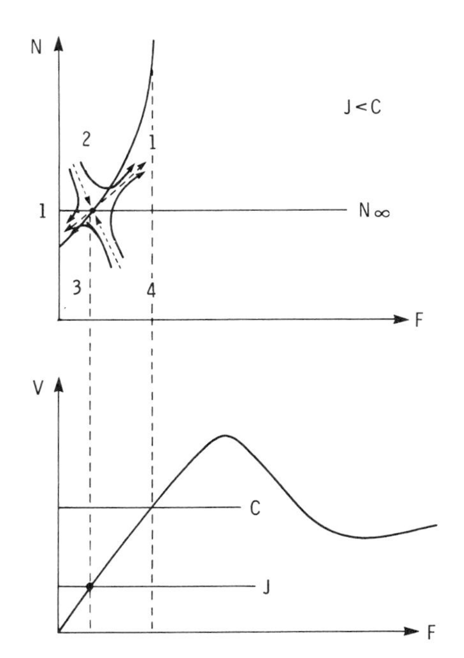

FIGURE 2-12. Same as Fig. 2-11, but for $C > J$.

singular point (dashed and dotted lines) is $dN/dF = \lambda$, where now (2-43) is modified as

$$\lambda = \frac{1}{2D}\left\{J - C \pm \left[(J - C)^2 + 4D\frac{dV}{dF}\right]^{1/2}\right\}$$

The interesting feature of these solutions is that the phase plane solutions of Figs. 2-11 and 2-12 are all propagating with a constant velocity C, as shown in Fig. 2-13. Of course, the question of whether a solution such as that of Fig. 2-13 is stable and actually exists physically cannot be determined from the phase plane study.

We next consider a solution near a singular point within the NDM region, as shown in Fig. 2-14 for $J > C$. For $J < C$, the phase plane separation near the singular point is shown in Fig. 2-15.

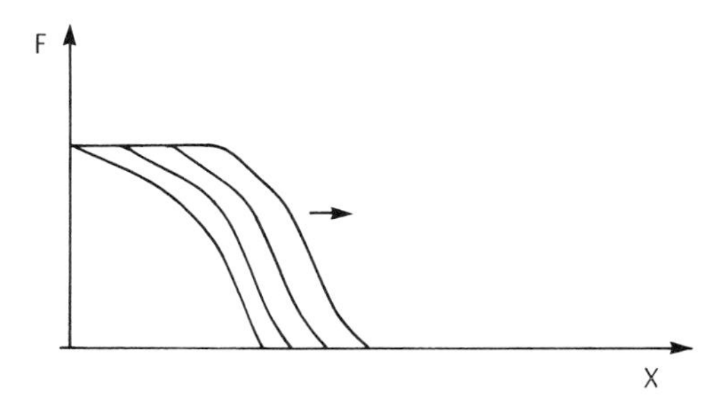

FIGURE 2-13. Field profiles propagating with constant velocity in the X direction. (Schematic.)

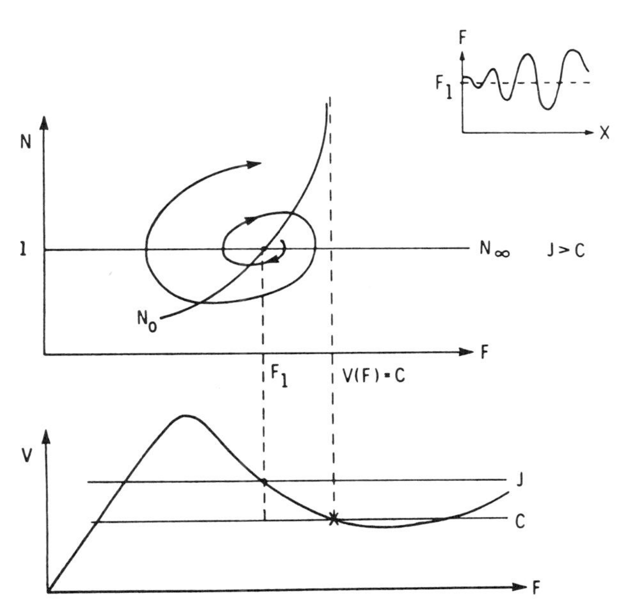

FIGURE 2-14. Phase portrait for $C < J$ with a singular point in the NDM region; the two singular points in the PDM region are not shown.

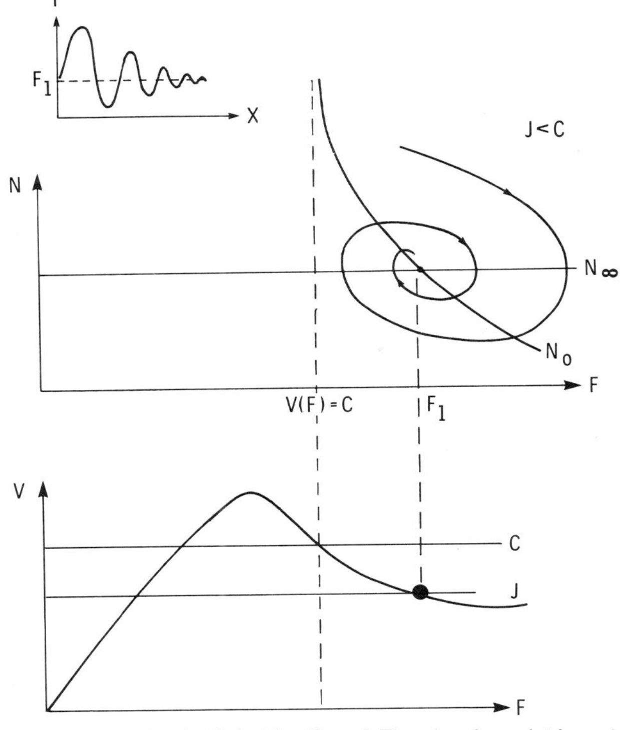

FIGURE 2-15. Same as Fig. 2-14, but for $C > J$. The singular point is a stable focus.

The characteristic features of the solutions of Figs. 2-14 and 2-15 are different. In Fig. 2-14 $J > C$ and, depending on the magnitude of T and $\underset{\sim}{D}$ [see Eqs. (1-24) and (1-25)], the singular point is either an unstable node or focus. For the unstable focus, the solution shows significant oscillatory spatial structure. In Fig. 2-15, where $J < C$, the singular point is stable. Thus, for a focus, a trajectory originating near the singular point either spirals outward or inward, as indicated.

A special case arises for $J = C$. Then, with $T = 0$ and $\underset{\sim}{D} > 0$ the eigenvalues of the singular point are purely imaginary. The singular point within the NDM region represents a center, and we can expect closed trajectories. As an introduction to this discussion we integrate Eq. (2-52). To do this we first rearrange it as

$$\frac{N - 1}{N}\, dN = \frac{1}{D} \frac{[N(V - C) - (J - C)]\, dF}{N} \tag{2-56}$$

Defining, as in Eq. (2-47a),

$$\psi_c(F, N) = \psi_0(F, N) + \frac{C}{D} \int dF \tag{2-57}$$

ψ_c satisfies the differential equation

$$d\psi_c = -\frac{1}{D}\left(\frac{J - C}{N}\right) dF \tag{2-58}$$

and two points in phase space connected by a solution of Eq. (2-56) satisfy

$$\psi_c(F'', N'') - \psi_c(F', N') = \frac{C - J}{D} \int_{F'}^{F''} \frac{dF}{N} \tag{2-59}$$

Equation (2-59) is extremely important; it indicates that for a nontrivial situation ($J \neq 0$) closed curves in phase space may occur. This occurs for $J = C$. For this case Eq. (2-59) reduces to

$$(N' - N'') - \ln\left(\frac{N'}{N''}\right) = \frac{1}{D} \int_{F''}^{F'} (V - C)\, dF \tag{2-60}$$

Let us consider the case where a closed phase plane trajectory surrounds the singular point (F_2, $N_2 = 1$), as shown in Fig. 2-16, and originates and ends at the singular point (F_1, $N_1 = 1$). Identifying (F', N') with (F_1, N_1) and (F'', N'') with ($F_{\max}$, 1) in Eq. (2-60), we obtain a high field propagating domain, subject to the condition

$$\int_{F_1}^{F_{\max}} V\, dF = C(F_{\max} - F_1) \tag{2-61}$$

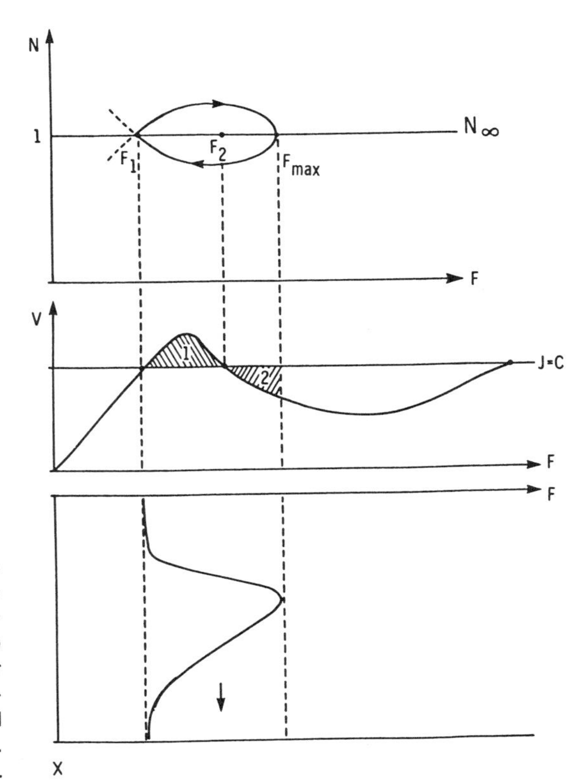

FIGURE 2-16. High-field domain propagating with velocity $C = J$. The top shows the closed phase space trajectory ("homoclinic orbit") surrounding the singular point (F_2, $N_2 = 1$). The center shows the velocity versus field characteristic, where the hatched areas 1 and 2 must be equal, thus determining F_{max}. The bottom shows the field profile $F(X)$.

A result often referred to as the "equal areas" rule. This rule states that areas "1" and "2" in Fig. 2-16 are equal and that equality at a given value of current determines the peak field F_{max}. In terms of phase trajectories, the high field domain represents a closed separatrix loop ("homoclinic orbit") originating at the saddle point in the low-field ohmic regime. Under more general field-dependent diffusion conditions Eq. (2-61) is generalized to read

$$\int_{F_1}^{F_{max}} \left(\frac{V - C}{D}\right) dF = 0 \tag{2-62}$$

Another closed configuration is shown in Fig. 2-17, corresponding to a low field propagating domain (Knight and Peterson, 1966, 1967).

For a given $V(F)$ characteristic, the equal areas rule for high field domains (Fig. 2-16) can be satisfied at higher values of current J, while the equal areas rule for low field domains (Fig. 2-17) applies at lower values of J. For the borderline value of J between these two cases, two propagating profiles of interest are shown in Fig. 2-18. They correspond to traveling accumulation and depletion layers.

The above discussion is purely mathematical. It is included here because virtually all field distributions discussed above have been seen in large signal

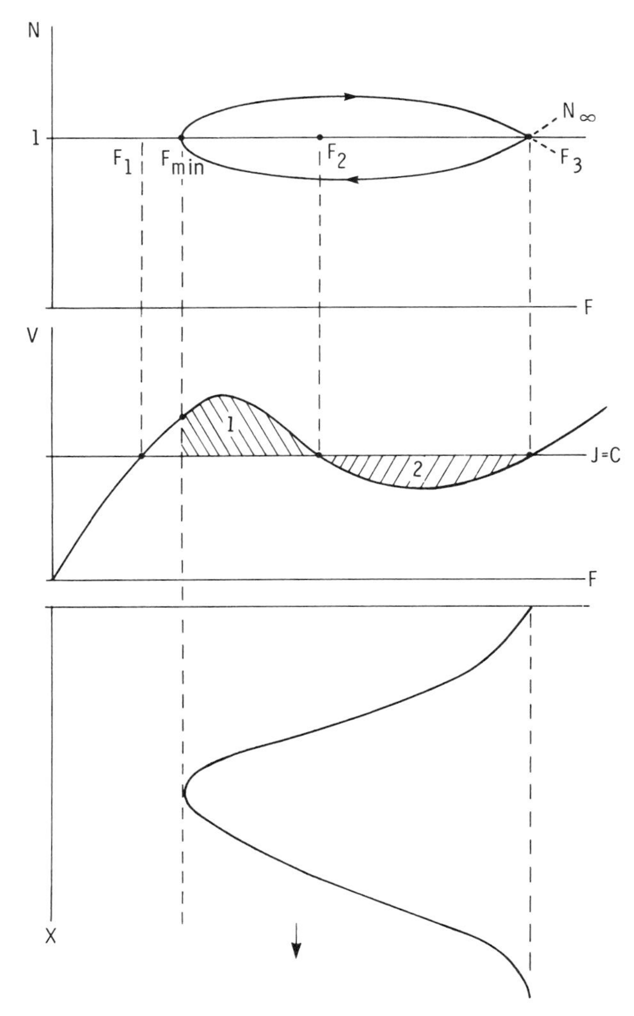

FIGURE 2-17. Low-field domain propagating with $C = J$. Plots as in Fig. 2-16. The hatched areas must be equal, thus determining F_{min}.

numerical studies, and have been tied to experiment. But the connection to physics awaits the next chapters. Some preliminary comments are, however, useful at this point. The field profiles of Figs. 2-7, 2-8, and 2-9 were sketched to emphasize that all solutions are dependent upon the boundary conditions, particularly those at the cathode boundary. This result has been critical in explaining high-field current instabilities in gallium arsenide and indium phosphide. Figures 2-14 and 2-15 show considerable structure in the $F(X)$ curves, which in turn is reflected in the $N(X)$ plots. Such profiles have been seen numerically and represent strong spatial fluctuations in propagating accumulation layers. They are responsible in part for limiting the performance of high-frequency sources. Figure 2-16 represents the well-known high-field propagating Gunn domain.

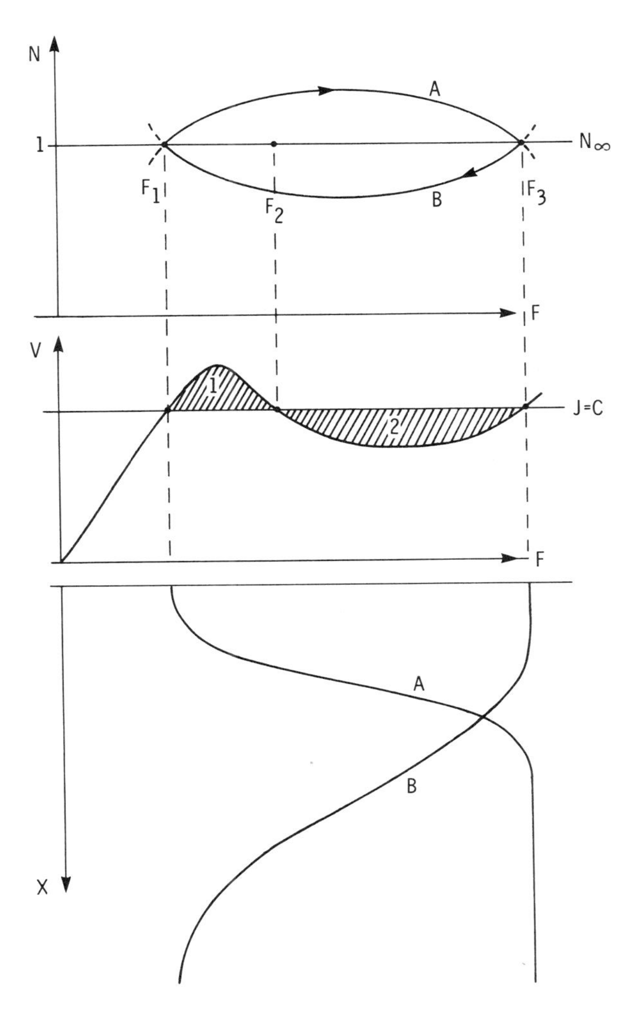

FIGURE 2-18. Traveling accumulation and depletion layers. Plots as in Fig. 2-16. The condition is given by equality of the areas 1 and 2.

The situation when $J = C$ is ideal. Ordinary fluctuations throughout a device preclude this from happening. Instead, adjustments in the domain velocity must be made continually in order for domains to propagate the required distance. This is not always the case, as we shall discuss in Chap. 5. The importance of domain propagation to solid state devices should not be underestimated. Because of this we digress for a moment and rederive the equal areas rules using an intuitive argument.

We begin with Eq. (2-56), integrating the left side from 1 to N and the right side from F_1 to F,

$$N - \ln N - 1 = \frac{1}{D} \int_{F_1}^{F} dF \left[(V - C) - \frac{(J - C)}{N} \right] \tag{2-63}$$

Note that the accumulation and depletion branches must neutralize ($N = 1$) both at $F = F_1$ and $F = F_{max}$. Since the left-hand side of Eq. (2-63) vanishes when $N = 1$, the right-hand side must vanish at $F = F_{max}$.

Thus, Eq. (2-63) yields

$$\int_{F_1}^{F_{max}} dF\left\{[V(F) - C] - \frac{1}{N}[J - C]\right\} = 0 \tag{2-64}$$

Since the contribution to the integral from the first term is independent of N, but the contribution from the second term depends on whether we are integrating along the accumulation or depletion branch, the integral can vanish only if each contribution vanishes separately. Furthermore, the second contribution vanishes for either $N \lessgtr 1$ only if $C = J$. We therefore conclude that for a field-independent diffusion coefficient a freely traveling domain will move at the same velocity as the carriers outside the domain. Note that with regard to the equal areas rule (the vanishing of the first contribution to the integral) for a particular $V(F)$ curve there is a minimum V below which the equal areas rule cannot hold. This implies that for a given $V(F)$ curve there is a maximum domain field that can be achieved. For velocities below the minimum value moving domains will be unstable (Grubin et al. 1971).

2.2.3. SNDC and Filaments

2.2.3.1. The Electrical Stability of SNDC Elements

We will again use the Poisson and continuity equations to ascertain the stability of SNDC elements against small fluctuations. The possibility of filaments introduces a two-dimensional aspect into the problem, which requires—unlike in the NNDC case—the use of all four of Maxwell's equations. First, we shall give a qualitative stability argument, idealizing the problem by introducing cylindrical symmetry. Important boundary effects may therefore be lost.

Consider the cylindrically symmetric configuration of Fig. 2-19, which possesses a homogeneous bulk $J(F)$ characteristic with an SNDC region. SNDC is produced by an as yet unstated nonmagnetic isothermal mechanism. Imagine biasing the device to the point where there are three intersection points, as in Fig. 2-19b. The stability of these points is identified below.

For a material with radius a, length l, and permeability μ_0, Faraday's law,

$$\text{curl}\,\mathbf{F} + \frac{\partial \mathbf{B}}{\partial t} = 0 \tag{2-65}$$

where $\mathbf{B}$ is the magnetic flux density, for the present geometry is

$$\frac{\partial F_z}{\partial r} = \frac{\partial B_\theta}{\partial t} \tag{2-66}$$

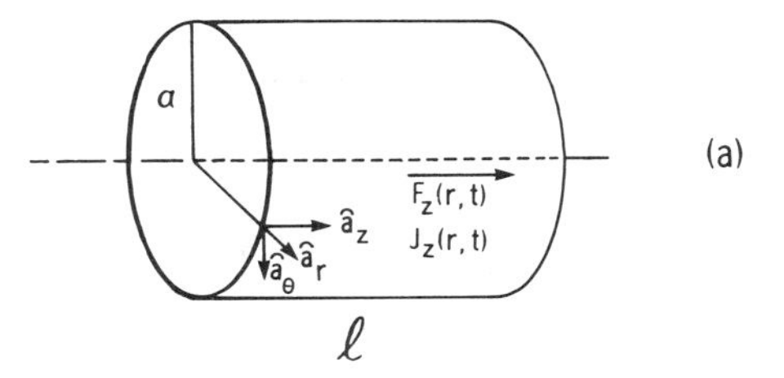

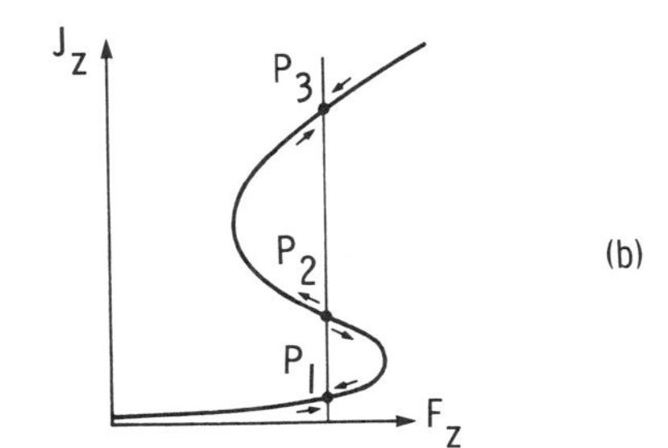

FIGURE 2-19. Schematic sketch of a cylindrically symmetric SNDC element of radius a (a), characterized by a current density–field characteristic with SNDC (b).

where we have introduced cylindrical coordinates r, θ, z. Defining

$$\Omega(r, t) = l \int_r^a B_\theta \, dr \tag{2-67}$$

as the magnetic flux contained in a cylindrical shell of length l and width $a-r$, an integration of Eq. (2-66) yields

$$lF_z(a, t) = lF_z(r, t) + \frac{\partial \Omega(r, t)}{\partial t} \tag{2-68}$$

As can be seen by applying Poynting's theorem (Shaw et al., 1973a), $lF_z(a, t) = \Phi(t)$ is the potential drop between the end faces of the cylinder; thus we have

$$\Phi(t) = lF_z(r, t) + \frac{\partial \Omega(r, t)}{\partial t} \tag{2-69}$$

All radial dependences in field are accompanied by transients in the magnetic flux. Under time-independent conditions $\Phi = lF_z$, and F_z is independent of r.

Since we are interested in determining the stability of points P_1, P_2, and P_3 in Fig. 2-19b we need at least a qualitative argument concerning the dependence of F on Ω. Application of the "curl H" equation, with $\mathbf{B} = \mu_0\mathbf{H}$, together with the constitutive $J_z(F_z)$ relation, provides this. (Here $\mathbf{H}$ is the magnetic field and μ_0 the permeability of free space.) First

$$\text{curl}\,\mathbf{B} = \mu_0\mathbf{J} \tag{2-70}$$

where $\mathbf{J}$ includes the conduction current $\mathbf{J}_c$ and the displacement current $\epsilon\partial\mathbf{F}/\partial t$.

For the problem of interest, this yields

$$\frac{1}{r}\frac{\partial}{\partial r} rB_\theta = \mu_0 J_z \tag{2-71}$$

(Note: Because B_θ is arbitrary to within the gradient of a scalar function, B_θ is not a unique function of J.) For a spatially uniform current

$$B_\theta = \frac{\mu_0 J_z r}{2} \tag{2-72}$$

and the flux is

$$\Omega(r) = \frac{\mu_0 l}{2} J_z(a^2 - r^2) \tag{2-73}$$

Thus, as F_z is a nonlinear function of J_z given by the inverted current density–field relation, it is also a nonlinear function of $\Omega(r)$, i.e.,

$$F_z(J_z) = F_z\left[\frac{2\Omega(r)}{\mu_0 l(a^2 - r^2)}\right] \tag{2-74}$$

Thus, e.g., at a point P_2 in the SNDC region, a small increase in J_z along the $J_z(F_z)$ characteristic, and hence an increase in Ω will result in a decrease in F_z by ΔF_z. Thus for fixed voltage $\Phi(t) = \Phi_0 = lF_z$,

$$\frac{\partial \Omega}{\partial t} = \Phi_0 - l(F_z - \Delta F_z) = l\,\Delta F_z \tag{2-75}$$

will be positive, and Ω will further increase. This means that the point P_2 is unstable. Similar arguments show that the points P_1 and P_3 in the PDC region are stable. We conclude that Lenz's law does not hold in the NDC region.

We shall now adopt a more systematic and analytical approach. Eliminating the magnetic field from Eqs. (2-65) and (2-70), we obtain

$$\text{curl}\,\text{curl}\,\mathbf{F} + \mu_0 \frac{\partial \mathbf{J}}{\partial t} = 0 \tag{2-76}$$

The other equation governing the electric field is Poisson's equation

$$\text{div}\,\mathbf{F} = \frac{1}{\epsilon}\rho \tag{2-77}$$

where ρ is the charge density.

These two equations have to be supplemented by constitutive relations for **J** and ρ. The usual form for single-carrier transport is

$$\mathbf{J}(n, \mathbf{F}) = e\mathbf{v}(\mathbf{F})n + eD \operatorname{grad} n + \epsilon \frac{\partial \mathbf{F}}{\partial t} \tag{2-78}$$

$$\rho(n, F) = e[N_D^* - N_{\text{tot}}(n, F)] \tag{2-79}$$

Here we have assumed negatively charged carriers (electrons) in contrast to Eq. (2-26), which applies to positively charged carriers. N_D^* is the background charge density provided by ionized donors and compensating acceptors, and

$$N_{\text{tot}}(n, F) = n + n_t(n, F) \tag{2-80}$$

is the sum of the free (n) and trapped (n_t) carrier concentration. Note that Eq. (2-79) generalizes the charge density used in Eq. (2-14) by including the possibility of trapping. n_t is given for time-independent states by the trapping kinetics as a function of the independent variables n and F, where the dependence upon F occurs through the field-dependent rate constants involved in the trapping and detrapping processes. Throughout the following we will assume that this explicit dependence upon F is on the magnitude of F, not the direction, and is monotonically decreasing, since an increase in field will generally decrease the number of trapped electrons through field-enhanced generation or impact ionization.

In Sec. 2.2.2 we considered a nonmonotonic constitutive $v(F)$ relation and neglected the trapped charge density $n_t(n, F)$, which is characteristic of a bulk-induced NNDC mechanism such as intervalley transfer. We shall now consider a nonmonotonic dependence of N_{tot} upon n in some range of F. Thus, there is a profound analogy between the constitutive relation $v(F)$ and $N_{\text{tot}}(n, F)$ for NNDC and SNDC elements, respectively. We shall further elaborate on the analogy between NNDC and SNDC elements when we study the circuit stability in Sec. 2.3, where we shall find a complete symmetry of NNDC and SNDC behavior.

A typical plot of N_{tot} as a function of n is shown in Fig. 2-20. It is representative of a variety of different bulk generation–recombination mechanisms that yield SNDC (Crandall, 1970; Kastalski, 1973; Zabrodskij and Shlimak, 1975; Pickin, 1978; Adler et al., 1980; Schöll, 1981; Proctor et al., 1982).

Physically, when the curves N_{tot} versus n at fixed F display a "falling" region in some intermediate field range, as shown in Fig. 2-20, this corresponds to states

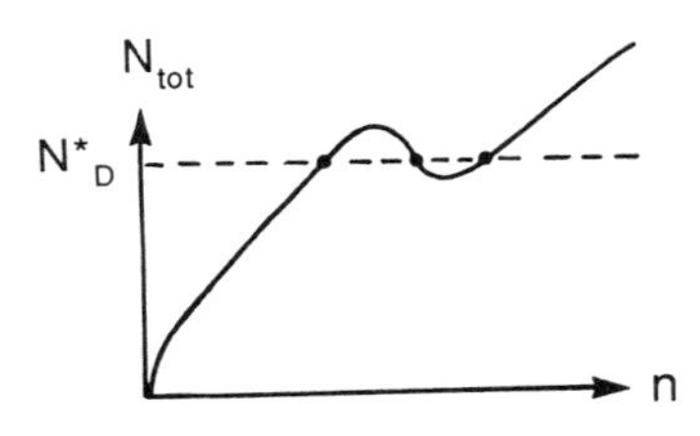

FIGURE 2-20. Total free and trapped electron density N_{tot} versus free electron density n for a generation–recombination-based SNDC mechanism. The spatially homogeneous steady states are given by the charge neutrality condition $N_{\text{tot}} = N_D^*$, where $N_D^* \equiv N_D - N_A$ is the effective donor density.

where a decrease in the total carrier concentration N_{tot} leads to an increase in the density of free carriers n. It is easy to see that this cannot happen if the distribution of carriers between free and trapped states is governed by a thermal equilibrium distribution, or more generally, a quasi-Fermi distribution

$$n = \frac{N_T}{1 + e^{(E_t - E_F)/kT}} \tag{2-81}$$

where the quasi-Fermi level E_F is given by

$$n = N_c e^{(E_F - E_c)/kT} \tag{2-82}$$

Here E_t is the trap depth, N_T the trap density, E_c the conduction band edge, and N_c the effective density of states in the conduction band. From Eqs. (2-81) and (2-82) it follows by eliminating E_F that $n_t = N_T n[n + N_c \exp(E_t - E_c)/kT]^{-1}$ and hence $N_{tot} = n + n_t$ is a monotonically increasing function of n, and vice versa.

A nonmonotonic $N_{tot}(n)$ relation can occur if the rates that describe the balance of trapping and detrapping depend strongly upon n, such that the presence of a few more free carriers strongly depletes the trapped carriers. This can be achieved with impact ionization. As a simple illustration, we consider a two-level impact ionization mechanism (Schöll, 1981). Here the concentration of trapped electrons n_t is the sum of those trapped in a ground level (n_{t1}) and in an excited level (n_{t2}). The rate equations for the generation–recombination processes shown in Fig. 2-21 are

$$\frac{dn_{t2}}{dt} = -X_1^S n_{t2} + T_1^S(N_T - n_{t1} - n_{t2})n - X_1^* n_{t2} n + X^* n_{t1} - T^* n_{t2}, \tag{2-83}$$

and

$$\frac{dn_{t1}}{dt} = -X^* n_{t1} + T^* n_{t2} - X_1 n_{t1} n \tag{2-84}$$

where T_1^S, X_1^S, T^*, X^*, X_1, and X_1^* denote the rate constants of capture into the excited state, (thermal or optical) ionization of the excited state, relaxation of the excited state into the ground state, excitation of the ground state, impact

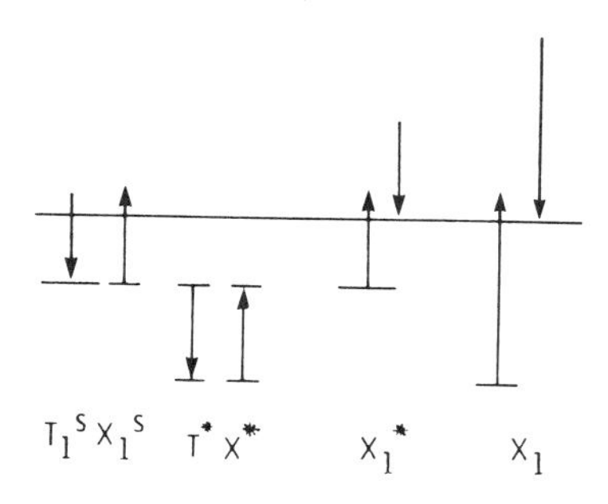

FIGURE 2-21. A simple generation–recombination (g–r) mechanism involving the conduction band and the impurity ground level and its first excited level. The various g–r processes, discussed in the text, are respective rate coefficients. (After Schöll, 1982a.)

ionization of the ground state and of the excited state, respectively, as defined in the figure. In the steady state n_{t1} and n_{t2} are readily obtained from Eqs. (2-83) and (2-84), as

$$n_{t1}(n, F) = N_T T^* T_1^S n / \Delta(n) \tag{2-85}$$

$$n_{t2}(n, F) = N_T (X^* + X_1 n) T_1^S n / \Delta(n) \tag{2-86}$$

with

$$\Delta(n) = (X_1^S + T_1^S n + X_1^* n)(X^* + X_1 n) + T^*(T_1^S + X_1)n$$

The occupation numbers of the two levels are plotted for an intermediate field range in Fig. 2-22. This leads to a nonmonotonic $N_{tot}(n)$ curve, as shown in Fig. 2-20. Note that the explicit F dependence of $n_{t1}(n, F)$ and $n_{t2}(n, F)$ occurs through the impact ionization coefficients X_1, X_1^*, which are increasing functions of F. In contrast to the situation shown in Fig. 2-20, for very low fields the depletion of the ground state n_{t1} is diminished, while for very high fields it is enhanced; in both cases the resulting $N_{tot}(n)$ curve is monotonic. The underlying physical mechanism is the following: For a small total number of carriers N_{tot}, impact ionization is negligible, and the distribution between free and trapped states is governed by the balance of capture and thermal ionization. For large N_{tot}, thermal ionization is negligible, and capture is balanced by impact ionization. In an intermediate range of N_{tot}, both states are possible, and additionally there exists an NDC state, where N_{tot} is decreasing in n, since impact ionization depletes the ground state and forms a highly nonthermal distribution of free and trapped carriers.

The intersection of the line $N_{tot} = N_{D^*}$ with the curve $N_{tot}(n, F)$ (see Fig. 2-20) gives the charge-neutral, spatially homogeneous steady states, by Eq. (2-79). If the background charge is chosen suitably, and N_{tot} as a function of n is N-shaped in some range of F, such as in Fig. 2-20, this leads to an S-shaped current-field characteristic

$$J_z(F_0) = e\mu F_0 n_0(F_0) \tag{2-87}$$

where the subscript 0 denotes the uniform steady state, and F_0 is the uniform field applied in the z direction.

We now proceed to discuss the stability of uniform SNDC states against small space- and time-dependent fluctuations (Schöll, 1987).

A small signal analysis of SNDC elements was described, e.g., by Shaw et al. (1973), Mitin (1977) and Bass et al. (1983). Our treatment here is different in that

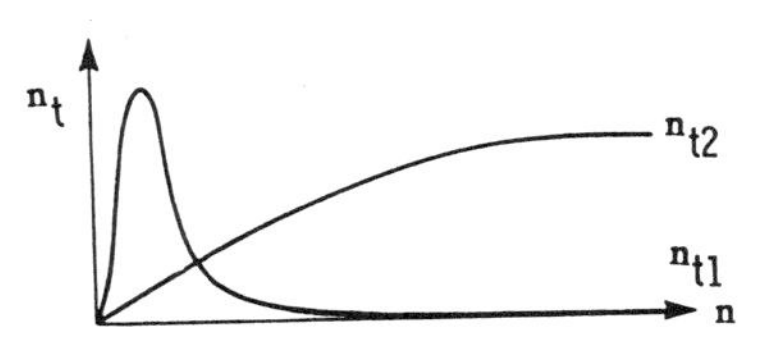

FIGURE 2-22. Trapped electron density in the ground state (n_{t_1}) and in the excited state (n_{t_2}) versus free-electron concentration. The sharp decrease in (n_{t_1}) may be due to impact ionization. (After Schöll, 1987.)

it also includes fluctuations of the carrier densities:

$$\delta\tilde{\mathbf{F}}(\mathbf{x}, t) = \mathbf{F}(\mathbf{x}, t) - \mathbf{F}_0 \tag{2-88}$$

$$\delta\tilde{n}(\mathbf{x}, t) = n(\mathbf{x}, t) - n_0 \tag{2-89}$$

$$\delta\tilde{n}_t(\mathbf{x}, t) = n_t(\mathbf{x}, t) - n_{t,0} \tag{2-90}$$

We substitute Eqs. (2-88)–(2-90) into the governing Eqs. (2-76) and (2-77), using Eqs. (2-78) and (2-79), and retain only terms linear in the fluctuations. For the time dependence we assume

$$\delta\tilde{\mathbf{F}}(\mathbf{x}, t) = \delta\mathbf{F}(\mathbf{x})e^{\lambda t} \tag{2-91}$$

$$\delta\tilde{n}(\mathbf{x}, t) = \delta n(\mathbf{x})e^{\lambda t} \tag{2-92}$$

$$\delta\tilde{n}_t(\mathbf{x}, t) = \delta n_t(\mathbf{x})e^{\lambda t} \tag{2-93}$$

This yields

$$\text{curl curl } \delta\mathbf{F} + \mu_0\lambda\left[e\mu n_0\, \delta\mathbf{F} + en_0\frac{d\mu}{dF_0}\mathbf{F}_0\, \delta F_z + e\mu\mathbf{F}_0\, \delta n + eD \text{ grad } \delta n + \epsilon\lambda\, \delta\mathbf{F}\right] = 0 \tag{2-94}$$

$$\text{div } \delta\mathbf{F} = \frac{1}{\epsilon}\left(\rho_N\, \delta n + \rho_F\, \delta F_z\right) \tag{2-95}$$

where we have allowed for a field-dependent mobility, $\mu(F_0)$, which is related to the drift velocity v by $v = \mu F_0$, and defined $\rho_N = \partial\rho/\partial n$, and $\rho_F = \partial\rho/\partial F$. Note that in our approximation ρ depends only on the magnitude and not on the direction of $\mathbf{F}$, and therefore in Eq. (2-95) only the field fluctuation δF_z parallel to $\mathbf{F}_0$ occurs.

In general, the concentrations of electrons trapped at levels $i = 1, 2, \ldots,$ are governed by rate equations of the type used in Eqs. (2-83) and (2-84). Substitution of Eqs. (2-88)–(2-90) and linearization in the concentrations yields a matrix equation that can be solved for δn_{ti} as a function of δn, δF, and λ. Therefore, δN_{tot} and ρ_N, ρ_F in Eq. (2-95) also depend upon λ. Thus, although Eq. (2-95) does not contain the trapped electron concentrations explicitly, it includes the temporal evolution of the occupancy of the individual trap levels implicitly through the λ dependence. We have allowed for independent fluctuation variation of the trap occupanices, which goes beyond the often invoked assumption of time-independent trapping, e.g., Shockley–Read–Hall kinetics. This generalization is, in fact, essential for the description of generation–recombination-induced instabilities. The λ dependence of ρ_N and ρ_F is in general complicated; for the mechanism of Fig. 2-21, for example, it is (Schöll, 1987)

$$\rho_N = e(-\lambda^2 + \lambda\beta_1 + \beta_2)/(\lambda^2 + \lambda\theta + \Delta) \tag{2-96}$$

$$\rho_F = e(\lambda\alpha_1 + \alpha_2)/(\lambda^2 + \lambda\theta + \Delta) \tag{2-97}$$

with constants $\beta_1 > 0$, $\beta_2 > 0$ (for the NDC state), $\theta > 0$, $\Delta > 0$, $\alpha_1 > 0$, $\alpha_2 > 0$, which depend on the generation–recombination coefficients.

Eliminating δN from Eqs. (2-94) and (2-95), we find

$$\text{curl curl}\,\delta\mathbf{F} + \mu_0\lambda\sigma(\lambda)\,\delta\mathbf{F} = 0 \tag{2-98}$$

where

$$\sigma(\lambda) = \epsilon\lambda + en_0\left(\mu + \frac{d\mu}{dF_0}\mathbf{F}_0 \otimes \hat{a}_z\right)$$
$$+ \frac{e}{\rho_N(\lambda)}(\mu\mathbf{F}_0 + D\,\text{grad}) \otimes [\epsilon\,\text{div} - \rho_F(\lambda)\hat{a}_z] \tag{2-99}$$

denotes a dynamic differential conductivity tensor, $\hat{a}_z$ is the unit vector in the z direction, and $\otimes$ denotes the tensor product. $\sigma(\lambda)$ describes the response of the current fluctuation to a space- and time-dependent field fluctuation. For homogeneous, time-independent fluctuations δF_z, Eq. (2-99) reduces to the static scalar differential conductivity

$$\sigma_{\text{diff}} = en_0\left(\mu + \frac{d\mu}{dF_0}F_0\right) - e\mu\frac{\rho_F(0)}{\rho_N(0)}F_0 = en_0\frac{dv}{dF_0} + ev\frac{dn}{dF_0} \tag{2-100}$$

Equation (2-100) shows that NDC can arise from $dv/dF_0 < 0$ (negative differential mobility), as well as from $dn/dF_0 < 0$ (carrier density decreasing with increasing field). The latter can be due to either $\partial\rho/\partial F_0 < 0$ (field-induced) or $\partial\rho/\partial n < 0$ (generation–recombination-induced). NNDC or SNDC require that $\sigma_{\text{diff}} \to 0$ or $\sigma_{\text{diff}} \to \infty$, respectively, at the turning points of the $J(F_0)$ characteristic. Thus, NNDC can occur if dv/dF_0 or $\partial\rho/\partial F_0$ become sufficiently negative, while SNDC is associated with a change of sign of $\partial\rho/\partial n$, and thus requires negative $\partial\rho/\partial n$. Equation (2-100) is a special case of the general relation (2-5), where the additional dynamic variables q are now given by the carrier densities n, n_{t1}, n_{t2}, ...

It is important to note that Eq. (2-100) can only describe changes in current along the static current–field characteristic of uniform states, whereas the more general expression Eq. (2-99) is needed in order to account for all possible responses to a general time- and space-dependent fluctuation of the field and the carrier concentration. The tensor character of Eq. (2-99) is important for filamentation, since this requires going beyond the one-dimensional analysis we used for moving field domains in the NNDC case.

We shall investigate next the spectrum of eigenvalues λ of Eqs. (2-98). Negative values of λ indicate, according to our assumption [Eqs. (2-91)–(2-93)], that fluctuations decay, and the uniform steady state is stable. Positive values of λ, however, indicate an instability, which will eventually lead to a new spatial structure such as current filaments. For different eigenmodes, i.e., different orientations and wavelengths of the initial spatial fluctuation $\delta\mathbf{F}(\mathbf{x})$ described by a wave vector $\mathbf{k}$, we anticipate different eigenvalues; this yields a whole set of dispersion relations $\lambda(\mathbf{k})$. For simplicity we confine attention to the case of a field-independent mobility μ.

For the cylindrical configuration of Fig. 2-19, the vector components of Eq. (2-98) are

$$0 = \left\{ -\frac{1}{r}\frac{\partial}{\partial r} r \frac{\partial}{\partial r} + \alpha\left[\tilde{V} + \rho_F\left(e\mu F_0 + eD\frac{\partial^2}{\partial z^2}\right) - e\mu F_0 \epsilon \frac{\partial}{\partial z}\right]\right\} \delta F_z + \left[\frac{\partial}{\partial z} - \alpha\epsilon\left(e\mu F_0 + eD\frac{\partial}{\partial z}\right)\right]\frac{1}{r}\frac{\partial}{\partial r} r\,\delta F_r$$

$$0 = \left[\frac{\partial}{\partial z} + \alpha\left(eD\rho_F - eD\epsilon\frac{\partial}{\partial z}\right)\right]\frac{\partial}{\partial r}\delta F_z + \left[-\frac{\partial^2}{\partial z^2} + \alpha\left(\tilde{V} - eD\epsilon\frac{\partial}{\partial r}\frac{1}{r}\frac{\partial}{\partial r} r\right)\right]\delta F_r \tag{2-101}$$

where we have defined

$$\alpha(\lambda) \equiv -\mu_0\lambda/\rho_n(\lambda) \tag{2-102}$$

$$\tilde{V}(\lambda) \equiv -(\epsilon\lambda + e\mu n_0)\rho_N(\lambda) \tag{2-103}$$

Note that

$$\alpha\tilde{V} = \mu_0\lambda(\epsilon\lambda + e\mu n_0)$$

Before we discuss the possible modes in detail (Schöll, 1987) it should be noted that the contact geometry induces important boundary conditions. If we assume metallic contacts in the planes $z = 0$ and $z = l$, then the total voltage fluctuation between these contacts,

$$\delta\Phi = \int_0^l \delta F_z\, dz \tag{2-104a}$$

must be independent of the transverse coordinates. Since we are interested in filamentary instabilities, where δF_z depends upon r, it must also depend upon z in order to satisfy Eq. (2-104a).

A further boundary condition at metallic contacts is

$$\delta F_r|_{z=0} = \delta F_r|_{z=l} = 0 \tag{2-104b}$$

which states that δF_r cannot be independent of z unless it vanishes everywhere. Finally, the condition of vanishing radial currents at the cylindrical surface requires

$$\delta J_r = eD\frac{\partial}{\partial r}\delta n + (\epsilon\lambda + e\mu n_0\,\delta F_r)|_{r=a} = 0 \tag{2-104c}$$

We attempt a solution of Eq. (2-101) in the form (see Fig. 2-23)

$$\begin{aligned} \delta F_z(r, z) &= \delta\tilde{F}_z J_0(k_\perp r)e^{ik_\| z} \\ \delta F_r(r, z) &= \delta\tilde{F}_r J_1(k_\perp r)e^{ik_\| z} \end{aligned} \tag{2-105}$$

where $\|$ and $\perp$ denote the components of $\mathbf{k}$ parallel and perpendicular to $\mathbf{F}_0$, and J_0, J_1 are Bessel functions of the first kind of order zero and one, respectively. They satisfy the relations

$$\begin{aligned} \frac{1}{r}\frac{\partial}{\partial r}r\frac{\partial}{\partial r}J_0(k_\perp r) &= -k_\perp^2 J_0(k_\perp r) \\ \frac{\partial}{\partial r}\frac{1}{r}\frac{\partial}{\partial r}rJ_1(k_\perp r) &= -k_\perp^2 J_1(k_\perp r) \\ \frac{\partial}{\partial r}J_0(k_\perp r) &= -k_\perp J_1(k_\perp r) \\ \frac{1}{r}\frac{\partial}{\partial r}rJ_1(k_\perp r) &= k_\perp J_0(k_\perp r) \end{aligned} \tag{2-106}$$

Substitution of Eq. (2-105) reduces the differential Eq. (2-101) to the algebraic matrix equation

$$\begin{pmatrix} A_{11} & A_{12} \\ A_{21} & A_{22} \end{pmatrix}\begin{pmatrix} \delta\tilde{F}_z \\ \delta\tilde{F}_r \end{pmatrix} = 0 \tag{2-107}$$

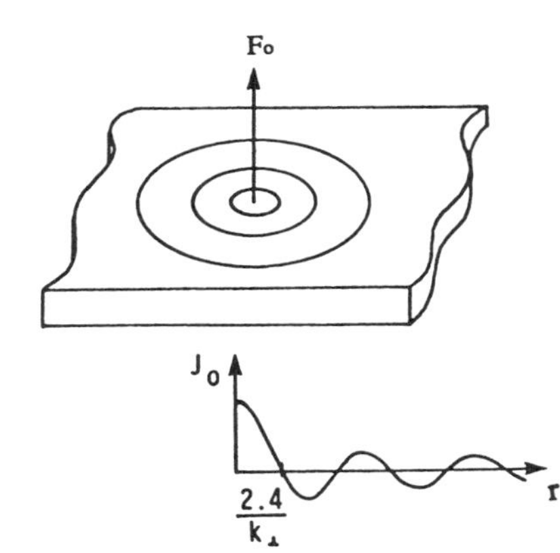

FIGURE 2-23. Schematic sketch of fluctuations leading to a cylindrical current filament, when a field F_0 is applied in the z direction. The inset shows the Bessel function $J_0(k_\perp r)$ that describes the radial field fluctuations $\delta F_z(r)$. (After Schöll, 1987.)

where

$$A_{11} \equiv k_\perp^2 + \alpha[\tilde{V} + \rho_F(e\mu F_0 + eDik_\parallel) - e\mu F_0 \epsilon ik_\parallel] \tag{2-108a}$$

$$A_{12} \equiv [ik_\parallel - \alpha(\epsilon e\mu F_0 + eD\epsilon ik_\parallel)]k_\perp \tag{2-108b}$$

$$A_{21} \equiv -[ik_\parallel + \alpha(eD\rho_F - eD\epsilon ik_\parallel)]k_\perp \tag{2-108c}$$

$$A_{22} \equiv [k_\parallel^2 + \alpha(\tilde{V} + eD\epsilon k_\perp^2)] \tag{2-108d}$$

The dispersion relation $\lambda(k_\parallel, k_\perp)$ is given by the condition that the determinant of the matrix in (2-107) vanishes:

$$A_{11}A_{22} - A_{12}A_{21} = 0 \tag{2-109}$$

A numerical solution of Eq. (2-109) shows that for NDC states there exists a region of unstable modes (Re $\lambda > 0$) in the $(k_\parallel, k_\perp)$ plane.

The boundary condition [Eq. (2-104c)] requires values of $k_\perp$ such that $J_1(k_\perp a) = 0$, which follows by using Eqs. (2-95) to eliminate δn, and Eqs. (2-105) and (2-106) to express all field fluctuations in terms of J_1. The permitted values of $k_\parallel$ are such that Im δF_z and Im δF_r satisfy Eqs. (2-104a) and (2-104b). For zero voltage fluctuations, $\delta\Phi = 0$, we obtain $k_\parallel = n(2\pi/l)$ with $n = 1, 2, 3, \ldots$.

In order to gain some analytical insight into the dispersion relation, we neglect $k_\parallel$ in Eq. (2-108), assuming that l is large. Equation (2-109) is then simplified:

$$\alpha(\lambda)\{[k_\perp^2 + \mu_0\lambda(\epsilon\lambda + e\mu n_0)][\tilde{V}(\lambda) + eD\epsilon k_\perp^2 + \rho_F(\lambda)e\mu F_0\mu_0\lambda(\epsilon\lambda + e\mu n_0)\} = 0 \tag{2-110}$$

For $\rho_F \equiv 0$, the curly bracket in Eq. (2-110) factorizes into two uncoupled modes:

(a) A stable transverse dielectric relaxation mode

$$k_\perp^2 + \mu_0\lambda(\epsilon\lambda + e\mu n_0) = 0 \tag{2-111}$$

The substitution of Eq. (2-111) into Eq. (2-107) shows that this corresponds to a purely transverse electromagnetic mode with $\delta F_r = 0$: $\mathbf{k} \perp \delta\mathbf{F} \parallel \mathbf{F}_0$. Hence, because div $\mathbf{F} = 0$, this mode does not couple to charge density fluctuations. Since the generation–recombination instability manifests itself essentially in charge density fluctuations, this mode is always stable (Re $\lambda < 0$):

$$\lambda = -\frac{e\mu n_0}{2\epsilon} \pm \left[\left(\frac{e\mu n_0}{2\epsilon}\right)^2 - \frac{k_\perp^2}{\mu_0\epsilon}\right]^{1/2} \tag{2-112}$$

The real part of the spectrum is plotted in Fig. 2-24. The damping (dielectric relaxation) is due to the conductivity $\sigma_0 \equiv e\mu n_0$, which is [in contrast to the *differential* conductivity σ_{diff}, Eq. (2-100)] always positive.

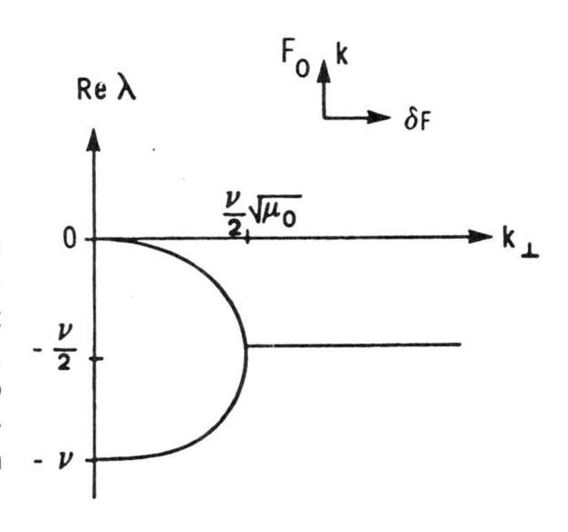

FIGURE 2-24. Transverse electromagnetic modes of an SNDC element (schematic). The real part of the eigenvalue λ is plotted versus the wave vector $k_\perp$. The inset shows the orientation of the applied static electric field $\mathbf{F}_0$, the field fluctuation $\delta\mathbf{F}$, and the wave vector $\mathbf{k}$ relative to each other. All modes are damped (Re $\lambda < 0$). The notation $\nu = e\mu n_0/\epsilon$ has been used for the dielectric relaxation frequency. (After Schöll, 1987.)

(b) The second factor in Eq. (2-110) gives the recombination–diffusion mode

$$\tilde{V}(\lambda) + eD\epsilon k_\perp^2 = 0 \tag{2-113}$$

From Eq. (2-107) we obtain for the eigenvector $\delta\mathbf{F}$:

$$\frac{\delta F_z}{\delta F_r} = \frac{\alpha(\lambda)\epsilon e\mu F_0 k_\perp}{k_\perp^2 + \mu_0\lambda(\epsilon\lambda + e\mu n_0)} \tag{2-114}$$

This mode can become unstable, and in fact does on SNDC points. The corresponding spectrum $\lambda(k_\perp)$ is given by

$$k_\perp = [(\epsilon\lambda + e\mu n_0)\rho_N(\lambda)/(eD\epsilon)]^{1/2} \tag{2-115}$$

where we have used Eq. (2-103). It can be shown (Schöll, 1987) that the solutions $\lambda_1 > \lambda_2 > \cdots > \lambda_M$ of the equation $\rho_N(\lambda) = 0$ describe spatially homogeneous, charge-neutral fluctuations of the free and the trapped carrier densities, and that in the NDC state the largest eigenvalue λ_1 is positive. In the long-wavelength limit $k_\perp \to 0$ the dispersion relation Eq. (2-113) may be expanded in terms of $\lambda - \lambda_1$. For NDC states we obtain a top branch with undamped recombination–diffusion modes

$$\lambda \approx \lambda_1 - \tilde{D}_\perp k_\perp^2 \tag{2-116}$$

where

$$\tilde{D}_\perp \equiv eD\epsilon\left[\frac{\partial}{\partial\lambda}\tilde{V}(\lambda)\right]_{\lambda=\lambda_1}^{-1} \tag{2-117}$$

is an effective transverse diffusion constant. It reflects the slowing down of laterally diffusing carriers by multiple trapping and detrapping. The dispersion relation is plotted in Fig. 2-25 for a typical g–r instability (Schöll, 1982b).

Now we consider the coupling of the two groups of modes Eqs. (2-111) and (2-113) through $\rho_F \neq 0$ in Eq. (2-110). The relevant part of the resulting spectrum is shown in Fig. 2-26. We first discuss the behavior of the spectrum near

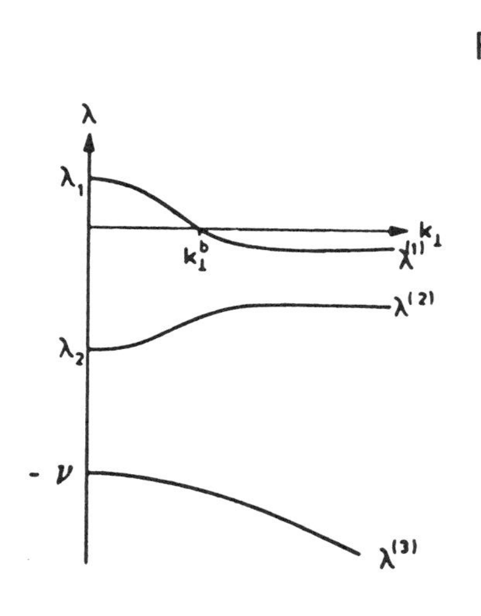

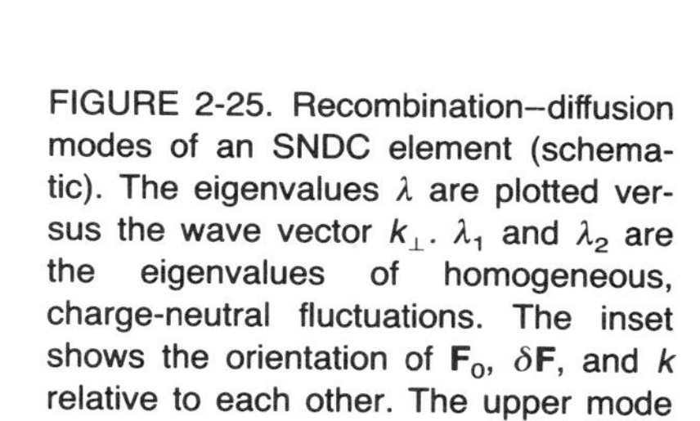

FIGURE 2-25. Recombination–diffusion modes of an SNDC element (schematic). The eigenvalues λ are plotted versus the wave vector $k_\perp$. λ_1 and λ_2 are the eigenvalues of homogeneous, charge-neutral fluctuations. The inset shows the orientation of $\mathbf{F}_0$, $\delta\mathbf{F}$, and k relative to each other. The upper mode is unstable for $k_\perp < k_\perp^b$. (After Schöll, 1982b.)

those points where $\lambda = 0$. Expanding Eq. (2-110) in terms of λ and $k_\perp$ and using (2-100), we obtain in the vicinity of $(\lambda = 0,\ k_\perp = 0)$:

$$\lambda \approx -k_\perp^2/(\mu_0 \sigma_{\text{diff}}) \tag{2-118}$$

which is undamped for NDC states ($\sigma_{\text{diff}} < 0$). The corresponding eigenvector is parallel to F_0 for small λ: $\delta F_r = 0$.

The eigenvalues of Eq. (2-110) cross from positive to negative values of λ (Fig. 2-26) at a finite value

$$k_\perp^b = [-\tilde{V}(0)/(eD\epsilon)]^{1/2} = [\mu n_0 \rho_N(0)/(D\epsilon)]^{1/2} \tag{2-119}$$

This crossing point $k_\perp^b$ is the same as for the pure recombination–diffusion mode Eq. (2-113). Substitution into Eq. (2-107) reveals that the eigenvector satisfies $\delta F_z = 0$: $\delta\mathbf{F} \parallel \mathbf{k} \perp \mathbf{F}_0$. All modes with $0 < k_\perp < k_\perp^b$ are unstable and grow in time. As $k_\perp$ increases, the electromagnetic modes change gradually from purely transverse ($\delta F_r = 0$) to purely longitudinal ($\delta F_z = 0$).

With increasing F_0 the wave vector $k_\perp^b$ changes as shown in Fig. 2-27. At both end points of the SNDC regime $k_\perp^b \to 0$, and the range of unstable wave vectors disappears altogether. As F_0 increases, a whole family of modes with $k_\perp^b(F_0)$ becomes sucessively undamped. As we have shown within the general framework

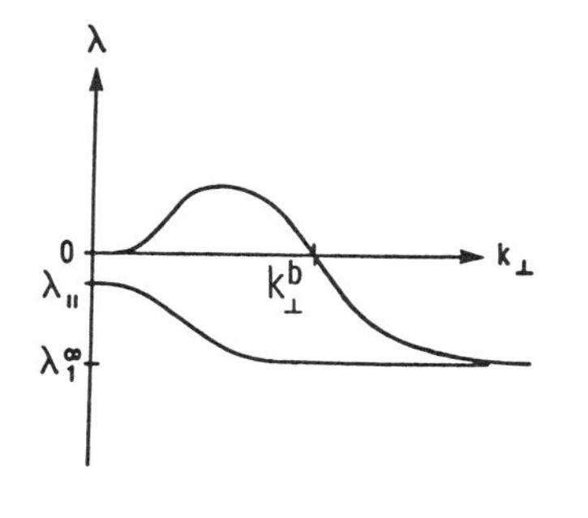

λ

FIGURE 2-26. Coupled electromagnetic filamentary mode of an SNDC element (schematic). The parameters are discussed in the text and Figs. 2-24 and 2-25. (After Schöll, 1987.)

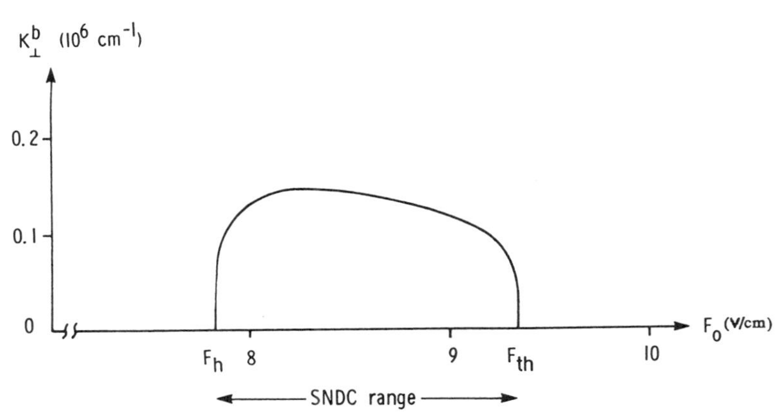

FIGURE 2-27. Bifurcation line $k_\perp^b$ as a function of the applied field F_0 for an SNDC element. The modes $k_\perp < k_\perp^b$ are undamped; all modes $k_\perp > k_\perp^b$ are damped. The bifurcation line gives the wave vectors of bifurcating spatial structures. The numerical parameters are for the g–r mechanism of Fig. 2-21. (After Schöll, 1982b.)

of bifurcation theory (Sec. 1.2.1 E), this leads to the successive bifurcation of a family of filamentary stationary solutions $\delta F_r(r) \propto J_1(k_\perp^b r)$ from the unstable homogeneous NDC state. These are modulated perpendicular to the static field F_0 on a characteristic length scale $1/k_\perp^b(F_0)$. At the bifurcation point $\delta F_z \equiv 0$ holds by Eq. (2-107). From Eq. (2-95) it then follows that the excess carrier density of the bifurcating solution is

$$\delta n(r) = [\epsilon k_\perp^b / \rho_N(0)]\, \delta \tilde{F}_r J_0(k_\perp^b r) \tag{2-120}$$

The current density $\delta \mathbf{J} = \sigma(\lambda = 0)\, \delta \mathbf{F}$ follows from Eq. (2-99):

$$\begin{aligned} \delta J_z(r) &= e\mu F_0[\epsilon k_\perp^b / \rho_N(0)]\, \delta \tilde{F}_r J_0(k_\perp^b r) \\ \delta J_r(r) &\equiv 0 \end{aligned} \tag{2-121}$$

Equation (2-121) shows that there is a modulation of the longitudinal current density δJ_z although $\delta F_z = 0$. This is because the transversally modulated carrier density induces a ripple in the drift current J_z.

The transverse current fluctuation δJ_r vanishes identically due to our approximation $k_\parallel = 0$ so that the boundary condition Eq. (2-104c) does not impose a restriction upon the permitted values of $k_\perp^b$.

The external circuit may, however, impose constraints. If we assume that the SNDC element is part of a heavily loaded resistive circuit, then the total current fluctuation

$$\delta I = 2\pi \int_0^a \delta J_z(r) r\, dr \propto J_1(k_\perp^b a) \tag{2-122}$$

must vanish, and the family of permitted bifurcation wave vectors $k_\perp^b$ is specified by the roots of $J_1(k_\perp^b a) = 0$, as also required by the boundary condition (2-104c) in case of $k_\parallel \neq 0$. The lowest root is $k_\perp^b \sim 3.8/a$. This boundary condition implies that the radial field $\delta F_r(a) \propto J_1(k_\perp^b a)$ vanishes on the cylindrical surface, and we have a charge-free surface.

It can be shown under quite general conditions (Schöll, 1987) that among all possible inhomogeneous solutions the one with the smallest $k_\perp^b$, corresponding to the profile with the least number of extrema, is the only stable and hence physical one. We note that for $k_\perp^b = 3.8/a$, $\delta J_z(r) \propto J_0(k_\perp^b r)$ changes sign once in going from $r = 0$ to $r = a$ [at $r \sim (2.4/3.8)a$], leading either to current pinching at the center or current spreading at the cylindrical surface: we have filamentation. The condition $k_\perp^b(F_0) = 3.8/a$ specifies the two limiting values of the field F_0 between which current filaments can exist (cf. Fig. 2-27). For large a this range includes almost the full SNDC range, but for small a it may be considerably shorter. If the radius a is so small that $k_\perp^b(F_0) < 3.8/a$ for any F_0, then the device is stable against all inhomogeneous fluctuations, and current filaments *cannot* form.

Equation (2-118) is the dispersion relation used in more intuitive filamentation arguments (Shaw et al., 1973; Bass et al., 1983). From our more detailed analysis, which includes fluctuations of the carrier concentrations, it follows that the dielectric mode Eq. (2-118) approximates the spectrum in the vicinity of $\lambda = 0$, $k_\perp$ only, and that the bifurcation of filamentary structures occurs, in fact, at a larger wave vector $k_\perp^b$, Eq. (2-119), which is determined by generation–recombination processes and not by the static differential conductivity σ_{diff}. Our bifurcation argument requires that the real part of λ crosses from negative to positive values at a finite $k_\perp$ which cannot happen with the dispersion relation of Eq. (2-118).

2.2.3.2. Phase Portrait Analysis and Equal Areas Rule for a One-Carrier Semiconductor

In the preceding section we have demonstrated that stable current filaments bifurcate at two field values F_0 from the homogeneous SNDC state. As the field is varied, the amplitude $\delta \tilde{J}_z$ of these filaments grows and the profiles cannot be described any longer by the Bessel functions arising from the linearized transport equations. Instead, the profiles have to be calculated from the nonlinear transport Eqs. (2-77)–(2-79) (Schöll, 1982b).

We assume a cylindrical stationary filamentary solution of the form $n(r)$, $\mathbf{F}(r) = \mathbf{F}_0 + \mathbf{F}_r(r)$, where $F_r(r)$ is the internal radial field generated by the radial carrier concentration gradient. Observing the boundary condition of vanishing radial current density, we obtain from Eqs. (2-77) and (2-78),

$$J_z = e\mu F_0 n \tag{2-123}$$

$$\frac{d}{dr} n = -\frac{\mu}{D} n F_r \tag{2-124}$$

$$\frac{d}{dr} F_r = -\frac{F_r}{r} + \frac{1}{\epsilon} \rho(n, F_0) \tag{2-125}$$

where we have assumed $F_r \ll F_0$, so that ρ depends essentially upon the external field F_0 only. Thus, F_0 may be considered as a control parameter.

We first consider current filaments where the wall of the filament is thin compared to the radius r_0 of the filament. In this case we can approximate Eq. (2-125) as

$$\frac{d}{dr} F_r \sim \frac{1}{\epsilon} \rho(n, F_0) \tag{2-126}$$

which corresponds to approximating the cylindrical wall by a plane interface, so that we are effectively dealing with a one-dimensional plane geometry. Equations (2-124) and (2-126) then represent an autonomous dynamic system, and we can proceed with a phase-space analysis similar to the one that we performed in the NNDC case in Sec. 2.2.2.2. [cf. Eqs. (2-34), (2-35)]. As pointed out earlier, the essential nonlinearity that was represented by $v(F)$ in the NNDC case is now contained in $\rho(n, F) = e[N_D^* - N_{\text{tot}}(n, F)]$. The trajectories of the system equations (2-124) and (2-126) in the (n, F_r) phase space are given by

$$\frac{dF_r}{dn} = -\frac{\rho(n, F_0)}{F_r n} \frac{D}{\epsilon \mu} \tag{2-127}$$

which can be solved by the exact first integral

$$F_r(n) = \pm\{2[C - \tilde{\Phi}(n, F_0)]\}^{1/2} \tag{2-128}$$

where

$$\tilde{\Phi}(n, F_0) = \frac{D}{\epsilon\mu} \int \rho(n, F_0) \frac{dn}{n} \tag{2-129}$$

is a closed analytical integral since the integrand is rational for all common (i.e., polynomial) g–r rates. Different trajectories belong to different values of the integration constant C. The trajectories are parametrized by the radial coordinate r. The null-isoclines, where either $dF_r/dn = 0$ or $dn/dF_r = 0$, are given by $\rho(n, F_0) = 0$ and $F_r n = 0$, respectively. If F_0 is chosen in the SNDC range, $\rho(n, F_0) = 0$ has three solutions $n_1(F_0)$, $n_2(F_0)$, $n_3(F_0)$. Therefore, the phase plane trajectories cross the straight lines $n = n_1$, $n = n_2$, and $n = n_3$ with zero slope dF_r/dn. When a trajectory crosses the n axis, its slope is infinite. The phase portraits for these different values of F_0, all of them in the SNDC range, are shown in Fig. 2-28. Figure 2-28a corresponds to the case $\tilde{\Phi}(n_1, F_0) > \tilde{\Phi}(n_3, F_0)$, which occurs at lower values of F_0, while Fig. 2-28c corresponds to $\tilde{\Phi}(n_1, F_0) < \tilde{\Phi}(n_3, F_0)$, which belongs to higher F_0. A phase portrait for a value of F_0 above the SNDC range is shown in Fig. 2-29. Here there exists one singular point only: a saddle point.

The three singular points of Fig. 2-28 where $dF_r/dr = dn/dr = 0$ correspond to the spatially homogeneous steady states $F_r = 0$, $n = n_1$, $n = n_2$, and $n = n_3$, defined by $\rho(n_i, F_0) = 0$, where $i = 1, 2, 3$. Since there exists a first integral, i.e., the system is conservative, the singular points may be saddles or centers.

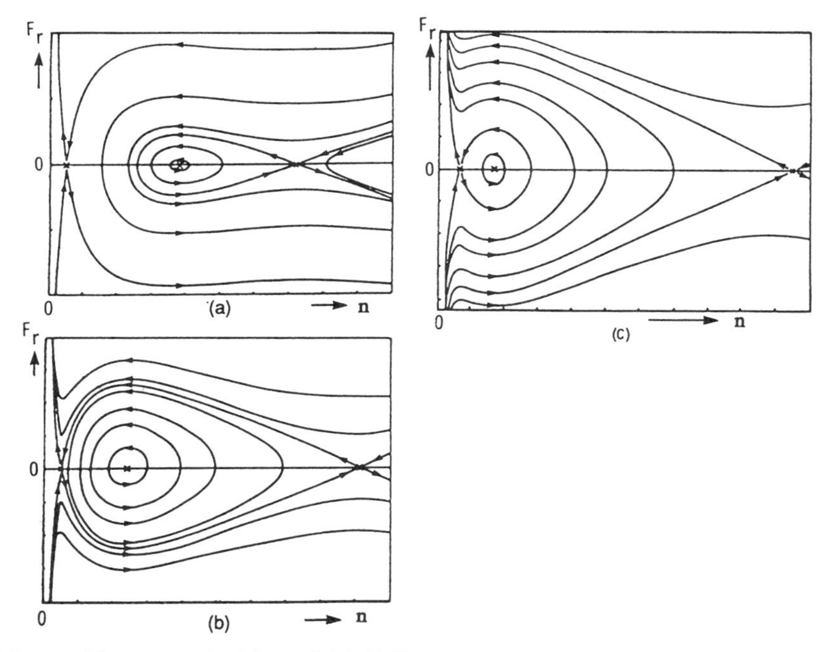

FIGURE 2-28. Phase portrait of the radial field F_r versus the electron density n for an SNDC element. The parameter varying along the phase trajectories is the radial coordinate r. The singular points (x) correspond to the homogeneous steady states n_1, n_2, n_3. Parts (a), (b), and (c) correspond to a sequence of increasing applied fields F_0 within the SNDC regime. (After Schöll, 1982b.)

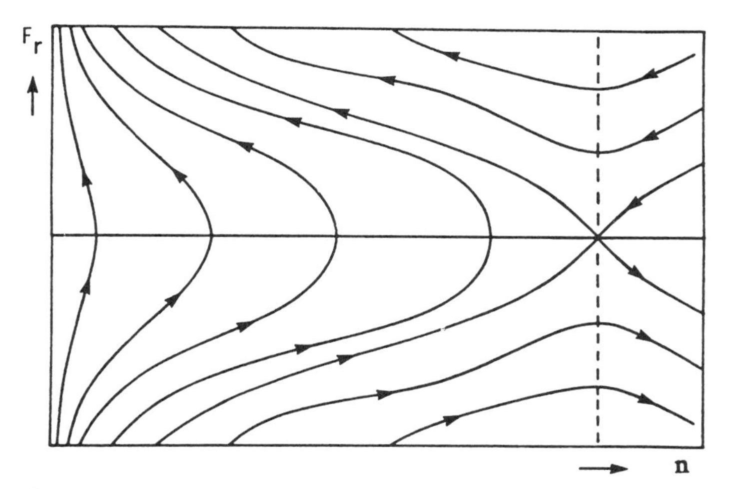

FIGURE 2-29. Phase portrait of the radial field F_r versus the electron density n as in Fig. 2-28, but for a higher field F_0 in the PDC regime, where only one singular point exists.

Linearization of Eqs. (2-124) and (2-126) around the singular points yields

$$\begin{pmatrix} \dfrac{d}{dr}\delta n \\ \dfrac{d}{dr}F_r \end{pmatrix} = \begin{pmatrix} 0 & -\dfrac{\mu}{D}n_i \\ \dfrac{1}{\epsilon}\rho_N & 0 \end{pmatrix} \begin{pmatrix} \delta n \\ F_r \end{pmatrix} \tag{2-130}$$

with $\delta n = n - n_i$, $i = 1, 2, 3$. The eigenvalues of the matrix in Eq. (2-130) are

$$K_i = \pm\left(-\frac{\mu}{\epsilon D} n_i \rho_N \right)^{1/2} \tag{2-131}$$

Hence [compare Fig. 2-20 and observe Eq. (2-79)] the singular points represent a saddle point if the corresponding homogeneous steady state is stable, i.e., $\rho_N < 0$, and a center if the homogeneous state is unstable, $\rho_N > 0$. In case of a saddle point the slope of the separatrices at the saddle point is given by the eigenvectors of Eq. (2-130)

$$dF_r/dn = \pm[-D\rho_N/(\epsilon\mu n_i)]^{1/2}$$

The slope is large for small n_i and large D. The center is surrounded by closed trajectories, corresponding to periodic oscillatory solutions. These are ellipses near the center, and represent small amplitude harmonic spatial oscillations with wave vector

$$k = iK_2 = \pm\left(\frac{\mu}{D\epsilon} n_2 \rho_N\right)^{1/2} \tag{2-132}$$

A comparison of Eqs. (2-132) and (2-119) shows that this is the bifurcation wave vector for the given value of F_0, and the small amplitude oscillations are the stationary solutions whose bifurcation we have analyzed in Sec. 2.2.3.1.

The phase plane region of oscillatory solution is confined by separatices originating at the saddle points. The three topologically different cases that occur are shown in Figs. 2-28a, 2-28b, and 2-28c. In Fig. 2.28a two separations of the saddle point at n_3 are joined together, forming a closed loop ("homoclinic orbit") with a minimum electron concentration n_{min}. In Fig. 2-28c, which corresponds to higher values of F_0, two separatrices of the saddle point at n_1 form a closed loop with maximum n_{max}. Figure 2-28b shows the limiting case between Figs. 2-28a and 2-28c, where n_1 and n_3 are connected by two symmetric separatrices ("heteroclinic orbits"). The corresponding inhomogeneous profiles connect the two stable steady states with a narrow interfacial layer.

All trajectories originating outside these separatrices are unbounded and represent profiles where either n or F_r tends to infinity for large $|r|$.

Which trajectories actually correspond to physical solutions depends upon the boundary conditions on the cylindrical axis and on the lateral surfaces of the SNDC elements. The regularity of F_r at $r = 0$ requires

$$F_r(0) = 0 \tag{2-133}$$

Thus, all trajectories originate on the n axis. Consider, for example, the phase portrait in Fig. 2-28a. A solution with a low electron density on the cylinder axis, $n(0) < n_1$, tends to $n \to 0$ and $F_r \to \infty$ for $r \to \infty$ (trajectory A in Fig. 2-30a). The radial field must be very large in order to balance the diffusion current at large r. For $n(0) = n_1$ the profile remains constant (B). If the electron concentration at the center satisfies $n_1 < n(0) < n_{\min}$, $n(r)$ increases monotonically (C). For $n(0) = n_{\min}$ the profile corresponds to part of the separatrix loop and tends asymptotically to the stable homogeneous steady state n_3 for $r \to \infty$ (D). For $n_{\min} < n(0) < n_3$ the solutions are periodic, and the corresponding profiles oscillate around the unstable steady state (E). For $n(0) = n_3$ the profile is constant (F), and for $n(0) > n_3$ the profiles again tend to $n \to \infty$ (G). Of all possible density profiles only solution D resembles a filamentary profile: a low density core and a monotonic increase in density up to the stable homogeneous value n_3, far from the core. The profiles of type D are shown for different, increasing values of the applied field F_0 in Fig. 2-31. As F_0 increases, the low-current filament becomes wider and deeper.

For values of F_0 corresponding to the phase portrait of Fig. 2-28c the profiles are inverted, corresponding to a high-density core and a low-density mantle that asymptotically reaches the stable homogeneous value n_1. Upon further increase

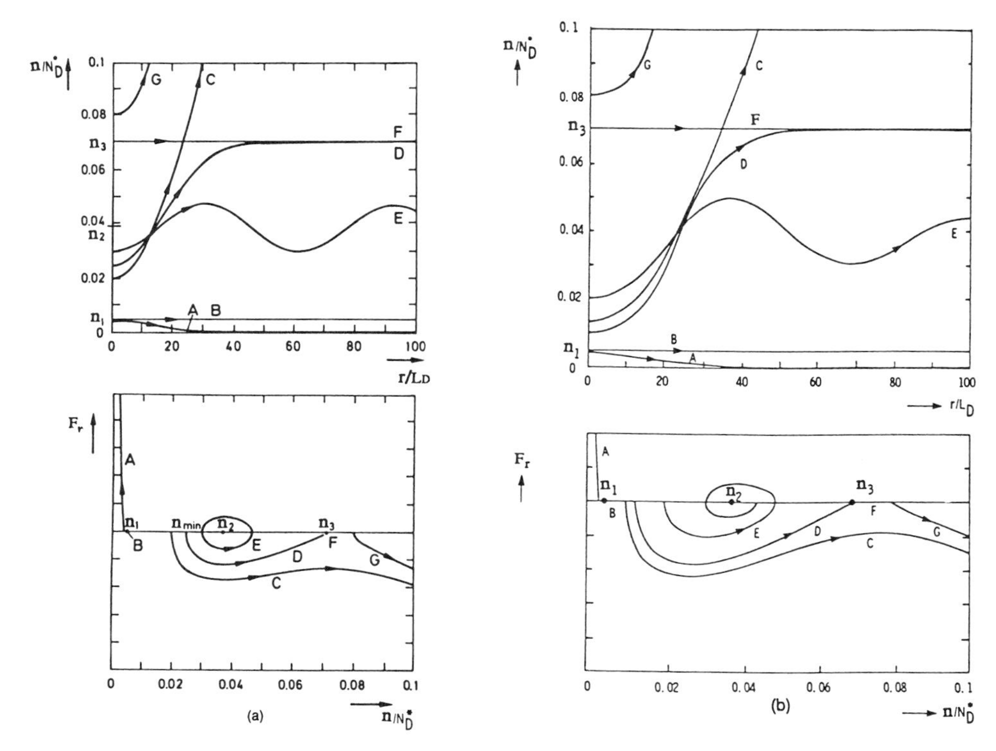

FIGURE 2-30. Electron density profiles n in units of N_D^* versus radial coordinate in units of the effective Debye length L_D (top), and associated phase trajectories in the (F_r, n) phase plane (bottom). (a) Corresponds to a planar and (b) corresponds to a cylindrical geometry. Solutions corresponding to different boundary conditions $n(0)$ are denoted by letters A, B, C, D, E, F, G. (After Schöll, 1987.)

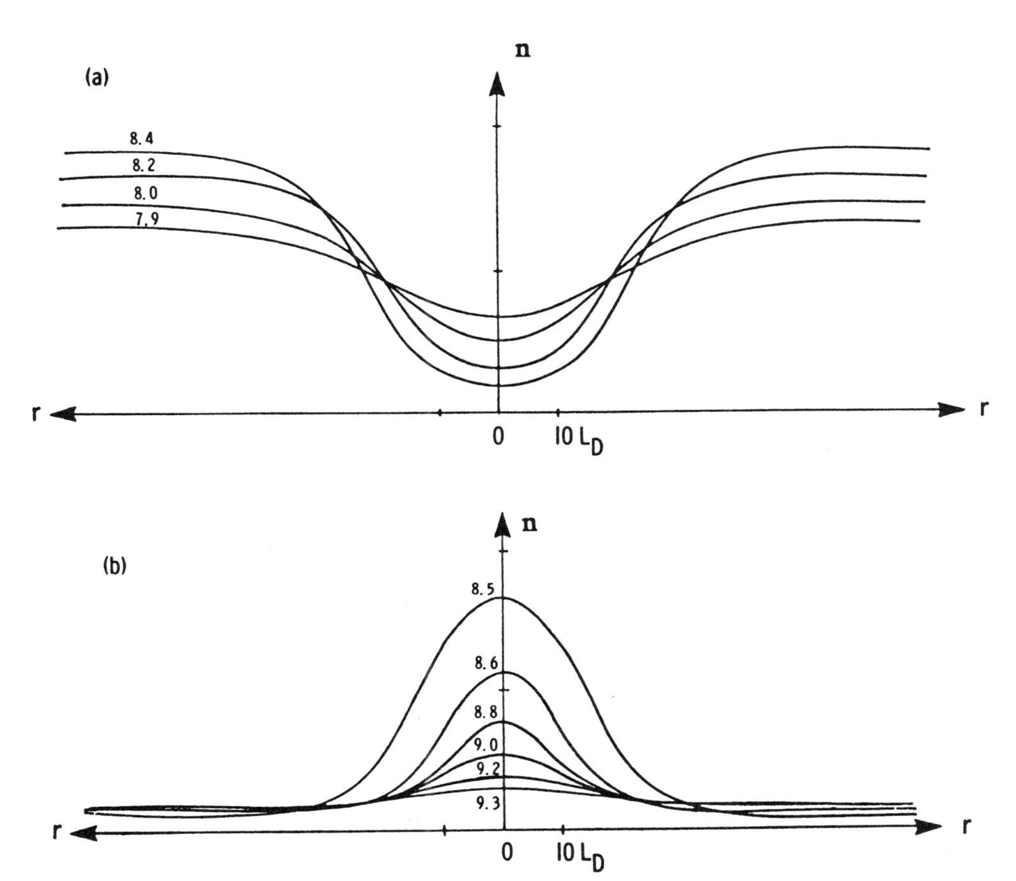

FIGURE 2-31. Spatially inhomogeneous electron density profiles $n(r)$ for different applied fields F_0 (the numbers at the profiles give F_0 in V cm^{-1}). (a) Low-current filaments; (b) high-current filaments. (After Schöll, 1987.)

of F_0 the high-current filament shrinks, and disappears as the end of the SNDC range is reached. The profiles shown in Fig. 2-31 lack a characteristic feature of low or high-current filaments in SNDC devices: They do not have a *wide flat* core of radius $r_0 \gg L_D$, where $L_D = [\epsilon D/(N_D^* e\mu)]^{1/2}$ is an effective Debye length, and a thin, charged filament wall several L_D's thick. It can also be shown (Schöll, 1983), by applying a small fluctuation to these stationary carrier profiles and linearizing the time-depending transport equations around them, that they are unstable, unless the total current,

$$I = e\mu F_0 2\pi \int_0^a n(r, t) r \, dr \tag{2-134}$$

is kept constant in time.

Wide, flat current filaments can be idealized in infinitely extended systems by the boundary conditions

0.25

$$n(0) \cong n_3, \qquad \lim_{r\to\infty} n(r) = n_1 \tag{2-135}$$

in the case of high-current filaments, and by

$$n(0) \cong n_1, \qquad \lim_{r\to\infty} n(r) = n_3 \tag{2-136}$$

in the case of low-current filaments. This situation occurs in the phase portrait of Fig. 2-28b, where we can identify the two saddle-to-saddle separatrices as the filamentary solutions. We shall show now that this condition singles out a specific value of the applied field $F_0 = F_{c0}$, where spatial coexistence between the stable homogeneous phases n_1 and n_3 is possible.

From Eq. (2-128) it follows that the singular points ($n = n_1$, $F_r = 0$) and ($n = n_3$, $F_r = 0$) lie on the same trajectory if and only if

$$\tilde{\Phi}(n_1, F_0) = \tilde{\Phi}(n_3, F_0) \tag{2-137}$$

Using the definition of $\tilde{\Phi}$, Eq. (2-129), we can rewrite Eq. (2-137) as the equal areas rule

$$\int_{n_1}^{n_3} \rho(n, F_0)\frac{dn}{n} = 0 \tag{2-138}$$

which determines F_0 such that the two hatched areas in Fig. 2-32 are equal. The equal areas rule for filamentation in SNDC elements, Eq. (2-138), is analogous to

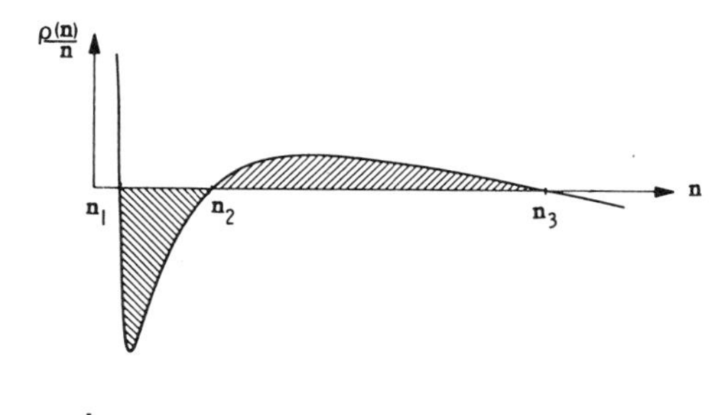

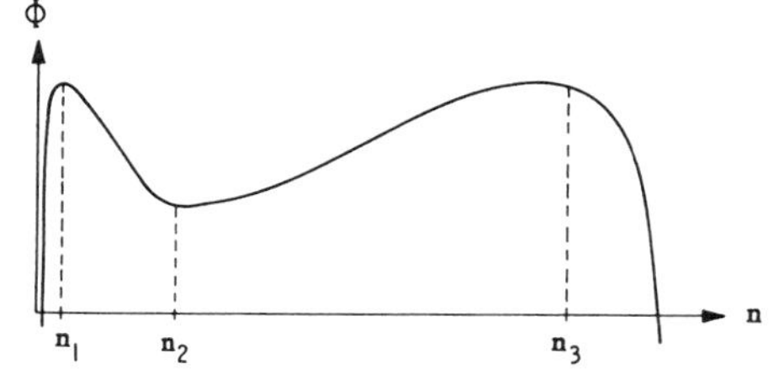

FIGURE 2-32. Condition for coexistence of the two phases n_1 and n_3 with a planar boundary layer; ρ is the charge density. The two hatched areas must be equal to ensure $\Phi(n_1) = \Phi(n_3)$. This is the equal areas rule for a planar geometry. (After Schöll, 1982a.)

the equal areas rule for domain formation in NNDC elements derived in Sec. 2.2.2.2. A similar equal areas rule has been derived for thermally induced SNDC (Volkov and Kogan, 1969). We also note the analogy with Maxwell's rule of phase coexistence in equilibrium thermodynamics.

The equal areas rule, Eq. (2-138), can also be derived in a more intuitive way from the formal analogy of Eqs. (2-124, 2-126) with the mechanical equation of motion of a unity mass particle in the potential $\tilde{\Phi}$ if we identify $\ln n$ with the spatial coordinate, and $r\mu/\epsilon D$ with the time: Eliminating F_r from Eqs. (2-124) and (2-126), we obtain

$$\frac{d^2}{dr^2}\ln n + \frac{\mu}{\epsilon D}\rho(n, F_0) = 0 \tag{2-139}$$

From the definition Eq. (2-129) it follows that the extrema of $\tilde{\Phi}$ are the homogeneous steady states where $\rho(n_i, F_0) = 0$. More precisely, stable homogeneous states correspond to maxima ($\rho_N < 0$), while the unstable homogeneous state corresponds to a minimum ($\rho_N > 0$). The filamentary profile corresponds to a motion where the particle is initially at rest at one of the peaks of the potential, then travels through the minimum in a relatively short time and comes asymptotically to rest at the other potential peak. Therefore the maxima must be of equal height (Fig. 2-32b), which gives Eq. (2-137). Near the peaks the particle moves slowly. Hence the profile $n(r)$ drops from n_3 to n_1 (or vice versa) within a thin interfacial layer and is essentially constant everywhere else.

Two electron density profiles corresponding to a high-current filament and a low-current filament are shown in Fig. 2-33 for $F_0 = F_{c0}$. From the phase portrait (Fig. 2-28b) it follows that the filament wall, say at r_0, can be shifted to arbitrarily large radii, if the initial point of the trajectory, i.e., the electron concentration value on the cylinder axis, is chosen arbitrarily close to one of the singular points. Thus, the total current can be widely varied between the high- and low-conductivity values corresponding to n_3 and n_1, respectively, at fixed field F_{c0},

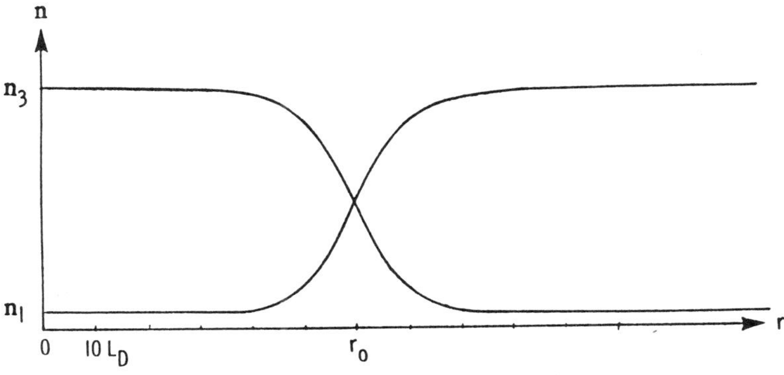

FIGURE 2-33. Electron density profiles $n(r)$ corresponding to high- or low-current filaments of radius r_0. (Schematic.)

simply by expanding r_0:

$$I = e\mu F_{c0}\pi[n_3 r_0^2 + n_1(a^2 - r_0^2)] \tag{2-140}$$

in the case of a high-current filament. (In the case of a low-current filament n_3 and n_1 are interchanged.) This means that an SNDC element containing a filament is a voltage limiter. This is analogous to the NNDC case where excess voltage applied to the system goes into (and widens) the domain while the current remains constant.

In the preceding analysis we have neglected the curvature of the filament by approximating Eq. (2-125) by Eq. (2-126). In a more rigorous treatment (Schöll, 1982b) we shall now retain Eq. (2-125) and combine it with Eq. (2-124), which gives

$$\frac{d^2}{dr^2}\ln n + \frac{1}{r}\frac{d}{dr}\ln n + \frac{\mu}{\epsilon D}\rho(n, F_0) = 0 \tag{2-141}$$

Equation (2-141) is analogous to the equation of motion for a particle in the double-peak potential $\tilde{\Phi}$ of Eq. (2-129), with a time-dependent friction force. This "friction force", which reflects the curvature of the filaments, leads to a modification of the phase portrait and the profiles of Fig. 2-30a, as shown in Fig. 2-30b. Depending upon the initial conditions, either damped oscillations around n_2 or overshooting unbounded solutions are possible. For small amplitudes A the oscillatory solutions are the Bessel functions obtained in the bifurcation analysis of Sec. 2.2.3.1.:

$$n(r) = n_2 + AJ_0(k_\perp^b r) \tag{2-142}$$

In the limit case between oscillatory and unbounded solutions there is a monotonic solution that satisfies

$$\lim_{r\to\infty} n(r) = n_i \qquad (i = 1, 3)$$
$$\lim_{r\to\infty} F_r(r) = 0 \tag{2-143}$$

for a suitable $n_0 \equiv n(0)$ in the range $n_1 < n_0 < n_3$. The case $i = 3$ applies to $F_0 < F_{c0}$, i.e., $\tilde{\Phi}(n_1) > \tilde{\Phi}(n_3)$, and leads to $n_0 < n_{\min} < n_2$; hence $n(r)$ represents a cylindrical current filament of low current density. The inverse case $i = 1$ applies to $F_0 > F_{c0}$, and leads to $n_0 > n_{\max} > n_2$; it represents a high-current filament.

With increasing $F_0 < F_{c0}$ the difference in the potential peaks $\tilde{\Phi}(n_1, F_0) - \tilde{\Phi}(n_3, F_0)$ decreases. Hence, n_0 tends to n_1, and the low-current filament becomes wider. For F_0 close to but smaller than F_{c0}, the filamentary solution rises from a value just above n_1 to n_3 within a thin cylindrical transition layer at r_0. Since $n(r)$ varies considerably in the transition layer only, we may approximately set $r \cong r_0$ in the second terms of Eq. (2-141).

We will now show that the equal areas rule has to be modified owing to the curvature of the filament wall. Formal integration of Eq. (2-141) over $d(\ln n)$ yields

$$\frac{1}{2}\left(\frac{d}{dr}\ln n\right)^2 + \int \frac{1}{r}\left(\frac{d\ln n}{dr}\right)^2 dr + \frac{\mu}{\epsilon D}\int \rho(n, F_0)\, d(\ln n) = C \qquad (2\text{-}144)$$

where C is an integration constant. A monotonic solution matching the filamentary boundary conditions (2-133) and (2-143) exists if and only if

$$\frac{\mu}{\epsilon D}\int_{n(0)}^{n(\infty)} \rho(n, F_0)\, d(\ln n) = -\Sigma \qquad (2\text{-}145)$$

where

$$\Sigma \equiv \int_0^\infty \frac{1}{r}\left(\frac{d\ln n}{dr}\right)^2 dr = \left(\frac{\mu}{D}\right)^2 \int_0^\infty \frac{1}{r} F_r^2\, dr > 0 \qquad (2\text{-}146)$$

represents a "surface tension" term. It is the analog of the radius-dependent coexistence pressure of droplets in equilibrium thermodynamics. Σ can be further evaluated in the approximation of a thin transition layer:

$$\Sigma \cong \frac{1}{r_0}\int_0^\infty \left(\frac{d\ln n}{dr}\right)^2 dr = \frac{1}{r_0}\int_{n(0)}^{n(\infty)} \left(\frac{d}{dr}\ln n\right) d(\ln n) \qquad (2\text{-}147)$$

which yields, if we approximate the integrand by its value at n_2,

$$\Sigma \cong \frac{m}{r_0}\ln\frac{n(\infty)}{n(0)} \qquad (2\text{-}148)$$

where

$$m = \left(\frac{d}{dr}\ln n\right)_{n=n_2} = (-F_r)_{n=n_2}$$

can be calculated self-consistently from Eq. (2-144):

$$-\frac{1}{2}m^2 + \frac{m}{r_0}\ln\frac{n(\infty)}{n_2} + \frac{\mu}{\epsilon D}\int_{n_2}^{n(\infty)} \rho(n, F_0)\, d(\ln n) = 0$$

For large r_0 this yields approximately

$$m \cong \left[\frac{2\mu}{\epsilon D}\int_{n_2}^{n(\infty)} \rho(n, F_0)\, d(\ln n)\right]^{1/2}$$

This approximation is reasonable for values of F_0 close to F_{c0}, as discussed above.

Equation (2-145) can again be interpreted as an equal areas rule, using the defining equation (2-129) (Fig. 2-34). The right-hand side of Eq. (2-145) is always

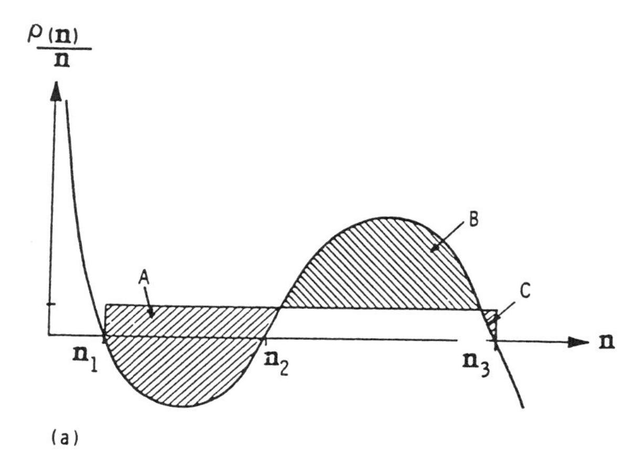

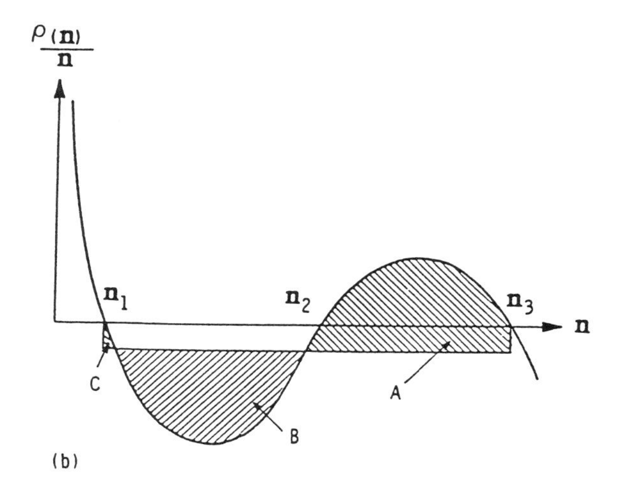

FIGURE 2-34. Equal-areas rule for cylindrical current filaments. The areas $A + C$ must be equal to B. (a) High-current filament; (b) low-current filament. ρ is the local charge density. (After Schöll, 1982a.)

positive, therefore $\tilde{\Phi}(n(0), F_0) > \tilde{\Phi}(n(\infty), F_0)$ must hold. For $F_0 < F_{c0}$ [i.e., $\tilde{\Phi}(n_1, F_0) > \tilde{\Phi}(n_3, F_0)$] this requires $n(\infty) = n_3$, which gives a filament with a low-density core. For $F_0 > F_{c0}$ high-current filaments are obtained [$n(\infty) = n_1$]. For a given field $F_0 < F_{c0}$, the equal areas rule [Eq. (2-145)] singles out a current filament with a certain radius r_0 and wall thickness Δr in accordance with Eq. (2-148). As F_0 approaches F_{c0}, i.e., $\tilde{\Phi}(n_3, F_0) - \tilde{\Phi}(n_1, F_0) \to 0$, and $n(\infty) = n_3$, $n(0) \to n_1$, the surface term Σ must tend to zero. Therefore, the filament radius $r_0 \to \infty$, corresponding to a plane interfacial layer. Hence, with increasing field F_0 the low-current core becomes wider and eventually fills the entire sample. This gives rise to a new, filamentary branch in the current–voltage characteristic. Conversely, with decreasing field $F_0 > F_{c0}$, a high current filament expands according to Eqs. (2-145) through (2-148), leading to a second filamentary branch (Fig. 2-35a). If a heavily loaded circuit is connected to the device and the bias is increased, the operating point first travels on the homogeneous low-conductivity branch up to point A, then switches to the filamentary branch through the

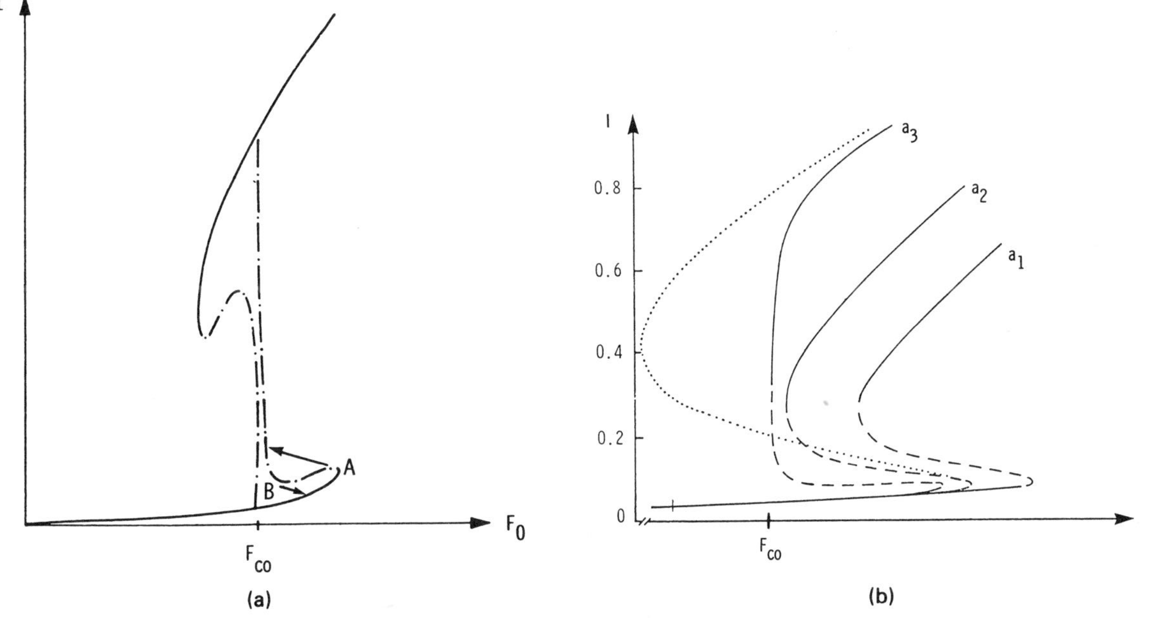

FIGURE 2-35. Current–field characteristic of an SNDC element. (a) Infinitely extended system. The solid lines denote stable homogeneous steady states, the dash-dotted lines denote low–(left branch) and high- (right branch) current filaments. A and B denote operating points where transitions between homogeneous and filamentary states occur. F_{co} makes the coexistence of two phases with a planar interface. (b) Influence of lateral boundaries (Dirichlet boundary conditions at $r = a_i$, $i = 1, 2, 3$). Homogeneous states are dotted. Solid and dashed lines correspond to stable and unstable filaments, respectively. The radius of the SNDC element is chosen as $a_1 = 50L_D$, $a_2 = 75L_D$, $a_3 = 200L_D$. (After Schöll, 1987.)

spontaneous nucleation of a high-current filament. Upon further increase of bias, the current rises by expanding the current filament until the sample is filled homogeneously by the high-conductivity state, while the voltage $F_0 l$ remains about constant (neglecting the slight decrease down to $F_{c0}l$). Upon further increase of bias the operating point moves up on the stable homogeneous high-conductivity branch. The reverse process of decreasing the bias can lead to hysteresis of the switch-back from the filamentary to the homogeneous low-conductivity branch (point B). The resulting current–voltage characteristic is very similar to that shown in Fig. 2-1d.

We have so far assumed an infinitely extended medium. Finite lateral boundaries can have substantial influence upon filamentation and upon the current–voltage characteristic (Schöll, 1985a). Figure 2-35b shows that the filamentary branch is rounded off and shifted to larger values of voltage if the carrier density at the lateral surface is pinned to a low value $n(a) < n_1$. This can be achieved, for example, by surface recombination into suitable surface states. The smaller the lateral dimension, the more the SNDC range is shifted to higher voltages. The shift is most pronounced for dimensions in the range of about 20–100 L_D, which typically corresponds to values in the near-micron and submicron range. It is remarkable that two different stable filamentary profiles are possible at the same value of F_0, even above the threshold value of homogeneous SNDC, i.e., in the regime of positive differential conductivity of the homogeneous steady state, provided the dimensions and boundary conditions are suitably chosen.

Another extension of the preceding analysis is to consider time-dependent spatiotemporal structures: "Breathing" current filaments were suggested as a possible mechanism for self-generated current oscillations (Schöll, 1987). Recent space- and time-resolved experimental investigations (Mayer et al., 1988; Brandl et al., 1989; Rau et al., 1989) have indeed demonstrated that in certain regimes of the negative conductance part of the filamentary $I(\Phi)$ characteristics in n-GaAs and p-Ge at helium temperatures, self-generated nonlinear oscillations arise that are localized at the filament boundaries, and which represent a breathing motion of the filament walls. They are associated with impact ionization of shallow donors or acceptors. In order to understand this, an analysis of the nonlinear time-dependent transport equations (2-77)–(2-79), supplemented by the carrier continuity equation, is required. This can be done by deriving nonlinear ordinary differential equations for the radius of the current filament $r_0(t)$ and other relevant dynamic variables, such as the longitudinal field $F_z(t)$ and the position of the peak of the radial field $F_r(r, t)$ (Schöll, 1991; Schöll and Drasdo, 1990). The method is similiar to the one used to derive the equal areas rule (2-145), and thus the result may be viewed as a "time-dependent" equal areas rule, or as a nonlinear mode expansion of the basic transport equations in terms of the *breathing mode*. A Hopf bifurcation of a limit cycle in the phase space $(r_0, F_z, F_r^{\text{peak}})$ is found, corresponding to a periodically breathing current filament. The physical mechanism underlying the breathing oscillations is similar to the spatially homogeneous model for dielectric relaxation oscillations discussed elsewhere (Schöll, 1987). The two models are simple approximations of two different modes of dielectric relaxation oscillations: breathing or bulk-dominated, respectively. Both types of oscillations are predicted to occur on the negative conductance branch of the (filamentary or homogeneous, respectively) current–voltage characteristic, generated by Hopf bifurcations when the applied current is decreased. This agrees well with the experiments of Rau et al. (1989), who found two different types of oscillations in different regimes of the negative conductance regime of the $I(\Phi)$ characteristics, associated with large-amplitude circuit-limited (CLO = bulk-dominated) modes and small-amplitude structure-limited (SLO = breathing) modes with lower frequency, followed by a regime of stable current filaments. It appears that this behavior can be consistently explained by a crossover between the filamentary and the spatially homogeneous NDC branches (cf. Fig. 2-35b). Thus, with increasing current four different regimes subsequently occur: a stable, homogeneous, weakly conducting state that persists up to the onset of impurity breakdown (solid line); bulk-dominated oscillations (dotted); a breathing current filament (dashed); and a stable current filament (solid line). It is interesting to note that the regime of the breathing mode occurs in the same regime that circuit-controlled relaxation oscillations are found (Shaw et al., 1973a). These oscillations are discussed in Sec. 2.3.

2.2.3.3. Filamentation in a Semiconductor Having Two Types of Charge Carriers

Results similar to those outlined above are obtained when two types of carriers (electrons and holes) are involved in an SNDC system. Consider the case where electron–hole pairs are produced by impact ionization in a cylindrical sample where quasineutrality holds (Adler et al. 1980; Schöll, 1986a). The equations describing the radial motion of the carriers in such a system under steady state condition are

$$\frac{\partial}{\partial r} r(J_p + J_n) = 0 \tag{2-149}$$

$$\frac{1}{er}\frac{\partial}{\partial r}(rJ_p) = \varphi_p(n, p; F_0) \tag{2-150}$$

$$J_n = e\left(\mu_n n F_r + D_n \frac{\partial n}{\partial r}\right) \tag{2-151}$$

$$J_p = e\left(\mu_p p F_r - D_p \frac{\partial p}{\partial r}\right) \tag{2-152}$$

$$n = N_D^* + p - n_t \tag{2-153}$$

where n and p are the electron and hole concentration, n_t is the concentration of trapped electrons, φ_p is the hole generation–recombination rate, and J_n, J_p are the radial electron and hole current densities. Assuming that n_t can be neglected in Eq. (2-153), we obtain the ambipolar transport equations:

$$J_p = -eD_A(p)\frac{\partial p}{\partial r} \tag{2-154}$$

and

$$\frac{\partial J_p}{\partial r} + \frac{J_p}{r} = ef(p, F_0) \tag{2-155}$$

where

$$D_A(p) \equiv \frac{p + (N_D^* + p)}{(N_D^* + p)/D_p + p/D_n} \tag{2-156}$$

is the ambipolar diffusion coefficient and

$$f(p, F_0) \equiv \varphi_p(N_D^* + p, p; F_0)$$

In two-carrier SNDC mechanisms, however, trapping is important (cf. Sec. 1.3.4), and n_t may not be neglected. The preceding analysis can be generalized by defining an appropriate *effective* ambipolar diffusion coefficient $D_A(p)$, which includes trapping (Schöll, 1987). Equations (2-154) and (2-155) then retain their form.

Equations (2-154) and (2-155) are the analogue of Eqs. (2-124) and (2-125), respectively, if we replace $(1/e)J_p$ with F_r, p with n, $D_A(p)$ with $D/(\mu n)$, and f with ρ/ϵ. Identical arguments as those following Eq. (2-125) apply in this case. For a filament with a wide flat core ($r_0 \gg L_D$), Eq. (2-155) can be approximated as

$$\frac{\partial J_P}{\partial r} = ef(p, F_0) \tag{2-157}$$

and a similar phase portrait analysis in the (J_p, p) phase plane can be performed. The homogeneous steady states are given by $J_p = J_n = 0$ and $f(p, F_0) = 0$. The first integral of Eqs. (2-154), (2-157) is

$$J_p(p) = \pm e\{2[C - \psi(p, F_0)]\}^{1/2} \tag{2-158}$$

where

$$\psi(p, F_0) \equiv \int D_A(p)f(p, F_0)\, dp \tag{2-159}$$

From Eq. (2-158) it follows that a current filament with $p(0) = p_1$ and $p(\infty) = p_3$, or vice versa, exists for a value of F_0 such that the equal areas rules

$$\int_{p_1}^{p_3} D_A(p)f(p, F_0)\, dp = 0 \tag{2-160}$$

is satisfied. The $I(F_0)$ characteristic that results is similar to the previous one:

$$I = eF_0\pi\{[\mu_p p_1 + \mu_n(p_1 + N_D^*)](a^2 - r_0^2) + [\mu_p p_3 + \mu_n(p_3 + N_D^*)]r_0^2\}. \tag{2-161}$$

The curvature of the filament walls leads to a similar modification of the equal areas rule as in Eq. (2-145). Retaining the second term in Eq. (2-155), we obtain

$$\frac{\partial}{\partial r}\left(D_A \frac{\partial p}{\partial r}\right) + \frac{1}{r} D_A \frac{\partial p}{\partial r} + f(p, F_0) = 0 \tag{2-162}$$

which can be integrated over $D_A(p)\,dp$:

$$\frac{1}{2}\left(D_A \frac{dp}{dr}\right)^2 + \int \frac{1}{r} D_A^2 \frac{dp}{dr}\,dp + \psi(p, F_0) = C$$

From this, the equal areas rule

$$\int_{p(0)}^{p(\infty)} D_A(p) f(p, F_0)\,dp = -\int_0^{\infty} \frac{1}{r}\left(D_A \frac{dp}{dr}\right)^2 dr \tag{2-163}$$

results.

The major difference in the properties of the current filaments in systems with one or two types of carriers lies in the concentration profile within the walls of the filament. In the latter case, the walls are generally wider, the width being determined by the effective ambipolar diffusion length.

2.3. STABILITY OF AN NDC ELEMENT IN A CIRCUIT CONTAINING REACTIVE COMPONENTS

2.3.1. Representation of the NDC Element and its Local Environment

An NDC element in a circuit will exhibit a variety of instabilities, some involving the resonant response of controlled or spurious reactive circuit elements (Solomon et al., 1972). Since these circuit oscillations are of fundamental importance in understanding the complete response of the NDC element to a specific excitation, it is vital that the important reactive elements be identified. Furthermore, the role of the contacts (boundaries) to the NDC element will also play a major role in determining the complete response (Shaw et al., 1969, 1979; Solomon et al., 1975). Therefore, both the circuit and contact conditions must be specified before an analysis of the response of the NDC element can be undertaken. In this chapter we introduce those aspects of the circuit problem that are required to solve the problem. Those aspects of the contact problem (Shaw, 1981) that are important will be discussed with regard to specific devices in later chapters.

To understand the electrical behavior of an NDC element we must first properly represent the NDC element and its local environment. With regard to its environment, which consists of the leads, contacts, and support components, we note the following.

1. The attachment of metallic leads to the NDC element introduces a lead resistance R_l and lead inductance L_l.
2. The contact regions themselves most often produce a nonlinear resistance, which we label R_c, and will also impose specific electric field conditions at the interface of the NDC material.
3. Supporting, mounting, or holding the NDC element in any way introduces package capacitance C_p and package inductance L_p.
4. An external voltage source (we treat only dc sources now) will contain its own internal resistance R_I.

These contributions are shown in Fig. 2-36 in a lumped element approximation of the circuit containing the NDC element, which we have represented as a cylindrical block of material. Also shown is a load resistor R_L, which may represent the actual load in the circuit.

Now that we have identified the important reactive components in the local environment of the NDC element, it remains to approximate the element itself in terms of lumped elements. To do this we first must realize that when current flows along the length of a cylindrical conductor of radius a and length l, energy flows into it radially through its surface sheath (Feynman, 1964). Conservation of energy results in

$$-\int (\mathbf{F} \times \mathbf{H}) \cdot d\mathbf{S}_r = \int \mathbf{J} \cdot \mathbf{F}\, d\tau + \frac{1}{2}\frac{d}{dt}\int (\mu_0^{-1}\mathbf{B} \cdot \mathbf{B} + \epsilon \mathbf{F} \cdot \mathbf{F})\, d\tau \quad (2\text{-}164)$$

where $d\mathbf{S}_r$ is a surface (sheath) element, $d\tau$ a volume element, and $\mathbf{B} = \mu_0\mathbf{H}$. The left-hand side of Eq. (2-164) represents the total power driving the electron

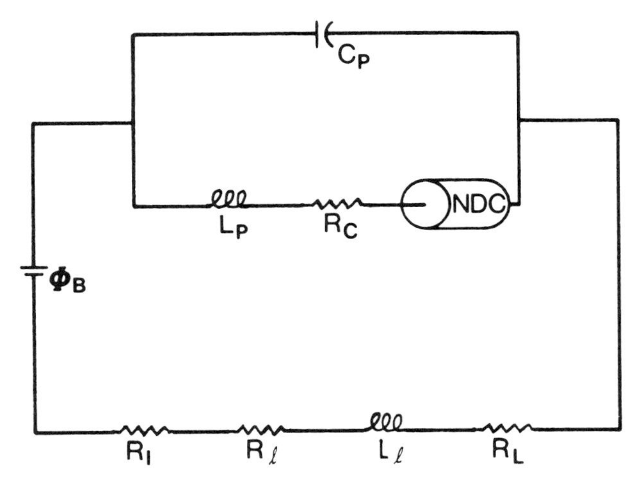

FIGURE 2-36. Lumped element approximation of a circuit containing an NDC element. Φ_B is the applied bias voltage, C_p and L_p are package capacitance and inductance, R_c is a contact resistance, R_ℓ and L_ℓ are lead resistance and inductance, R_L and R_I are load resistance and internal resistance of the voltage source, respectively.

stream. From the symmetry of the problem, we have

$$-\int (\mathbf{F} \times \mathbf{H})\, d\mathbf{S}_r = F_z(r = a)l \int H_\theta\, ds = F_z(r = a)lI = \Phi I \qquad (2\text{-}165)$$

where ds is a path element on the circumference of the cylinder, Φ is the total voltage drop across its length l, and I is the total current flowing down its length parallel to the z axis. Defining the intrinsic inductance L_i as

$$L_i = \frac{1}{I^2} \int \mu_0^{-1} \mathbf{B} \cdot \mathbf{B}\, d\tau \qquad (2\text{-}166)$$

and intrinsic capacitance C_i as

$$C_i = \frac{1}{\Phi^2} \int \epsilon \mathbf{F} \cdot \mathbf{F}\, d\tau \qquad (2\text{-}167)$$

Equation (2-164) becomes

$$I\Phi = \int \mathbf{J} \cdot \mathbf{F}\, d\tau + \frac{1}{2} \frac{d}{dt}(L_i I^2 + C_i \Phi^2) \qquad (2\text{-}168)$$

We now examine Eq. (2-168) in two limits: no displacement current $[(d/dt)C_i\Phi^2 = 0]$; no inductive voltage $[(d/dt)L_iI^2 = 0]$. In the first case we may write

$$\Phi = \frac{1}{I} \int \mathbf{J} \cdot \mathbf{F}\, d\tau + \frac{1}{2I} \frac{d}{dt}(L_i I^2) \qquad (2\text{-}169)$$

and in the second case

$$I = \frac{I}{\Phi} \int \mathbf{J} \cdot \mathbf{F}\, d\tau + \frac{1}{2\Phi} \frac{d}{dt}(C_i \Phi^2) \qquad (2\text{-}170)$$

If we define an *effective conductive voltage*, $\Phi_c(t)$, as

$$\Phi_c(t) \equiv \frac{1}{I} \int \mathbf{J} \cdot \mathbf{F}\, d\tau$$

then Eq. (2-169) becomes

$$\Phi(t) = \Phi_c(t) + \frac{1}{2I} \frac{d}{dt}(L_i I^2) \qquad (2\text{-}171)$$

This represents the voltage drop across the circuit shown in Fig. 2-37a, where the conductive element is our $\Phi_c(t)$ characteristic. That is, $\Phi_c(t)$ represents the

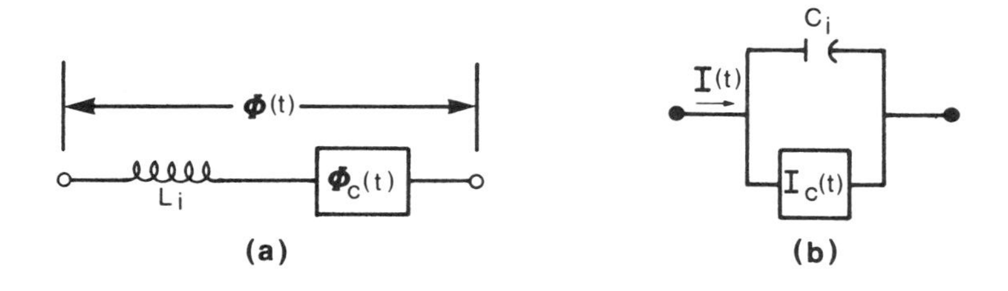

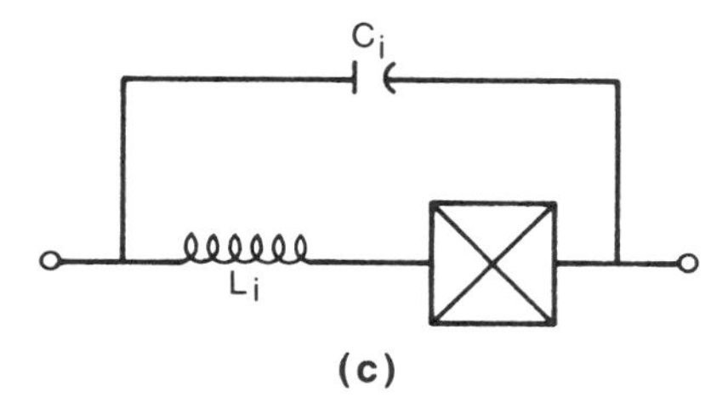

FIGURE 2-37. Lumped element approximation of an NDC element. (a) No displacement current; (b) no inductive voltage; (c) general case. L_i and C_i are the internal (intrinsic) inductance and capacitance, respectively. Φ is the total voltage across the NDC element, and Φ_c is its conductive part. I is the total current through the NDC element, and I_c is its conductive part.

voltage drop across the sample when the time derivatives vanish. It is that voltage component that drives carriers through the system in a steady dc mode.

In the second case, if we define an *effective transport* or *conduction current* $I_c(t)$, as

$$I_c(t) = \frac{1}{\Phi}\int \mathbf{J}\cdot\mathbf{F}\,d\tau \tag{2-172}$$

then Eq. (2-170) becomes

$$I(t) = I_c(t) + \frac{1}{2\Phi}\frac{d}{dt}(C_i\Phi^2) \tag{2-173}$$

This represents the total current flowing in the circuit shown in Fig. 2-37b, where the conductive element is our $I_c(t)$ characteristic. That is, $I_c(t)$ represents the current flowing through the sample when the time derivative vanishes. It is that current component that represents carriers moving through the system in a steady dc mode (in response to potential, concentration, and temperature gradients).

When both intrinsic inductive voltage drops and displacement currents are present, the sample can be approximated at low frequencies by the model shown in Fig. 2-37c. In this model we assume that the inductance is defined only via the transport current; the displacement current produces magnetic energy that radiates away. As we shall see, the limiting cases are most often the important ones for NDC elements, so the models shown in Figs. 2-37a and 2-37b will be emphasized. However, independent of which model for the NDC element is chosen, by inserting it into Fig. 2-36 we see that we are dealing with a system that contains *at least* four reactive components and one highly nonlinear resistive component. The time-dependent solutions will generally be complicated and numerical assistance is often called for in the analysis. There are, however, a large number of common situations that arise that allow further simplification of the circuit shown in Fig. 2-36. These will be discussed in the next section. But before we do this we must also realize that in the absence of current density filamentation (or high-field domains) $\Phi_c(t)$ [or $I_c(t)$] is a single-valued function of I (or Φ) and can be written as $\Phi_c(I)$ [or $I_c(\Phi)$], and L_i (or C_i) is a constant. In this case we can apply standard techniques to solve for the circuit response. However, when filamentation (or domain formation) is present, $\Phi_c(I)$ [or $I_c(\Phi)$] is multivalued and L_i (or C_i) varies over an oscillatory cycle. We will show how $\Phi_c(I)$ and $I_c(\Phi)$ are obtained for the inhomogeneous SNDC case in Chap. 7. The NNDC case has been described by Shaw et al. (1979), and is simply the dual of the SNDC case. We first consider the case where L_i and C_i are constants and the samples remain electrically homogeneous (*diodic* NDC elements sometimes exhibit this type of behavior). Here the $I_c(\Phi)$ and $\Phi_c(I)$ curves scale the $J(F)$ curves. This analysis will prove quite useful when we analyze the more common and important inhomogeneous cases. In what follows we will derive the governing differential equations and note their duality features. We will then discuss the solutions in a semiquantitative manner. More detailed quantitative and numerical solutions will be developed with regard to specific cases in later chapters.

2.3.2. The Circuit Response for an NNDC Element

When an NNDC element undergoes time-dependent NDC-driven processes it is generally the case that the relative changes in voltage across it are large compared to the relative changes in current through it. Thus, here the capacitive effect is much larger than the inductive effect. Because of this a convenient approximation is to neglect L_i and L_p, and include R_c into the $I_c(\Phi)$ characteristic. The circuit of Fig. 2-36 then reduces to that shown in Fig. 2-38, where we have modeled the NNDC element by the circuit of Fig. 2-37b (Solomon et al., 1972). Here $R = R_I + R_l + R_L$, $L = L_l$, and $C = C_p + C_i$. Kirchoff's laws yield

$$\Phi_B = IR + L\frac{dI}{dt} + \Phi \tag{2-174}$$

and

$$I = I_c(\Phi) + C\frac{d\Phi}{dt} \tag{2-175}$$

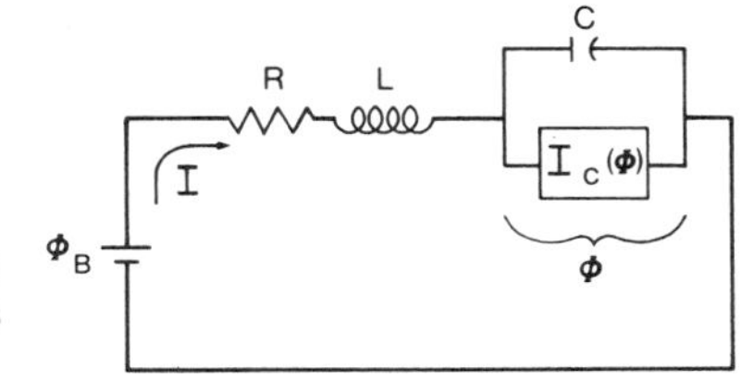

FIGURE 2-38. Lumped element approximation of an NNDC element in its primary circuit environment, as given by Figs. 2-36 and 2-37b.

Solutions of these equations come in many forms. We may obtain the current–time (or voltage–time) profiles and the current–voltage Lissajous-type figures. We will require both. We therefore transform these two equations to produce the governing differential equations of the NNDC circuit problem for the voltage as a function of time:

$$\Phi_B = \left[1 + \frac{I_c(\Phi)}{\Phi} R\right]\Phi + \frac{Z_0}{R_0}\left[R_0 \frac{dI_c(\Phi)}{d\Phi} + \frac{R_0 C}{L/R}\right]\frac{d\Phi}{dT} + \frac{d^2\Phi}{dT^2} \qquad \text{(2-176)}$$

and for the trajectories in the $I(\Phi)$ phase plane:

$$\frac{dI}{d\Phi} = \frac{\Phi_B - \Phi - RI}{I - I_c(\Phi)} \frac{1}{Z_0^2} \qquad \text{(2-177)}$$

where $Z_0 = (L/C)^{1/2}$ and $T = t/(LC)^{1/2}$. Z_0 has the following significance: in the case of a circuit consisting of two reactive elements, an inductor and capacitor, Z_0 is the ratio of voltage to current amplitude across either element. Thus, large values of Z result in large-amplitude voltage oscillations. The low-voltage resistance of the NNDC element, $R_0 \equiv [(dI_c/d\Phi)(\Phi \rightarrow 0)]^{-1}$ has been introduced in Eq. (2-176) in order to make the bracketed part of the damping term a dimensionless factor, which is in most practical cases of the order unity.

Before we explicitly solve Eqs. (2-176) and (2-177), it is useful to elaborate some general features of the dynamic system equations (2-174) and (2-175) in the $I(\Phi)$ phase plane. In terms of the scaled time $T = t/(LC)^{1/2}$ these two equations can be recast in the form

$$\frac{d}{dT} I = \frac{1}{Z_0}(\Phi_B - \Phi - RI) \qquad \text{(2-174')}$$

$$\frac{d}{dT}\Phi = Z_0[I - I_c(\Phi)] \qquad \text{(2-175')}$$

The steady states are given by the intersections of the load line $\Phi = \Phi_B - RI$ with the NNDC characteristic $I = I_c(\Phi)$. If the magnitude of the slope of the load line is larger than that of the $I_c(\Phi)$ characteristic at any point of the NDC region,

$$R < R_n \equiv -\left[\min\left(\frac{dI_c}{d\Phi}\right)\right]^{-1} \qquad \text{(2-178)}$$

there always exists one unique steady state (I_0, Φ_0). For larger values of R there exist three steady states for a range of bias Φ_B. The stability of the steady states is readily determined from linearization of Eqs. (2-174'), (2-175') around the steady state, as outlined in Chap. 1. We obtain with $\xi_1 = I - I_0$, $\xi_2 = \Phi - \Phi_0$:

$$\begin{pmatrix} \dfrac{d}{dT}\xi_1 \\ \dfrac{d}{dT}\xi_2 \end{pmatrix} = \begin{pmatrix} -\dfrac{R}{Z_0} & -\dfrac{1}{Z_0} \\ Z_0 & -Z_0\dfrac{dI_c}{d\Phi} \end{pmatrix} \begin{pmatrix} \xi_1 \\ \xi_2 \end{pmatrix} \tag{2-179}$$

where $dI_c/d\Phi$ is to be evaluated at the steady state. The stability parameters defined by Eqs. (1–22) and (1–26) are

$$\underset{\sim}{T} = -\left(\frac{R}{Z_0} + Z_0\frac{dI_c}{d\Phi}\right) \qquad \text{(trace)} \tag{2-180}$$

and

$$\underset{\sim}{D} = R\frac{dI_c}{d\Phi} + 1 \qquad \text{(determinant)} \tag{2-181}$$

The different regimes of stability are shown in the $(R, dI_c/d\Phi)$ plane in Fig. 2-39. The PDC states are always stable because of $\underset{\sim}{D} > 0$, $\underset{\sim}{T} < 0$. *It is remarkable that for $R > 0$ some of the NDC states also become stable.*

For $Z_0 < R_n$ (Fig. 2-39a) the two curves $\underset{\sim}{D} = 0$ and $\underset{\sim}{T} = 0$ do not intersect. When the $\underset{\sim}{T} = 0$ line is crossed, $\underset{\sim}{T}$ changes sign, while $\underset{\sim}{D}$ remains positive, and $\underset{\sim}{T}^2 - 4\underset{\sim}{D} < 0$. Hence the steady state, which is unique for $R < R_n$, changes from stable focus to unstable focus. On the $\underset{\sim}{T} = 0$ line a Hopf bifurcation occurs in which a stable limit cycle bifurcates from the focus. This limit cycle exists in the whole regime where $\underset{\sim}{T} > 0$. Only for $R = 0$ does the limit cycle bifurcate right at

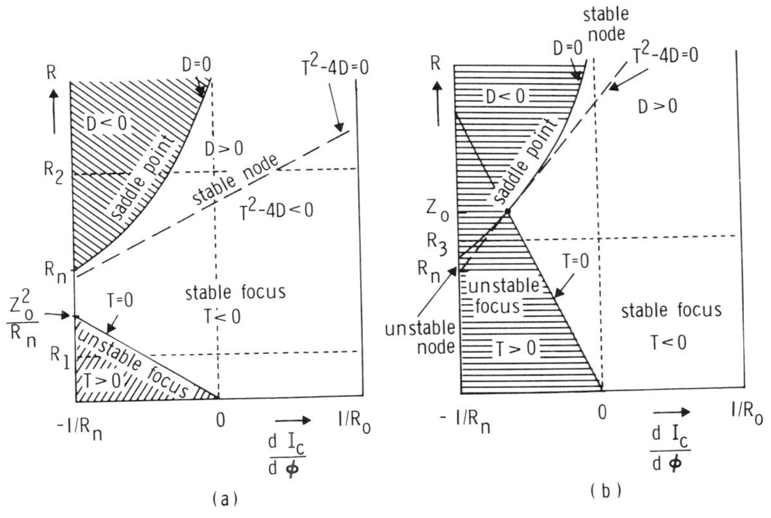

FIGURE 2-39. Regimes of stability of the steady state for an NNDC element in a reactive circuit. (a) $Z_0 < R_n$; (b) $Z_0 > R_n$. Hopf bifurcations of limit cycles occur on the line $T = 0$; saddle-node bifurcations occur on the curve $D = 0$. Regions of instability of the operating point are hatched.

the beginning of the NDC region. With increasing load R the bifurcation point is shifted further into the NDC regime. For $R_n > R > Z_0^2/R_n$, e.g., for small values of Z_0, all NDC points of the $I_c(\Phi)$ characteristic are stable, and no self-sustained oscillation is generated. When the bias Φ_B is slowly increased at a constant value of the load $R = R_1 < Z_0^2/R_n$, then $dI_c/d\Phi$ decreases from $1/R_0$ to its minimum value $-1/R_n$ and then increases again to zero, as the intersection point of the load line and the $I_c(\Phi)$ characteristic travels along the $I_c(\Phi)$ curve. Hence, the line $\underset{\sim}{T} = 0$ is crossed twice, which marks the two boundaries of the regime of bias where a limit cycle exists. The associated $I(\Phi_B)$ plot is shown in Fig. 2-40a.

Now let us consider a larger value $R_2 > R_n$, such that the curve $\underset{\sim}{D} = 0$ can be reached, while $\underset{\sim}{T}$ is negative throughout. For $\underset{\sim}{D} > 0$ and $\underset{\sim}{T}^2 - 4\underset{\sim}{D} > 0$ the steady state is a stable node; for $\underset{\sim}{D} < 0$ it is a saddle point, i.e., unstable. It follows from Eq. (1-181) that $\underset{\sim}{D} < 0$ is equivalent to $dI_c/d\Phi < -1/R$. Hence (for $R > Z_0$) exactly those points are unstable for which the slope of the $I_c(\Phi)$ characteristic is steeper than that of the load line. In this case there exist two more intersections of the load line with the $I_c(\Phi)$ curve satisfying $dI_c/d\Phi > -1/R$; hence these are stable steady states, no matter whether they are NDC or PDC states. For values on the curve $\underset{\sim}{D} = 0$ the load line becomes tangent to the $I_c(\Phi)$ characteristic. This corresponds to a saddle-node bifurcation of the type shown in Fig. 1-9, case (A1), where two steady states (operating points on the load lines) coalesce and disappear. This is illustrated by the corresponding $I(\Phi_B)$ plot in Fig. 2-40b. The current exhibits hysteresis as a function of bias, but no Hopf bifurcation.

Finally, we consider the case $Z_0 > R_n$ (Fig. 2-39b). Now a combination of limit cycle and saddle-node bifurcations may occur if the load R_3 satisfies $R_n < R_3 < Z_0$. On the curve $\underset{\sim}{D} = 0$ a saddle point and an *unstable* node coalesce and disappear. Figure 2-40c shows a typical $I(\Phi_B)$ plot, in which a limit cycle surrounds up to three singular points.

The analysis that we now perform is equivalent to that for the tunnel diode (Schuller and Gartner, 1961)—the "uniform field" case. The first thing to note is that Eq. (2-176) is an oscillator equation with a nonlinear damping term (the second term on the right-hand side) and a nonlinear force term. When biased in

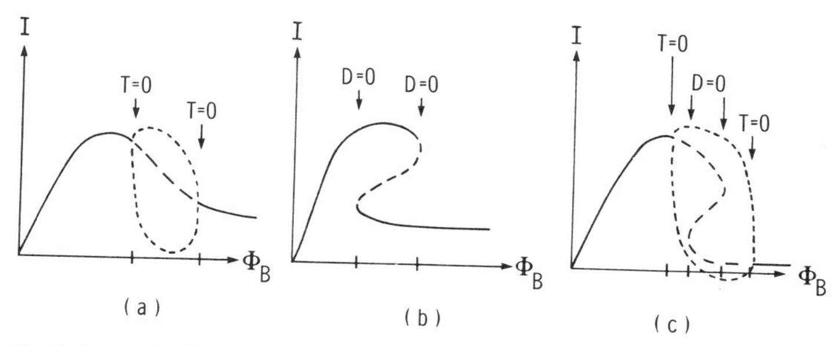

FIGURE 2-40. Schematic plot of the current I through the NDC element versus bias voltage Φ_B for an NNDC element in a reactive circuit (a) at fixed $R = R_1 < Z_0^2/R_n$ for $Z_0 < R_n$; (b) at $R = R_2 > R_n$ for $Z_0 < R_n$; (c) at $R = R_3$ with $R_n < R_3 < Z_0$ for $Z_0 > R_n$. Solid lines denote stable steady states; dashed lines, unstable steady states; dotted lines, amplitudes of the limit cycle. The bifurcation points are marked by $T = 0$ and $D = 0$.

the region of negative slope ($dI_c(\Phi)/d\Phi < 0$) we see that the damping can be negative, leading to growth rather than decay. In general the bracketed part of the damping term is of order unity and the strength of the damping term is determined by Z_0/R_0. Equation (2-176) is a generalization of the Van der Pol equation [Eq. (1-54)] for a free-running oscillator, and the general techniques for its solution have been outlined in Chap. 1. For small Z_0/R_0 the damping term is a small perturbation and the limit cycle solutions for $\Phi(T)$ are nearly sinusoidal. This is shown in Fig. 2-41a for small R and $Z_0/R_0 = 3$, where we plot $\Phi(T)$, $I(T)$, and $I(\Phi)$, obtained numerically. The current oscillations are nearly sinusoidal while the voltage oscillations show evidence of the nonlinear damping term. For large Z_0/R_0 the damping term is important and the solutions become well-defined relaxation oscillations. Figure 2-41b illustrates a case for $Z_0/R_0 = 12$. Here the current oscillations are almost sawtooth and the voltage oscillations exhibit sharp spikes. This is due to a time scale separation between the variables I and Φ for large Z_0, as evident from Eqs. (2-174′) and (2-175′), such that Φ very rapidly attains a pseudo-steady state given by $I = I_c(\Phi)$, and I changes slowly. Therefore, the relaxation oscillation cycle consists of a slow rise close to the PDC portion of the $I_c(\Phi)$ characteristic, until the maximum current I_p is reached, and a fast, spiky second part. Uniform field circuit oscillations for a given value of Z_0/R_0 and Φ_B have similar shapes and amplitudes and differ only in frequency due to the time scale $T = t/(LC)^{1/2}$.

Rather than use numerical solutions at the present time we can employ a

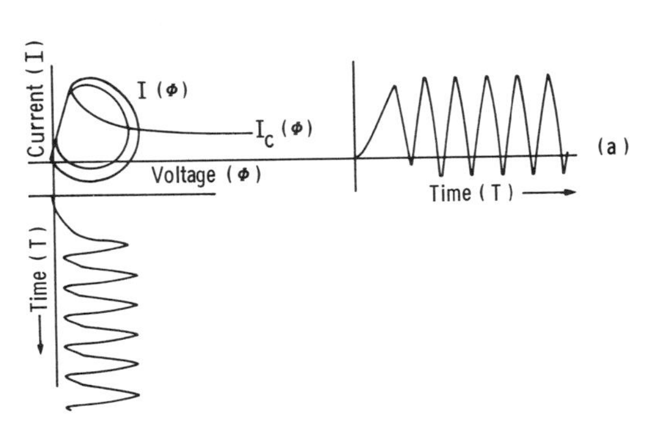

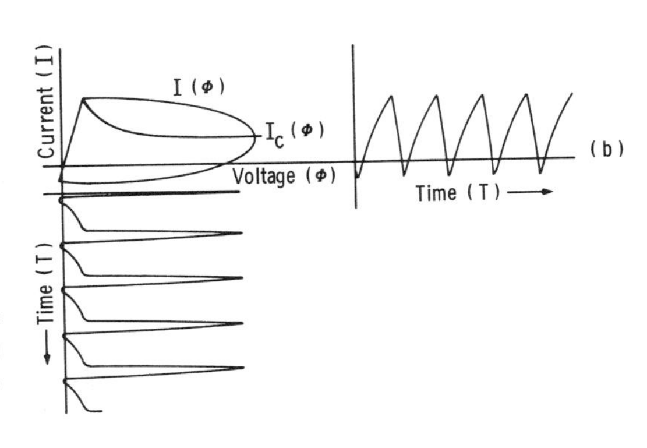

FIGURE 2-41. Numerical solutions of the phase trajectories $I(\Phi)$ and the associated time series $I(T)$ and $\Phi(T)$ for an NNDC element in the circuit of Fig. 2-38. (a) $Z_0/R_0 = 3$, (b) $Z_0/R_0 = 12$. The static NNDC characteristic $I_c(\Phi)$ is also shown in the phase plane. (After Solomon et al., 1972.)

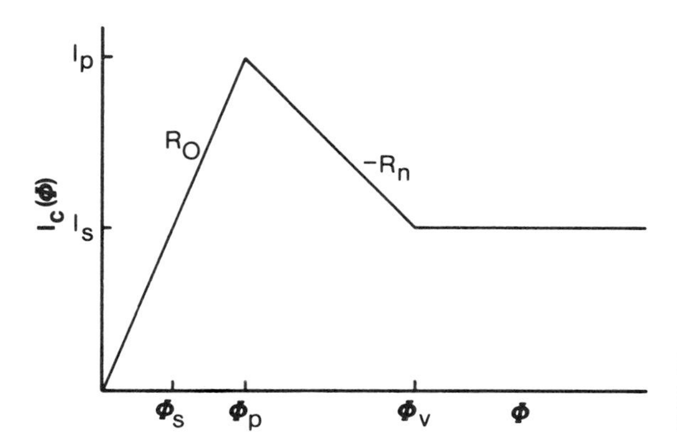

FIGURE 2-42. Three-piece linear approximation of the NNDC characteristic $I_c(\Phi)$. (After Solomon et al., 1972.)

three-piece linear approximation for the $I_c(\Phi)$ curve to obtain analytic solutions of the limit cycle and therefore more insight into the problem. A three-piece linear approximation is shown in Fig. 2-42. The problem is solved in the three regions $\Phi < \Phi_p$, $\Phi_p < \Phi < \Phi_v$, and $\Phi_v < \Phi$, which correspond to $dI_c(\Phi)/d\Phi = 1/R_0$, $-1/R_n$, and 0, respectively, where R_n is the magnitude of the negative differential resistance. The equations and solutions are shown in Table 2-1. The individual solutions are joined smoothly from one region to the next. The composition of the solution is evident in the relaxation oscillation shown in Fig. 2-41b. The voltage waveform begins with a slow exponential rise with time constant $L/(R_0 + R)$ and changes to a sharp spike in voltage when $\Phi = \Phi_p$. The time needed to reach Φ_p is bias dependent and if, for example, the minimum voltage is arbitrarily set equal to zero, the rise time to threshold is given approximately by

$$\Delta T = \frac{L}{R_0 + R} \ln\left[1 - \frac{\Phi_p}{\Phi_B}\left(1 + \frac{R}{R_0}\right)\right]^{-1}$$

Increasing Φ_B, all other things being equal, ΔT decreases and the frequency of oscillation increases. Bias tuning is a characteristic of NNDC relaxation oscillators.

The spike above threshold is composed of a fast exponential transit through the region of negative slope ($\Phi_p < \Phi < \Phi_v$), with time constant R_nC, followed by a damped sine wave for $\Phi > \Phi_v$ and another exponential transit for $\Phi_p < \Phi < \Phi_v$. An exponential decay for $\Phi < \Phi_p$ completes the limit cycle.

As Fig. 2-41b indicates, the $I(\Phi)$ limit cycle trajectory, obtained by eliminating T from $I(T)$ and $\Phi(T)$, has the appearance of a truncated ellipse, closed by the low-voltage part of the $I_c(\Phi)$ characteristic. Since this Lissajous-type figure provides an indication of the magnitudes of the current and voltage swings and their relation to the parameter Z_0/R_0, we consider it in some detail.

A useful approximation of the $I(\Phi)$ trajectory is obtained by solving Eq. (2-177) in the phase space region $\Phi > \Phi_v$, where $I_c(\Phi) \sim I_s$ (cf. Fig. 2-42) and then extending the solution until it crosses the "below threshold" section $I_c = \Phi/R_0$ of the $I_c(\Phi)$ characteristic. Integration yields

$$Z_0^2[I(\Phi) - I_s]^2 + (\Phi - \Phi_B)^2 = K^2 \tag{2-182}$$

TABLE 2-1. Differential Equations and Solutions for the Circuit of Fig. 2-38, Where the NNDC Element is Replaced by a Nonlinear Resistor in Parallel with the Capacitor C_i. $I_c(\Phi)$ for the Nonlinear Resistor is the Three-Piece Curve in Fig. 2-42.[a]

$\Phi < \Phi_p$:

$$\frac{d^2\Phi}{dT^2} + \frac{Z_0}{R_0}\left(1 + \frac{R_0C}{L/R}\right)\frac{d\Phi}{dT} + \left(1 + \frac{R}{R_0}\right)\Phi = \Phi_B$$

$$\Phi = \frac{\Phi_B}{1 + R/R_0} + \Phi_1^+ \exp(\alpha_1^+ T) + \Phi_1^- \exp(\alpha_1^- T)$$

$$\alpha_1^\pm = -\frac{Z_0}{2R_0}\left(1 + \frac{R_0C}{L/R}\right) \pm \left[\left(\frac{Z_0}{2R_0}\right)^2\left(1 + \frac{R_0C}{L/R}\right)^2 - \left(1 + \frac{R}{R_0}\right)\right]^{1/2}$$

$\Phi_p < \Phi < \Phi_v$:

$$\frac{d^2\Phi}{dT^2} + \frac{Z_0}{R_0}\left(-\frac{R_0}{R_n} + \frac{R_0C}{L/R}\right)\frac{d\Phi}{dT} + \left(1 - \frac{R}{R_n}\right)\Phi = \Phi_B - \Phi_p\left(\frac{R}{R_n} + \frac{R}{R_0}\right)$$

$$\Phi = \frac{\Phi_B - \Phi_p[(R/R_n) + (R/R_0)]}{1 - R/R_n} + \Phi_2^+ \exp(\alpha_2^+ T) + \Phi_2^- \exp(\alpha_2^- T)$$

$$\alpha_2^\pm = -\frac{Z_0}{2R_0}\left(-\frac{R_0}{R_n} + \frac{R_0C}{L/R}\right) \pm \left[\left(\frac{Z_0}{2R_0}\right)^2\left(-\frac{R_0}{R_n} + \frac{R_0C}{L/R}\right)^2 - \left(1 - \frac{R}{R_n}\right)\right]^{1/2}$$

$\Phi > \Phi_v$:

$$\frac{d^2\Phi}{dT^2} + \frac{Z_0}{R_0}\left(\frac{R_0C}{L/R}\right)\frac{d\Phi}{dT} + \Phi = \Phi_B - RI_s$$

$$\Phi = \Phi_B - RI_s + \Phi_3^+ \exp(\alpha_3^+ T) + \Phi_3^- \exp(\alpha_3^- T)$$

$$\alpha_3^\pm = -\frac{Z_0}{2R_0}\left(\frac{R_0C}{L/R}\right) \pm \left[\left(\frac{Z_0}{2R_0}\right)^2\left(\frac{R_0C}{L/R}\right)^2 - 1\right]^{1/2}$$

[a] After Solomon et al. (1972).

where we have also taken $R \ll |\Phi_B - \Phi|/I$. In Eq. (2-182), K is an integration constant. Extending Eq. (2-182) until it crosses the positive resistance portion of $I_c(\Phi)$ determines the constant, K^2. At threshold $\Phi = \Phi_p$, $I(\Phi) \sim I_c(\Phi_p) \equiv I_p$, so that the constant in Eq. (2-182) may be evaluated to give

$$K^2 = Z_0^2(I_p - I_s)^2 + (\Phi_p - \Phi_B)^2 \tag{2-183}$$

Over its region of validity a plot of $Z_0 I(\Phi)$ versus Φ generates a circle. A plot of $R_0 I$ versus Φ generates a family of ellipses whose ratio of semimajor axis to semiminor axis is Z_0/R_0. For $Z_0 > R_0$ the amplitude of Φ is large while that of $R_0 I$ is small. The reverse occurs for $Z_0 < R_0$. In other words, large voltage swings occur at the expense of small current swings and vice versa.

The ellipse equation, (2-183), shows that besides the circuit parameters Z_0 and Φ_B, the circuit response is determined by the NDC element parameters I_p, Φ_p, and I_s. This point is illustrated in Fig. 2-43 in which the circuit behavior is

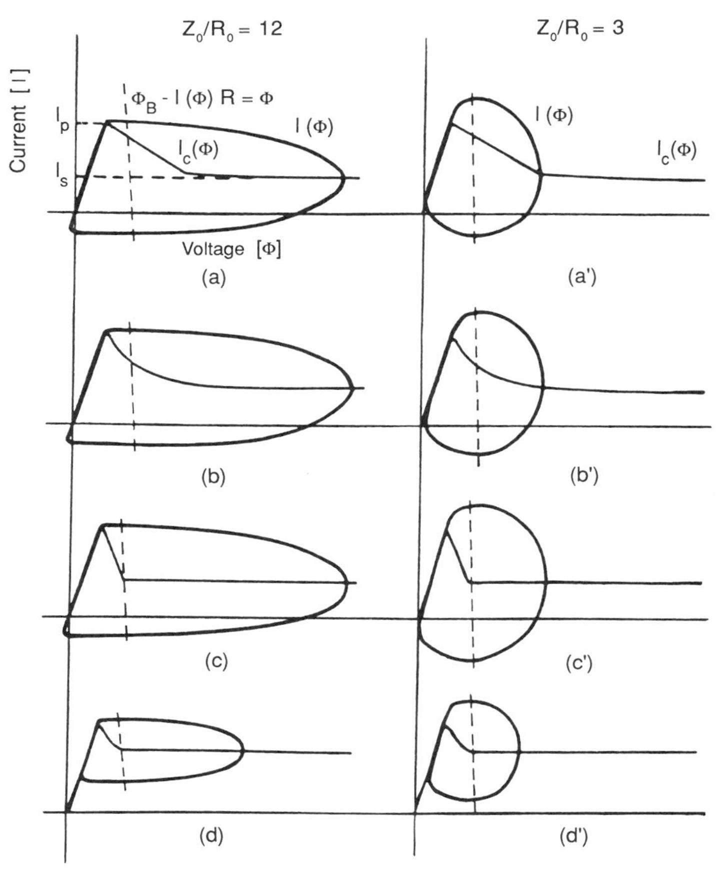

FIGURE 2-43. Numerical solutions of the limit cycle phase trajectories for four different $I_c(\Phi)$ characteristics $Z_0/R_0 = 12$ (a)–(d) and $Z_0/R_0 = 3$(a')–(d'). (After Solomon et al., 1972.)

obtained for four different $I_c(\Phi)$ curves and two values of Z_0/R_0. Figures 2-43a–2-43d show relaxation oscillations with $Z_0/R_0 = 12$ and Figs. 2-43a′–2-43d′ show nearly sinusoidal oscillations with $Z_0/R_0 = 3$. The $I_c(\Phi)$'s in Figs. 2-43a–2-43c and in Figs. 2-43a′–2-43c′ differ only in the shape of the region of negative slope, and we see that the $I(\Phi)$ trajectories for a given value of Z_0/R_0 are almost congruent. Figures 2-43d and 2-43d′ have a higher saturation current I_s. This has a substantial influence on $I(\Phi)$, reducing both the current and voltage amplitudes considerably.

Equation (2-183) provides qualitative information about the relative values of the current and voltage amplitudes. However, as we shall see in Chap. 5, to examine the formation and quenching of field nonuniformities (domains) it is important to obtain a more careful determination of the maximum current I_M and the minimum voltage Φ_m for a particular $I(\Phi)$ trajectory. In the case of Φ_m its value is a good measure, in long samples, of whether a domain will be quenched. If Φ_m exceeds the domain sustaining voltage, quenching will not occur. The values I_M and Φ_m may be obtained by numerical calculation, as shown in Fig. 2-44

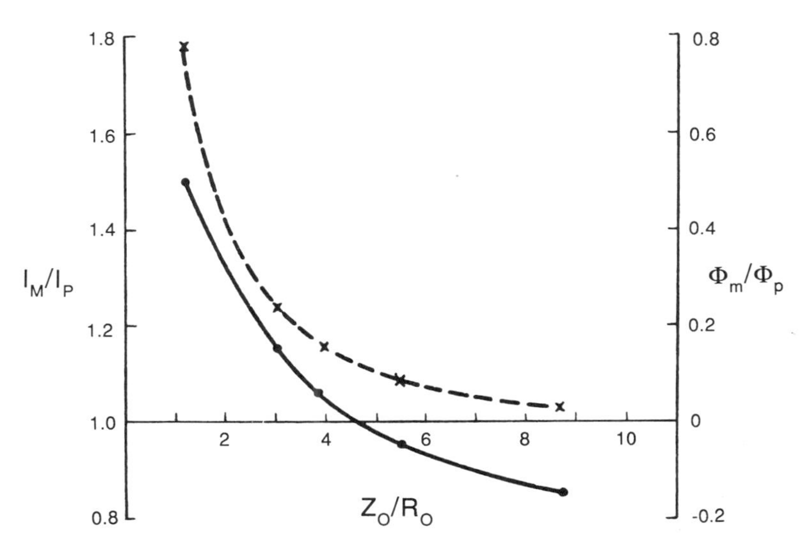

FIGURE 2-44. Current maxima I_M in units of I_p(dashed) and voltage minima Φ_m (solid line) in units of Φ_p versus Z_0/R_0 for $\Phi_B = 2\Phi_p$ and $R_0/R = 10$. For the circuit a small resistor (1 Ω) is included in series with the NNDC element to assure numerical stability. (After Solomon et al., 1972.)

for one value of bias. Here Φ_m and I_M are plotted as functions of Z_0/R_0. Φ_m varies from $0.6\Phi_p$ for small Z_0/R_0 to negative values for large Z_0/R_0. I_M varies from I_p for large Z_0/R_0 to $\sim 1.8 I_p$ for small values of Z_0/R_0. Generally, the parameters I_M and Φ_m both increase with decreasing Z_0/R_0.

In summary, the most important results of the uniform field (or tunnel diode) analysis are follows:

1. The form of the oscillation (relaxation or sinusoidal) is determined by Z_0/R_0.
2. The detailed shape of the oscillation is controlled by Z_0/R_0 and the NNDC element characteristics Φ_p, I_p, and I_s, and is relatively insensitive to the shape of the region of negative slope.
3. The parameters I_M and Φ_m both increase with decreasing Z_0/R_0.

We might now ask what application do these tunnel diode results have to the behavior of NNDC elements that become electrically inhomogeneous and form moving high-field domains? It has been found (Shaw et al., 1979) that as long as the circuit controls the formation and quenching of space-charge nonuniformities (i.e., fully formed domains do not transit to the anode), $I_c(\Phi)$ will be approximately defined by a single-valued trajectory in the $I(\Phi)$ plane, whose value of I_p, Φ_p, and I_s will closely correspond to their respective values in the uniform field case (see the dual SNDC analysis in Chap. 7). The important conclusion here is that the oscillatory behavior predicted by the simple tunnel diode analysis applies reasonably well to all cases where the oscillatory behavior is dominated by the circuit and not by domain transit or other space-charge effects. Furthermore, the simple tunnel diode analysis can also be used to define the competing limits of circuit and domain domination by providing an understanding of the formation and quenching of space-charge nonuniformities. With regard to quenching, this process occurs when Φ becomes small and the NNDC element returns to its positive differential mobility region. Values of Φ_m, the minimum value of Φ, are smallest for large Z_0/R_0 (see Fig. 2-44) and indicate that relaxation oscillations are more likely to quench space-charge nonuniformities than near-sinusoidal oscillations. On the other hand, we will learn from the dual SNDC analysis in Chap. 7, that I_M, the maximum value of current reached during the first cycle, is crucial in determining whether nonuniformities will form.

2.3.2. The Circuit Response for an SNDC Element

When an SNDC element undergoes time-dependent NDC driven processes it is generally the case that the relative changes in current through it are large compared to the relative changes in voltage across it. Thus, here the inductive effect is much larger than the capacitive effect. Furthermore, values of R substantially greater than L_l/t_s, where t_s is the transition time from one conductive state to another, are usually employed so that the system is heavily loaded. Under these conditions a convenient approximation is to neglect L_l and C_i, and include R_c in the $\Phi_c(I_c)$ characteristics. The circuit of Fig. 2-36 then reduces to that shown in Fig. 2-45, where we have modeled the SNDC element by the circuit of Fig. 2-37a (Shaw et al., 1973). Here $R = R_l + R_l + R_L$, $L = L_i + L_p + M$, where M is the mutual inductance that may exist between L_p and L_i, and $C = C_p$. Note also that we have written the effective conductive voltage drop as $\Phi_c(I_c)$ to specify that the conductive voltage drop across the SNDC element is a function of the *transport* current. Kirchoff's laws yield

$$\Phi_B = IR + \Phi \tag{2-184}$$

$$I = I_c + C\frac{d\Phi}{dt} \tag{2-185}$$

and

$$\Phi = \Phi_c(I_c) + L\frac{dI_c}{dt} \tag{2-186}$$

Again, Eqs. (2-184)–(2-186) can be cast in the form of a time-scaled dynamic

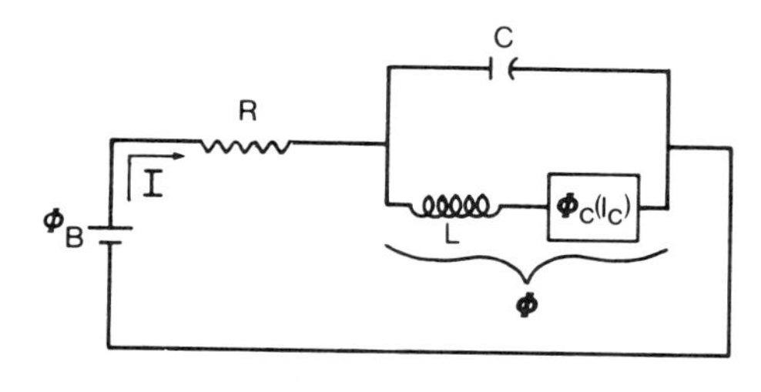

FIGURE 2-45. Lumped element approximation of an SNDC element in its primary circuit environment, as given by Figs. 2-36 and 2-37a.

system

$$\frac{d}{dT}\Phi = Z_0\left(\frac{\Phi_B - \Phi}{R} - I_c\right) \tag{2-187}$$

$$\frac{d}{dT}I_c = \frac{1}{Z_0}[\Phi - \Phi_c(I_c)] \tag{2-188}$$

where Z_0 and T have been defined in the NNDC circuit analysis. Comparison of Eqs. (2-187) and (2-188) with Eqs. (2-174′) and (2-175′) shows that Eq. (2-187) is the dual of Eq. (2-174′), and Eq. (2-188) is the dual of Eq. (2-175′). In going from the NNDC to the SNDC system we simply have to replace Φ, I, I_c, Z_0, R, Φ_B with I_c, Φ, Φ_c, $1/Z_0$, $1/R$, Φ_B/R, respectively. That is, in the context of the approximations noted, which are the common situations, the circuit theory transforms directly between the S and N cases when the important reactive components are identified for each NDC type. The important circuit parameters in each case produce a dual circuit system where the voltage across the NNDC element in its "primary" circuit behaves exactly as the current through the SNDC element in its primary circuit. Therefore, the arguments following Eq. (2-177) are similar to the ones we now present. However, the occurrence of limit cycle oscillations is now favored by large R and small Z_0.

The differential equation for the current I_c as a function of time is

$$\frac{\Phi_B}{R} = \left[1 + \frac{\Phi_c(I_c)}{I_c R}\right]I_c + \frac{R_0}{Z_0}\left[\frac{d\Phi_c(I_c)}{R_0\, dI_c} + \frac{L/R}{R_0 C}\right]\frac{dI_c}{dT} + \frac{d^2 I_c}{dT^2} \tag{2-189}$$

and the equation for the trajectories in the $\Phi(I_c)$ phase space is

$$\frac{d\Phi}{dI_c} = \frac{(\Phi_B - \Phi)/R - I_c}{\Phi - \Phi_c(I_c)} Z_0^2 \tag{2-190}$$

Equation (2-189) is the dual of Eq. (2-176), and Eq. (2-190) is the dual of Eq. (2-177).

We note that the strength of the damping term in Eq. (2-189) is determined by R_0/Z_0, since the bracketed part of the damping term is of order unity. [In Eq. (2-176) the controlling parameter was Z_0/R_0. Note the several inversions in comparing the equations.] Here, for large Z_0 the damping term is small and the solutions are near sinusoids. Small Z_0 now produces relaxation oscillations. We again employ a three-piece linear approximation to analyze the problem; this is shown in Fig. 2-46. $I_c(T)$ solutions can be obtained in the three regions $I_c < I_p$, $I_p < I_c < I_v$, and $I_v < I_c$, which correspond to $d\Phi_c/dI_c = R_0$, $-R_n$, and 0, respectively. A typical relaxation oscillation, displayed via a $\Phi(I_c)$ trajectory, is shown in Fig. 2-47. In the $I_c(T)$ plane the current waveform begins with a slow exponential rise with time constant $\sim(R_0/L + 1/RC)^{-1}$, followed by a sharp spike in current after $I_c = I_p$ is reached. The spike is composed of a fast exponential transit through the region of negative slope ($I_p < I_c < I_v$) with time constant $\sim(1/RC - R_n/L)^{-1}$, where $|R_n/L| > |1/RC|$, followed by a damped sine wave for $I_c > I_v$ and another exponential transit for $I_p < I_c < I_v$. An

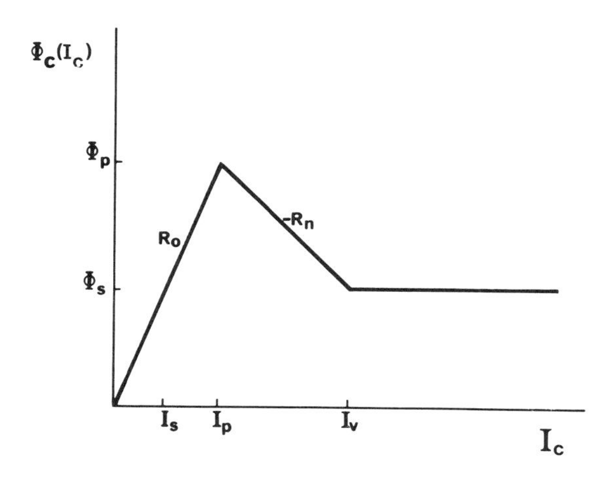

FIGURE 2-46. Three-piece linear approximation of the SNDC characteristic $\Phi_c(I_c)$.

exponential decay for $I_c < I_p$ completes the cycle. The time required to reach $I_c = I_p$ during the intial slow exponential rise depends on the applied bias Φ_B. Thus, the frequency of the relaxation oscillation is voltage tunable. Figure 2-41 shows the dual oscillation in an NNDC element. The voltage waveforms displayed there are analogues of the current waveforms just described. The voltage solutions shown in Table 2-1 are the duals of the current solutions.

We next integrate Eq. (2-190) over that portion of the cycle where

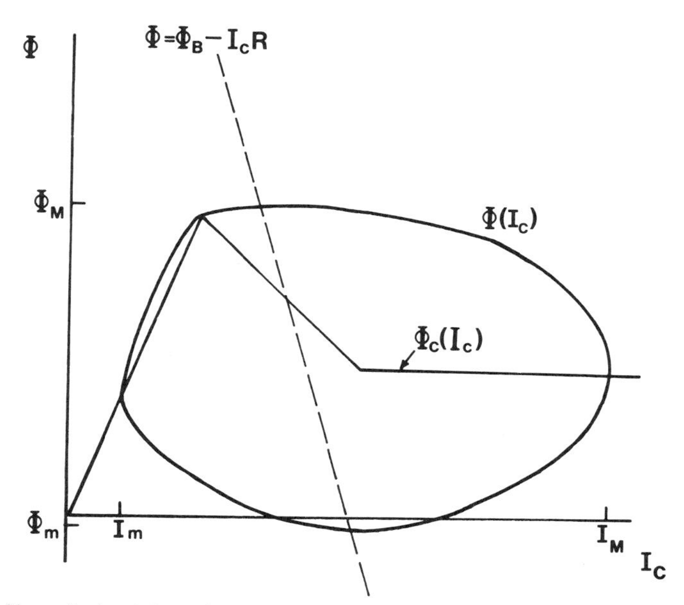

FIGURE 2-47. Numerical solution of the limit cycle phase trajectory $\Phi(I_c)$ for an SNDC element in the circuit of Fig. 2-45. The static SNDC characteristic $\Phi_c(I_c)$ is also shown in the phase plane. The dashed line is the dc load line. (After Shaw et al., 1973.)

$\Phi_c(I_c) = \Phi_s$, assuming $\Phi_B/R \gg \Phi/R$:

$$\frac{R_0^2}{Z_0^2}[\Phi(I_c) - \Phi_s]^2 + R_0^2\left(\frac{\Phi_B}{R} - I_c\right)^2 = \bar{K}^2 = \text{const} \tag{2-191}$$

which is the SNDC ellipse equation. For a particular SNDC element the shape of the ellipse is determined by Z_0/R_0. Plotting $\Phi(I_c)$ vs. $R_0 I_c$, the trajectory is a circle for $Z_0/R_0 = 1$, an ellipse with the major axis along $R_0 I_c$ for $Z_0/R_0 < 1$, and an ellipse with the major axis along Φ for $Z_0/R_0 > 1$. For small Z_0/R_0 the current amplitude is large and for large Z_0/R_0 the voltage amplitude is large. If we again assume the validity of extending the ellipse below I_v until it intersects the positive differential resistance part of $\Phi_c(I_c)$, then the $\Phi(I_c)$ curve is determined back to this point. At threshold ($I_c = I_p$), $\Phi(I_p) \cong \Phi_c(I_p) = \Phi_p$. The constant $\bar{K}^2$ in Eq. (2-191) may then be evaluated to give

$$\bar{K}^2 = \frac{R_0^2}{Z_0^2}(\Phi_p - \Phi_s)^2 + R_0^2\left(\frac{\Phi_B}{R} - I_p\right)^2 \tag{2-192}$$

Under the assumptions leading to Eq. (2-192), the complete $\Phi(I_c)$ trajectory is obtained by joining the ellipse Eq. (2-191) to the positive differential resistance segment of $\Phi_c(I_c)$. This approximation is best for small Z_0/R_0. An elliptical $\Phi(I_c)$ trajectory is plotted in Fig. 2-47. In the $\Phi(I)$ plane (replace the abscissa in Fig. 2-47 by $I = I_c + C\,d\Phi/dT$) the trajectory collapses to the load line (Shaw and Gastman 1971, 1972).

In analogy with the NNDC case, Eq. (2-192) shows that besides the parameters Z_0/R_0 and Φ_B/R, the circuit response is determined by the SNDC element parameters R_0, I_p, Φ_p, and Φ_s. As we shall show in Chap. 7, the form and nature of the circuit response is a dominant factor in the formation and quenching of current density filaments. In particular, the maximum voltage Φ_M and minimum current I_m (Fig. 2-47) reached during the first cycle are of major importance. From the ellipse Eq. (2-191) it follows that for large Z_0/R_0 (sinusoidal oscillations) Φ_M and I_m are high, whereas for small Z_0/R_0 (relaxation oscillations) Φ_M and I_m are low.

2.4. *SUMMARY*

In this chapter we have shown from analyses of Maxwell's equations that NDC points are unstable against the formation of both current density filamentation and high electric field domains. We have shown that the preferred form of the inhomogeneous distribution (current density or field) depends upon the detailed shape of the NDC curve (SNDC or NNDC). The inhomogeneous steady state solutions for both the SNDC and NNDC cases were shown to be governed by "equal areas" type rules.

Besides being unstable against inhomogeneous current density and field distributions, NDC elements can also be unstable against circuit-controlled oscillations. We have shown that these can vary from near sinusoids to spiky relaxation type oscillations, and that the SNDC element in its primary circuit is the dual of the NNDC element in its primary circuit.

In any real sample a common occurrence is that the material or device becomes unstable against both the formation of inhomogeneous distributions of charge and the development of circuit oscillations. Here the circuit and charge distribution problem must be solved simultaneously, and numerical assistance is often required. Specific cases will be disscussed in detail throughout the book. However, simple subelement-type models can also provide further insight; one of these is discussed in Chap. 7.

3

Tunnel Diodes

3.1. THE p-n JUNCTION

In Chap. 2 we analyzed the behavior of a circuit containing an NNDC element (see Fig. 2-38) for the case of a uniform electric field. We stressed that the theory of Eqs. (2-174)–(2-183) can be applied directly to a tunnel diode. Hence, the uniform field or tunnel diode theory is applicable for the circuit response of an NNDC element that does not form space charge waves such as high-field domains. In reality, the electric field is never homogeneous in tunnel diodes, and the only meaning of this terminology here is that the electric field distribution does not show domain structure. (We consider domain formation effects in Chap. 5).

The absence of domain formation is an attractive feature of tunnel diode analysis; this is the reason that we treat it before any other NNDC device. As we shall see in the next section (Sec. 3.2), the tunnel current will depend on the electric field, F, or electric potential, Φ, distribution in the diode. Since the tunneling effect manifests itself in a heavily doped p-n junction (Esaki, 1958, 1969), let us first consider the band diagram of a p-n junction.

The spatial distribution of electric field can be found by solving the Poisson equation (2-14). A relatively narrow transition region between p and n regions can be achieved by diffusion at high temperatures or by ion implantation. For such an ideal abrupt junction, the fixed background doping density, N_0, in (2-14) can be represented by the model steplike charge density

$$N_0(x) = +N_A\theta(-x) - N_D\theta(x) \tag{3-1}$$

as shown in Fig. 3-1. Here N_A is the concentration of charged acceptors, N_D is the concentration of charged donors, $\theta(x)$ is a theta function, and x is a coordinate directed from the p to the n region. The simplest approach for solving Eq. (2-14) is to neglect the effect of the mobile carriers in the depletion regions (depletion approximation), so that Eq. (2-14) can be written in the form

$$\frac{\epsilon}{e}\frac{\partial F}{\partial x} = -N_0 = -N_A\theta(-x) + N_D\theta(x) \qquad \text{for } -l_p < x < l_n \tag{3-2}$$

$$F = 0, \qquad n = N_D \qquad \text{for } x > l_n \tag{3-3}$$

$$F = 0, \qquad p = N_A \qquad \text{for } x < -l_p \tag{3-4}$$

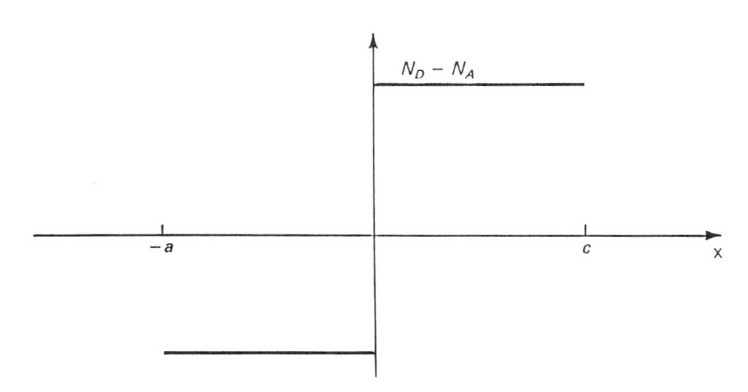

FIGURE 3-1. Doping concentration in an ideal abrupt *p-n* homojunction. The *p* side extends from $-a$ to 0, and the *n* side from 0 to *c*. (After Hess, 1988.)

where l_p and l_n are introduced as the lengths of the depletion regions in the p- and n-type sides of the junction, respectively.

After integration of (3-2)–(3-4) we have

$$F(x \leq -l_p) = 0, \qquad F(-l_p \leq x \leq 0) = -N_A(l_p + x)e/\epsilon \tag{3-5}$$

$$F(x \geq l_n) = 0, \qquad F(l_n \geq x \geq 0) = -N_D(l_n - x)e/\epsilon \tag{3-6}$$

Taking into account that both Eqs. (3-3) and (3-4) must provide the same value of electric field at $x = 0$,

$$F(x = 0) = -N_A l_p e/\epsilon = -N_D l_n e/\epsilon \tag{3-7}$$

we have a relation between l_p and l_n

$$N_A l_p = N_D l_n \tag{3-8}$$

This relation is the requirement of charge neutrality in the *p-n* junction, i.e., the total concentration of positively charged donors equals the total concentration of negatively charged acceptors in the depletion region.

If the effect of mobile carriers in the depletion region is taken into account, exponential tails on $n(x)$ and $p(x)$ in the depletion region arise, as shown in Fig. 3-2a. The dependence of the fixed charge density (3-8) and the electric field (3-7) on x are shown in Figs. 3-2b and 3-2c.

At this point, the width of the depletion region is still undetermined. To obtain l_n and l_p it is necessary to use one more equation in addition to (3-8). This is an equation analogous to Eq. (2-12), in which we define the potential difference between the p and n regions, the "built-in" potential Φ_{bi}

$$\Phi_{bi} = -\int_{-l_p}^{l_n} F\,dx \tag{3-9}$$

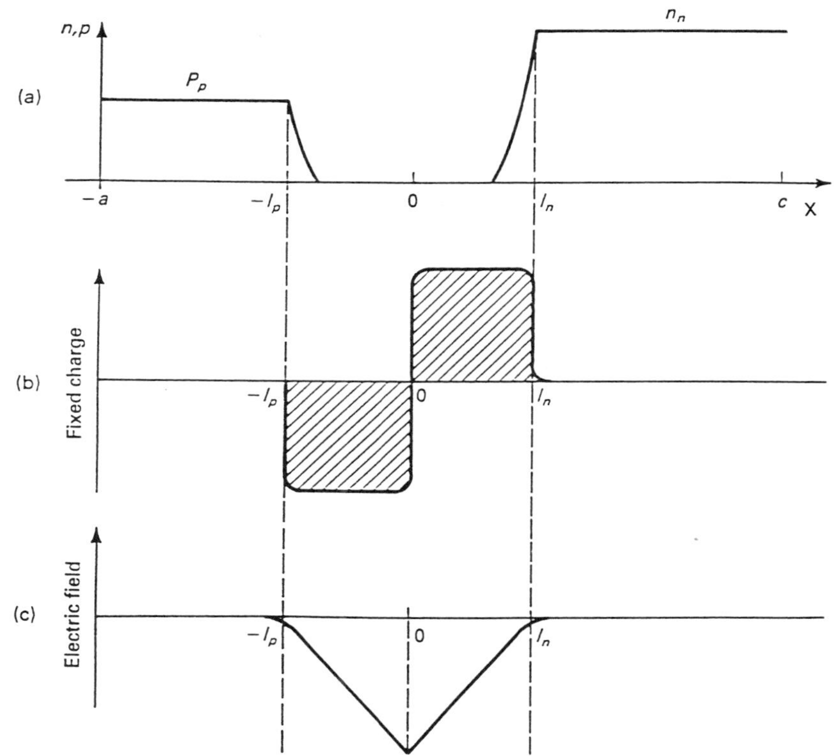

FIGURE 3-2. (a) Concentration of free carriers (electrons and holes) in a *p-n* junction at equilibrium; (b) distribution of the fixed charge in the depletion layer (from $-l_p$ to l_n); (c) electric field in the depletion region (after Hess, 1988).

The built-in potential is positive because F is negative in the depletion region. We choose $\Phi = 0$ in the n region, so

$$\Phi(x) = -\int_x^{l_n} F\,dx \tag{3-10}$$

Integration of (3-9) and (3-10), with Eqs. (3-5)–(3-7) taken into account, gives

$$l_n = \left(\frac{2\epsilon N_A}{eN_D}\frac{\Phi_{\text{bi}}}{N_A + N_D}\right)^{1/2} \tag{3-11}$$

$$l_p = \left(\frac{2\epsilon N_D}{eN_A}\frac{\Phi_{\text{bi}}}{N_A + N_D}\right)^{1/2} \tag{3-12}$$

$$\Phi(x \leq -l_p) = \Phi_{\text{bi}} \tag{3-13}$$

$$\Phi(-l_p \leq x \leq 0) = \Phi_{\text{bi}} - N_A(l_p + x)^2\frac{e}{2\epsilon} \tag{3-14}$$

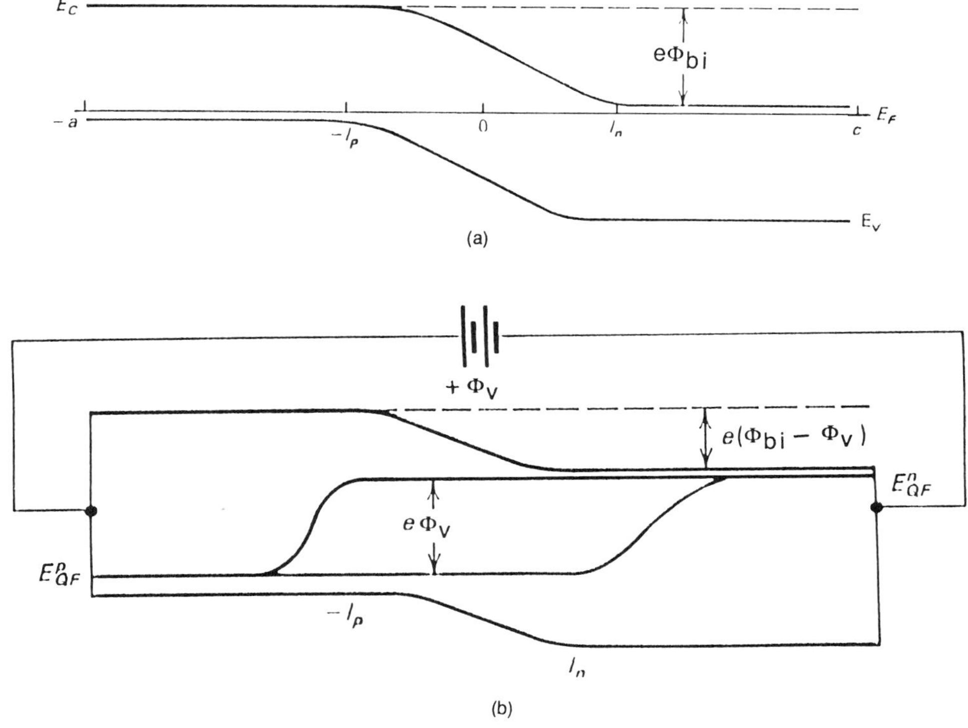

FIGURE 3-3. Band diagram for abrupt *p-n* junction at equilibrium (a) and with external voltage $\Phi_V < 0$ applied (b) (after Hess, 1988).

$$\Phi(0 \leq x \leq l_n) = N_D(l_n - x)^2 \frac{e}{2\epsilon} \tag{3-15}$$

$$\Phi(x \geq l_n) = 0 \tag{3-16}$$

The dependence of the potential on x is presented in Fig. 3-3a, where the straight horizontal line E_F corresponds to a Fermi level that is continuous throughout the p-n junction at equilibrium. If an external potential Φ_V is applied, Eqs. (3-1)–(3-8) remain the same, but in (3-9)–(3-14) we must replace Φ_{bi} by $\Phi_{\text{bi}} - \Phi_V$. For $\Phi_V > 0$, forward bias, the band diagram is shown in Fig. 3-3b, with constant quasi-Fermi levels for electrons, E^p_{QF}, and holes, E^n_{QF}, in the depletion region:

$$E^n_{\text{QF}} - E^p_{\text{QF}} = e\Phi_V \tag{3-17}$$

The concept of quasi-Fermi levels was not exploited in (3-1)–(3-16) because it was assumed that $n \cong p \cong 0$ in the depletion region. To determine $n(x)$ and $p(x)$ in that region, it is necessary to take into account the energy of the carriers as referred to the corresponding quasi-Fermi level. Equations (3-11) and (3-12) determine the dependence of the width, w, of the depletion region,

$$w = l_n + l_p = \left[\frac{2\epsilon}{e} \cdot \frac{N_A + N_D}{N_A N_D} (\Phi_{\mathrm{bi}} - \Phi_V)\right]^{1/2} \tag{3-18}$$

on the built-in and applied potentials and on the doping concentration. If both N_A and N_D increase, the depletion width decreases and the maximum value of electric field, which is at $x = 0$ (3-7), increases:

$$F_{\mathrm{max}} = -F(x = 0) = 2(\Phi_{\mathrm{bi}} - \Phi_V)/w \tag{3-19}$$

In Eqs. (3-18) and (3-19), the built-in potential

$$\Phi_{\mathrm{bi}} = (E_G + \xi_n + \xi_p)/e \tag{3-20}$$

also depends on doping through the electron, ξ_n, and hole, ξ_p, quasi-Fermi energies. For wide gap semiconductors we have

$$|\xi_n|, |\xi_p| \ll E_G \tag{3-21}$$

That is why

$$\Phi_{\mathrm{bi}} \cong E_G/e. \tag{3-22}$$

In spite of the fact that variations of ξ_n and ξ_p only make for small changes in Φ_{bi}, their variation with doping is important, because a qualitatively new phenomenon, NNDC, can occur in heavily doped *p-n* junctions. A band diagram for a lightly doped *p-n* junction is shown in Fig. 3-3, where $\xi_n, \xi_p < 0$ and the Fermi levels on both sides of the junction lie in the forbidden energy gap. For sufficiently heavy doping on both sides, the electrons on the *n* side and the holes on the *p* side are degenerate ($\xi_n, \xi_p < 0$). Instead of the diagram of Fig. 3-3a, the one shown in Fig. 3-4 results. For reverse bias there is no qualitative difference between degenerate and nondegenerate *p-n* junctions, but for forward bias, there is a great difference. The difference was first observed and explained by Esaki (1958): this was the invention of the tunnel diode. In the next section (Sec. 3.2) the current–voltage characteristic of this diode will be calculated, and a qualitative explanation of the NNDC will be given.

In the direct tunneling process electrons penetrate through the gap without changing their energy, as shown in Fig. 3-5. There is no current at zero bias (points a–d on the current–voltage characteristic of Fig. 3-6 correspond to the

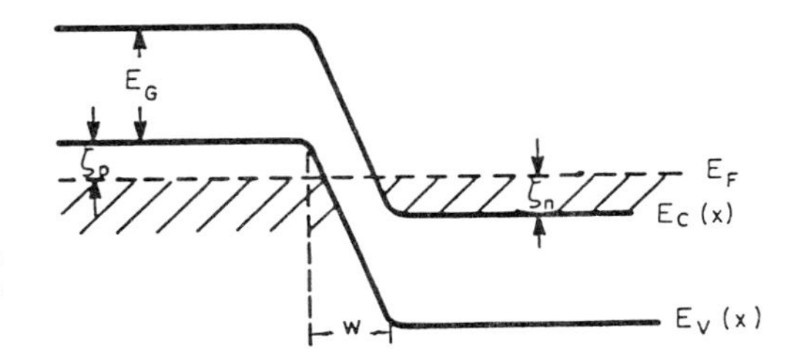

FIGURE 3-4. Equilibrium band diagram for a *p-n* junction degenerate on both sides (after Duke, 1969).

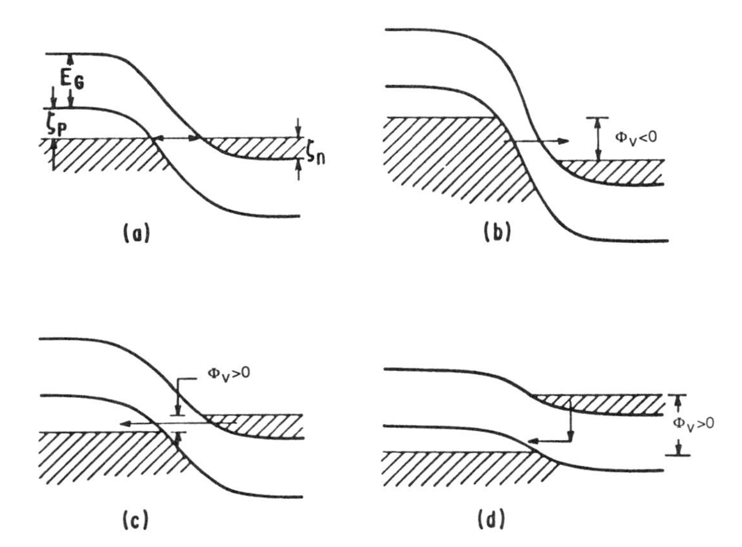

FIGURE 3-5. Schematic diagram of the potential energy as a function of position in a *p-n* tunnel diode at (a) zero bias; (b) reverse bias ($\Phi_V < 0$); (c) small forward bias ($\Phi_V > 0$); and (d) large forward bias ($\Phi_V > 0$). The arrows schematically indicate the tunneling path. (After Duke, 1969.)

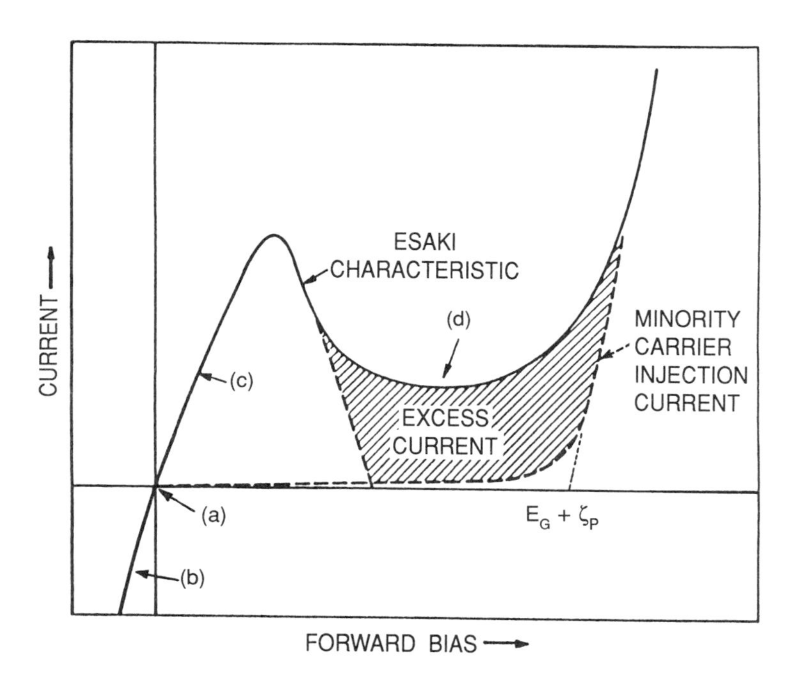

FIGURE 3-6. Schematic diagram of the current–voltage characteristic of the *p-n* tunnel diode described in Fig. 3-5. The notations (a)–(d) indicate the regions of the characteristic associated with the different parts of Fig. 3-5. The generic labels of the various regions of the characteristic are given in the figure. (After Duke, 1969.)

energy band diagrams shown in Figs. 3-5a–3-5d). Electrons can tunnel from the valence band to free states in the conduction band if a reverse bias is applied (Fig. 3-5b). Under forward bias, electrons from conduction states on the n-type side tunnel to hole states in the valence band on the p-type side. For these conditions the current increases as Φ_V increases. But this occurs only up to $\Phi_V \cong \xi_p/e$ (Fig. 3-5c). For $\Phi_V > \xi_p/e$, some electrons from the conduction states on the n-type side have energy states that lie inside the energy gap on the p-type side. These electrons cannot tunnel; here the current decreases as Φ_V increases. Ideally, for $\Phi_V > (\xi_p + \xi_n)/e$, all states in the conduction band on the n-type side correspond to energy gap states on the p-type side; here the tunnel current equals zero. An NNDC region occurs over the applied voltage range

$$\xi_p/e < \Phi_V < (\xi_n + \xi_p)/e, \qquad \xi_n > 0, \qquad \xi_p > 0 \tag{3-23}$$

In addition to the tunneling current, other additional current components also exist, providing excess current. The actual current–voltage characteristic looks like the continuous line shown in Fig. 3-6, where the major contribution for $\Phi_V > E_G + \xi_p$ comes from the normal thermionic current. In Chap. 2 we showed that the amplitudes of the current and voltage fluctuations in the circuit increase if the ratio of the maximum current I_p to valley current I_s increases (Fig. 2-42). This ratio is an important feature in an NNDC element.

For a nondegenerate junction ξ_n and ξ_p are negative. Here, Eq. (3-23) is not true and tunnel currents might occur only under reverse bias, $\Phi_V < (\xi_n + \xi_p)/e$. We see that NNDC can occur in tunnel diodes when both sides of the junction are degenerate.

3.2. TUNNELING IN A p-n JUNCTION

The theory of tunneling has been presented in detail by Duke (1969) (see also Kane, 1969; Kane et al., 1969; Sah, 1969; and others in the book edited by Burstein and Lundqvist, 1969). Hence, we outline here only those major features of tunneling showing that the current–voltage characteristic of a p-n junction has the form shown in Fig. 3-6, and identify those parameters which define the dependence of $I(t)$ and $\Phi(t)$ in a circuit containing a tunnel diode (Fig. 2-38).

3.2.1. Direct Tunnel Diodes

In our qualitative explanation of NNDC (Fig. 3-5), we assumed without any explanation that electrons from allowed states in the valence band can go through the forbidden energy gap (tunnel) to allowed states in the conduction band. The inverse process is also possible. We now consider tunneling in more detail (a comprehensive description can be found in Duke, 1969; and Burstein and Lundqvist, 1969).

Tunneling through the energy gap in a p-n junction is similar to Zener tunneling (Zener, 1943) in a uniform electric field. The only difference is that the electric field is inhomogeneous (3-5), (3-6) in a p-n junction. Figure 3-7

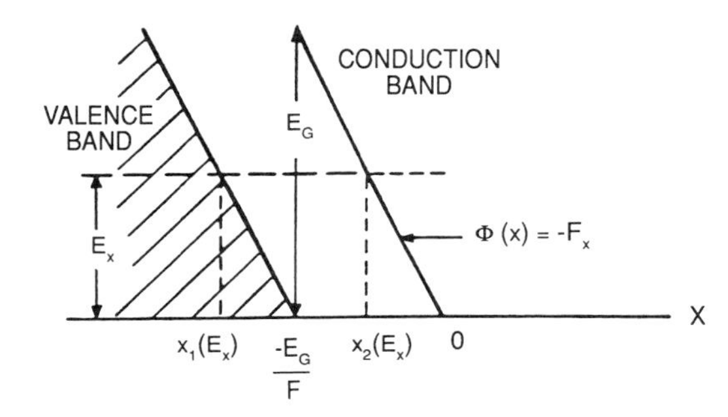

FIGURE 3-7. Schematic representation of the semiclassical barrier to be used for calculating the Zener tunneling probability in the uniform-field model (after Duke, 1969).

demonstrates Zener tunneling, which we consider now via the WKB approximation using a semiclassical approach (Zener, 1943; Duke, 1969; Franz, 1969). The tunneling probability in the WKB approximation is

$$D = \exp\left(-2\int_{x_1}^{x_2} \varkappa(x, E_x)\, dx\right) \tag{3-24}$$

where x_1 and x_2 define the classical turning points for the electrons with energy E_x and $\varkappa(x, E_x)$ is the local decay constant of the wave function of the electron inside the energy gap (Fig. 3-7). The decay constant is defined ordinarily by setting

$$k = i\varkappa \tag{3-25}$$

where k is the wave number of the electron, and by using the dispersion relation $E(k)$ inside the energy gap from the known relation in the allowed bands. The simplest dispersion relation has the form:

$$\hbar^2 k^2/(2m^*) = E(E_G - E)/E_G \tag{3-26}$$

where $E > 0$ denotes the conduction band and $E < -E_G$ denotes the valence band. The effective masses of electrons and holes near the band edges are considered to be the same (the conduction and the valence bands are completely symmetric). This approach is known as the quasirelativistic dispersion law because it is like the dispersion relations for electrons and positrons in the relativistic case if we replace m^* by m and E_G by $2mc^2$ (Kane, 1957; Amrachov et al., 1986).

Taking E to be the local kinetic energy, we obtain from Fig. 3-7,

$$E = e\Phi(x) - E_x = -eFx - E_x \tag{3-27}$$

If we put (3-25)–(3-27) into the integral (3-24) and note that

$$x_2 = \frac{-E_x}{Fe} \tag{3-28}$$

$$x_1 = \frac{-(E_G + E_x)}{Fe} \tag{3-29}$$

we obtain

$$D = \exp\left(\frac{-\pi E_G^{3/2} m^{*1/2}}{2^{3/2} e\hbar F}\right) \tag{3-30}$$

This expression was obtained via a number of approximations (Duke, 1969). First of all, it is known as the one-dimensional approach because the local decay constant of Eq. (3-24) must be determined [see Eq. (3-25)] by the k_x component of the vector **k** only. With $k_\parallel$, the component of momentum **k** parallel to the plane of the junction, the energy $E_\parallel$ is such that

$$E = E_\parallel + E_x \tag{3-31}$$

If (3-31) is taken into account in the calculation of the tunneling probability, the result is known as the three-dimensional approach; instead of (3-30) we obtain (Duke, 1969)

$$D = \exp\left(\frac{-E_G}{4\bar{E}_\parallel}\right)\exp\left(\frac{-E_\parallel}{\bar{E}_\parallel}\right) \tag{3-32}$$

where

$$\bar{E}_\parallel = \frac{e\hbar F}{\pi(2m^* E_G)^{1/2}} \tag{3-33}$$

An additional exponential factor, $\exp(-E/\bar{E}_\parallel)$ arises in D. This factor reduces the penetration probability for electrons having energy $E_\parallel$ comparable to, or larger than, $\bar{E}_\parallel$.

A second serious restriction is the equality of the effective masses in both bands (3-26). More rigorous calculations (Keldysh, 1958) prove that combinations of effective masses enter into Eq. (3-33):

$$m^* = \frac{2m_e m_h}{m_e + m_h} \tag{3-34}$$

where m_e and m_h are the effective masses of electrons and holes near the corresponding band edges. Even with this modification, Eq. (3-32) can be used only in direct gap semiconductors where both bands are at the center of the Brillouin zone. Figure 3-8 shows experimentally observed current–voltage characteristics of a GaAs tunnel diode at different temperatures. In this diode the NNDC is due to direct tunneling from the conduction to the valence band.

3.2.2. Phonon-Assisted Tunneling

A very important consequence of WKB theory (3-32) is that the component of the wave vector parallel to the plane of the junction, $k_\parallel$, is conserved when an electron tunnels from one side of the junction to the other. This prohibits, for

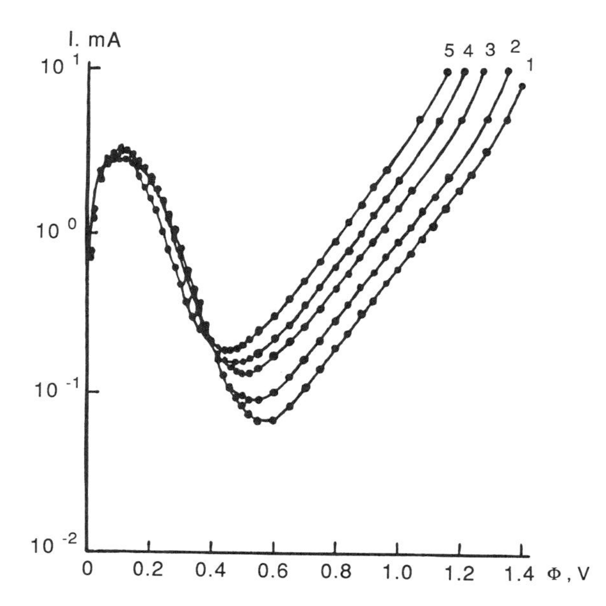

FIGURE 3-8. The current–voltage characteristics of a GaAs tunnel diode for forward bias at various temperatures T: 1, 77 K; 2, 196 K; 3, 292 K; 4, 359 K; 5, 427 K (after Imenkov et al., 1965).

example, tunneling through a junction lying in (100) planes in Ge from states near the conduction band minima. However, the first observation of tunnel-induced NNDC (Esaki, 1958) was in a Ge *p-n* junction. The answer is that tunneling occurred because the change in wave vector can be taken up by a phonon, or by impurities, or by the electron–electron interaction. These processes are known as the KPK (Keldysh–Price–Kane) mechanism of tunneling (Duke, 1969), because the theory was developed by Keldysh (1958a, b) and implemented by Price and Radcliffe (1959) and Kane (1961).

If the electron–phonon interaction is sufficiently strong, then the phonon-assisted tunneling contribution might also be important in tunnel diodes fabricated from direct band gap materials. The essential distinction between phonon-assisted tunneling in direct and indirect diodes lies in the consideration of momentum conservation. In direct diodes the sharp threshold behavior is associated with the required momentum transfer, $\mathbf{k}_0$. The energy of the phonon emitted, $\hbar\omega(\mathbf{k}_0)$, is finite, so phonon-assisted tunneling can only occur if $|e\Phi| > \hbar\omega(\mathbf{k}_0)$, i.e., threshold occurs for *both* acoustic and optic phonons. In direct diodes $\mathbf{k}_0 = 0$; here the sharp threshold effect can result only from the emission of optical phonons, for which $\hbar\omega$ is finite.

A review of theoretical and experimental results for phonon-assisted tunneling is presented in the books of Duke (1969), Scanlan (1966), and Roy (1986), and that edited by Burstein and Lundqvist (1969). They identify other modifications of the tunneling probabilities (3-32) which we will not discuss here. As far as indirect tunneling is concerned, we stress that the conduction and valence band extrema might not only be different in their values of $\mathbf{k}$ in the Brillouin zone, but they can even be different materials, as shown for the *p-n* heterojunction in Fig. 3-9.

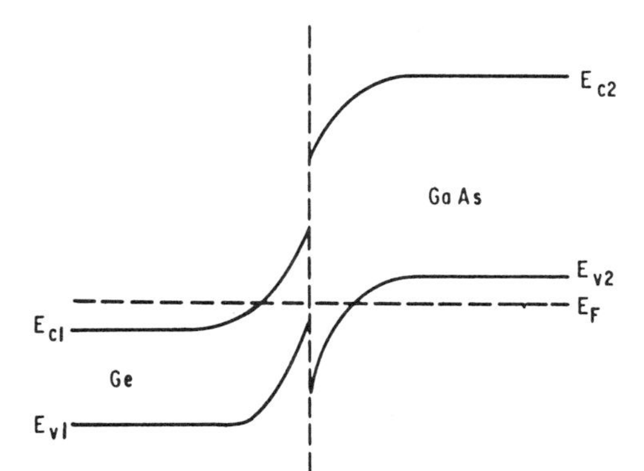

FIGURE 3-9. Schematic potential energy versus distance diagram for a Ge–GaAs *n-p* heterojunction (after Duke, 1969).

3.2.3. Current–Voltage Characteristics

A rigorous, contemporary calculation of the current–voltage characteristics based on new tunneling concepts can be found in the work of Roy (1986). Qualitatively, these concepts do not impact the major features of the current–voltage characteristics, hence we present here calculations based on the WKB method. This means that in the ordinary expression for the electron current density,

$$J = 2\frac{e}{(2\pi)^3}\int v_x f(\mathbf{k})\,d^3k \tag{3-35}$$

it is only necessary to multiply by the tunneling probability D and by the distribution function of the empty states on the opposite side of the junction. Here

$$v_x = \frac{\partial E}{\hbar\,\partial k_x} \tag{3-36}$$

is velocity, and $f(\mathbf{k})$ is the electron distribution function, which can be taken as the equilibrium Fermi distribution outside of the depletion region:

$$f(\mathbf{k}) = f(E) = \{1 + \exp[(E - E_F)/kT]\}^{-1} \tag{3-37}$$

where E_F is the Fermi energy.

Taking into account currents from the n to the p side and vice versa, we obtain

$$J = 2\frac{e}{(2\pi)^3\hbar}\int dk_x \frac{\partial E}{\partial k_x}\,d^2k_{\|}[f(E) - f(E + e\Phi)]D \tag{3-38}$$

The following simplifications can now be performed. In our approach, D depends only on $E_{\|}$; that is why instead of k_x and $k_{\|}$, the integration is easier to do over the variables E and $E_{\parallel}$

$$J = \left[\frac{2\pi m^* e P_0}{(2\pi\hbar)^3}\right] \int_{-\infty}^{\infty} dE[f(E) - f(E + e\Phi)] \int_0^{E_{\max}} \exp\left(\frac{-E_{\parallel}}{\bar{E}_{\parallel}}\right) dE_{\parallel} \quad (3\text{-}39)$$

where

$$P_0 = \exp\frac{-E_G}{4\bar{E}_{\parallel}} \quad (3\text{-}40)$$

$\bar{E}_{\parallel}$ and m^* are defined by Eqs. (3-33) and (3-34),

$$E_{\max} = \min\begin{Bmatrix} E \\ (\xi_n + \xi_p - E - e\Phi)m_h/m_e \end{Bmatrix} \quad (3\text{-}41)$$

The second integral in (3-39) is easy to calculate and

$$J = J_0 e^2 \Phi \bar{E}_{\parallel} 2\pi m^* P_0/(2\pi\hbar)^3 \quad (3\text{-}42)$$

where

$$J_0 = \frac{1}{e\Phi}\int_{-\infty}^{\infty} dE[f(E) - f(E + e\Phi)]\left[1 - \exp\left(\frac{E_{\max}}{\bar{E}_{\parallel}}\right)\right] \quad (3\text{-}43)$$

The factor J_0 is chosen in such a way that

$$J_0 \cong 1 \qquad \text{for } kT \ll e\Phi \ll \xi_n, \xi_p \quad (3\text{-}44)$$

when a small voltage is applied, because

$$f(E) - f(E + e\Phi) \cong -e\Phi\frac{\partial f}{\partial E} = e\Phi\delta(E - \xi_n) \quad (3\text{-}45)$$

and $E_{\max} \to \infty$.

We can estimate the dependence of J_0 on Φ by taking into account that for degenerate electrons and holes the limits of integration are not $-\infty$ and ∞, but rather 0 and $(\xi_n + \xi_p - e\Phi)$ (see Fig. 3-5). If (3-44) is true, it again yields $J_0 = 1$, but (Roy, 1986; Karlovsky, 1962)

$$J_0 \cong A(\xi_n + \xi_p - e\Phi) \qquad \text{for } e\Phi \sim \xi_n, \xi_p \quad (3\text{-}46)$$

Equations (3-42)–(3-46) yield the current–voltage characteristics as shown in Fig. 3-6. The current reaches its maximum at $\Phi = \Phi_p$:

$$e\Phi_p = (\xi_n + \xi_p)/3 \quad (3\text{-}47)$$

and its minimum at $\Phi = \Phi_n$:

$$e\Phi_n = (\xi_n + \xi_p) \quad (3\text{-}48)$$

Equation (3-46) is very approximate because it results from an expansion of the exponential in (3-43) using $E_{\max} \ll \bar{E}_{\parallel}$. This is the reason that we find other analytical expressions for the $I(\Phi)$ characteristics in the literature, e.g.,

$$J = J_p\left(\frac{\Phi}{\Phi_p}\right)\exp\left(\frac{1-\Phi}{\Phi_p}\right) + J_n\left(\frac{\Phi}{\Phi_n - 1}\right) + J_t \exp\left(\frac{e\Phi}{kT}\right) \tag{3-49}$$

where the first term represents the tunnel current, the second is the excess current, and the third is the thermal current, with J_t as the reverse saturated current density (Roy, 1986). The parameters Φ_p, Φ_n, J_p, J_n, J_t are chosen for a best fit of Eq. (3-49) and the experimental curve. Hence, such dependencies are phenomenological, and are convenient because of their simplicity, in contrast to Eq. (3-42), where it is necessary to integrate and also add the excess and thermal currents (Sah, 1969).

3.3. RESONANT TUNNELING

In the previous section we demonstrated that electrons can tunnel through the classically forbidden region into allowed states. The current decreases (see Fig. 3-6) when the number of allowed states decreases (Figs. 3-5c and 3-5d). This results in NNDC, which can be observed not only in the Esaki diode discussed in the previous section, but also in other structures where the number of allowed states that can be reached by tunneling decreases when the applied voltage increases. A review of such tunneling phenomena has been given, for example, by Capasso et al. (1986). We will consider some of these mechanisms here because they are generically related to the Esaki diode.

3.3.1. Resonant Tunneling between Quantum Wells

Resonant tunneling was first proposed in 1963 as a method of measuring the distances between energy levels in a quantum well via the separation between peaks observed in the current–voltage characteristics (Demikhovskii and Tavger, 1963, 1966). The phenomenon was described in a review article (Tavger and Demikhovskii, 1968), with some reference to experimental results; it was rediscovered in 1971 by Kazarinov and Suris (1971, 1972). Historically, it is commonly accepted that the publication of Tsu and Esaki (1973) is referred to as the one where the effect was rediscovered. In that paper, Tsu and Esaki emphasized double (and multiple barrier) structures, which we will discuss in the next subsection.

To describe the effect, let us consider the energy spectrum of a quantum well, which is similar to the problem of a "particle in a box." The simplest model of a quantum well is a one-dimensional structure where the potential $\Phi(x)$ is zero over the distance $(0, L)$ on the x axis and infinite at the boundaries 0 and L (Tavger and Demikhovskii, 1968; Ando et al., 1982). Owing to the limited width, L, of the well, the projection of the wave number, k_x, perpendicular to the plane of the well is indeterminate, that is why a discrete quantum number n replaces k_x.

As a result, the energy $E_\parallel$ in (3-31) depends on $k_\parallel$ in the same manner as in three dimensions. But, for E_x we have

$$E_x = \frac{(\pi\hbar n)^2}{2mL^2} \tag{3-50}$$

Since n is an integer, the electron can assume only certain discrete energies E_x (3-50), while $E_\parallel$ is a continuous function of $k_\parallel$. The discrete energies E_x are called quantum states or quantum levels, or more commonly quantum subbands, and they are characterized by the quantum number n. The subbands usually overlap, as shown in Fig. 3-10, because the energy interval $E_x(n + 1) - E_x(n)$ is, as a rule, smaller than the allowed values (or magnitudes) of $E_\parallel$. As a matter of fact, the relation (3-31) can be rewritten in the form

$$E_n(k_\parallel) = E_\parallel(k_\parallel) + \frac{(\pi\hbar n)^2}{2m^*L^2} \tag{3-51}$$

where n is a subband number.

We emphasize that the dispersion relation in the form of (3-51), shown in Fig. 3-10, is an idealized model because of the spread of the levels due to both the

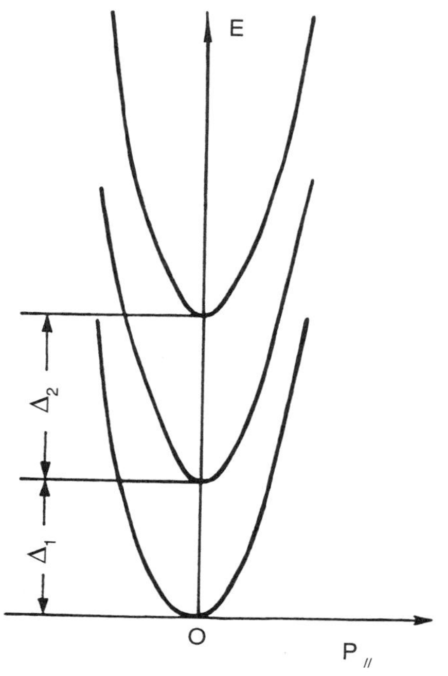

FIGURE 3-10. Subband structure for a quantum well showing the energy dependence of $p_\parallel = \hbar k_\parallel$ parallel to the quantum well component of the momentum. Δ_1 and Δ_2 are the energy distances between subbands.

finite temperature and scattering. This is the reason that the requirements

$$kT < E_{n+1} - E_n \tag{3-52}$$

and

$$\frac{\hbar}{\tau} \ll E_{n+1} - E_n \tag{3-53}$$

must be satisfied, where τ is the electron relaxation time. In our analysis we assume that the length of the quantum well L is small enough so that both conditions (3-52) and (3-53) are fulfilled. In addition to the imposed conditions there should also exist a limitation on the concentration of electrons in order that the number of populated subbands be small. In order for only one subband to be populated, it is necessary that the inequality

$$nL^3 < \frac{3\pi}{2} \tag{3-54}$$

be satisfied, which follows from (3-50) and dimensionality considerations.

In reality, the potential depth Φ_L of the well is not infinite, as assumed above, but is finite. This is the reason that another inequality,

$$\Phi_L \gg E_2 - E_1 \tag{3-55}$$

is essential, in order to assume that electrons exist in the deep quantum well.

Let us consider a one-dimensional superlattice having quantum wells of equal length L separated by potential barriers of height Φ_L. The period of the superlattice is d. Additional broadening Δ arises due to the degeneracy of the levels E_n through equal quantum wells. For the sake of simplicity, let us restrict ourselves to the case where, in addition to inequalities (3-52)–(3-55), one more condition,

$$\Delta \ll E_{n+1} - E_n \tag{3-56}$$

is satisfied. In the previous section we stressed that $E_{\parallel}$ is continuous during the tunneling. This is the reason why tunneling can take place only when allowed states exist on the other side of the barrier, i.e., when the applied potential drop over one period of superlattice, Φ_d, coincides with $\Delta_i = E_i - E_1$, so that the ground state in the quantum well, which supplies tunneling electrons, becomes degenerate with one of the excited levels on the other side of the barrier. Figure 3-11a demonstrates tunneling from the ground state in a quantum well to the first excited level in the next quantum well, when

$$e\Phi_{d,1} = E_2 - E_1 \tag{3-57}$$

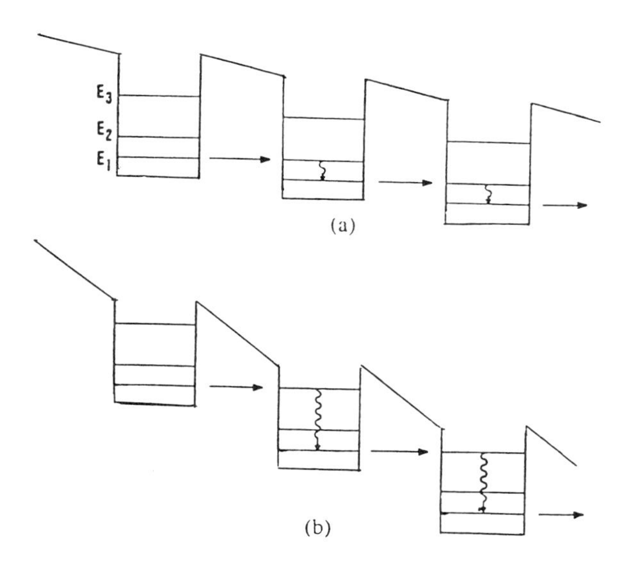

FIGURE 3-11. Schematic illustration of sequential resonant tunneling of electrons for a potential energy drop across the superlattice period equal to the respective energy difference between the first excited state and the ground state (3-57) of the wells (a) and to the energy difference between the second excited state and the ground state of the wells (3-58) (b) (after Capasso et al., 1986).

The next current peak corresponds to

$$e\Phi_{d,2} = E_3 - E_1 \tag{3-58}$$

when tunneling takes place onto the second excited level, as demonstrated in Fig. 3-11b. With an applied voltage Φ_V in Fig. 3-11, we assume that the field is uniformly distributed over the low conductivity superlattice regions and there is no voltage drop inside of the quantum wells.

It is important to note here that we considered tunneling between two adjacent wells. If the superlattice structure consists of more than two wells, as shown in Fig. 3-11, relaxation from the excited level to the ground level must take place before an electron can tunnel to the next adjacent well. This relaxation can be due to phonon or photon emission, or due to electron–electron interactions. (There presently is no reliable calculation of the current–voltage characteristics that treats different relaxation mechanisms.)

In accordance with (3-57) and (3-58) tunneling is possible only for a certain value of voltage $\Phi_{d,i}$, hence the term "resonant tunneling." (This is sequential resonant tunneling because relaxation to the ground state follows each tunneling between adjacent wells.) In reality, each subband is broadened, (3-52)–(3-56), so direct tunneling is realized for $\Phi_{d,i}$ sufficiently close to resonance. Figure 3-12 illustrates the same phenomenon via a momentum space band diagram. The voltage Φ_d is chosen out of resonance:

$$E^- = E_2 - E_1 - e\Phi_d, \qquad E^+ = E_3 - E_1 - e\Phi_d \tag{3-59}$$

From this presentation it is obvious that direct tunneling is possible if E^- is less than or equal to Δ or h/τ [compare (3-53) and (3-56)]. But phonon-assisted tunneling is possible even out of resonance (Franz, 1969; Logan, 1969; Kleinman, 1969), in analogy with the phenomena discussed in the previous section. As a

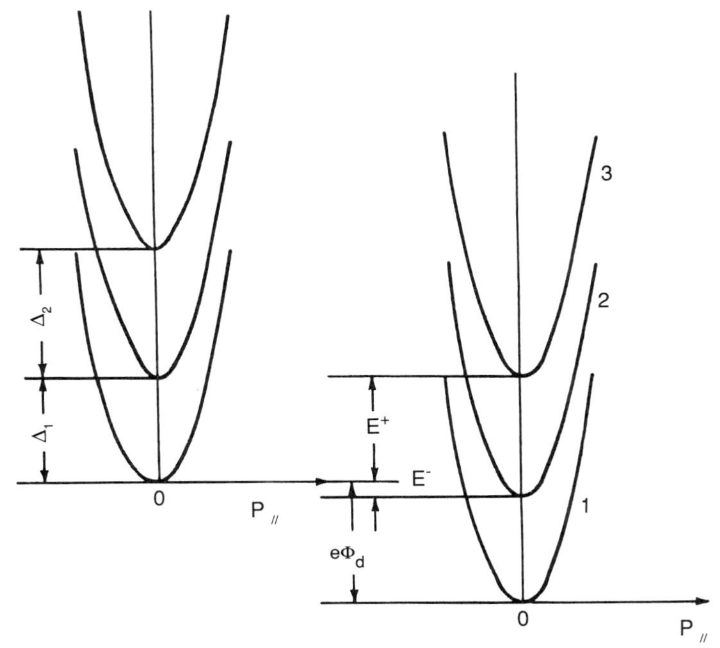

FIGURE 3-12. Momentum space representation of resonant tunneling between two adjacent wells (n and $n + 1$) in a superlattice. Shown is the subband structure; $p_{\parallel}$ is the momentum in the plane of the layers. Φ_d is the potential drop across the superlattice period; E^- and E^+ are the energy detunings from resonance. (After Capasso et al., 1986.)

result, the current–voltage characteristic has a finite current between the peaks rather than zero current, as demonstrated in Fig. 3-13. These are experimentally measured characteristics, demonstrating resonant tunneling in superlattices.

Let us now describe the experiment of Fig. 3-13 in a bit more detail (Capasso et al., 1986). The undoped superlattice consists of 36 periods of undoped (concentration of impurities N_D less than $10^{14}\,\mathrm{cm}^{-3}$) $Al_{0.48}In_{0.52}As$ and $Ga_{0.47}In_{0.53}As$, both having the same length, 139 Å, grown by molecular beam epitaxy. There is also a heavily doped ($N_D \cong 10^{17}\,\mathrm{cm}^{-3}$) $Al_{0.48}In_{0.52}As$ n^+ buffer layer. On this superlattice structure a heavily doped ($N_A \cong 2 \times 10^{18}\,\mathrm{cm}^{-3}$) $Al_{0.48}In_{0.52}As$ p^+ layer is grown. As a result, p^+-i-n^+ diodes with the superlattice as the i layer were formed. The i layer was completely depleted at zero bias. To achieve tunneling of the electrons through the superlattice i layer, the p^+ layer was illuminated with visible light that was completely absorbed in the p^+ layer. A reverse bias was applied so that only electrons were moving (tunneling) through the i superlattice layer. Below 50 K two peaks and two NNDC regions were observed in the current–voltage characteristics. The difference between the bias voltages corresponding to these peaks was in excellent agreement with 35 $(\Phi_{d,2} - \Phi_{d,1})$, where 35 is the number of periods in the supperlattice, and $\Phi_{d,2}$ and $\Phi_{d,1}$ are the corresponding resonant voltages (3-58) and (3-57). Note that $e(\Phi_{d,2} - \Phi_{d,1}) = E_3 - E_1 \cong 143\,\mathrm{meV}$. Taking into account the period of the superlattice $d = 2 \times 139$ Å, we find that the electric field is close to 10^5 V/cm.

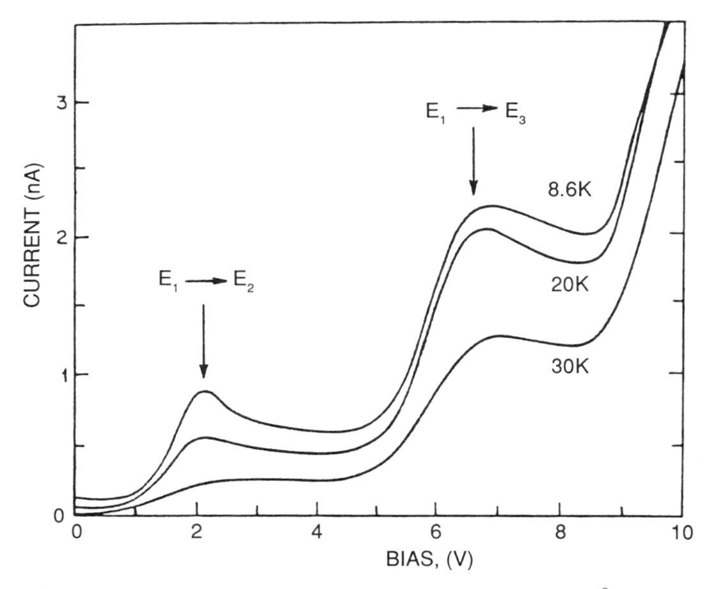

FIGURE 3-13. Current–voltage characteristics for a superlattice with 139-Å-thick wells and barriers and 35 periods. The arrows indicate that the peaks correspond to resonant tunneling between the ground state and the first two excited states as shown in Fig. 3-11. (After Capasso et al., 1986.)

This is why this d was optimal for the observation of NNDC. If d decreases, $\Phi_{d,i}$ increases as d^2 and electrical breakdown might occur before the first current peak. If d increases, the distance $E_i - E_1$ decreases and the effect can be observed only at low temperatures.

This dependence of current on applied voltage can be found in the papers of Kazarinov and Suris (1971), Tsu and Esaki (1973), Lee (1984), and Capasso et al. (1986). We will not reproduce them here because they underestimate the current between two peaks, where the main contribution is due to phonon- or photon- (Kazarinov and Suris, 1971) assisted tunneling and other excess current contributions, in analogy with the Esaki diode. For our considerations here it is important that the resonant tunneling results in NNDC. Because of the small dimensions of a superlattice structure, the operating frequency is higher than in Esaki diodes, persisting up to perhaps 500 GHz (Solner et al., 1987).

Superlattices with identical quantum wells were considered above. In a series of publications by Summers et al. (1987a,b, and references therein) resonant tunneling was studied in quantum well structures with the quantum well width progressively decreased through the superlattice, in order to create variably spaced superlattice energy filter structures, as shown in Fig. 3-14a. This is an energy filter structure because under appropriate bias the confined quantum levels in adjacent wells become degenerate, as shown in Fig. 3-14b. Electrons resonantly tunnel through the supperlattice, and a nearly mononergetic stream of high-energy electrons is produced. This is coherent resonant tunneling if more than two quantum wells are in the structure, in contrast to the case we treated above of sequential resonant tunneling, where tunneling to the vacant excited

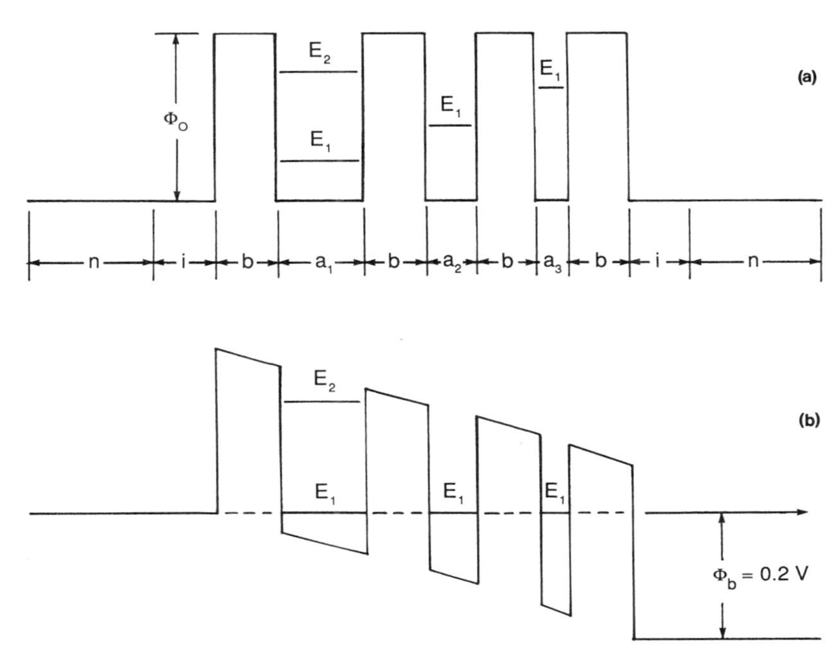

FIGURE 3-14. Conduction band profile of a three-well variably spaced superlattice energy filter structure designed for electron injection at 0.20 V, (a) for zero bias and (b) for $\Phi_b = 0.2$ V (after Summers et al., 1987b).

state is followed by relaxation to the ground state, etc. (compare Figs. 3-14 and 3-11). The advantage of the variably spaced superlattice is that the current–voltage characteristics show little temperature dependence and the ratio of peak-to-valley current is large.

The current–voltage characteristics for the three-well structure of Fig. 3-14 are shown in Fig. 3-15. The peak-to-valley ratio is 6:1 below 100 K. It is important that the device exhibits little temperature dependence in the region $10\,\text{K} < T < 100\,\text{K}$. It is also necessary to stress that fabrication of variably spaced superlattices is more complicated because they must be designed with a specific sequence of quantum well lengths in order to achieve resonance. For example, the current–voltage characteristics of Fig. 3-15 were obtained by Summers et al. (1987) in a structure using $Al_{0.35}Ga_{0.65}As$ as barriers of length 50.88 Å, and GaAs as quantum wells of lengths 67.84, 39.57, and 25.44 Å, respectively. A variation of one of these lengths will result in the shifting of the structure out of resonance, as shown in Fig. 3-14b. For the equal length quantum well structure (Fig. 3-11) the resonance between only two adjacent quantum wells is required.

To conclude this subsection we note that Esaki and Tsu (1970) (see also Lebwohl and Tsu, 1970) proposed NNDC in superlattice structures via the Bragg reflections experienced by an electron accelerated in an electric field. This NNDC mechanism was experimentally observed (see, e.g., Capasso et al., 1986), and in

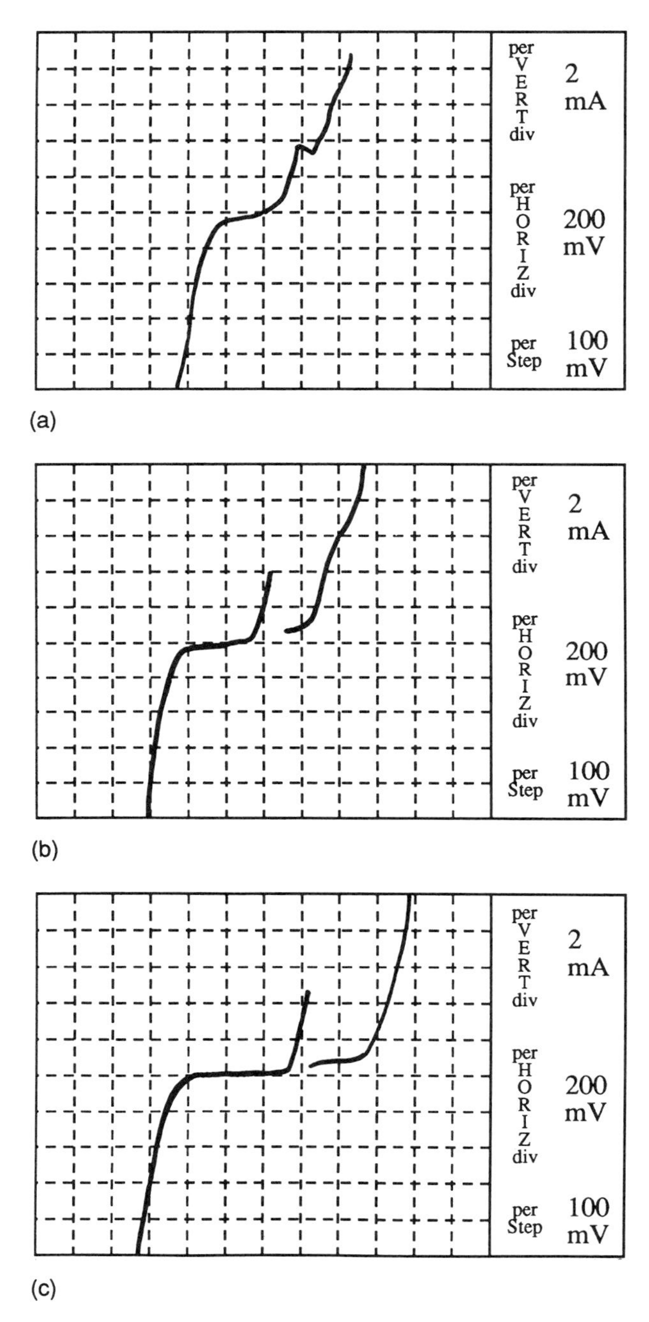

FIGURE 3-15. Current–voltage characteristics of a three-well 0.2-eV variably spaced superlattice energy filter structure at temperatures of (a) 200 K, (b) 100 K, (c) 10 K (after Summers et al., 1987b).

contrast to (3-57) and (3-58) it occurs in the essentially smaller electric field

$$\frac{\hbar}{\tau} < e\Phi_d \ll \Delta \tag{3-60}$$

where Δ is the width of the ground state.

Davies et al. (1985, 1986) (see also Kelly et al., 1986) proposed a tunnel diode with a moderately thick tunneling barrier between two superlattices. The difference between the case considered above and the one proposed is the fact that the subband structure presented in Fig. 3-10 must be replaced by minisubbands for the superlattice. As soon as the minisubband has finite length, tunneling through the barrier can only occur when there are available states on the other side of the barrier. Current–voltage characteristics with NNDC are possible, as shown in Fig. 3-16. We will not present here a detailed explanation of the origin of the NNDC because all of the details in Fig. 3-16 are understandable from our discussion of the analogous phenomena for tunneling between two quantum wells (compare Figs. 3-11 and 3-13). We want to stress that the above authors experimentally observed the proposed NNDC for structures with two ten period superlattices, with each period containing 3 nm of $Al_{0.25}Ga_{0.75}As$ and 6 nm of GaAs doped with silicon at $4 \times 10^{17}\,cm^{-3}$. Between these superlattices a $Al_{0.25}Ga_{0.75}As$ barrier was inserted. The width of the barrier was 6 nm and 8 nm in different experiments. Oscillations as high as 340 GHz were achieved in these diodes.

3.3.2. Resonant Tunneling through Double Barriers

In the previous subsection we discussed resonant tunneling between two adjacent quantum wells. These two wells could represent part of a superlattice. It is important to note that interfaces between the superlattice and the terminal electrodes are unavoidable. As a matter of fact, GaAs–$Ga_{1-x}Al_xAs$ superlattices grown on GaAs substrates are usually sandwiched between two GaAs regions to which low-resistance "ohmic" contacts are attached (Ando et al., 1982). This means that these regions must also be shown in Fig. 3-11. As a result, the multibarrier system with an emitter and collector must be treated self-consistently, as was done for the first time in 1973 by Tsu and Esaki. The simplest structure consists of a quantum well of length L, two barriers (each of length L_b and of height Φ_L in the symmetrical structure), and two n^+ preterminal regions with the Fermi level E_F above the conduction-band edge. This is a double barrier structure, and it is shown in Fig. 3-17 for three different applied bias voltages. For the sake of simplicity, only the ground level E_1 is shown in the quantum well. Tunneling from quantum well level E_1 to the collector is always possible because the collector has available states. Tunneling from the emitter to the quantum well level is analogous to the situation considered in the previous subsections. If the bias voltage is too small (as shown in Fig. 3-17a) or too large (Fig. 3-17c), no states are available for tunneling into the quantum well. Resonant tunneling can take place if level E_1 is in the energy interval between the conduction-band edge

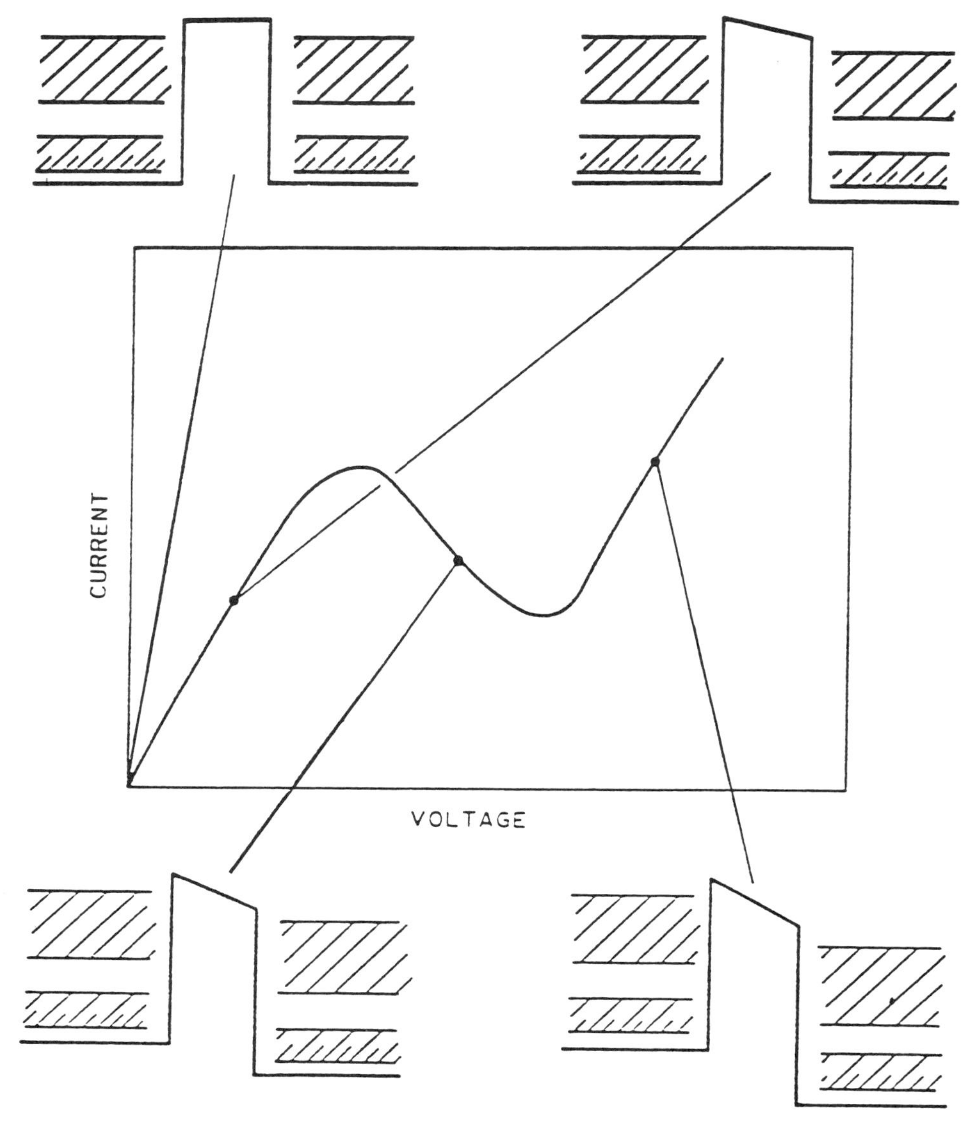

FIGURE 3-16. A sketch of the current–voltage characteristic expected for a tunnel barrier when superlattices filter the energy states on either side (after Kelly et al., 1986).

and the Fermi level, as shown in Fig. 3.17b,

$$E_c < E_1 < E_F \tag{3-61}$$

This means that instead of the resonant level of the previous subsection we now have the continuous spectrum on the left-hand side of Fig. 3-11 (i.e., all k_x are allowed up to k_F, where k_F is the Fermi momentum). Under the same

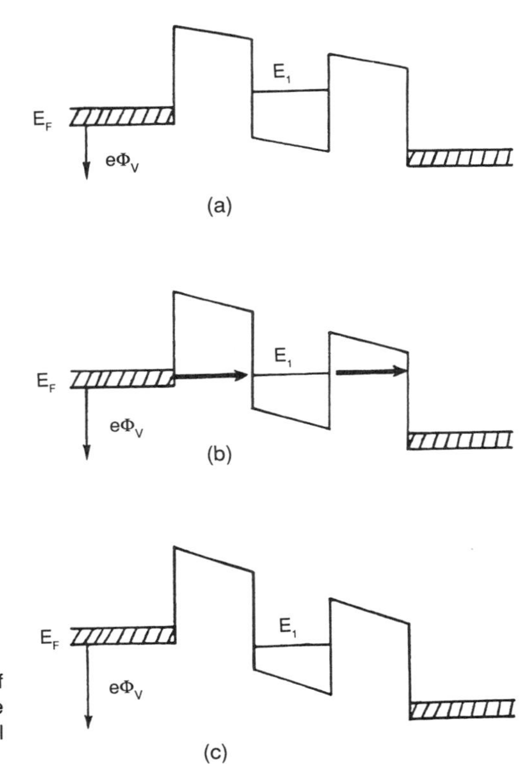

FIGURE 3-17. Schematic physical pictures of (a), (c) off-resonance and (b) on-resonance conditions in a double barrier quantum well diode (after Liu and Coon, 1987).

assumptions as before, the parallel component $k_{\parallel}$ of the momentum and energy $E_{\parallel}$ is conserved in tunneling. By comparing (3-31) and (3-51) we can determine the wave vector $k_x = k_0$ of the electrons in the emitter, which can tunnel to the quantum well (Luryi, 1985, 1987; Capasso et al., 1986), which for the simple parabolic dispersion law is

$$\hbar^2 k_0^2/(2m^*) = E_1 - E_c \tag{3-62}$$

That is, only those electrons whose wave vectors lie in a disk corresponding to $k_x = k_0$, shaded in Fig. 3-18, have isoenergetic $E_{\parallel}$ states in the quantum well. As the applied voltage increases, the number of electrons that can tunnel also increases, because the shaded disk on Fig. 3-18 moves downward to the equatorial plane of the Fermi sphere and the current increases to its peak value. For $k_0 = 0$ [E_1 coincides with the conduction-band edge (3-62)], the number of tunneling electrons per unit area equals $m^* E_F/\pi\hbar^2$. When E_1 moves below E_c, there are no electrons in the emitter that can tunnel into the quantum well at $T = 0\,\text{K}$. The current drops abruptly for this idealized model. In reality, the current decreases (NNDC occurs) up to the lowest valley current, and then increases again because tunneling to the first excited level in the quantum well occurs.

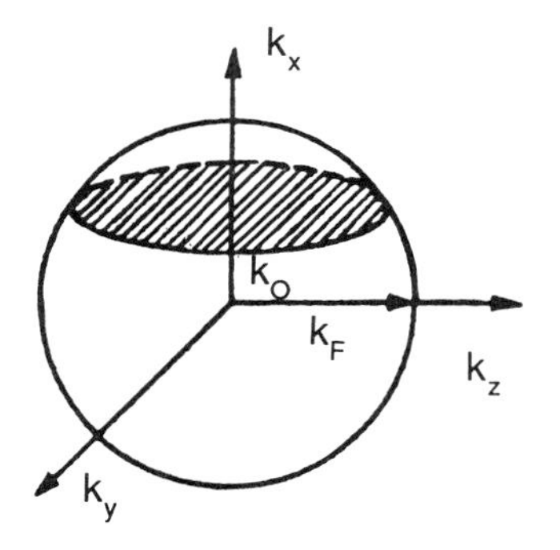

FIGURE 3-18. The Fermi surface for a degenerately doped emitter. Assuming conservation of the lateral momentum during tunneling, only those emitted electrons whose momenta lie on a disk $k_x = k_0$ (shaded disk) are resonant. The energy separation between E_0 and the bottom of the conduction band in the emitter is given by $\hbar^2 k_0^2/2m^*$. In an ideal diode at zero temperature, the resonant tunneling occurs in a voltage range during which the shaded disk moves down from the pole to the equatorial plane of the emitter Fermi sphere. At higher Φ_V (when $k_0^2 < 0$), resonant electrons no longer exist. (After Capasso et al., 1986.)

Wu et al. (1989) calculated the peak current, J_P, valley current, J_V, and peak-to-valley ratio (PTV), taking into account the finite temperature in the Fermi distribution. Figure 3-19 qualitatively explains the temperature dependence of J_P, J_V, and PTV shown in Fig. 3-20. Figure 3-19a shows the condition when the peak current occurs.

Wu et al. (1989) treated the case of temperature T_1 close to zero. When the temperature rises to T_2 and then to T_3, more and more electrons will move upward to higher energy states, causing a decrease in the amount of electrons having energies below E_F. As a result, when the temperature increases, the peak current decreases, as shown in Fig. 3-19, at the expense of the broadening of $J(\Phi)$ near the peak. The explanation of the temperature dependence of the valley current is given in Fig. 3-19b, where the major contribution to the valley current was taken via the excited state, which is out of the resonant condition. This is the reason that the valley current increases when the temperature increases.

The PTV decreases when the temperature increases (Fig. 3-20; see also Reed, 1986). The PTV ratio decreases even more if scattering of the electrons inside of the quantum well is taken into account (Goldmann et al. 1987a;

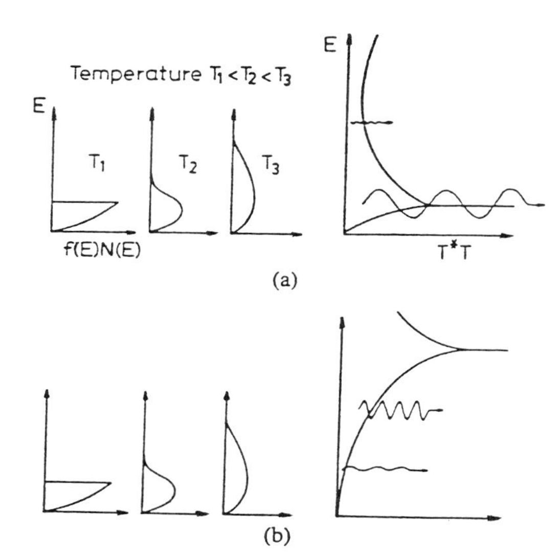

FIGURE 3-19. Energy distributions of electrons at different temperatures when (a) the peak current and (b) the valley current occur. Transmission coefficients shown on the right are not scaled. (After Wu, J. S., et al., 1989.)

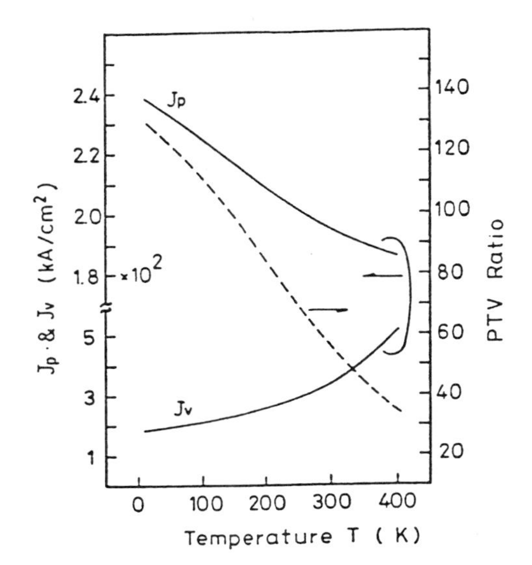

FIGURE 3-20. Calculated temperature dependences of peak and valley currents and peak-to-valley (PTV) ratio for a $GaAs/Ga_{1-x}Al_xAs$ structure with parameters $L = 50$ Å, $L_b = 10$ Å, $m^* = 0.067\ m_0$, $E_F = 0.0777$ eV, $\Phi_L = 1$ eV. (After Wu, J. S. et al., 1989.)

Wingreen et al., 1988; Jonson, 1989; Chevoir and Vinter, 1989; Frensley, 1989a; Eaves et al., 1989). It is also necessary to stress that for tunneling through the second level of the well the major part of the current, even in the peak region, is due to intersubband scattering inside the well (Chevoir and Vinter, 1989).

A simplified model of tunneling through a double barrier structure was presented above. The phenomenon is actually more complicated owing to formation of the accumulation and depletion layers in the emitter and the collector, respectively, and the formation of space charge in the quantum well in the biased device (Goldman et al., 1987b, 1988; Rousseau et al., 1988; Wolak, 1988; Brown, 1988, 1989; Frensley, 1989b). Hence, a more exact energy diagram under applied bias appears as shown in Fig. 3-21; it replaces Fig. 3-17. The voltage drops across the different regions of the structure shown in Fig. 3-21 must add up to the applied bias Φ_b:

$$e\Phi_b = E(\Phi_1 + \Phi_2 + \Phi_L) + \Delta_1 + \Delta_2 \tag{3-63}$$

In our previous considerations we had $\Delta_1 = \Delta_2 = 0$ and $\Phi_1 = \Phi_2$. If the areal concentration of electrons in the well is n_w, the areal space-charge density is

$$\Sigma = en_w. \tag{3-64}$$

This space-charge results in a difference in voltage drops

$$\Phi_2 = \Phi_1 + \Sigma L_b/\epsilon \tag{3-65}$$

in the collector, Φ_2, and emitter, Φ_1, barriers. The electron density in the quantum well can be estimated in the following way (Goldman et al., 1987b, 1988). If the transmission coefficient of the collector barrier is T_c, the lifetime of electrons in the well is $\tau \cong \hbar/(T_c E_1)$. The number flux of electrons passing through the well in the steady state is $-J/e$, and the areal concentration in the

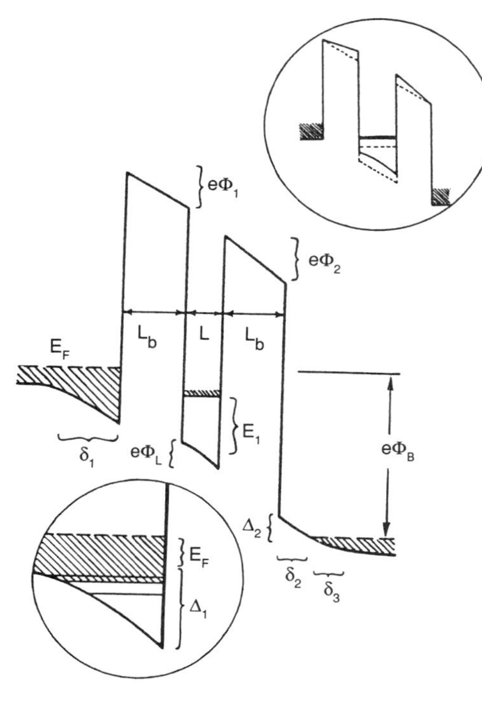

FIGURE 3-21. The conduction band energy diagram of a double barrier resonant tunneling structure under bias. Δ_1 and δ_1 are the parameters of the accumulation layer in the emitter (shown schematically in the lower inset); Δ_2, δ_2, and δ_3 describe the depletion layer in the collector. The electrostatic potential drops across the emitter barrier; the well and the collector barrier are, respectively, Φ_1, Φ_L, and Φ_2. The high- and low-J states of the structure are shown schematically in the top inset. (After Goldmann et al., 1988.)

well is

$$n_w = \hbar J/(eT_c E_1) \tag{3-66}$$

The voltage drop in the quantum well is related to Φ_1 and Φ_2 by the equations

$$\frac{\Phi_L}{L} = \frac{\Phi_1}{L_b} + \frac{\Sigma}{2\epsilon} = \frac{\Phi_2}{L_b} - \frac{\Sigma}{2\epsilon} \tag{3-67}$$

which follows from the boundary conditions for the electric field on the interfaces between the barriers and quantum wells. The boundary conditions on the interfaces between the barrier and the emitter (the barrier and the collector) together with an approximate solution of the Poisson equation [see Eqs. (2-14) and (3-2)] yield

$$\frac{\Delta_1}{\delta_1} \cong \frac{e\Phi_1}{L_b} \tag{3-68}$$

$$\sigma_A \cong \frac{\epsilon\Phi_1}{L_b} \cong eN(E_F)\Delta_1\delta_1 \tag{3-69}$$

$$\frac{\delta_2}{e} = \frac{\Phi_2\delta_1}{L_b} - \frac{eN_D\delta_2^2}{2\epsilon} \tag{3-70}$$

$$\delta_3 \cong \frac{\delta_2 E_F}{\Delta_2} \tag{3-71}$$

where σ_A is the areal charge density in the accumulation layer of the emitter and $N(E_F)$ is the density of states at the Fermi energy. If the neutrality of the structure,

$$\sigma_A + \Sigma = eN_D(\delta_2 + \delta_3/2) \tag{3-72}$$

and the expression for T_c in the quasiclassical approximation (Goldman and Krivchenkov, 1961),

$$T_C \cong 16\frac{E_1(U - E_1)}{U^2}\exp\left\{-\frac{4L_b(2m^*)^{1/2}}{3\hbar e\Phi_2}[(U - E_1 - e\Phi_L/2)^{3/2} - (U - E_1 - e\Phi_2 - e\Phi_L/2)^{3/2}]\right\} \tag{3-73}$$

are taken into account, Eqs. (3-63)–(3-73) establish the complete set of equations for the potential distribution in the structure for given applied bias Φ_b and current J. Here U is the conduction band discontinuity between the well and the barrier, so that $U - E_1$ is the barrier height for electrons tunneling out of the well at zero bias. Goldman et al. (1987b, 1988) used these equations, together with experimentally measured current–voltage characteristics, to verify that the set of equations is self-consistent.

The transmission coefficient T_c depends exponentially on Φ_2 and Φ_L (3-73), and both of them depend on T_c through the areal concentration n_w of electrons in the quantum well (3-66) via the areal charge density (3-64), (3-65), (3-67). As a result, for the applied voltage, Eqs. (3-63)–(3-73) give solutions with two different stable concentrations n_w in the quantum well, corresponding to two different stable distributions of voltage drops in the structure, as shown schematically in the upper insert of Fig. 3-21, and two different currents. The current bistability was experimentally observed (Goldman et al., 1987b,c; 1988) and is shown in Fig. 3-22. Calculations were also done by Sheard and Toombs (1988).

The measurement shown in Fig. 3-22 was performed on a structure grown by molecular-beam epitaxy on an $n^+\langle 100\rangle$GaAs substrate. It had a 56-Å GaAs well sandwiched between two 85 Å-thick $Al_{0.40}Ga_{0.60}As$ barriers. The GaAs emitter and collector regions had net concentrations $\cong 2 \times 10^{17}$ cm^{-3}. If the capacitor C is connected parallel to the structure, as show in the lower part of Fig. 3-22, the constant Φ_b is maintained to avoid circuit oscillations (Solner, 1987).

In addition to Eqs. (3-63)–(3-73), Goldman et al. (1987c) added an equation for the tunneling current through the first barrier

$$J = \frac{em^*}{2\pi^2\hbar^3}T_e\Delta E(E_F + \Delta_1 - \Delta E) \tag{3-74}$$

where T_e is the transmission coefficient of the emitter barrier, and

$$\Delta E \cong \Delta_1 + e(\Phi_1 + \Phi_L/2) - E_1 \tag{3-75}$$

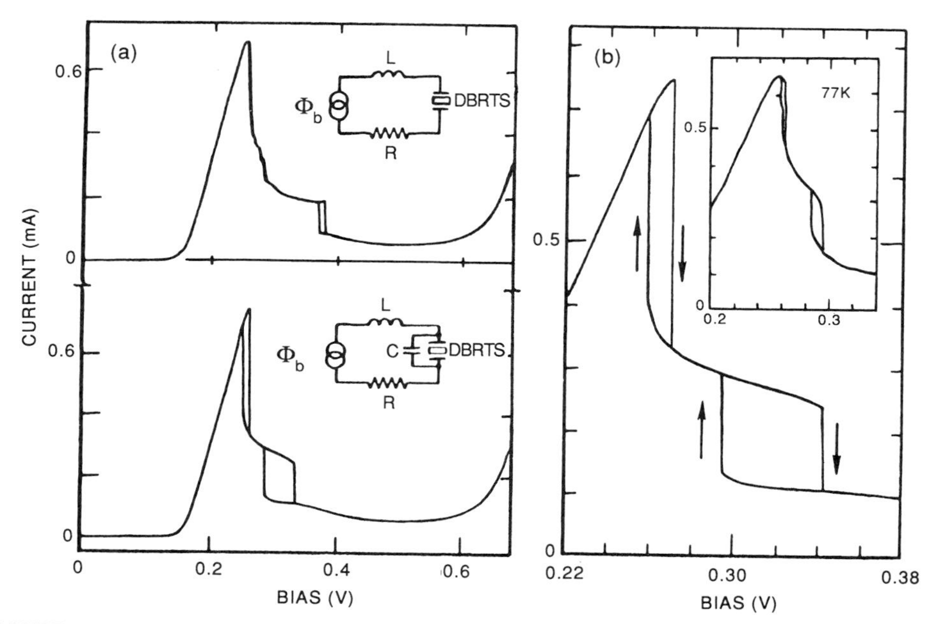

FIGURE 3-22. (a) the current–voltage characteristics of the double barrier resonant tunneling structure (DBRTS) (area $4.5 \times 10^{-6}\,\mathrm{cm}^2$) at 4.21 K measured with and without a 10-nF capacitor in parallel with the device. Equivalent circuits consist of a voltage source Φ_b, series resistance R (1.6 Ω), and inductance L; the upper circuit oscillates in the bias range $0.25\,\mathrm{V} < \Phi < 0.37\,\mathrm{V}$. I is measured on a 10-Ω resistor (included in R), and Φ_b is measured with use of a pseudo-four-terminal technique. (b) Same as the lower trace in (a); clearly seen are the two bistable regions (arrows indicate the direction of the voltage sweep, $\sim 0.1\,\mathrm{mV\,s^{-1}}$). Inset: The $I - \Phi$ curve of the same device at 77 K. (After Goldman et al., 1987c.)

is the energy separation between the emitter and the well, and

$$0 < \Delta E \leq E_F + \Delta_1 \tag{3-76}$$

The inequality (3-76) is the condition for the direct inelastic tunneling from emitter to well. Solutions of Eqs. (3-63)–(3-76) were found by Goldman et al. (1987c); we reproduce their results for the electron concentration in the well, Fig. 3-23. Note that at the current peak the concentration in the well is close to $10^{11}\,\mathrm{cm}^{-2}$. Electron–electron scattering becomes important for such concentrations at 4 K, and the characteristic time of this scattering is on the order of 10^{-13} s. (Yang et al., 1987; Goldman et al., 1987b). The charge accumulation in the quantum well was studied by Vodjdani et al. (1989).

To emphasize the importance of the electron accumulation in the quantum well on the bistability, Zaslavsky et al. (1988) designed an asymmetric double barrier structure with one barrier significantly higher than the other (Goldman et al., 1988). The 56-Å undoped GaAs quantum well was sandwiched between two 90-Å-thick undoped barriers of different heights. The low barrier was created using $Al_{0.42}Ga_{0.58}As$ and the higher barrier using $Al_{0.58}Ga_{0.42}As$. Under forward bias, the higher collector barrier of the device enhances charge storage up to

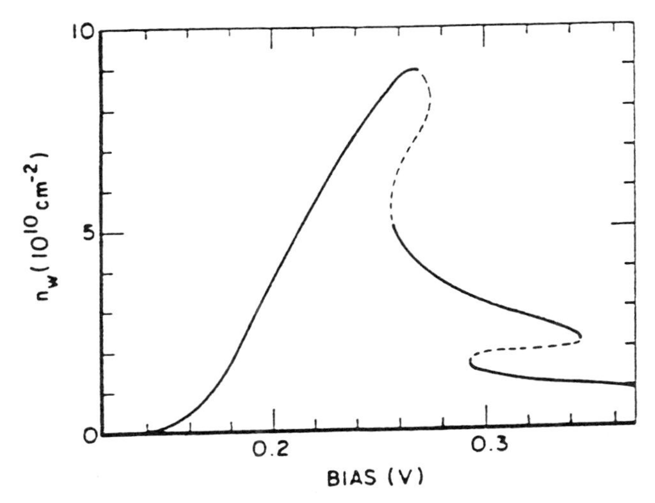

FIGURE 3-23. The steady state electron concentration in the well, n_w, vs. Φ_b, calculated with use of the current–voltage curve of Fig. 3-21b (after Goldman et al., 1-987c).

$4.5 \times 10^{11}\,cm^{-3}$ (compare with Fig. 3-23) at the voltage Φ_P corresponding to the current peak. The space charge, Σ, in the well becomes appreciable: this is the reason that the bistability is observed and only a fraction of the applied bias contributes to the increase of ΔE (3-75). Hence, Φ_P shifts out to 565 mV (Fig. 3-24).

Under reverse bias the tunneling current is small and the collector barrier is low. Here the space charge is negligible. No screening of the external electric field and a much lower value of the bias $\Phi_P = 265\,mV$ at the current peak (Fig. 3-24) were observed. There is no bistability under reverse bias.

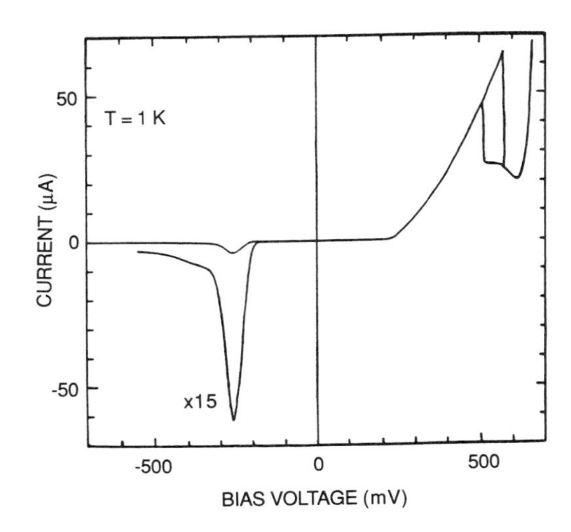

FIGURE 3-24. The current–voltage characteristic of the asymmetric double barrier structure described in the text (after Zaslavsky et al., 1988).

The experimental results presented in Figs. 3-22–2-24 were analyzed from the point of view of sequential tunneling in the double barrier structure. Recently, additional experimental evidence has been uncovered in favor of this model (see, e.g., Morkoc et al., 1986; Eaves et al., 1988; Heiblum et al., 1989; Frensley, 1989b; Goldmann et al., 1987a, b, c; 1988), as opposed to the coherent tunneling view. This is the reason that in this chapter we only discussed sequential resonant tunneling. The theory of coherent (Fabry–Perot-type) resonant tunneling can be found in the papers of Tsu and Esaki (1973), Capasso et al. (1986), and Ricco and Azbel (1984). We also note that Weil and Vinter (1987) proved that both interpretations of tunneling lead to the same predictions for the dc current–voltage characteristics. As far as these characteristics are concerned, double barrier structures produce very large PTV. For example, the PTV is 16:1 for forward bias and 20:1 for reverse bias, for the dependence shown in Fig. 3-24. A peak-to-valley ratio exceeding 10:1 is now readily obtained at room temperature by using, for example, InGaAs/AlAs pseudomorphic structures (Inata et al. 1987; see also Huang et al., 1987, and references therein).

In the NNDC region of the current–voltage characteristics it is possible to see a well-pronounced "plateaulike" structure (Figs. 3-22, 3-24), which was observed in the above work and also by Muto et al. (1989); Young et al. (1988); Huang et al. (1987), Eaves et al. (1989) (they also observed the bistability), and others. This plateaulike structure and charge accumulation in the quantum well make the circuit response analysis more difficult (Young et al., 1988), but as we have seen in Chap. 2, the detailed shape of the oscillation is relatively insensitive to the shape of the region of negative slope (2-174)–(2-183).

3.3.3. Multibarrier Structures

Progress in molecular beam epitaxy, and the resulting high-frequency response of double barrier structures (e.g., Solner et al., 1983, 1987), has led to detection and mixing at frequencies as high as 2.5 THz. Brown (1989) has reported the highest oscillation frequency, 420 GHz. Quantum transport calculations suggest that NNDC persists up to 5 THz (Frensley, 1987), which has initiated great activity in this and related areas. For example, $Si–Si_{1-x}Ge_x$ n-type

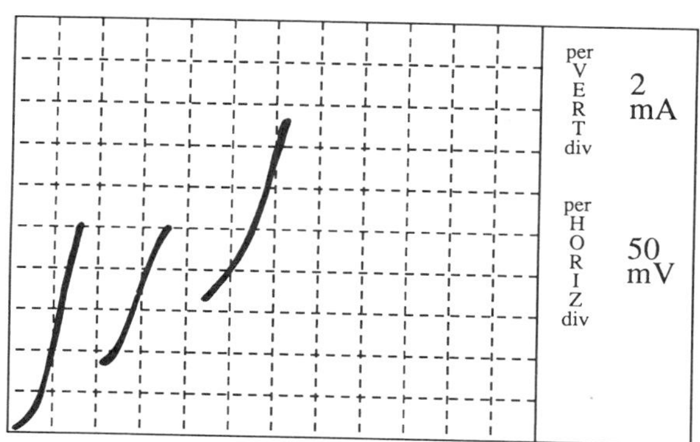

FIGURE 3-25. Typical current–voltage characteristics of a triple-well resonant tunnel diode at 220 K (after Mizuta et al., 1988).

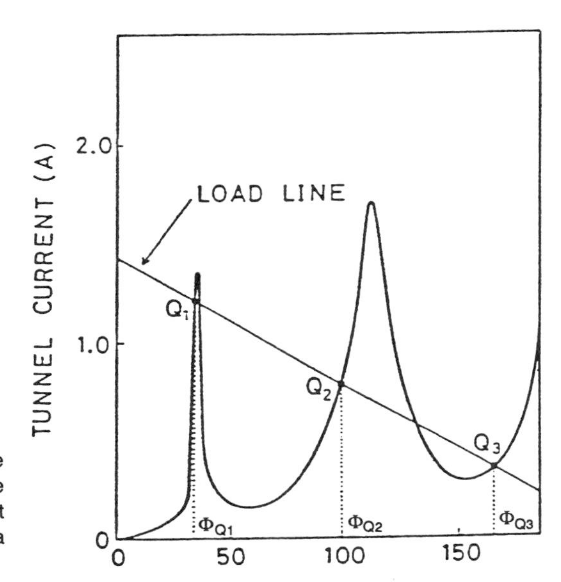

FIGURE 3-26. Schematic current–voltage characteristics and the load line. Three stable points Q_1, Q_2, and Q_3 are located at the voltages Φ_{Q1}, Φ_{Q2}, Φ_{Q3}. (After Mizuta et al., 1988.)

resonant tunnel structures as electron filters were studied by Rajakarunanayake and McGill (1989). Resonant tunneling with 16 resonances in the current–voltage characteristics for double barrier structures with compositionally graded parabolic quantum wells were reported by Sen et al. (1987a).

If the number of quantum wells in the structure increases, new opportunities for the variation of the current–voltage characteristics appear. The PTV increases if, instead of one quantum well, two are introduced into the structure (Sawaki et al., 1988, 1989; Collins, 1989). Double NNDC can easily be realized, with two peak voltages Φ_P independently controllable with well width, in the triple-well resonant tunnel diode (Mizuta et al. 1988; Sen et al., 1987b; Capasso et al., 1987). Both peak currents are nearly equal, as shown in Fig. 3-25. In contrast to the cases we treated in Chap. 2 (see Figs. 2-1 and 2-4), three stable points for a given load line are now possible, as shown in Fig. 3-26. Applications as a three-state memory cell (Capasso et al., 1987) become possible.

Other novel applications of resonant tunneling structures have recently been discussed by, among others, Leng (1987); Van Hove et al. (1989); Heiblum et al. (1987); and Woodward et al. (1987, 1988).

3.4. SUMMARY

In this chapter we first reviewed conventional *p-n* junction theory, and then the degenerate *p-n* system that produced the NNDC tunnel diode device. Direct and indirect tunneling mechanisms were discussed, with the latter involving phonon-assisted processes. The current–voltage characteristics were then developed. Resonant tunneling between quantum wells was discussed, with emphasis on double-well and multiple-well (superlattice) structures. In the next chapter we again focus on the *p-n* diode system, this time under reverse bias breakdown conditions.

4

The Avalanche Diode

4.1. INTRODUCTION

In the preceding chapter we analyzed the behavior of a *p-n* junction (diode) under forward bias. Let us now reconsider Fig. 3-5b under reverse bias conditions ($\Phi_V < 0$). Here, the tunneling current increases when $|\Phi_V|$ increases because the electric field $F_{\max}$ increases in proportion to $(\Phi_{\text{bi}} - \Phi_V)^{1/2}$ [Eqs. (3-18) and (3-19)], resulting in an enhanced tunneling probability [Eqs. (3-30) and (3-32)]. However, this is only one consequence of the increased $F_{\max}$. In addition, after tunneling to the *n*-side of the barrier, the electron finds itself in a high electric field region (see Fig. 3-2). When a sufficiently large reverse bias [$e(\Phi_{\text{bi}} - \Phi_V) > E_G$] is applied to the diode, the energy received by the electron from the electric field becomes so large that it can reach an energy higher than E_G. Now the electron can cause impact ionization and create an electron–hole pair. The primary and secondary electrons are accelerated by the field, but they are now leaving the high-field region (Figs. 3-5b and 3-2c). However, the secondary hole moves further into the high-field region, where it can reach sufficiently high energy to produce another electron–hole pair. An avalanche "breakdown" develops, which results in junction breakdown. The current then increases abruptly at the avalanche voltage Φ_B, as shown in Fig. 4-1. Under certain conditions it is sometimes possible for the avalanche process to be self-supporting at a lower bias $|\Phi_V|$ than the breakdown voltage Φ_B. Here the dc $I(\Phi)$ characteristics of a reverse-biased diode can exhibit SNDC, as shown in Figs. 1-37 and 2-35. In Sec. 4.3 we develop an understanding of the avalanche process because it is vital for the class of solid state microwave oscillators and amplifiers discussed in this chapter—IMPATT (impact avalanche transit time) diodes (see, e.g., Bauhahn and Haddad, 1977). However, in the next section, Sec. 4.2, we first obtain the electric field and potential distribution for different structures, because preferential *p-n* junctions for IMPATT diodes are not the same as those discussed in Chap. 3.

4.2. POTENTIAL AND ELECTRIC FIELD DISTRIBUTION IN A READ-TYPE AVALANCHE DIODE

We have already seen that the energy gained by an injected electron is the most important parameter in impact ionization. To increase the energy optimally, it is necessary to have a *p-n* junction where, in contrast to the situation displayed

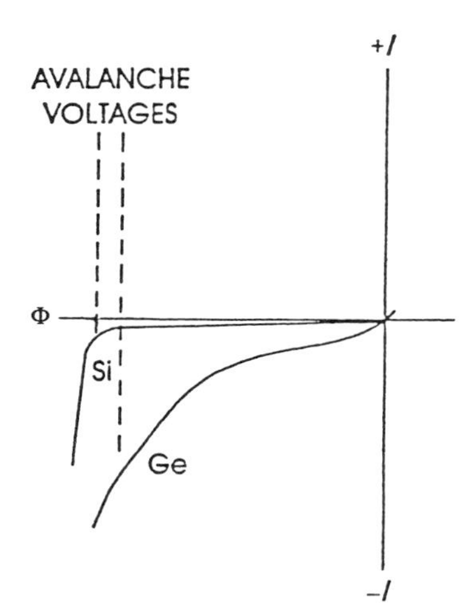

FIGURE 4-1. Current–voltage characteristic of a *p-n* junction under reverse bias.

in Fig. 3-2, the entire voltage drop, $\Phi_0 = \Phi_{bi} - \Phi_V$, in the structure occurs on only one side of the *p-n* junction. From Eqs. (3-11)–(3-16) we can see that if

$$N_A \gg N_D \tag{4-1}$$

the entire depletion region is on the lower doped *n* side of the junction [see Eq. (3-18)]

$$w \cong l_n = [2\epsilon\Phi_0/(eN_D)]^{1/2}, \qquad l_p \ll l_n \tag{4-2}$$

and the potential drop occurs primarily across the *n* side:

$$\Phi(x \leq 0) \cong -\Phi_0 \equiv -(\Phi_{bi} - \Phi_V) \tag{4-3}$$

$$\Phi(0 \leq x \leq w) = -N_D(w - x)^2 e/(2\epsilon) \tag{4-4}$$

$$\Phi(x \geq w) = 0 \tag{4-5}$$

where w is the width of the depletion region.

For the electric field in the region $x > 0$ and for its maximum at $x = 0$, Eqs. (3-6) and (3-19) are valid. The dependence on x of the fixed charge and of the electric field on the *n* side of the junction are the same, as shown in Fig. 3-2 for $x > 0$. The only difference between this and the previous case is the fact that l_p approaches 0 because of condition (4-1).

A p^+-n junction was considered above; for a n^+-p junction the reverse inequality in Eq. (4-1) is valid. Here l_n approaches 0 and almost the entire voltage, $\Phi_0 = \Phi_{bi} - \Phi_V$, drops across the p side of the junction. Equations (3-5)–(3-19) can be used, with $N_D \gg N_A$. If we assume that the n^+ side is at

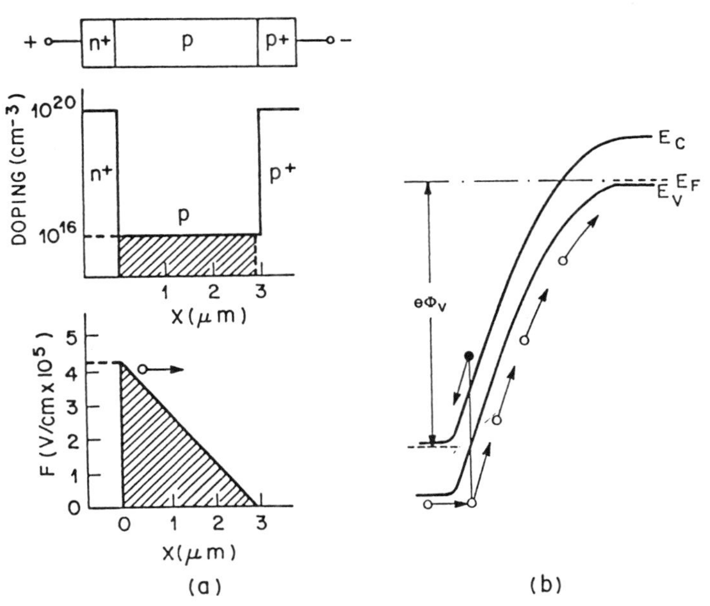

FIGURE 4-2. (a) Doping profile and electric-field distribution for a one-sided abrupt n^+-p diode at avalanche breakdown. (b) Energy band diagram of the diode. (After Sze, 1985.)

$x > 0$, as shown in Fig. 4-2, instead of (4-2)–(4-5) we have for the n^+-p junction

$$w \cong l_p = [2\epsilon\Phi_0/(eN_A)]^{1/2} \tag{4-6}$$

$$\varphi(x \leq 0) \cong \Phi_0 = (\Phi_{bi} - \Phi_V) \tag{4-7}$$

$$\Phi(0 \leq x \leq w) = N_A(w - x)^2 e/(2\epsilon) \tag{4-8}$$

$$\Phi(x \geq w) = 0 \tag{4-9}$$

$$F_{max} = F(x = 0) = 2\Phi_0/w = (2eN_A\Phi_0/\epsilon)^{1/2} = N_A we/\epsilon \tag{4-10}$$

$$F(0 \leq x \leq w) = F_{max} - N_A xe/\epsilon \tag{4-11}$$

where we again take the potential equal to zero in the region we are interested in, i.e., in this case the p region. From Eqs. (4-6) and (4-10), it follows that the width of the depletion region decreases and the maximum electric field increases when the doping N_A increases.

The attachment of metallic leads to heavily doped regions of a semiconductor usually produces low resistance metal–semiconductor contacts (Shaw, 1981; Rhoderick and Williams, 1988). This is one of the major reasons that a lightly doped region of a junction is always followed by a heavily doped region. Hence, the simplest n^+-p structure for an IMPATT diode is shown in Fig. 4-2, along with its electric field distribution and the corresponding energy band diagram. (A

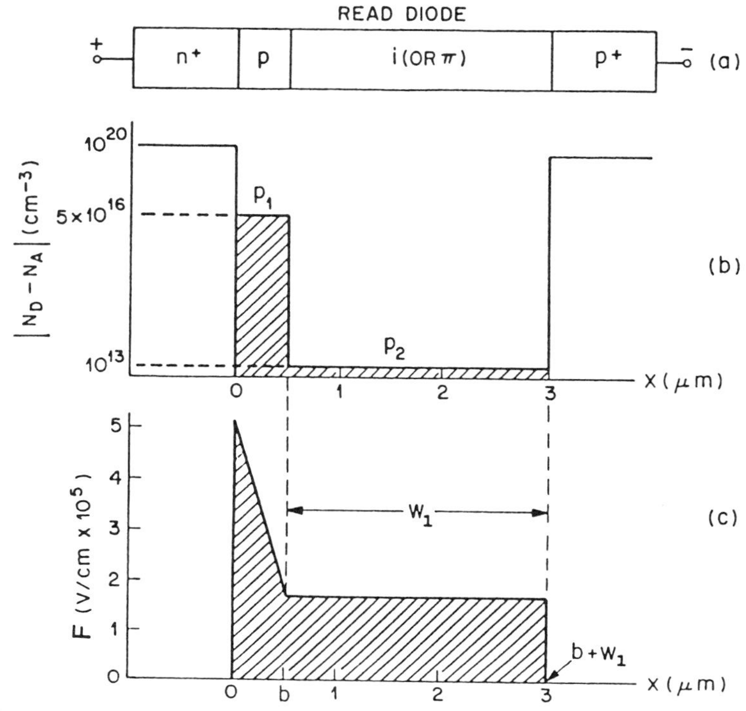

FIGURE 4-3. (a) Read diode; (b) doping profile; (c) electric field distribution (after Sze, 1985).

negative sign is shown on the p^+ side to stress that the diode is under reverse bias.) Microwave oscillations of the IMPATT type have been observed and exploited in the p-n diode shown in Fig. 4-2 (see, e.g., DeLoach, 1976; Johnston et al., 1965; Val'd-Perlov et al., 1966). However, Read (1958) originally proposed a structure of the type n^+-p-i-p^+, or n^+-p-π-p^+, as shown in Fig. 4-3, for the generation of microwaves (see also Sec. 4.4). The Read diode consists of two regions of essentially different doping : region $p_1(0 < x \leq b)$ of relatively high doping and, as we shall see below, high field, and a region $p_2(b \leq x \leq b + w_1 = d)$ of doping close to intrinsic, having a low electric field. To determine the potential (and the field) distribution in a Read diode, it is necessary to solve Poisson's Eq. (3-2), taking into account that doping N_{A1} is in the region $0 \leq x \leq b$, doping N_{A2} is in $b \leq x \leq b + w_1 = d$, and the n^+ region is highly doped so that $N_D \gg N_{A1}, N_{A2}$; the entire voltage drops across the p side of the junction. The solution yields

$$F_{max} = F(x = 0) = [N_{A1}b + N_{A2}(w - b)]e/\epsilon \tag{4-12}$$

$$F(0 \leq x \leq b) = F_{max} - N_{A1}xe/\epsilon \tag{4-13}$$

$$F(b \leq x \leq w) = F_{max} - [N_{A1}b + N_{A2}(x - b)]e/\epsilon \tag{4-14}$$

where w is the width of the depletion region:

$$w = [2\epsilon\Phi_0/(eN_{A2}) - b^2(N_{A1}/N_{A2} - 1)]^{1/2} \tag{4-15}$$

The potential distribution in the region $0 \le x \le w$ is easy to obtain by integrating (4-12)–(4-14):

$$\Phi(x \le 0) = \Phi_0 \tag{4-16}$$

$$\Phi(0 \le x \le b) = [N_{A1}(b - x)^2 + N_{A2}(w - b)(w + b - 2x)]e/(2\epsilon) \tag{4-17}$$

$$\Phi(b \le x \le w) = N_{A2}(w - x)^2 e/(2\epsilon) \tag{4-18}$$

$$\Phi(x \ge w) = 0 \tag{4-19}$$

For $N_{A1} \gg N_{A2}$ the electric field in the heavily doped region (4-13) decreases faster than in the lightly doped region. Equations (4-12) to (4-14) are written for the case

$$b \le w < b + w_1 = d \tag{4-20}$$

i.e., when the width of the depletion region is larger than b but smaller than the total length of the p region, d. This is not a "punched-through" diode. For the case where w reaches d, the diode becomes punched-through because the entire p region is depleted of charge. This condition is shown in Fig. 4-3.

Now let us rewrite the inequalities (4-20) for the potential rather than the length. Using Eq. (4-15) we have

$$\Phi_{p1} \le \Phi_0 \le \Phi_{p2} \tag{4-21}$$

where

$$\Phi_{p1} = b^2 N_{A1} e/(2\epsilon) \tag{4-22}$$

and

$$\Phi_{p2} = [d^2 N_{A2} + b^2(N_{A1} - N_{A2})]e/(2\epsilon) \tag{4-23}$$

We have already mentioned that if Φ_0 became larger than the right-hand part of Eq. (4-21), then the diode is punched-through. That is why Φ_{p2} is called the punch-through "spiking" voltage of the diode. If

$$\Phi_0 < \Phi_{p1} \tag{4-24}$$

the size of the depletion region, w, is less than b, so Eqs. (4-6) to (4-11) must be used instead of Eqs. (4-12)–(4-19). This means that for small voltages [Eq. (4-24)] the depletion layer lies only in the heavily doped region. The width w increases when Φ_0 increases, and $w = b$ when Φ_0 reaches Φ_{p1}. Hence, Φ_{p1} is the spiking voltage of the heavily doped region. For $\Phi_0 > \Phi_{p1}$ this region is punched-through and depletion of the lightly doped region becomes important.

The high-field region plays a major part in the avalanche process, and inhomogeneities in this field [Fig. 4-3c, Eq. (4-13)] introduce essential difficulties in governing the avalanche process. This is why the modified Read structure,

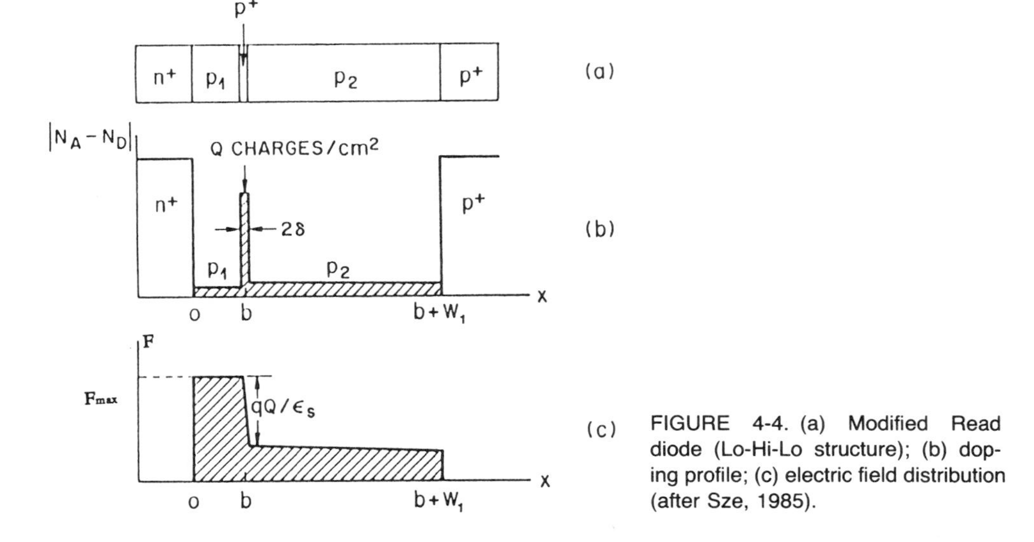

FIGURE 4-4. (a) Modified Read diode (Lo-Hi-Lo structure); (b) doping profile; (c) electric field distribution (after Sze, 1985).

shown in Fig. 4-4, generally has advantages in comparison to the two previous structures. This is a "Lo-Hi-Lo" structure, because the p region in the diode is a lightly doped region of thickness b, having a concentration N_{A1} close to the intrinsic level. The heavily doped region of thickness 2δ has a high concentration N_{A2}. The lightly doped region of thickness w_1 has concentration N_{A3}, which again is close to intrinisc. It is always the case that

$$2\delta \ll w_1 \tag{4-25}$$

so for the total device length we can take, as before, $d = b + w_1$.

The solution of Poisson's equation can be readily corrected as above. For small voltages (4-24) the solutions are given by Eqs. (4-6)–(4-11). For intermediate voltages, when only the first Lo region is punched-through, the solutions are given by Eqs. (4-18) and (4-19). (Note that now $N_{A2} \gg N_{A1}$). For a higher voltage, when the Hi region is punched-through, and the depletion layer spreads through the second Lo region,

$$b + 2\delta < w < d = b + w_1 \tag{4-26}$$

the results are (Ivastchenko et al., 1984)

$$F_{\max} = F(x = 0) = [N_{A1}b + N_{A2}2\delta + N_{A3}(w - b - 2\delta)]e/\epsilon \tag{4-27}$$

$$F(b + 2\delta \leq x \leq w) = F_{\max} - [N_{A1}b + N_{A2}2\delta + N_{A3}(x - b - 2d)]e/\epsilon \tag{4-28}$$

The electric field for $0 \leq x \leq b$ is defined by Eq. (4-13) and for $b \leq x \leq b + 2\delta$ by Eq. (4-14). The width of the depletion region is

$$w = [2\epsilon\Phi_0/(eN_{A3}) - b^2(N_{A1}/N_{A3} - N_{A2}/N_{A3}) - (2\delta + b)^2(N_{A2}/N_{A3} - 1)]^{1/2} \tag{4-29}$$

If we put (4-29) into (4-26) the voltage interval is defined by

$$\Phi_{p2} \le \Phi_0 \le \Phi_{p3} \tag{4-30}$$

when the depletion layer spreads through the second Lo region. Here

$$\Phi_{p3} = [d^2 N_{A3} + (b + 2\delta)^2 (N_{A2} - N_{A3}) + b^2 (N_{A1} - N_{A2})] e/(2\epsilon) \tag{4-31}$$

is the spiking voltage of the diode, and Φ_{p2} is the spiking voltage of the Hi region, which can be taken from Eq. (4-23) if we substitute $(b + 2\delta)$ for d, i.e., by the actual size of the first two regions.

If the voltages increases so that

$$\Phi_0 > \Phi_{p3} \tag{4-32}$$

the diode is punched-through and the electric field distribution is described by Eq. (4-28) over the entire second Lo region, and by Eq. (4-13) over the first Lo region. The only difference is that F_{max} is defined by the expression

$$F_{max} = F(x = 0) = \{\Phi_0 + [w_1^2 N_{A3} + 4\delta(w_1 + \delta) N_{A2} + b(d + w_1 + 2\delta) N_{A1}] e/\epsilon\}/d \tag{4-33}$$

At the critical value, when $\Phi_0 = \Phi_{p3}$, Eq. (4-33) yields the same value for F_{max} as Eq. (4-27). The essential difference between these two cases is in the fact that the electric field differs from zero everywhere, including the boundary $x = d$, where Eq. (4-28) gives

$$F_{min} = F(x = d) = F_{max} - (N_{A1} b + N_{A2} 2\delta + N_{A3} w_1) 2e/\epsilon = F_{max} - N_s 2e/\epsilon \tag{4-34}$$

where

$$N_s = N_{A1} b + N_{A2} 2\delta + N_{A3} w_1 \tag{4-35}$$

is the total concentration of impurities in the diode per unit area.

The doping and the sizes of the aforementioned three regions are ordinarily chosen such that the inequality (4-32) is true for characteristic operating voltages Φ_V of the IMPATT diode. It is also often true even for $\Phi_V = 0$, because the built-in potential Φ_{bi} is sufficient (4-32) to punch through the structure. As a matter of fact, the doping satisfies not only the aforementioned condition and $N_{A2} \gg N_{A1}$, but also

$$N_{A2} 2\delta \gg N_{A1} b, \quad N_{A3} w_1 \tag{4-36}$$

i.e., the total concentration of acceptors in the Hi region is essentially larger than that in both Lo regions. Under these conditions the second terms on the right-hand side of Eqs. (4-13) and (4-14) are small in comparison with the first terms. That is, the electric field is practically constant, $F(x) = F_{max}$, in the first Lo region (see Fig. 4-4); at $x = b$ it reduces very abruptly (4-14) to the value $F(x) \approx F_{min}$ and remains almost constant throughout the second Lo region. In this case (4-36), Eq. (4-33) simplifies to

$$F_{max} = [\Phi_0 + N_s w_1 e2/\epsilon]/d \tag{4-33'}$$

or

$$\Phi_0 = F_{max} b + F_{min} w_1 \tag{4-37}$$

Equation (4-37) implies that we have a low field region (we will show in Sec. 4.4 that this is the "transit" region) and a high-field region with a uniform field. We first treat the high-field region, where impact ionization occurs.

4.3. THE AVALANCHE PROCESS

The avalanche process is initated by impact ionization, wherein a free carrier (electron or hole) gains sufficient energy from the electric field so that it can create an additional electron–hole pair by exciting an electron from the valence band to the conduction band. The main parameter characterizing impact ionization is the threshold energy E_{th}.

4.3.1. Threshold Energy for Impact Ionization

Referring to the primary electron as the one that provides the energy capable of producing the secondary electron and secondary hole, let us consider a simple, direct gap semiconductor with isotropic parabolic bands (see Chap. 3)

$$E_{e,h}(k) = \hbar^2 k^2/(2m_{e,h}) \tag{4-38}$$

where e and h denote electrons and holes, respectively. As long as the Golden Rule can be applied, we expect both the energy and momentum to be conserved. In the initial state, i, we have a primary electron with vector momentum $\mathbf{k}_i$; and energy $E_e(\mathbf{k}_i)$, and in the final state, f, after ionization, we have a primary electron $(f, \mathbf{k}_f, E_e(\mathbf{k}_f))$, a secondary electron $(f', 4\mathbf{k}_{f'}, E_e(\mathbf{k}_{f'}))$, and secondary hole $(i', \mathbf{k}_{i'}, E_h(\mathbf{k}_{i'})$:

$$E_e(\mathbf{k}_i) = E_h(\mathbf{k}_{i'}) + E_e(\mathbf{k}_f) + E_e(\mathbf{k}_{f'}) + E_G \tag{4-39}$$

$$\mathbf{k}_i = \mathbf{k}_{i'} + k_f + \mathbf{k}_{f'} \tag{4-40}$$

To find the threshold energy, E_{th}, it is necessary to minimize Eq. (4-39) taking into account Eqs. (4-38) and (4-40). The result is (for details see Anderson and Crowell, 1972; Hauser et al., 1979; Ivastchenko and Mitin, 1990):

$$E_{th} = E_G[1 + m_e/(m_e + m_h)] \tag{4-41}$$

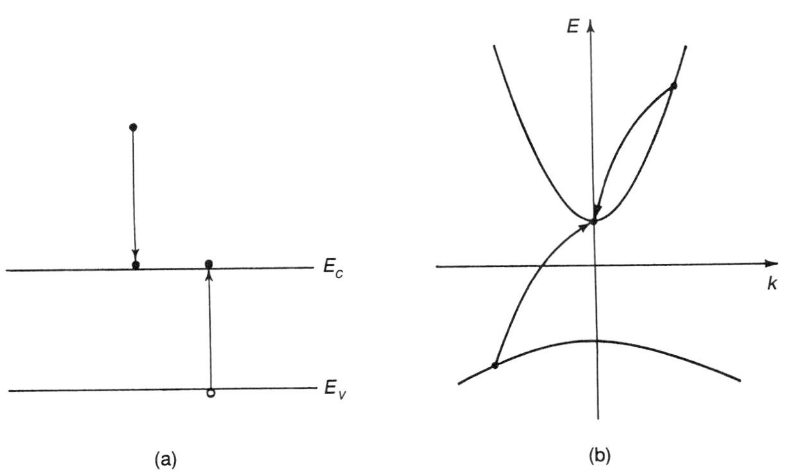

FIGURE 4-5. The impact ionization process in (a) real space and (b) **k** space (after Hess, 1989).

At the threshold energy the momenta of the particles after ionization are

$$\mathbf{k}_f = \mathbf{k}_{f'} = \mathbf{k}_i m_e/(2m_e + m_h), \qquad \mathbf{k}_{i'} = \mathbf{k}_i m_h/(2m_e + m_h) \tag{4-42}$$

The impact ionization process for $m_h \gg m_e$ is shown in Fig. 4-5. Here E_{th} approaches E_G, $\mathbf{k}_f = \mathbf{k}_{f'}$ approaches 0, and the entire momentum of the primary electron is given to the hole. In the case where $m_e = m_h$, we obtain from Eq. (4-41) a threshold energy $E_{\text{th}} = 3E_G/2$. E_{th} increases when m_h/m_e decreases. For $m_h \ll m_e$ we have E_{th} approaching $2E_G$; the momentum of the primary electron is distributed equally between the primary and secondary electrons.

In actuality, most semiconductors are indirect, so that the conduction band mimima are shifted by $\mathbf{K}_0$ from the center of the Brillouin zone. It is necessary to change the dispersion relation (4-38) in accordance with the band structure; the vector $\mathbf{K}$ is added to the right-hand side of Eq. (4-40). For the direct processes considered above, $\mathbf{K} = 0$. For Umklappp processes, which include Bragg reflection, $\mathbf{K}$ equals the primitive translation wave vector of the reciprocal lattice (Kane, 1967; Dmitriev et al., 1987). The solution for Si gives $E_{\text{th}} = 1.3E_G$ (Yadau et al., 1976). Because of the anisotropy of the electron effective mass, the threshold energy is a function of $\mathbf{k}_i$. In III–V semiconductors E_{th} is also a function of $\mathbf{k}_i$. (In addition to the above process, an alternative exists. Here a primary electron is in the upper X or L valley, and the secondary electron will be in the lower Γ valley). In these cases a cumbersome procedure is required to obtain $E_{\text{th}}(\mathbf{k}_i)$, which very often includes a numerical solution. The review by Robbins (1980) provides extensive information on threshold energies.

4.3.2. The Probability of Impact Ionization

As long as the Golden Rule is applicable, the probability of impact ionization can be defined via the Born approximation. This is the probability P_i of the transition of two electrons from the state $|i, i'\rangle$ into the state $\langle f, f'|$. As a result of this transition, the secondary hole will be in state i' and the secondary electron in f'.

$$P_i = (2\pi/\hbar)\,|M|^2\,\delta(E_i + E_{i'} - E_f - E_{f'}) \tag{4-43}$$

where $|M|^2$ is the absolute value of the matrix element that connects the initial and final states:

$$M = \langle f, f' |\, I \,| i, i' \rangle \tag{4-44}$$

The interaction potential I, is the screened Coulomb potential

$$I = [e^2/(\epsilon\,|r|)]\exp(-|r|/\lambda_D) \tag{4-45}$$

where r is the distance between electrons, and λ_D is the screening length.

Taking into account momentum conservation (4-40) and using Bloch wave functions,

$$\Psi_{\mathbf{k}_i} = U_{\mathbf{k}_i}(\mathbf{r})\exp(i\mathbf{k}_i\mathbf{r}) \tag{4-46}$$

we obtain from Eq. (4-43)

$$M = \frac{e^2}{\epsilon V_0}\frac{I_{ee}(\mathbf{k}_i, \mathbf{k}_f)I_{eh}(\mathbf{k}_{i'}, \mathbf{k}_{f'})}{|\mathbf{k}_i - \mathbf{k}_f|^2 + \lambda_D^{-2}}, \tag{4-47}$$

where V_0 is the volume of crystal, and I_{ee} and I_{eh} are the overlap integrals of the amplitude of the Bloch functions:

$$I_{ee}(\mathbf{k}_i, \mathbf{k}_f) = \int U^*_{\mathbf{k}_f e}(\mathbf{r}_1)U_{\mathbf{k}_i e}(\mathbf{r}_1)\,d\mathbf{r}_1 \tag{4-48}$$

$$I_{eh}(\mathbf{k}_{i'}, \mathbf{k}_{f'}) = \int U^*_{\mathbf{k}_f' e}(\mathbf{r}_2)U_{\mathbf{k}_i' h}(\mathbf{r}_2)\,d\mathbf{r}_2 \tag{4-49}$$

In the isotropic parabolic bands (4-38) the overlap integral for two states in the conduction band is unity, but $I_{eh} = 0$, because the valence and conduction band wave functions taken at the top of the valence band and at the bottom of the conduction band are orthogonal to each other. Hence, even if we assume that I_{ee} is close to unity for all possible bandstructures, the details of the dependence of I_{eh} on $\mathbf{k}_{i'}$ and $\mathbf{k}_{f'}$ are very important.

The probability, $W_i = 1/\tau_i$, that the primary electron will produce an electron–hole pair with any $\mathbf{k}_{i'}$, and $\mathbf{k}_{f'}$ is actually

$$W_i = [V_0/(8\pi)^3]^2 \int P_i\,d\mathbf{k}_f\,d\mathbf{k}_{f'}, \tag{4-50}$$

where the integral of P_i is over all possible final states. If we assume that both I_{ee} and I_{eh} are independent of momentum and neglect the screening effect, integration of Eq. (4-50) can be performed, resulting in (Ridley, 1987)

$$W_i(E_i) = W_0 \frac{m_e/m_0}{\epsilon^2} \frac{I_{ee}^2 I_{eh}^2}{(1 + 2m_e/m_h)^{3/2}} \left(\frac{E_i - E_{th}}{E_G} \right)^2 \tag{4-51}$$

where

$$W_0 = e^4 m_0/\hbar^3 = 4.17 \times 10^{16}\,\mathrm{s}^{-1} \tag{4-52}$$

m_0 is the free electron mass and $W_i = 0$ for $E_i < E_{th}$. We must stress, however, that there is no reasonable situation where Eq. (4-51) can be used, because I_{eh}^2 always depends upon $\mathbf{k}_{i'}$ and $\mathbf{k}_{f'}$; otherwise it is zero. Band nonparabolicity and warping must be taken into account in narrow gap semiconductors. These produce $I_{ee}^2 \neq 1$ and $I_{eh}^2 \neq 1$. Rather than Eq. (4-51), Avramenko and Strikha (1986) obtained for narrow gap semiconductors such as InSb:

$$W_i(E_i) = W_0 \frac{m_e/m_0}{\epsilon^2} \frac{2E_G^3}{E_i(E_i + E_G)(2E_i + E_G)} \left(\frac{E_i - E_{th}}{E_G} \right)^2 K \tag{4-53}$$

where K is a numerical coefficient that is very close to unity everywhere except for a narrow energy region near threshold, $E_i - E_{th} = 0.01\,\mathrm{eV}$. In (4-53) it was the case that $m_h \gg m_e$ and $k_f \ll k_{f'}$. This is the reason that Eq. (4-53) is strictly valid only near threshold. But, near threshold ($E_i - E_{th} \ll E_{th}$) all possible dependencies can be written in the form (Keldysh, 1965; Hess, 1988)

$$W_i(E_i \geq E_{th}) = W_1 \left(\frac{E_i - E_{th}}{E_G} \right)^p; \qquad W_i(E_i \leq E_{th}) = 0 \tag{4-54}$$

where W_1 and p are adjustable parameters, and the exponent p is typically between 1 and 2.

In the more general case the dependence of W_i on E_i can be obtained only by numerical calculation. Kane (1967) calculated $W_i(E_i)$ (see Fig. 4-6) for electrons in silicon, taking into account screening and using wave functions, U_i, of first-order perturbation theory. The ordinate of Fig. 4-6 is the impact ionization probability $W_i = 1/\tau_i$, in units of $\hbar/\tau_i$. This is the quantum mechanical uncertainty of energy ΔE related to the fact that the lifetime (τ_i) is not infinite:

$$\Delta E \approx \hbar/\tau_i \tag{4-55}$$

This uncertainty reaches the order of 0.1 eV at kinetic energies of electrons of the order of 5 eV (note that in Fig. 4-6 the energy is zero at the top of the valence band), and this corresponds to $\tau_i \approx 10^{-14}\,\mathrm{s}$. This introduces broadening of the energy band at high energies.

In addition to impact ionization, scattering processes also take place. Hence, in Eq. (4-55) it is necessary to replace τ_i by τ_{tot}, where τ_{tot} is time between and two subsequent collisions, including ionization:

$$\Delta E \approx \hbar/\tau_{tot} \tag{4-56}$$

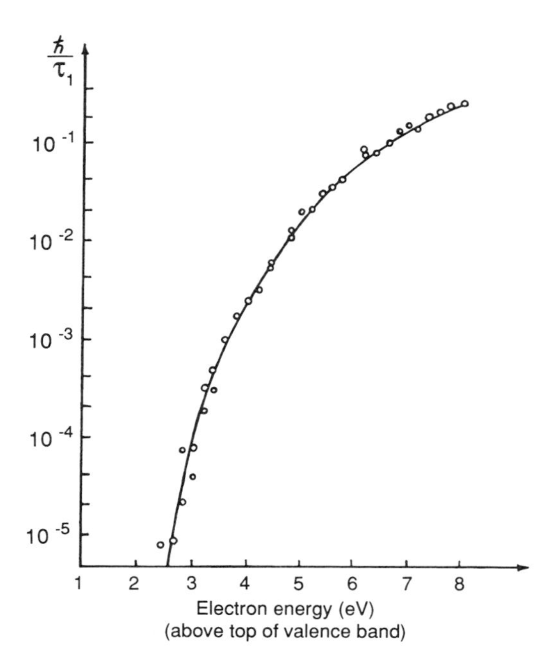

FIGURE 4-6. Impact ionization rate for electrons in silicon as a function of energy (after Kane, 1967).

In GaAs, we can see from Fig. 5-22, τ_{tot} at energies $E_i > E_G$ is close to 10^{-14} s, but in InP (Wu et al., 1991, Fig. 12) it is even less than 10^{-14} s. Thus, the collision broadening, ΔE, reaches values comparable to the energy scale that is important for impact ionization. Conservation of energy becomes questionable and the δ function in the Golden Rule (4-43) must be replaced by a more general expression. (A more detailed discussion of collision broadening can be found, for example, in the publications of Reggiani et al., 1987, 1988; Mahan, 1972, 1981; Barker, 1973; and others.)

One more effect must then be taken into account in performing a detailed analysis of impact ionization. A simple estimation, on the basis of the equations discussed in Sec. 4.2, shows that for voltages Φ_0 on the order of 1 V, electric fields in Read diode structures can reach values on the order of 10^6 V cm^{-1}. The electron wave functions cannot be presented in the form of Eq. (4-46) as independent of electric field. In other words, an impact ionization event cannot occur between free electron states. Rather, it must occur between states of an electron in an electric field (see, e.g., Levinson et al., 1972; Herbert et al., 1982; Seminozhenko, 1982; Khan et al., 1987; Reggiani et al., 1988 and references therein). Accounting for the electric field through electron states in the transition matrix element (4-44) accounts for the intracollisional field effect.

There is no theory of impact ionization that takes into account the intracollisional field effect and collisional broadening, except by Quade et al., 1991. In addition, the band structures, as mentioned in the previous subsection, are very complicated in most semiconductors at energies high enough for impact ionization to occur. This is the reason that Eq. (4-54) can be used for the probability of impact ionization in a semiquantitative theory. For actual semiconductor band structures numerical calculations must be performed. But, the results of modeling impact ionization processes in wide gap semiconductors (Chwang et al., 1979; Shichijo and Hess, 1981) show that the rate of impact ionization for large **k** is not strongly dependent on **k**, i.e. the process actually has a threshold character.

For $E > E_{th}$, the probability of impact ionization exceeds that of any other scattering probability. As it is for GaAs and other semiconductors, the threshold energy E_{th} is the most important parameter. This is why the probability of impact ionization in wide gap semiconductors can be taken as the simple form of Eq. (4-54), even for numerical calculations that take into account real band structures (see, e.g., Shichijo and Hess, 1981).

We will restrict ourselves here to the simple model of impact ionization described by Eq. (4-54) because the complexity of the theory of IMPATT diodes is also enhanced by the nonlocality of carrier multiplication, as we will see in the next subsection, in addition to the phenomena discussed above.

4.3.3. Rates of Impact Ionization and Nonlocality of Carrier Multiplication

The probability of impact ionization in the form of Eq. (4-54), or any other approach discussed in the previous subsection, involves the microscopic character of the process. As we have already seen in Chaps. 1 and 2, macroscopic equations must be solved to determine the oscillatory behavior of the circuit. It is then necessary to introduce macroscopic parameters of impact ionization, which depend upon the scattering and dynamic behavior of the carriers, as well as on their distribution in real and momentum space. The ionization process in solids is described by ionization rates for electrons (α) and holes (β). α and β are generally different, demonstrating that these carriers have different dynamic properties (Pearsall et al., 1977). It is possible to define α and β in different ways. They are usually introduced as an average rate of ionization per unit distance:

$$\alpha = \int W_{ie}(\mathbf{k}) f_e(\mathbf{k})\, d\mathbf{k} \bigg/ \int v_e(\mathbf{k}) f_e(\mathbf{k})\, d\mathbf{k} \tag{4-57}$$

$$\beta = \int W_{ih}(\mathbf{k}) f_h(\mathbf{k})\, d\mathbf{k} \bigg/ \int v_h(\mathbf{k}) f_h(\mathbf{k})\, d\mathbf{k} \tag{4-58}$$

where $v_{e,h}(\mathbf{k})$, $W_{ie,h}(\mathbf{k})$, and $f_{e,h}(\mathbf{k})$ are the group velocities, impact ionization probabilities, and distribution functions for electrons (e) and holes (h), respectively.

It follows from Eqs. (4-57) and (4-58) that the rate of ionization is the ratio of the average ionization probability to the drift velocity:

$$v_{de} = \int v_e(\mathbf{k}) f_e(\mathbf{k})\, d\mathbf{k} \bigg/ \int f_e(\mathbf{k})\, d\mathbf{k} \tag{4-59}$$

$$v_{dh} = \int v_h(\mathbf{k}) f_h(\mathbf{k})\, d\mathbf{k} \bigg/ \int f_h(\mathbf{k})\, d\mathbf{k} \tag{4-60}$$

where v_{de} and v_{dh} are the electron and holes drift velocities, respectively.

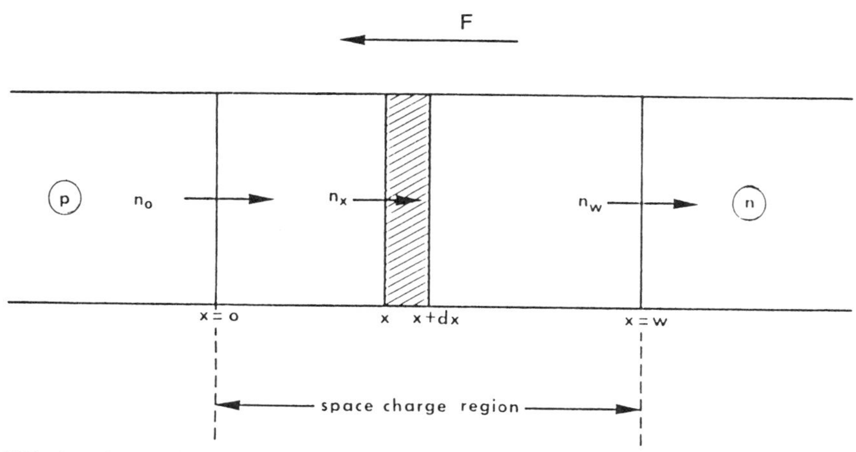

FIGURE 4-7. A *p-n* junction under reverse bias where carrier multiplication occurs in the space-charge region. n_0 electrons enter the space-charge region at $x = 0$ and n_w leave at $x = w$. n_x enter the volume element at x. (After Shaw, 1981.)

To calculate the dependence of α and β on electric field F, we must know the distribution functions $f_{e,h}$. We will discuss the details of these calculations shortly, but first we shall assume that $\alpha(F)$ and $\beta(F)$ are known. Once $\alpha(F)$ is known the carrier multiplication factor, M, can be determined. For a *p-n* junction we can define M as the number of pairs collected by the junction as the result of the insertion of a free carrier into the high-field region. Consider, e.g., the plane parallel geometry shown in Fig. 4-7. By way of example we first assume $\alpha = \beta$, $v_{de} = v_{dh}$ and insert n_0 electrons at $x = 0$. If n_x is the number of electrons entering the volume element at x, there will be $n_x\alpha(F)\,dx$ pairs formed in the element from x to $x + dx$ due to the passage of these electrons. In addition, since there are $(n_w - n_x)$ holes in this region as well, these holes will lead to the creation of $(n_w - n_x)\alpha(F)\,dx$ pairs in the element. Therefore, the total number of electrons formed in this element (from x to $x + dx$) is

$$dn_x = n_x\alpha(F)\,dx + (n_w - n_x)\alpha(F)\,dx = n_w\alpha(F)\,dx \tag{4-61}$$

Therefore, the total number of secondary electrons in the sample is

$$\int_{n_0}^{n_x} dn_x = n_w - n_0 = n_w \int_0^w \alpha(F)\,dx \tag{4-62}$$

or

$$1 - n_0/n_w = \int_0^w \alpha(F)\,dx \tag{4-63}$$

But since $M_n = n_w/n_0$, we have

$$1 - 1/M_n = \int_0^w \alpha(F)\,dx \tag{4-64}$$

and a knowledge of $\alpha(F)$ and $F(x)$ will then give us M.

Equation (4-64) is valid even if $v_{de} \neq v_{dh}$, but for $\alpha \neq \beta$ Eq. (4-64) is replaced by the more general result,

$$1/M_n = 1 - \int_0^W \alpha(F(x)) \exp\left\{ - \int_0^x [\alpha(F(x')) - \beta(F(x'))]\, dx' \right\} dx \quad (4\text{-}65)$$

where M_n is the electron multiplication factor. A similar equation holds for M_p, the hole multiplication factor.

Breakdown results when $M \to \infty$, i.e.,

$$\int_0^W \alpha(F(x)) \exp\left\{ - \int_0^x [\alpha(F(x')) - \beta(F(x'))]\, dx' \right\} dx = 1 \quad (4\text{-}66)$$

is required for breakdown, which for a known distribution of $F(x)$ (see the previous section) defines the breakdown voltage Φ_B. But before actual breakdown of the junction occurs, there will be some carrier multiplication within the space-charge region. If without any breakdown mechanism the reverse current is I_0, then the actual reverse current will be MI_0. In fact, a reasonable empirical representation of M as a function of Φ_0 has been found to be (Miller, 1955)

$$M = 1/[1 - (\Phi_0/\Phi_B)^n] \quad (4\text{-}67)$$

where Φ_B is the junction breakdown voltage and $3 \leq n \leq 6$ depending on the specific semiconductor and type of doping.

We see that the avalanche breakdown voltage Φ_B is a very important parameter of impact ionization. To determine Φ_B it is necessary to solve Eq. (4-66). Recall that the dependence of F on x in the different types of diodes was discussed in Sec. 4.2. It is now necessary to specify the dependence of α on F. This is a major problem of the theory of IMPATT diodes. $\alpha(F)$ was calculated for a uniform electric field distribution, $F(x) = \text{const}$, by many workers in the field (see, e.g., Baraff, 1962, 1964; Ridley, 1983; Burt, 1985; Shockley, 1961; Wolff, 1954, Keldysh, 1960, 1965). From their results it follows that the field dependence of the ionization rates can be expressed as

$$\alpha(F) = A' \exp(-b/F^m) \quad (4\text{-}68)$$

where A', b, and m are parameters. For most semiconductors m is close to unity. For example, for electrons in Si $A' = 3.8 \times 10^6\ \text{cm}^{-1}$, $b = 1.75 \times 10^6\ \text{V cm}^{-1}$, and $m = 1$.

The simplest approach to multiplication phenomena in IMPATT diodes includes two major assumptions: (i) α depends only on the electric field F; despite the fact that F varies with x, no specific dependence of α on x is introduced, so Eq. (4-68) is used with $F = F(x)$; (ii) the number of electron–hole pairs formed in the element from x to $x + dx$ due to impact ionization by electrons is $n_x \alpha(F(x))\, dx$, i.e., it depends on the number of electrons in this element from x to $x + dx$ and the rate $\alpha(F(x))$. This means that double locality is assumed. The second assumption we explained in deriving Eq. (4-61) as obvious.

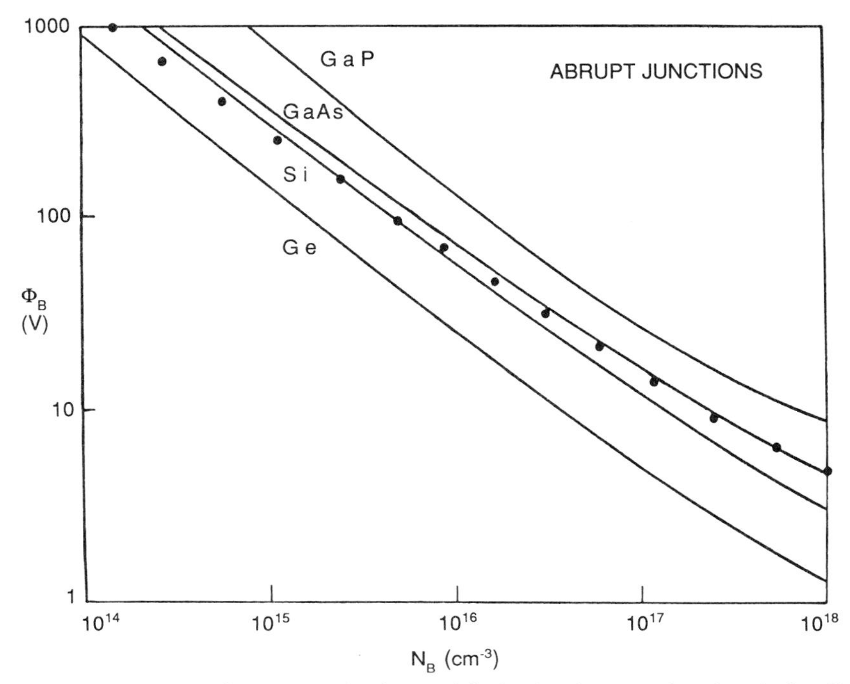

FIGURE 4-8. Breakdown voltage versus background doping for abrupt *p-n* junctions in Ge, Si, GaAs, and GaP. Points are experimental results for Si. (After Sze and Gibbons, 1966.)

But it actually is not obvious, and, furthermore, it is not always true. But if both of these assumptions are taken into account, Eqs. (4-66) and (4-67) can be solved for a given $F(x)$ and the dependence of the breakdown voltage Φ_B on parameters of the p-n junction can be obtained. Figure 4-8 demonstrates the dependence of Φ_B on doping level N_B for the case of an abrupt junction (in Ge, Si, GaAs, and GaP) with homogeneous doping, when the electric field and potential distributions are described by Eqs. (4-2)–(4-11). These results can be approximated by a universal expression that holds for the four semiconductors investigated:

$$\Phi_B = 60(E_G/1.1)^{3/2}(N_B/10^{16})^{-3/4} \text{ V} \tag{4-69}$$

N_B is taken as the concentration at the base of the diode.

Despite the reasonable agreement between the calculated and experimental results for Si in Fig. 4-8, there is no proof that the model is really good, because there is no exact experimental definition for Φ_B. As we can see from typical current–voltage characteristics, there is no real steplike dependence with a well-defined breakdown voltage. The same fact is reflected in Eq. (4-67). Furthermore, an increasing discrepancy between calculation and experiment appears with increasing concentration N_B. When N_B increases, the width of the depletion region decreases [Eq. (4-6)] and the steepness of the electric field profile increases [Eq. (4-11)]. Hence, the double locality approximation fails, and it is necessary to obtain a more rigorous definition for the multiplication factor [Eq. (4-65)] and for α [Eq. (4-68)].

It is now time to stress the importance of Eq. (4-65). Equation (4-66) follows from Eq. (4-65) for $M \to \infty$; it includes both α and β, and it defines Φ_B, which is difficult to determine experimentally. The multiplication factor is in accordance with its definition, n_w/n_0, measured experimentally over a quite broad range of voltage Φ_0. Furthermore, it follows from Eq. (4-65) that

$$M_n - 1 \approx \int_0^w \alpha\, dx \tag{4-70}$$

for

$$M_n - 1 \ll 1 \tag{4-71}$$

The same is true for holes:

$$M_p - 1 \approx \int_0^w \beta\, dx \tag{4-72}$$

for

$$M_p - 1 \ll 1 \tag{4-73}$$

This means that for small multiplication factors (4-71), (4-73) only impact ionization by primary carriers is important. Hence, if the experimental dependence of M on Φ_0 is known, Eq. (4-70) [or Eq. (4-72)] can be solved to obtain the α (or β) dependence on x and on the electric field. For example, the parameters of Eq. (4-68) can be readily determined if we put that dependence into Eq. (4-70) and take into account the dependencies of w on Φ_0 [Eq. (4-6)] and of F on x and Φ_0 [Eqs. (4-10), (4-11)] for a given $M(\Phi_0)$. Let us therefore analyze what is wrong with the local approach in Eq. (4-68).

In Sec. 4.3.1 we discussed that for impact ionization an electron needs to have an energy higher than E_{th}, the threshold energy. The electron is injected into the space charge region at $x = 0$ with energy E_0 essentially less than E_{th}, and needs to drift a distance d_n to reach an energy equal at least to E_{th}. If the electron reaches threshold without scattering, the distance d_n is obtained from

$$E_{\text{th}} = e \int_0^{d_n} F(x)\, \mathrm{d}x \tag{4-74}$$

If an electron is in the region $0 < x < d_n$ it cannot ionize and

$$\alpha = 0 \qquad \text{for} \quad 0 < x < d_n \tag{4-75}$$

Here d_n is a "dark space" region where the electron cannot ionize. This distance must be taken into account, not only for initially injected electrons, but also for secondary electrons and for those primary electrons that produce electron–hole pairs and fall back to lower energies. Equation (4-74) can be modified to take into account possible optical phonon absorption within a distance d_n, as was done by Okuto and Crowell (1974), who first introduced the dark space:

$$E_{\text{th}} = e \int_0^{d_n} F(x)\, dx + N_r E_r \qquad \text{(4-74')}$$

Here N_r is the net number of optical phonons having energy E_r absorbed by the carrier in the distance d_n. This correction to (4-74) can be important only in narrow gap semiconductors. In wide gap semiconductors E_{th} exceeds E_r by a factor essentially larger than 10 and N_r is of the order of unity if a reasonably large electric field is considered.

The dark space effect can obviously be neglected if

$$\alpha d_n \ll 1, \qquad \beta d_p \ll 1 \qquad \text{(4-76)}$$

$$d_n, d_p \ll w \qquad \text{(4-77)}$$

The inequality (4-76) means that the average distance $(1/\alpha)$ that an electron needs to traverse to impact ionize an electron–hole pair is essentially larger than the dark space. If the electric field increases, so does α, and d_n^{-1} is the upper limit of α, which is reached near breakdown.

Inequality (4-77) is true for lightly doped diodes. But it fails in heavily doped diodes where w becomes small by (4-6).

Because of the existence of the dark space discussed above, the second assumption of locality fails. Inside the dark space the product $n_x\alpha$ equals zero because $\alpha = 0$ (4-75). But, outside the dark space electrons n_x cannot ionize at x, they can ionize only at $x + d_n$. Hence, Eq. (4-65) must be modified to take into account nonlocality. All nonlocal effects can be included in α and β, as proposed by Gribnikov, Ivastchenko, and Mitin (1981). The continuity equations for electrons and holes

$$\partial n(x)/\partial t = \frac{1}{e} \operatorname{div} J_n(x) + G(x) - R(x) \qquad \text{(4-78)}$$

$$\partial p(x)/\partial t = -\frac{1}{e} \operatorname{div} J_p(x) + G(x) - R(x) \qquad \text{(4-79)}$$

should be taken within the framework of the drift and diffusion approximation, where G is the generation rate for the creation of electron–hole pairs

$$G(x) = \alpha(x) n(x) v_{\text{de}}(x) + \beta(x) p(x) v_{\text{dh}}(x) \qquad \text{(4-80)}$$

and R is the electron–hole recombination rate, which includes the relevant recombination processes not considered here. So we simply put $R = 0$.

Only the drift component of the current densities are important in J_n and J_p because of the very strong electric field in the range of impact ionization processes:

$$J_n(x) = -en(x)v_{de}(x) \qquad J_p(x) = ep(x)v_{dh}(x) \tag{4-81}$$

Equations (4-78)–(4-81), together with the definitions of α, β and v_{de}, v_{dh} (4-57)–(4-60), represent the complete set that provides all the information about the problem in question. Nonlocality is included in the definitions of α and β self-consistently through the distribution function and the limits of integration in the numerators of Eqs. (4-57) and (4-58) because $W_i = 0$ for $E < E_{th}$ (4-54).

If the conditions (4-76), (4-77) are not fulfilled, it is clear that the impact ionization process affects not only the carrier distribution functions at high energies ($E > E_{th}$), but determines all their energy and spatial structure. Therefore, a spatial inhomogeneity of the ionization rate leads to an inhomogeneity in the distribution function itself. Thus, when conditions (4-76), (4-77) are not fulfilled, the phenomenological attempts to take into account nonlocal effects (Okuto and Crowell, 1974) without calculating the distribution functions cannot be considered as quite correct. For example, let us demonstrate the importance of the nonlocality of α related to an electric field inhomogeneity along the length d_n (Gribnikov et al., 1981), which is not taken into account by the dark space (4-75). If the distribution function is highly anisotropic over most of the energy interval from the mean energy $\bar{E}$ up to threshold E_{th}, and its energy dependence is determined by just precollision drift in the electric field, then the asymptotic form (for $E \gg \bar{E}$) of the distribution function in the case of a homogeneous field is given (Keldysh, 1965; Baraff, 1962; Gribnikov, 1978) by

$$f(E) = n \exp\left\{-\int_0^E dE' \Big/ [e\,|F|\,l(E')]\right\} \tag{4-82}$$

Here $l(E)$ is the mean free path of electrons drifting with energy E along the field F, and n is a normalization concentration. Equation (4-82) leads to $\alpha(F)$ in the form of Eq. (4-68) with parameter $m = 1$. It is important that $m = 1$ for the dependence of W_i on E (4-54) with any parameter p. This is due to the fact that for $E_{th} \gg \bar{E}$ integration of Eq. (4-57) for different p's yields the same exponential dependence with different preexponential factors A' that depend upon F. The dependence of A' on F can be neglected in the range of fields important for impact ionization, in comparison with the very strong dependence of $\exp(-b/F)$ on F.

In the case of a spatial inhomogeneity, asymptotics analogous to that presented by Eq. (4-82) may be obtained from the equation

$$v_x\, \partial f/\partial x - eF(x)\, \partial f/\partial p_x + f/\tau_{\mathbf{p}} \approx 0, \qquad f = f(\mathbf{p}, x) \tag{4-83}$$

Here $\mathbf{p}$ and p_x are the momentum and its projection on the field, $\mathbf{v}$ and v_x are the velocity of the electron with momentum $\mathbf{p}$ and its projection upon the field, and $\tau_{\mathbf{p}}$ is the scattering time from the state $\mathbf{p}$. In Eq. (4-83) there is no term describing electrons coming into the state **p**, since it can be neglected when high-energy asymptotics are considered. Such an approach corresponds to the drift asymptotics because for the homogeneous case Eq. (4-83) follows from Eq. (4-82).

An approximate solution of Eq. (4-83) is given by (Gribnikov et al., 1981);

$$f(E, x) = n(\Phi(x) - E/e) \exp\left\{ - \int_0^E dE' \Big/ [e\, |F(\Phi(x) - (E - E')/e)|\, l(E')] \right\} \tag{4-84}$$

Here $n(\Phi)$ and $F(\Phi)$ are the carrier concentration and the electric field at the point of potential Φ. The difference between Eqs. (4-84) and (4-82) lies in the fact that in Eq. (4-84) the carrier concentration $n(\Phi(x))$ at the point x is replaced by, generally speaking, another $n(\Phi(x) - E/e)$, and instead of the field $F(\Phi(x))$ all the fields are present between $F(\Phi(x))$ and $F(\Phi(x) - E/e)$. Equations (4-82) and (4-84) are valid if $E \gg \bar{E}$, i.e., if the exponents are large enough. In this case conditions (4-76), (4-77) are fulfilled so that

$$n(\Phi(x) - E_{\text{th}}/e) \approx n(\Phi(x)) \tag{4-85}$$

The continuity equations (4-78) and (4-79) remain unchanged, and in these equations we put

$$\alpha(x) \cong \alpha_\infty \exp\left\{ - \int_0^{E_{\text{th},e}} dE' \Big/ [e\, |F(\Phi(x) - (E_{\text{th},e} - E')/e)|\, l_e(E')] \right\} \tag{4-86}$$

and

$$\beta(x) \cong \beta_\infty \exp\left\{ - \int_0^{E_{\text{th},h}} dE' \Big/ [e\, |F(\Phi(x) - (E_{\text{th},h} - E')/e)|\, l_h(E')] \right\} \tag{4-87}$$

instead of

$$\alpha(x) \cong \alpha_\infty \exp\left[- \frac{1}{e\, |F(x)|} \int_0^{E_{\text{th},e}} dE' \Big/ l_e(E') \right] \tag{4-88}$$

and

$$\beta(x) \cong \beta_\infty \exp\left[- \frac{1}{e\, |F(x)|} \int_0^{E_{\text{th},h}} dE' \Big/ l_h(E') \right] \tag{4-89}$$

where different threshold energies $E_{\text{th},e}$ and $E_{\text{th},h}$ and mean free paths l_e and l_h were introduced for electrons and holes. α_∞ and β_∞ are independent of coordinate parameters, corresponding to $F \to \infty$. Equations (4-88) and (4-89) are being used in a strictly local approach.

The correction terms connected with the use of Eqs. (4-86) and (4-87) instead of Eqs. (4-88) and (4-89) are essential if the field $F(x)$ varies significantly in the vicinity of its maximum value over a length on the order of d_n, d_p. The latter condition means that in intervals of the order of d_n, d_p (with voltages

$E_{th,e}/e$, $E_{th,h}/e$ across them) practically all of the impact ionization must take place for voltages not too low compared to the breakdown voltage Φ_B. The latter takes place only in a diode with a rather thin region of strong electric field (see Figs. 4-2 and 4-3), and if the diode has a low breakdown voltage (a low breakdown means that Φ_B exceeds $E_{th,e,h}/e$ by less than an order of magnitude). Of special interest in this case is the Read diode with a narrow heavily doped layer, as shown in Fig. 4-3, because the field varies rapidly with x and contributes to α and β [Eqs. (4-86) and (4-87)].

With α and β defined by Eqs. (4-86) and (4-87), the multiplication factors M_n and M_p can be calculated using Eqs. (4-70) and (4-72). If the electric field dependence on x is taken from Eq. (4-11), the integration can be done for the case when the exponents in Eqs. (4-86) and (4-87) are assumed to be large enough. As a result, the quantities $M_{n,p} - 1$ decrease compared to $M^0_{n,p} - 1$, calculated using a local approach. At $E_{th,e,h} \ll e\Phi_0$ Gribnikov et al. (1981) obtained

$$M_{n,p} - 1 \approx (M^0_{n,p} - 1) \exp[-E^2_{th,e,h}/(4e^2\Phi_0 l_{e,h} |F_{max}|] \tag{4-90}$$

Note that $l_{e,h}$ in Eqs. (4-88) and (4-89) were assumed energy independent, as is the case of most semiconductors (including Ge, Si, GaAs) at an energy E on the order of E_{th}, where $l_{e,h}$ is determined by optical phonon emission (Keldysh, 1965; Baraff, 1962; Gribnikov, 1978).

The most important result of Eq. (4-90) is that both M_n and M_p decrease compared to M^0_n and M^0_p. In the case of complete symmetry of the electron–hole parameters ($E_{th,e} = E_{th,h}$, $l_e = l_h$, $\alpha_\infty = \beta_\infty$) the approximate equation (4-90) gives $M_n = M_p$. More accurate equations maintain the inequality between M_n and M_p due to the asymmetry of $F(x)$ for electrons and holes within the depletion layer. We will discuss this in more detail at the end of this section. But, by analyzing the approximate equation (4-90), we see that the decreases of M_n and M_p are due to different reasons. Let us consider a diode with an n-type base. The multiplication factor M_n is measured by the injection of electrons at $x = 0$ from the high-field region, and M_p by the injection of holes from the low-field region at $x = w$. This asymmetry is reflected in Eqs. (4-86) and (4-87). The decrease of M_p is due to the fact that $\beta(x)$ in the nonlocal approach is less than the local $\beta(x)$, because holes come from the low-field region and their distribution function (4-84) is cooler than that in the local approach. This nonlocality in the distribution function accounts completely for the decrease in M_p and cannot be deduced from the phenomenological approach of Okuto and Crowell (1974).

The explanation for the behavior of M_n is somewhat more complicated. The dark space exists, which acts to decrease M_n in comparison with the local approach. But, electrons at point x came from the high-field region; that is why they are "hotter" [Eq. (4-84)] than might be expected from the local approach. Because of this, α and M_n should increase. This second phenomenon was not taken into account by Okuto and Crowell (1974). Both of these effects result in a smaller decrease of M_n than what would follow from the phenomenological picture of Okuto and Crowell (1974).

The nonlocality of the distribution function (4-84) is important when the quantities $d_{n,p}(F)$ vary because of the field inhomogeneity over the distances $d_{n,p}$ more than they vary over the mean free path $l_{e,h}$, i.e., if

$$|(d_{n,p}/F)(dF/dx)| \geq l_{e,h}/d_{n,p} \tag{4-91}$$

It is easily seen that condition (4-91) is fulfilled in the heavily doped region. The analytical results for α, β [Eqs. (4-86) and (4-87)] and $M_{n,p}$ [Eq. (4-90)] were presented here to provide us with an understanding of nonlocal multiplication. If the inequalities (4-76) and (4-77) fail, nonlocal effects become pronounced, but in this case only numerical solutions can be obtained. A Monte Carlo simulation was used by Gribnikov et al. (1981) and Higman et al. (1988) to calculate impact ionization coefficients and multiplication factors for nonlocal multiplication (also see Lippens et al., 1984, 1983; Ridley and El-Ela, 1989; Chen and Tang, 1988). The Monte Carlo results demonstrate (Gribnikov et al., 1981) that in thin heavily doped layers the distribution function for electrons and holes and their impact ionization coefficients are nonlocal, i.e., they depend not only upon the electric field strength at a specific point, but also on the fields in the vicinity of the point. Furthermore, in these nonlocal situations introduction of the impact ionization coefficients is justified only for small multiplication coefficients. For large M_n and M_p, the coefficients α and β have no meaning. M_n and M_p can be calculated in a direct Monte Carlo procedure without reintroducing α and β.

If we assume that the holes and electrons differ from each other only by the sign of their charge, the nonlocality effect changes not only the values of α and β obtained in the local approximation, but also provides the difference between the electron and hole multiplication coefficients (because of the electric field asymmetry in the depletion layer). Hence, the local approach for α and β [Eqs. (4-88) and (4-89)] can only be used for the simplified theory of avalanche diodes presented in what follows. A self-consistent numerical simulation of device performance, with nonlocal multiplication taken into account, is required for devices with narrow heavily doped regions (see, e.g., Lippens et al., 1984). Such calculations are also of importance in the understanding of novel modern devices such as the avalanche photodiode. Here the enhancement of impact ionization in periodic heterolayers is employed to have the device act as a photomultiplier with superior noise properties (Capasso, 1987).

4.4. MICROWAVE GENERATION USING AVALANCHE DIODES

4.4.1. The Read Diode Oscillator

Figure 4.3 shows the doping profile of a standard Read Diode (Read, 1958), and its associated field profile. As an oscillator it makes use of the transit time of electrons through the i (or π) drift region to elicit the optimum phase delay ($\theta = \pi$) between the applied voltage and resulting current in order to achieve the maximum gain. The device is reverse biased and mounted in a microwave cavity whose impedance is primarily inductive. This impedance is matched to the mainly capacitive impedance of the diode so as to form a resonant system. Frequencies above 100 GHz can be achieved with devices of this type.

For sufficiently high fields at the n^+–p interface, both electrons and holes are produced; the drift of the holes across the space-charge region produces a current $I_e(t)$ in the external circuit. Since the hole current density $j = pev$, and since the hole drift velocity v is constant, then

$$I_e = jA = \frac{v}{w} epwA = pVe\frac{1}{\tau} \tag{4-92}$$

Here pVe is the total charge of the holes that move across the drift region in a time $\tau = w/v$. V is the volume of the device, w is the width, and A is the cross-sectional area of the uniform structure.

Suppose a pulse of hole charge δq is suddenly generated at the n^+–p junction. A constant current $I_e = \delta q/\tau$ immediately begins to flow in the external circuit, and continues to flow during the time, τ, that the holes are moving across the space-charge region from the n^+–p interface to the p^+ region. If the pulse occurs at time t_0 then current will flow in the external circuit from time t_0 till $t_0 + \tau$. Thus, on the average, the external current $I_e(t)$ due to the moving holes is delayed by $\tau/2$ relative to the current $I_0(t)$ generated at the n^+–p interface.

We will also show, in the discussion of multiplication, that the current $I_0(t)$ is delayed by $\pi/2$ relative to the ac voltage impressed. Thus, to obtain a total delay of π, we want the delay $\tau/2$ to also provide a $\pi/2$ delay, i.e., we want the phase delay due to drift to be $\omega\tau/2 = \pi/2$. Therefore,

$$\omega = \frac{\pi}{\tau} \tag{4-93}$$

and the cavity should thus be tuned to have a resonant frequency of $(2\tau)^{-1}$.

In normal operation the diode is biased so that $F > F_c$ during the positive half of the voltage cycle and $F < F_c$ during the negative half (F_c is the breakdown field). This produces a $\pi/2$ phase lag between Φ and $I_0(t)$, and therefore a π phase lag between Φ and $I_e(t)$. Under these conditions, the dissipated power $P = -I\Phi$.

In treating multiplication the simplest case to consider is the unphysical situation where the field is constant, F_0, in the narrow multiplication region. (See Fig. 4.9.) Here we have

$$\int \alpha\,dx = \int_0^{x_1} \alpha(F_0)\,dx = x_1\alpha(F_0) \tag{4-94}$$

Experiments indicate that (roughly) $\alpha = aF^m$. Therefore,

$$\int \alpha\,dx = x_1 a F_0^m = (F_0/F_c)^m \tag{4-95}$$

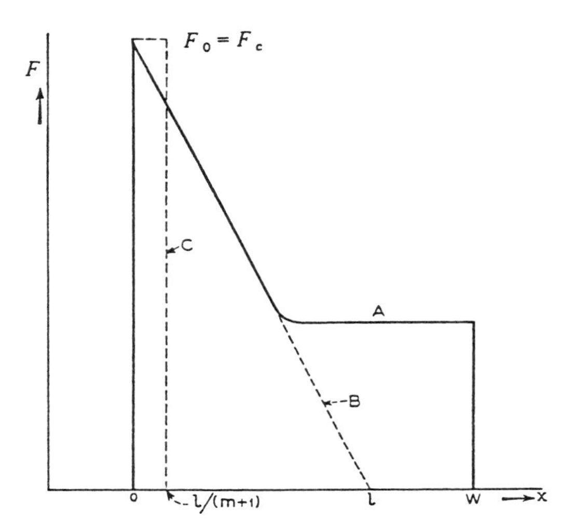

FIGURE 4-9. The electric field profile in a Read diode showing the approximation of a uniform field in the high field region. The parameters are explained in the text. (After Read, 1958.)

where $aF_c^m = 1/x_1$; F_c is the critical field for breakdown given by $\alpha(F_c) = 1/x_1$ [since the breakdown condition (4-66) is $\int \alpha\, dx = 1$].

Next, consider the linear field distribution that would correspond to a one-sided step junction. In Fig. 4-9, A is the field distribution at breakdown. Take $F = F_0 - kx$ in the multiplication region. The flat section contributes negligibly to the multiplication, so that we can replace A by B. Now

$$\int \alpha\, dx = \int_0^{x_0} a(F_0 - kx)^m\, dx$$

where x_0 is the zero field intercept of the tangent curve. Therefore, we have

$$\int \alpha\, dx = (F_0/F_c)^{m+1} \tag{4-96}$$

Here F_c is determined by the definition $aF_c^{m+1} = (m + 1)k$. Since $\alpha(F_c) = aF_c^m$, then $\alpha(F_c) = (m + 1)k/F_c = (m + 1)/l$, where $l \equiv F_c/k$ is the zero intercept of the tangent curve (B in Fig. 4-9) at breakdown. Thus, the peak field F_c at breakdown is equal to the breakdown field for a constant field region of width $l/(m + 1)$, as shown by curve C in Fig. 4-9. [For a linearly graded junction the field varies parabolically with x and $\int \alpha\, dx = (F_0/F_c)^{m+1/2}$.] In what follows, we use a linear field approximation, i.e., $\int \alpha\, dx = (F_0/F_c)^{m+1}$.

Our task is to derive a differential equation for the current as a function of time. If we assume that in Eq. (4-81)$v_{de} = v_{dh} = v$ and v is field independent and $\alpha = \beta$, then hole–electron pairs are being generated at rate $\alpha v(n + p)$ [see Eq. (4-80)]. We also assume that all the multiplication occurs in a relatively narrow multiplication region from $x = 0$ to $x = x_1$, $x_1 \ll w$, and that the total current for unit area $[I(x, t) = I_p(x, t) + I_n(x, t)]$ in the multiplication region is a function of time only $[I(x, t) = I_0(t)]$.

Adding the two continuity equations (4-78) and (4-79) under the above restrictions yields

$$\frac{\partial}{\partial t}(n + p) = \frac{1}{e}\frac{\partial}{\partial x}(I_n - I_p) + 2\alpha v(n + p) \tag{4-97}$$

which can be integrated to produce

$$\tau_1 \frac{dI_0}{dt} = -(I_p - I_n)_0^{x_1} + 2I_0 \int_0^{x_1} \alpha\, dx \tag{4-98}$$

where $\tau_1 = x_1/v$ is the transit time across the multiplication region, as shown in Fig. 4-10.

To determine the boundary conditions, note that

$$I_n = I_0 - I_p \tag{4-99}$$

Thus, at $x = 0$ in Fig. 4-10, $I_p = I_{ps}$; i.e., the hole current at $x = 0$ consists entirely of the reverse saturation current I_{ps} of holes thermally generated in the n^+ region that have moved to the n^+–p junction by diffusion. Therefore, at $x = 0$, we have

$$I_p - I_n = 2I_{ps} - I_0 \tag{4-100}$$

At x_1 in Fig. 4-10 the electron current consists of the reverse saturation current I_{ns} of electrons thermally generated in both the space-charge and p^+ regions. Therefore, we have

$$I_p - I_n = -2I_{ns} + I_0 \tag{4-101}$$

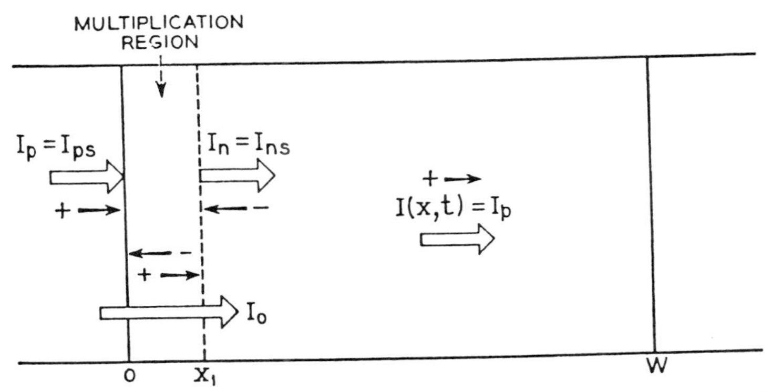

FIGURE 4-10. Current in an n^+-p-i-p^+ Read diode. The parameters are defined in the text (after Read, 1958).

which allows us to write Eq. (4-98) as

$$\frac{\tau_1}{2}\frac{dI_0}{dt} = I_0\left(\int_0^{x_1} \alpha\, dx - 1\right) + I_s \tag{4-102}$$

where $I_s = I_{ns} + I_{ps}$.

Note that in the dc case $I_0 = I_{dc}$ and the left-hand side of Eq. (4-102) vanishes. We are left with the result stated before.

At breakdown $\int_0^{x_1} \alpha\, dx = 1$, and we see that I_0 will increase linearly at a rate $2I_s/\tau_1$ and become infinite. If a larger field is applied, I_0 will approach infinity exponentially. For a smaller field I_0 will approach a finite value.

For an abrupt one-sided step junction we obtain

$$\frac{\tau_1}{2}\frac{d}{dt}\ln I_0 = \left(\frac{F_0}{F_c}\right)^{m+1} - 1 + \frac{I_s}{I_0} \tag{4-103}$$

which relates the current in the multiplication region to the peak field $F_0 = F_0(t)$.

For most practical cases $I_s/I_0 \ll 1$; we can neglect the last term in Eq. (4-103). Furthermore, for low enough amplitudes of oscillation ($F_0/F_c \approx 1$), we can expand F_0/F_c to obtain

$$\frac{d}{dt}\ln I_0 \cong \frac{2(m+1)}{\tau_1}\left(\frac{F_0}{F_c} - 1\right) \tag{4-104}$$

If I_0 is to be periodic, then F_0 must be periodic and the dc bias must be such that the average F_0 is F_c. Suppose we apply a periodic voltage with the proper bias so that $F_0 = F_c + F_a \sin \omega t$, where ω is the optimum frequency π/τ [Eq. (4-93)]. Then, integration yields

$$\ln \frac{I_0(t)}{I_0(0)} = \frac{2(m+1)}{\pi}\frac{\tau}{\tau_1}\frac{F_a}{F_c}(1 - \cos \omega t) \tag{4-105}$$

Since $F_0 = F_c + F_a \sin \omega t$, we see that at $t = 0$, $\ln I_0(t)$ is at its minimum value and F_0 is at its average value. F_0 reaches its maximum value at $\omega t = \pi/2$, whereas $\ln I_0(t)$ reaches its maximum value at $\omega t = \pi$; they are 90° out of phase. Note also that even if the amplitude of F_a is as small as 1% or 2% of F_c, I_0 will vary by a factor of 10 over a cycle. Thus we can have small signals in field and voltage, but large signals in current. Figure 4-11 shows that I_0 has a sharp peak in the middle of the cycle. Therefore, if I_0 varies by a large factor, the current is generated mainly by a pulse in the middle of the ac voltage cycle.

We have now dealt with the multiplication region and obtained an equation relating I_0 and F_0. We next consider the rest of the space-charge region, where current generation is negligible. Here we neglect I_{ns} compared to the total current $I(x, t)$.

We assume that the multiplication region has zero width so that all the

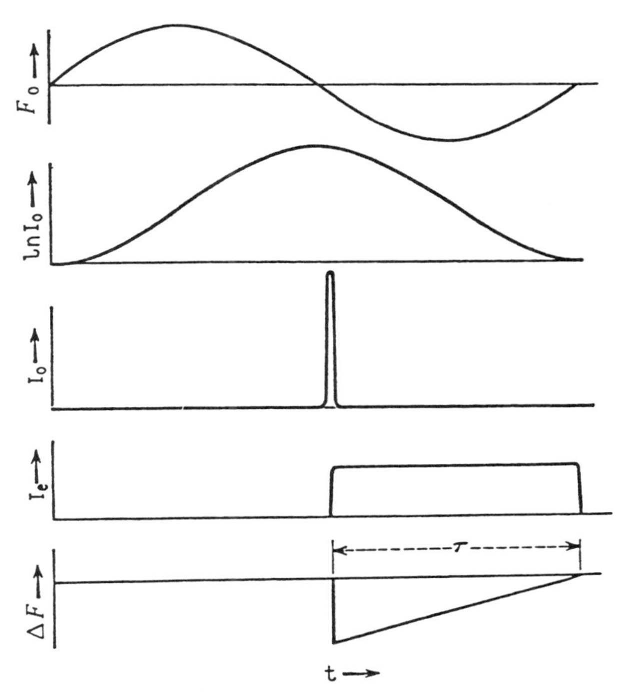

FIGURE 4-11. The phase relation of the field and current in a Read diode (after Read, 1958).

current is generated at $x = 0$. $I_0(t)$ is then a current of holes flowing out of the point $x = 0$ in Fig. 4-10. The current $I(x, t)$ at any point x and time t is, by current continuity,

$$I(x, t) = I(0, t - x/v) = I_0(t - x/v) \tag{4-106}$$

These holes traversing the space-charge region give rise to a current $I_e = I_e(t)$ in the external circuit which is just the average current flowing in the space charge region. Therefore,

$$I_e(t) = \frac{1}{w}\int_0^w I(x, t)\, dx = \frac{1}{\tau}\int_{t-\tau}^{t} I_0(t')\, dt' \tag{4-107}$$

where $t' = t - x/v$. That is, as emphasized before, the current in the external circuit is the total charge in the space charge region divided by the transit time τ.

In addition to I_c, which arises from carriers moving through the space-charge region, there is a displacement current $I_c = C\, d\Phi/dt$ flowing in the external circuit. I_c is just the current required to charge and discharge the diode regarded as a capacitor with capacitance C. It furnishes the variation in charge at the edges of the space-charge region.

Also note that the space charge of the holes acts to reduce the peak field for a given voltage. The stability of the device comes from the fact that current

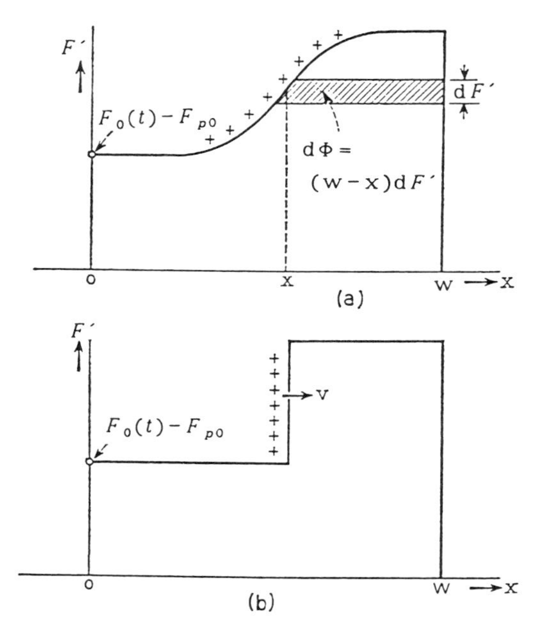

FIGURE 4-12. The difference in field between the current carrying and non-current-carrying cases as a function of position (after Read, 1958).

multiplication increases as F_0 increases, but the current carriers provide a space charge that reduces F_0. We now show how the space charge of the holes reduces the peak field for a given voltage.

If there were no current flowing, any increase in Φ above Φ_p (punch-through voltage) would simply raise the entire field distribution by an amount $(\Phi - \Phi_p)/w$. Figure 4-12 is a plot of the difference $F'(x, t) = F(x, t) - F_p(x)$ at a given time when there is current flowing.

The slope of the F' curve is determined entirely by the hole space charge; the effect of the fixed negative space charge is already included in $F_p(x)$, the field at punch-through. The holes provide a charge density $I(x, t)/v$. Thus, integration of Poisson's equation (see Chap. 3)

$$\frac{\partial F'(x, t)}{\partial x} = I(x, t)/\epsilon = I_0(t - x/v)/\epsilon \tag{4-108}$$

gives

$$F_0(t) = F_{p0} + \frac{\Phi(t) - \Phi_p}{w} - \frac{1}{\epsilon\tau}\int_{t-\tau}^{t} I_0(t')(\tau - t + t')\,dt' \tag{4-109}$$

The last term on the right-hand side of Eq. (4-109), which we call $\Delta F(t)$, is the effect of the current on the field at $x = 0$; it is always negative. Equations (4-105) and (4-109) relate the current $I_0(t)$, the field $F_0(t)$, and the voltage $\Phi(t)$. Thus, in principle, the current can be found for any applied voltage.

Read's original paper (1958) continues the analysis via dimensionless variables, calculating the average values, and discussing the linear small signal case. He determines the small signal equivalent circuit and Q value, which is the energy stored per cycle divided by the time average power dissipation. He finds that the Q of the diode varies linearly with τ_1/I_d and is negative for I_d less than $\pi^2\tau_1/2(m + 1)$. When I_d is equal to this value, the diode becomes an open circuit for this optimum frequency. This means that none of the ac current generated in the multiplication region flows out of the diode. Rather, it flows into the edges of the space charge region and provides the current that charges and discharges the diode regarded as a capacitor. In other words, at this frequency, the unit acts as a capacitor generating its own charge internally. Hence the diode voltage can vary with no external alternating ac voltage impressed; it is an oscillator.

4.4.2. The p-n Diode Avalanche Oscillator

Most modern avalanche diode microwave oscillators are basically Si or GaAs *p-n* or *p-i-n* diodes that are referred to as IMPATT diodes, the acronym standing for impact avalanche transit time. They are useful as millimeter wave dynamic devices (see, e.g., Lippens et al., 1984). Their upper frequency limits have been ascertained (see, e.g., Doumbia et al., 1975; Lippens and Constant, 1981) as well as their noise characteristics (see, e.g., Okamoto, 1975). One of the first calculations of the small signal characteristics of such structures was provided by Misawa (1966), and we will adhere to this development in our calculation of their behavior.

Figure 4-13 shows the critical parameters in a *p-n* diode under avalanche conditions. To understand this diagram, imagine that we have an electron density perturbation that varies sinusoidally in space in a field high enough to produce scattering limited velocities. The density perturbation produces the charge and, hence, the field perturbation. Since electrons drift at the scattering limited velocity regardless of field, the perturbation does not spread out but propagates without attenuation in the direction of electron drift. The field wave lags the electron density wave by $\pi/2$. We see that the "stiffened" electron–hole plasma can convey a nonattenuating space-charge wave.

The generation rate of electron–hole pairs is larger both when the electric field is stronger and when there are more carriers. Thus, the generation rate peaks somewhere between the place where the field is strongest and where the density is largest. Thus, the generation rate lags the electron density wave by less than $\pi/2$. This increased generation rate gives rise to an excess electron density that lags the rate by $\pi/2$. In the positive half cycle the excess electron density keeps increasing and in the negative half cycle the excess density decreases. The resultant total electron density provides a current that lags the field by more than $\pi/2$.

One of the easiest ways to produce an avalanching electron–hole plasma is to break down a *p-n* junction. Consider Fig. 4-14, where electrons enter from the left and holes from the right. We consider the voltage transient after an impulsive current is applied to the avalanching *p-n* junction. First, the applied impulsive current produces charge spikes at both edges of the space-charge region. This corresponds to a momentary widening of the space-charge region, as shown in Fig. 4-15. Since the saturation current of the *p-n* junction is very small, negligible numbers of electrons and holes are emitted. Therefore, almost all the charges in

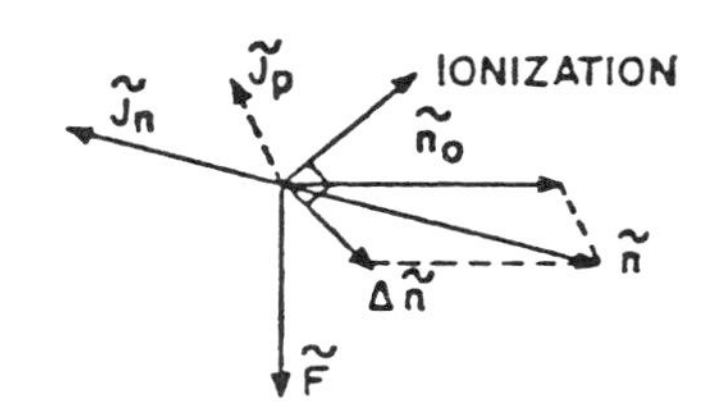

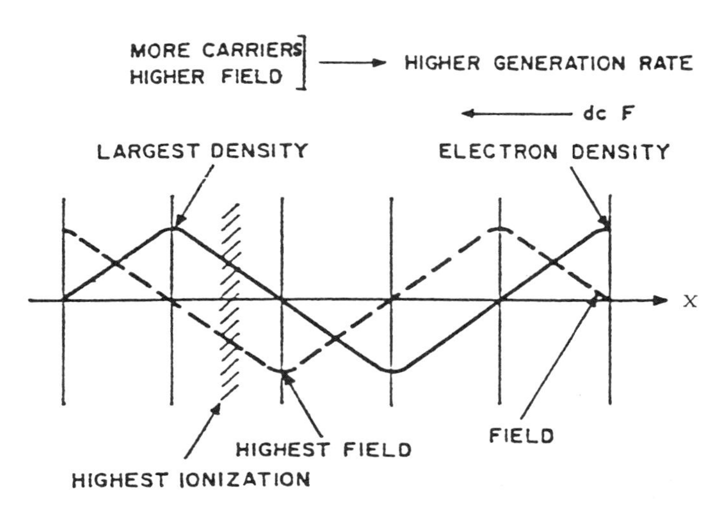

FIGURE 4-13. Field, electron density, and ionization rate versus distance in an avalanche *p-n* diode. The inset shows the phase relationship. (After Misawa, 1966.)

the spikes are not the charge of mobile carriers but the immobile charge of those impurities that were neutralized by carriers. The charge spikes have to wait to be neutralized by carriers generated by avalanche. These charge spikes raise the field in the space-charge region by a constant amount. This increased field causes more avalanching and generated electrons and holes neutralize the charge spikes. However, the situation is such that we still have a supply of electrons and holes

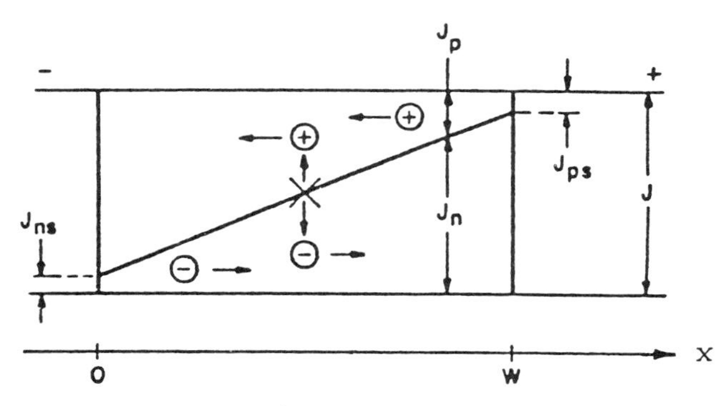

FIGURE 4-14. Space-charge region of a p^+-n junction under avalanche breakdown conditions (after Misawa, 1966).

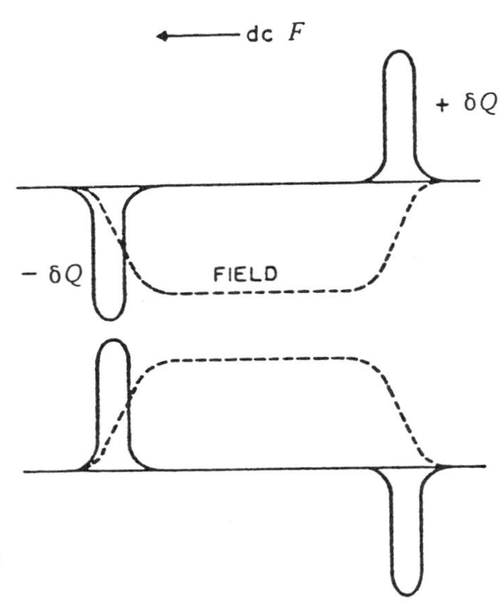

FIGURE 4-15. Two phases of the transit after an impulsive current is applied (after Misawa, 1966).

even after the spikes are neutralized. Subsequently, we have too many holes on the left end and too many electrons at the right end. This is a momentary narrowing of the space-charge region. Now these extra charges decrease the field in the space charge region and we have a situation just opposite what has just been considered (see Fig. 4-15). Since we now have a negative, or decreased, supply, these spikes die away and again the "pendulum" swings beyond the steady state. The situation is repetitive. We may have an instability. To determine the criteria for such an event we consider the fundamental equations governing the dynamics of the avalanching electron–hole plasma in an infinite medium. We use the continuity Eqs. (4-78), (4-79), Eq. (4-80), Poisson's equation, and neglect diffusion. We then divide the quantities into dc parts and small ac parts, the latter changing as $e^{i\omega t}$ ($\partial/\partial t \rightarrow i\omega$), and retain only first-order terms. With a tilde representing ac components, we obtain in units of field normalized by weN_0/ϵ and units of current normalized by eVN_0

$$\frac{\partial \tilde{F}}{\partial x} = \tilde{J}_n - \tilde{J}_p \tag{4-110}$$

and

$$\frac{\partial \tilde{J}_n}{\partial x} = (\alpha - i\omega)\tilde{J}_n + \alpha\tilde{J}_p + \alpha'\tilde{F}\tilde{J} \tag{4-111}$$

where $\tilde{J}$ is the total current and $\alpha' = \partial\alpha/\partial\tilde{F}$.

Similarly we have

$$\frac{\partial \tilde{J}_p}{\partial x} = -\alpha'\tilde{J}\tilde{F} - \alpha\tilde{J}_n - (\alpha - i\omega)\tilde{J}_p \tag{4-112}$$

Note that since α is an increasing function of $|F|$ then $\alpha'\tilde{J}$ is always positive regardless of the sign of $\tilde{J}$.

Adding (4-111) to (4-112) yields

$$\frac{\partial \tilde{J}_n}{\partial x} + \frac{\partial \tilde{J}_p}{\partial x} = i\omega(\tilde{J}_p - \tilde{J}_n) \tag{4-113}$$

and substituting into (4-110) yields after a spatial integration

$$\tilde{J} = \tilde{J}_n + \tilde{J}_p + i\omega\tilde{F} = \text{const} \tag{4-114}$$

For constant avalanche α and α' are independent of x. Then the solution of (4-110)–(4-112) is given by a superposition of solutions of the form e^{-ikx} with three different values of k. The dispersion relation is obtained as follows. First, we spatially differentiate (4-110), and set it equal to the difference between (4-111) and (4-112). This yields

$$\frac{\partial^2 \tilde{F}}{\partial x^2} = 2\alpha'\tilde{J}\tilde{F} + (2\alpha - i\omega)(\tilde{J} - i\omega\tilde{F}) \tag{4-115}$$

Another differentiation yields

$$\frac{\partial^3 \tilde{F}}{\partial x^3} = 2\alpha'\tilde{J}\frac{\partial \tilde{F}}{\partial x} - (2\alpha - i\omega)i\omega\frac{\partial \tilde{F}}{\partial x} \tag{4-116}$$

Putting $F = F_0 e^{ikx}$ provides the dispersion relation

$$k^2 + 2\alpha'\tilde{J} - 2\alpha i\omega - \omega^2 = 0 \tag{4-117}$$

The major assumption above actually has been that F_{dc} is independent of x so that the avalanche occurs uniformly all over, and thus a plane wave ac solution of the form $e^{i(\omega t - kx)}$ is possible. We obtain from (4-117)

$$i\omega = \alpha \pm (\alpha^2 - 2\alpha'\tilde{J} - k^2)^{1/2} \tag{4-118}$$

For real k we have a spatially sinusoidally varying perturbation. The perturbation varies as $e^{i\omega t}$, and grows exponentially with time, which means that it is unstable. When the square root in (4-118) is negative, the perturbation oscillates with time and the time constant for the exponential growth is $1/\alpha$.

When (4-117) is solved for k, we obtain

$$k = \pm [\omega^2 - 2\alpha'\tilde{J} + (2i\omega\alpha)]^{1/2} \tag{4-119}$$

We see that since the imaginary part of the square root is positive, one of the k's in the first quadrant and the other is in the third quadrant (for positive ω). When k is in the first quadrant, the wave propogates in the positive x direction because the real part of k is positive and its amplitude increases toward the positive x axis because the imaginary part is positive. When k is in the third quadrant the wave propogates and grows in the negative x direction. This means that if it is possible to excite this growing wave by an input probe, an output probe, which is placed in a position distant from the input probe, picks up the amplified wave. This is traveling-wave tube type amplification.

One of the easiest ways to produce an avalanching electron–hole plasma in a finite medium is to break down a p-n junction. Since we have boundaries in this case the situation is different. However, we expect a negative resistance owing to the above-mentioned instability of the plasma.

Misawa's (1966) detailed analysis of the p-n junction proceeds with the same equations as for the avalanche in the infinite media except for the boundary conditions $\tilde{J}_n(0) = \tilde{J}_{ns}$, $\tilde{J}_p(w) = \tilde{J}_{ps}$. He obtains expressions for $\tilde{F}$, $\tilde{J}_n$, and $\tilde{J}_p$. These values are used to calculate the Q of the avalanching diode. Since there is gain in the system, the Q is inherently negative (the power dissipation is negative).

The time average of the power dissipation is given by

$$-\left\langle \frac{dP}{dt} \right\rangle = \frac{1}{2}\mathrm{Re}[(\tilde{J}_n + \tilde{J}_p)\tilde{F}] \tag{4-120}$$

The energy stored is the field energy

$$\langle P \rangle = \frac{1}{2}\mathrm{Re}\frac{|\tilde{F}|^2}{2} \tag{4-121}$$

and the Q is

$$Q = \frac{\omega \int_0^w \langle P \rangle \, dx}{-\int_0^w \langle dP/dt \rangle \, dx} \tag{4-122}$$

Misawa (1966) considered a specific numerical example of a 5-μm-thick junction region in Si.

4.5. SUMMARY

In this chapter we have described different structures (specifically the Read and p-n diodes) used for IMPATT diode devices. The electric field distribution in these devices has been presented and its coordinate dependence stressed. The impact ionization processes were described and the threshold energy for impact ionization introduced and discussed. We emphasized the impact ionization coefficients of holes and electrons, with emphasis on the importance of the effect of nonlocality on impact ionization. The discrepancy between the phenomenological approach in describing impact ionization and a more rigorous microscopic treatment was examined. Microwave generation in IMPATT diodes was presented via a description of the oscillatory properties of both a Read structure and a conventional p-n diode.

5

The Gunn Diode

5.1. INTRODUCTION

Detailed experimental and theoretical treatments of the Gunn diode, whose operation depends on the transferred electron, or Gunn–Hilsum effect, have been presented by many authors (see, e.g., Shaw et al., 1979). Most of these have centered on sample and device lengths greater than 10 μm. Because recent emphasis has been placed on structures that have near-micrometer or submicrometer lateral dimensions (Wu et al., 1991), we will focus on them in this chapter.

Negative differential mobility (NDM) elements such as the Gunn diode (often referred to as a transferred electron device, TED) operate (1) by virtue of controlled large-amplitude instabilities (as discussed in Chap. 2) and (2) as negative conductance elements characterized by small signal instability. The distinction between large- and small-signal operation is a meaningful one when treating NDM elements and can be described as follows. For small-signal operation an ac source is generally superimposed on a stationary state of the element, and the ac output may be varied continuously by varying the input. For large-signal operation, which is far richer and includes conversion from dc to ac, the output and input are not necessarily continuously related. For example, in the case of a cathode-to-anode transit-time NDM element subjected to a dc bias that is insufficient to cause an instability, the superposition of an ac source will result in a discontinuous relation of output to input when the input amplitude is sufficiently high to cause a domain to nucleate and propagate from the cathode to the anode (Chap. 2).

The analytical description of small-signal operation is more restrictive than that just discussed and assumes that the steady state field profiles, which are nonuniform and vary according to constraints imposed by boundary and space-charge requirements, undergo only negligible alterations during time-dependent operation. The equations of motion of small- and large-signal instabilities are also different. For the former, time-dependent phenomena are described by equations linearized about a stationary solution of the dynamic equations and calculations are amenable to analytical methods. The dynamic equations describe the large-signal behavior of the NDM element; numerical techniques are required here.

While there are more approximations associated with small-signal calculations than with a large-signal analysis, the former are studied because devices are often operated as small-signal amplifiers and oscillators. Furthermore, while numerical methods have been applied to solve the time-dependent small-signal equations, analytical techniques are often used because the explanations of small-signal behavior are sometimes very subtle. Also, the question of large-signal stability can be examined using small-signal analysis. Systems that do not exhibit small-signal instabilities will not exhibit large-signal instabilities. In this chapter we, therefore, examine the small-signal time-dependent behavior of an NDM element. We concentrate on evaluating the small-signal impedance to determine the criteria for small-signal negative resistance and discuss space-charge- and boundary-dependent problems associated with NDM element stability.

One of the earliest small-signal calculations for NDM elements was that of Mahrous and Robson (1966). They derived a closed form expression for the small-signal impedance of an NDM element with extrinsic parameters (e.g., length and doping concentration) similar to those of GaAs specimens for which negative conductance was measured (Thim et al., 1965). A region of small-signal negative resistance was predicted to occur over frequency ranges similar to those experimentally observed.

The procedures used by Mahrous and Robson (1966) in the small-signal calculations have since been used by others (see, e.g., McWhorter and Foyt, 1966; Grubin and Kaul, 1975) and involve three steps:

(1) For a prescribed time-independent value of the cathode boundary field (see Chap. 2), the current density equation [cf. Eq. (2-27)]

$$J_0(t) = v(F)\left[eN_0(x) + \epsilon\frac{\partial F(x,t)}{\partial x}\right] - D(F)\left[\epsilon\frac{\partial^2 F(x,t)}{\partial x^2} + e\frac{\partial N_0(x)}{\partial x}\right] + \epsilon\frac{\partial F(x,t)}{\partial t} \tag{5-1}$$

with $N_0(x) = \text{const}$, $D(F) = 0$, and $\partial F/\partial t = 0$, is solved for a velocity–electric field relation represented by three linear pieces (Fig. 5-1).

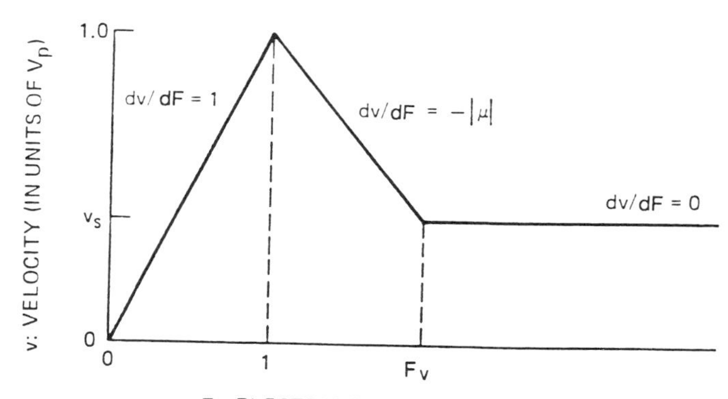

FIGURE 5-1. Three-piece linear representation of an NDM velocity–electric field relation. The axes are labeled in accordance with the dimensionless parameters of Table 5-A1. (After Shaw et al., 1979.)

(2) Perturbations of the stationary electric field profile, $\delta F(x, t)$, are found as solutions to a differential equation obtained by linearizing Eq. (5-1) subject to the above restrictions (zero diffusion, etc.).

(3) The small-signal potential $\delta\Phi(t)$ is calculated by integrating $\delta F(x, t)$ over the length of the NDM element and the small-signal impedance is obtained thereof. An example of the computation, taken from Mahrous and Robson (1966), is displayed in Fig. 5-2. The axes are labelled in accordance with the dimensionless parameters of Table 5-A1.

Calculations similar to those of Fig. 5-2 were performed by McWhorter and Foyt (1966), who, in addition, presented parallel experimental results. By assuming a peak carrier velocity of 2×10^7 cm/s, a value of $F_p = 4000$ V/cm, a low-field mobility of 5000 cm^2/V s, and allowing the NDM to be an adjustable parameter, they were able to fit theory to experiment with an NDM of 2500 cm^2/V s. These results are shown in Fig. 5-3, along with the relevant experimental parameters.

An important point to note from the preceding results, particularly those of Fig. 5-2, is that the theoretical small-signal negative resistance is obtained from NDM elements in which the dc voltage exceeds the threshold voltage for NDM, namely, $F_p \times$ sample length (F_p is approximately 3.2 kV/cm for GaAs having $\mu \approx 6000$ cm^2/V s). The experiments (Thim et al., 1965) revealed that the negative conductance began at applied fields (bias/voltage divided by sample length) of 3.1 kV/cm and increased with increasing applied field until the onset of current oscillations. While there was experimental and theoretical correlation, the dilemma present at that time was the thought that an NDM device would not be dc stable when operated at applied fields in excess of F_p.

The interpretive difficulties were thought to be resolved by another small-signal calculation, by McCumber and Chynoweth (1966), who calculated the small-signal impedance assuming a dc uniform average field within the NDM element. They demonstrated that the NDM element with uniform fields is small-signal stable provided the product of doping concentration and sample length was less that $2.7 \times 10^{10}/cm^2$. Bott and Fawcett (1968), using a more representative $v(F)$ relation than that used by McCumber and Chynoweth (1966), computed a critical $N_0 l$ product for uniform fields of $0.76 \times 10^{11}/cm^2$. In the experiments of McWhorter and Foyt (1966), for example, the operating background carrier concentration was $1.6 \times 10^{13}/cm^3$ and the sample length was 50 μm. Thus $N_0 l = 0.8 \times 10^{11}/cm^2$.

The results of McCumber and Chynoweth (1966) provided a dividing line between small-signal stable and unstable samples. The low $N_0 l$ product samples were dc stable and called "subcritical"; the high $N_0 l$ samples, which were not dc stable at high bias levels, were referred to as "supercritical." But there were still difficulties in interpretation; for example, in one of the early experiments by Gunn (1969), prethreshold negative conductance was observed from a nominally supercritical sample. In a later set of experiments, Perlman et al. (1970) demonstrated that supercritical devices could be operated as dc stable amplifiers at applied bias fields substantially in excess of F_p.

Two separate small-signal studies were presented later that resolved the conflict associated with the presence of dc stable small-signal amplification with

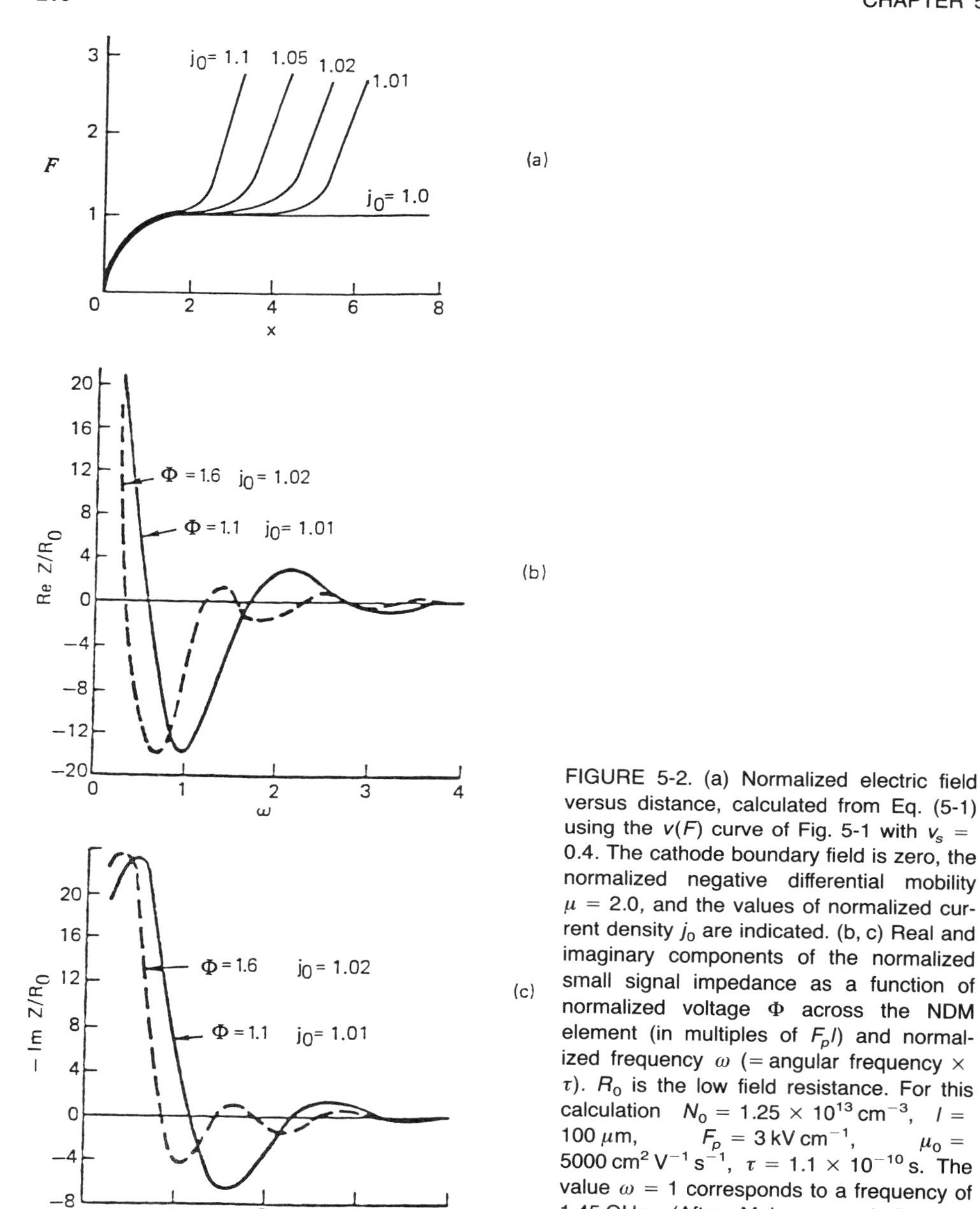

FIGURE 5-2. (a) Normalized electric field versus distance, calculated from Eq. (5-1) using the $v(F)$ curve of Fig. 5-1 with v_s = 0.4. The cathode boundary field is zero, the normalized negative differential mobility μ = 2.0, and the values of normalized current density j_0 are indicated. (b, c) Real and imaginary components of the normalized small signal impedance as a function of normalized voltage Φ across the NDM element (in multiples of $F_p l$) and normalized frequency ω (= angular frequency × τ). R_0 is the low field resistance. For this calculation $N_0 = 1.25 \times 10^{13}\,\mathrm{cm}^{-3}$, l = 100 μm, $F_p = 3\,\mathrm{kV\,cm}^{-1}$, $\mu_0 = 5000\,\mathrm{cm}^2\,\mathrm{V}^{-1}\,\mathrm{s}^{-1}$, $\tau = 1.1 \times 10^{-10}$ s. The value $\omega = 1$ corresponds to a frequency of 1.45 GHz. (After Mahrous and Robson, 1966.)

supercritical samples. In one case numerical evaluation of the small-signal impedance was presented along with experiments (Spitalnik et al., 1973); in another, analytical computations were performed (Grubin and Kaul, 1975). Both studies, which were for nonuniform depletion and accumulation layers beginning within the NDM region, indicated that the length of the NDM region rather than the length of the NDM element was as critical a factor in determining the

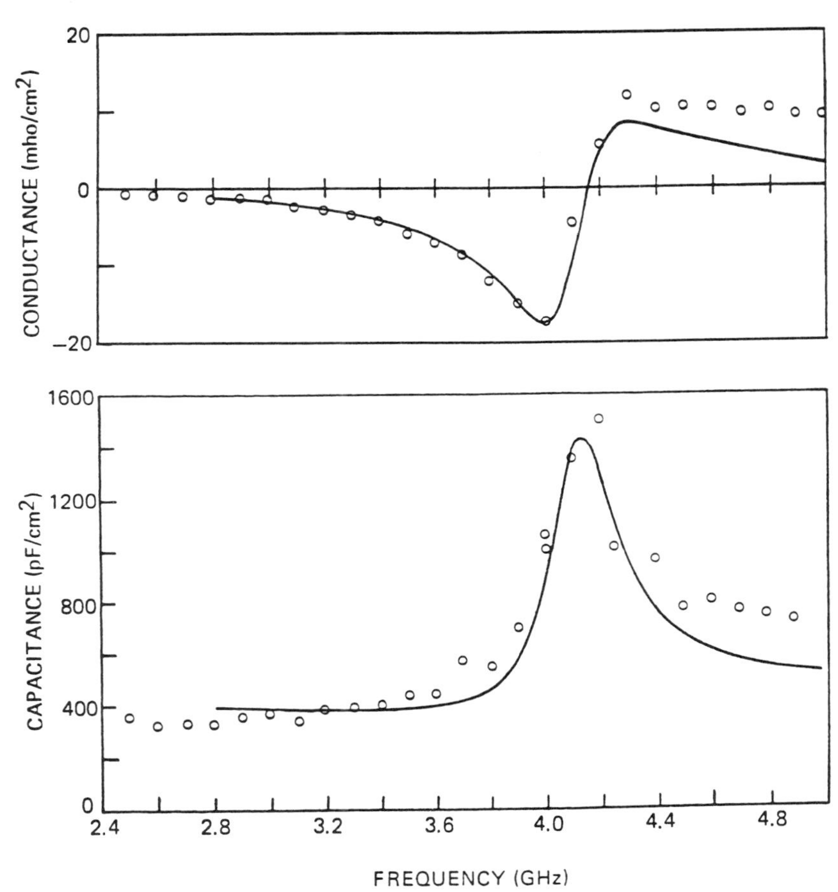

FIGURE 5-3. Comparison of the measured small-signal conductance and capacitance per unit area (○) with calculations based on the $v(F)$ curve of Fig. 5-1 with $v_s = 0.4$, $v_p = 2 \times 10^7$ cm s^{-1}, and $F_p = 4$ kV cm^{-1}. The sample, which was n-GaAs, was 50 μm long, and at the operating temperature of −40°C, $N_0 = 1.6 \times 10^{13}$ cm^{-3} and $\mu_0 = 6000$ cm^2 V^{-1} s^{-1}. The bias was 29 V, which is above $F_p l$ for an assumed $F_p = 4$ kV cm^{-1}. (After McWhorter and Foyt, 1966.)

small-signal stability of an NDM element as the $N_0 l$ product. Supercritical samples were shown to be dc stable under certain conditions. Noting that most samples operate under nonuniform field conditions, it should not be surprising that the stability criterion depends on the length of the NDM region.

The preceding arguments indicate that small-signal calculations have played an important role in developing concepts associated with the operation of NDM elements. But there is also another point that should be noted: small-signal negative resistance occurs only for a finite range of frequencies. If it were possible to obtain, in nature, a true negative resistance element, we would expect the gain to be limited only by the properties of the external circuit. Yet it is found (Fig. 5-3) that small signal negative resistance occurs only over a limited frequency range. The origin of this effect has to do with the fact that NDM elements contain moving space-charge layers and that a perturbation in electric field at a point x_1 will be carried downstream from the cathode and contribute to a change in the value of the electric field at the point $x_2(>x_1)$. The field perturbation at x_2 thus receives two contributions: a nonlocal contribution from x_1 and a local contribution from x_2. Consequently, there is a transit-time delay between the origin of a perturbation and its effect at different points along its trajectory. This nonlocal transit-time delay, which is basically a consequence of the fact that the carriers travel at finite velocities, introduces phase changes between the current through the NDM element and the voltage across it. For a certain range of frequencies there will be gain and for another range of frequencies there will be loss.

This chapter is concerned primarily with analytically elaborating on the concepts discussed in the preceding paragraphs. The order in which we do this is similar to that of Mahrous and Robson (1966).

In Sec. 5.2 we obtain analytical expressions for stationary depletion and accumulation layers beginning within the NDM region. The profiles are the zeroth-order solutions from which the time-dependent perturbations are obtained. The current–voltage characteristics associated with these layers are also obtained. The sensitivity of the results to the value of the NDM and cathode field is discussed. All computations are in terms of the dimensionless variables listed in Table 5-A1.

The time-dependent calculations have been presented by Shaw et al. (1979). They derived the space and time dependence of the perturbations in the stationary field profiles. They isolated, by specific examples, the significance of the nonlocal contributions to the perturbed field, which contributes to transit-time delays and phase differences between the current and voltage contributions. They then defined the small-signal impedance, relating it to the Fourier coefficients of the small-signal potential. They computed the small-signal impedance of an NDM element for a variety of dc electric field profiles, including depletion and accumulation layers beginning within the NDM region. Then, using approximate expressions, they examined the stability of depletion and accumulation layer profiles, seeking zeros of the total impedance. They extended the arguments of McCumber and Chynoweth (1966) and included the effects of a finite load resistance.

The calculations of this chapter are specific to both long and short supercritical NDM elements, but the analytical results are also applicable to subcritical NDM elements. The differential equation describing the stationary field profile is of first order and no details of the anode boundary are included. In the computations the anode region is simply that point downstream from the cathode where the computations cease. However, some of the approximations, particularly the approximate stability calculations, are specific to long samples and are so identified in the text.

5.2. TIME-INDEPENDENT CALCULATIONS

The steady state characteristics of the NDM element may be obtained from the time-independent electric field versus distance profile, $F(x)$. $F(x)$ is obtained as a solution to Eq. (5-1) for $\partial F/\partial t = D(F) = \partial N_0(x)/\partial x = 0$ and $N_0 = 1$:

$$J_0 = v(F)(eN_0 + \epsilon\, dF/dx) \tag{5-2a}$$

Then with $J_0 = N_0 e v_p j_0$, $F = F_p F$, $x = x\epsilon F_p/N_0 e$, where v_p is the peak velocity and F_p is the field at peak velocity, the current equation is conveniently written as

$$j_0 = v(F)(1 + dF/dx) \tag{5-2b}$$

Equation (5-2) is a first-order differential equation requiring one boundary condition for its solution. Since $v(F)$ is represented by three linear pieces, $F(x)$ is computed separately in each of the three regions. The boundary condition to Eq. (5-2) is taken to be the value of $F(x)$ at the beginning of the region of interest. At the cathode, $F(x = 0)$ is denoted by F_c and the value of $v(F_c)$ is denoted by v_c.

The $F(x)$ profiles are obtained by integrating Eq. (5-2),

$$x_2 - x_1 = \int_{F(x_1)}^{F(x_2)} \frac{v(F')\, dF'}{j_0 - v(F')} \tag{5-3}$$

which transforms to

$$x_2 - x_1 = F(x_1) - F(x_2) + j_0 t(x_1, x_2) \tag{5-4}$$

Here,

$$t(x_1, x_2) = \int_{x_1}^{x_2} dx''/v[F(x'')] \tag{5-5}$$

is the transit time of a carrier between the points x_1 and x_2 normalized to the dielectric relaxation time $\tau = \epsilon F_p/(N_0 e v_p)$. For transit within either the NDM or ohmic regions, we have

$$t(x_1, x_2) = -\mu^{-1} \log\{j_0 - v[F(x_2)]\}/\{j_0 - v[F(x_1)]\} \tag{5-6a}$$

where $\mu = dv/dF$ and all logarithms are to the base e. Within the saturated drift velocity (SDV) region, we have

$$t(x_1, x_2) = [F(x_1) - F(x_1)]/[j_0 - v_s] \tag{5-6b}$$

The quantity $t(x_1, x_2)$ is a derived quantity that is used in much of the following discussion. Transit-time periods, amplification bandwidths, etc., are related to the transit times across specific regions of the NDM element. When space-charge layers are uniformly distributed, transit time and "distance traveled" are related by a constant; when the space charge is nonuniformly distributed, as is usually the case in NDM elements, they are not.

The electric field versus distance profile (see Chap. 2) may be computed from the preceding equations. Within a given region, depletion layers form if $v(F) > j_0$ is satisfied everywhere within the region. Accumulation layers form if $v(F) < j_0$. Figure 5-4 displays two groups of depletion layer profiles beginning within the

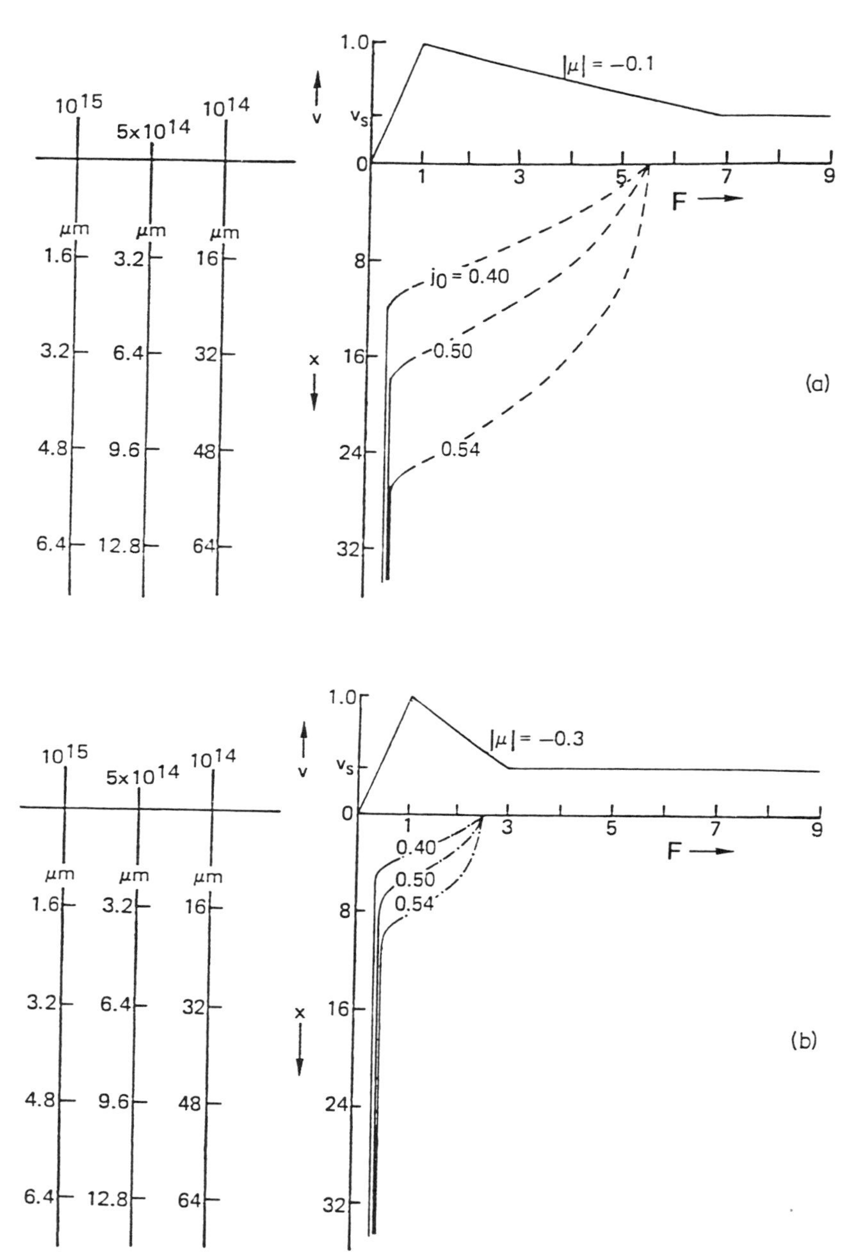

FIGURE 5-4. The electric field versus distance profile as obtained from Eq. (5-1) for the indicated $v(F)$ curve. j_0 is displayed, and (a) $\mu = -0.1$ and (b) $\mu = -0.3$. Four sets of axes are displayed for each computation: the dimensionless axes, and those corresponding to finite doping densities. (Note: In these computations $X = L_0x$, and $L_0 = 2$, 0.4, and 0.2 μm at 10^{14}, 5×10^{14}, and 10^{15} cm^{-3}, respectively.) (After Grubin et al., 1973.)

NDM region. In Fig. 5-4a $F(x)$, for $F_c = 5.5$ and $v_c = 0.55$, is plotted for three different values of j_0. The NDM for this case is -0.1. In Fig. 5-4b a similar calculation is performed for NDM of -0.3. Here $F_c = 2.5$ and $v_c = 0.55$. In dimensionless units the threshold current density for cathode originated instabilities is v_c, and both sets of calculations displayed in Fig. 5-4 are for the same v_c. The calculations demonstrate that for NDM elements with similar values of threshold current density the voltage drop across the cathode region is greater for the NDM element with the shallower NDM region. Another point worth emphasizing is also displayed in Fig. 5-4. Here we show four separate distance axes. One is in dimensionless units, while the others are for doping levels of 10^{14}–$10^{15}\,\text{cm}^{-3}$. We see that the length of the depletion layer is greatest for the NDM element with the lowest doping. Thus, as found in the case of stably propagating domains (Butcher, 1967), the width of the cathode adjacent depletion layer depends on the carrier concentration.

Figure 5-5 shows $F(x)$ for F_c within the SDV region. As in Fig. 5-4b, the NDM is -0.3. But, even for the smaller value of j_0, the depletion layer is significantly greater than that associated with the former calculation.

As the two preceding figures indicate, the length of the depletion layer is approximately determined by that portion of the sample within the NDM region (Fig. 5-4) or that portion within the SDV and NDM regions (Fig. 5-5). The portion of the depletion layer within the ohmic region can be obtained from Eqs. (5-3)–(5-6). The length, in dimensionless units, of that portion of the depletion layer within the NDM region extending from $x = 0$ to that point where $F(x) = 1$ (see Fig. 5-1) is

$$\Delta_{d,\text{NDM}} = |\mu^{-1}|\,[1 - v_c + j_0 \log(1 - j_0)/(v_c - j_0)] \tag{5-7}$$

where within the NDM region we have used the relation

$$v = 1 - |\mu|\,(F - 1) \tag{5-8}$$

Equation (5-7) indicates that $\Delta_{d,\text{NDM}}$ is smallest for v_c approximately equal to unity and is greatest for $v_c \cong v_s$. For solutions starting within the SDV region at $x = 0$ and extending to the point where $F(x) = F_v$, the depletion length is

$$\Delta_{d,\text{SDV}} = v_s(F_c - F_v)/(v_s - j_0) \tag{5-9}$$

For the case in which the length of the depletion layer extends across the SDV and NDM regions, we add Eq. (5-7) to (5-9) with v_c in Eq. (5-7) replaced by v_s. Note that in real units the length of the depletion layer is obtained as follows:

$$\text{Depletion layer length in centimeters} = \Delta_d v_p \tau \tag{5-10}$$

The gross properties of the cathode region voltage drop can be obtained experimentally by voltage probing. The experimental current–voltage characteristics provide additional information (Shaw et al., 1979). The current–voltage

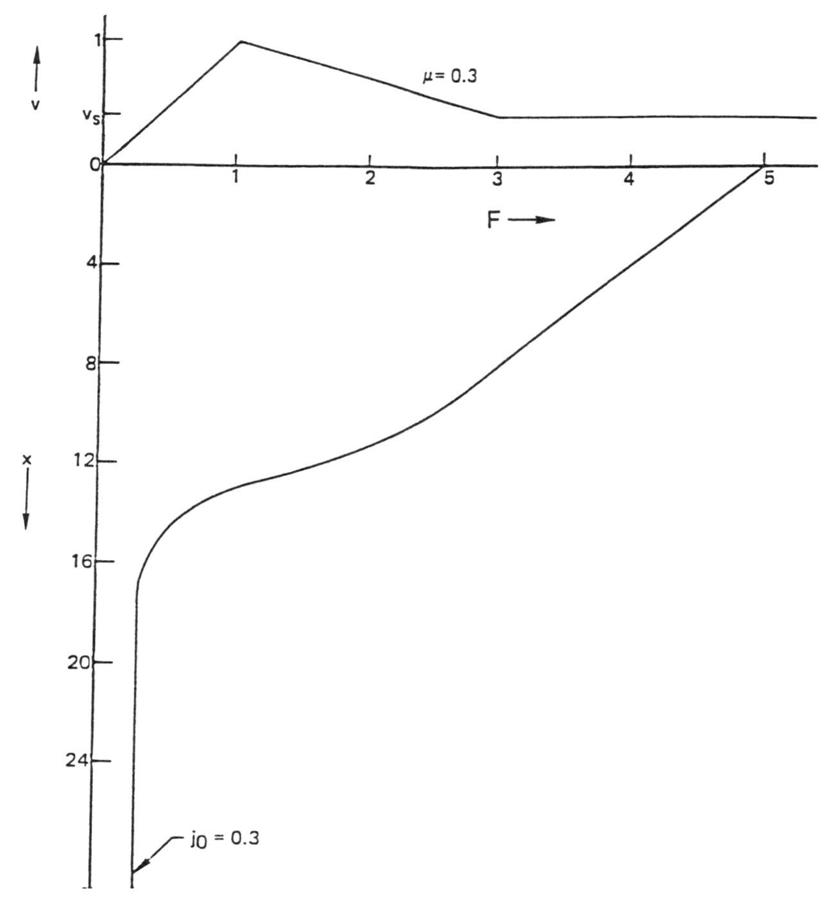

FIGURE 5-5. The electric field versus distance profile as in Fig. 5-4. Here $F/F_p = 5.0$, $j_0 = 0.3$, and $\mu = -0.3$. (After Grubin et al., 1973.)

characteristics can be obtained analytically from the preceding equations. In doing this we find it convenient to write the potential as

$$\Phi = \frac{1}{l'} \int_0^{l'} F(x)\, dx \cong j_0 + \langle F_{\rm ex} \rangle \equiv \langle F \rangle \tag{5-11}$$

where $\langle F_{\rm ex} \rangle$ denotes the excess voltage across the depletion layer and $l' \equiv l/(v_p \tau)$. $\langle F \rangle$, which is the normalized voltage, also denotes an average field. Integration of the electric field profiles is direct, and for solutions beginning within the NDM region, we have

$$\langle F_{\rm ex} \rangle l' = \tfrac{1}{2}(F_c^2 - F_a^2) + j_0(1 + |\mu^{-1}|)[t(0, \Delta_{d,\rm NDM}) - \Delta_{d,\rm NDM}] \tag{5-12}$$

where $\Delta_{d,\rm NDM}$ is given by Eq. (5-7) and $t(0, \Delta_{d,\rm NDM})$ is given by Eq. (5-5). Also, the anode field, F_a, is approximately equal to j_0. For solutions beginning within the SDV region, we have

$$\langle F_{\text{ex}} \rangle l' = [\tfrac{1}{2}(F_c^2 - F_v^2) + j_0(F_v - F_c)[\tfrac{1}{2}(F_c + F_v) - v_s]/(v_s - j_0)]$$
$$+ [\tfrac{1}{2}(F_v^2 - F_a^2) + j_0(1 + |\mu^{-1}|)$$
$$\times \{t(\Delta_{d,\text{SDV}}, \Delta_{d,\text{SDV}} + \Delta_{d,\text{NDM}}) - \Delta_{d,\text{NDM}}\}] \tag{5-13}$$

where $\Delta_{d,\text{NDM}}$ is given by Eq. (5-7) with v_c replaced by v_s and F_c replaced by F_v. The first part of Eq. (5-13) is the excess voltage across the SDV region.

An important consequence of the preceding result is that $\langle F_{\text{ex}} \rangle$ is greater than zero. Thus, two points follow:

(1) In the absence of a depleted cathode region (the uniform field case), the preinstability current–voltage relation scales the preinstability velocity–electric field relation.

(2) For a linear preinstability $v(F)$ the presence of a sublinear preinstability current–voltage relation indicates the presence of a cathode depletion region. Several illustrations follow.

Figure 5-6 displays the current density versus average electric field for two $5 \times 10^{14}\,\text{cm}^{-3}$ doped 100 μm-long NDM elements with an NDM of -0.1. The elements are distinguished by their v_c values (hence F_c values). (The choice of average field rather than voltage as abscissa has the merit of emphasizing the relative contribution of the cathode region.) Note the sublinear characteristics with an enhanced sublinearity for the higher cathode field element. (See Fig. 5-4a for the field profiles.) The two sets of calculations are presented to draw attention to the fact that a reduction in cathode field generally results in a decrease in the cathode voltage drop. Also, for comparison, we include the linear current versus average field relation. (Note a shift in the curves away from the uniform field case. This is an unphysical feature peculiar to the fixed cathode field model where F_c is independent of current.)

Figure 5-7 displays the current–average field curves for two 100-μm-long NDM elements with $v_c = 0.55$ and $\mu = -0.1$. These elements differ in their

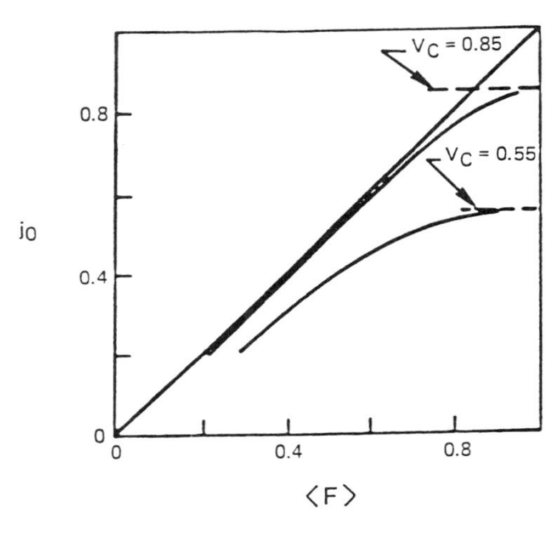

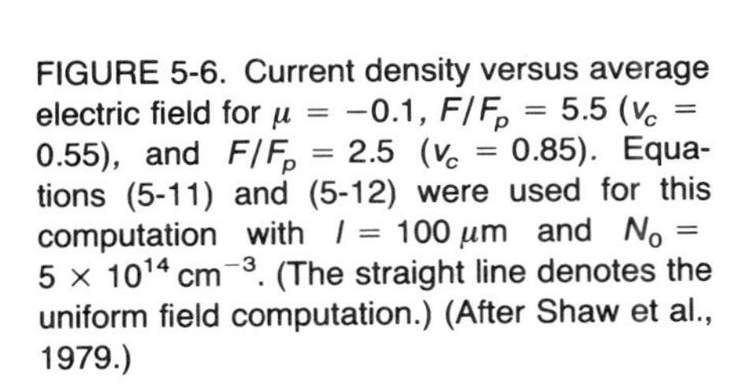
FIGURE 5-6. Current density versus average electric field for $\mu = -0.1$, $F/F_p = 5.5$ ($v_c = 0.55$), and $F/F_p = 2.5$ ($v_c = 0.85$). Equations (5-11) and (5-12) were used for this computation with $l = 100\,\mu$m and $N_0 = 5 \times 10^{14}\,\text{cm}^{-3}$. (The straight line denotes the uniform field computation.) (After Shaw et al., 1979.)

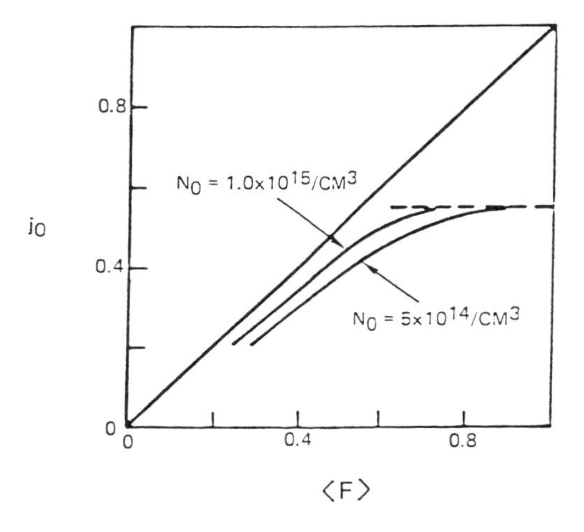

FIGURE 5-7. Current density versus average electric field for $\mu = -0.1$, $F/F_p = 5.5$, and $l = 100\ \mu$m. Two values of N_0 are chosen for computations and are indicated. Equations (5-11) and (5-12) were used. (After Shaw et al., 1979.)

doping level; the calculations are displayed to demonstrate the dependence of the characteristics on the doping. (See the discussion associated with Fig. 5-4.)

Figure 5-8 displays the current–average field characteristic for a cathode field well into the SDV region. The saturated current density approaches v_s.

The preceding results indicate that in NDM elements where the velocity–field relation is linear prior to NDM, a sublinear current–voltage relation is a signature of a cathode field in excess of the threshold field for NDM. (Shaw et al., 1969; 1979). When the sublinear current–voltage relation is followed in a resistive circuit by a time-dependent instability, these results also indicate that the cathode boundary field is in the NDM region. (The results are true even when the cathode field is current dependent (Grubin, 1976).) For GaAs, where $v(F)$ is approximately linear prior to NDM (Ruch and Kino, 1967), a prethreshold current–voltage measurement is therefore a useful tool. When the $v(F)$ relation is sublinear prior to NDM, as in n-Ge (Chang and Ruch, 1968) and n-InP (Kaul et al., 1972; Fawcett and Herbert, 1972), a sublinear current–voltage relation requires voltage versus distance probe measurements for clarification or com-

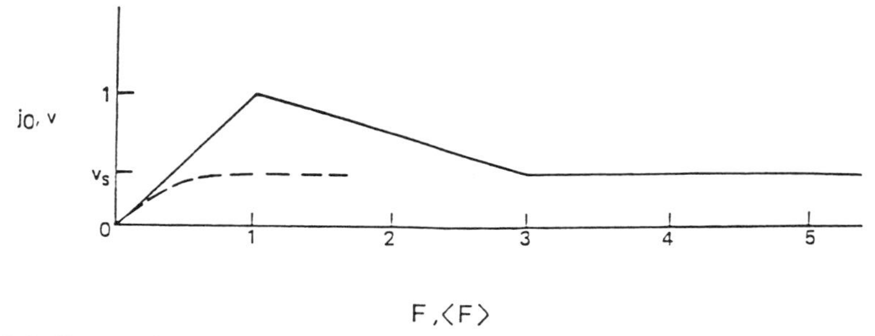

FIGURE 5-8. Current density (---) versus average electric field for F/F_p within the SDV region. Here $F/F_p = 5$ and $\mu = -0.3$. Equations (5-11) and (5-13) were used for this computation with $l = 100\ \mu$m and $N_0 = 10^{15}\ \text{cm}^{-3}$. Also shown for reference is the $v(F)$ curve. (After Grubin et al., 1973.)

parison with a known uniform field case. However, with respect to these semiconductors, there are also features that influence the extent to which each exhibits sublinearity in the current–voltage characteristic. For example, in the case of n-GaAs, an averaged μ obtained from the relation

$$\mu = \mu_0[1 - v(@10\,\text{kV/cm})v_p]/[(1 - 10\,\text{kV/cm})/F_p]$$

is -0.2, and for n-Ge at 33 K, $\mu = -0.007$ (Chang and Ruch, 1968). Thus, for a given range of F_c, the length of the depletion layer will be greatest for n-Ge, yielding the most sublinear prethreshold current–voltage curves due to contact effects.

Another point is worth mentioning: The "bias at threshold" includes the voltage across the cathode depletion region as well as the voltage across the quiescent part of the sample. At threshold, the relative cathode contribution to the voltage decreases as the length of the NDM element increases. Therefore, for NDM elements with similar values of F_c, as the length increases, the bias at threshold divided by sample length decreases. This behavior has been observed in n-GaAs by, among others, Foyt and McWhorter (1966) and Gunn (1964).

The preceding discussion has so far been confined to depletion layer profiles. But Eqs. (5-1)–(5-6) are also applicable to the study of accumulation layer profiles. An accumulation layer of particular interest is that which begins within the region of NDM and increases into the SDV region, where it continues to grow until the anode is reached; the field subsequently decreases. This type of profile is unstable in long samples but is thought to be responsible for amplification in short, supercritical (Grubin and Kaul, 1975; Spitalnik et al., 1973) devices, i.e., 10-μm-long, $10^{15}\,\text{cm}^{-3}$ NDM elements. A sketch of an accumulation layer profile as derived from the preceding equations for $j_0 > v_c$ is shown in Fig. 5-9 for a value of $\mu = -0.2$ and $N_0 = 10^{15}\,\text{cm}^{-3}$. Note that $v_c = 0.5$ and $F_c = 3.5$. In Fig. 5-9, $\Delta_{a,\text{NDM}}$ is the length of that portion of the accumulation layer that begins at $x = 0$ and ends at a point where $F(x) = F_a$. It is calculated from Eq. (5-14). Unlike depletion layer profiles, here the length of the accumulation layer within the NDM region decreases with increasing current. It is important to note that a requirement for an accumulation layer to form is that j_0 exceed v_c. It is not necessary that the current exceed the NDM threshold current density, which in dimensionless units is equal to unity.

The length of the accumulation layer within the NDM region is obtained from Eqs. (5-3) and (5-6) and is equal to

$$\Delta_{a,\text{NDM}} = |\mu^{-1}|\,[v_s - v_c + j_0 \log(v_s - j_0)/(v_c - j_c)] \tag{5-14}$$

The potential across the NDM element, including contributions across the NDM and SDV regions is

$$\begin{aligned}\langle F_{\text{ex}}\rangle l' = \tfrac{1}{2}(F_c^2 - F_v^2) + (F_a - F_v)[v_s/(j_0 - v_s)][\tfrac{1}{2}(F_a + F_v) - j_0] \\ + j_0(1 + |\mu^{-1}|)[t(0, \Delta_{a,\text{NDM}}) - \Delta_{a,\text{NDM}}]\end{aligned} \tag{5-15}$$

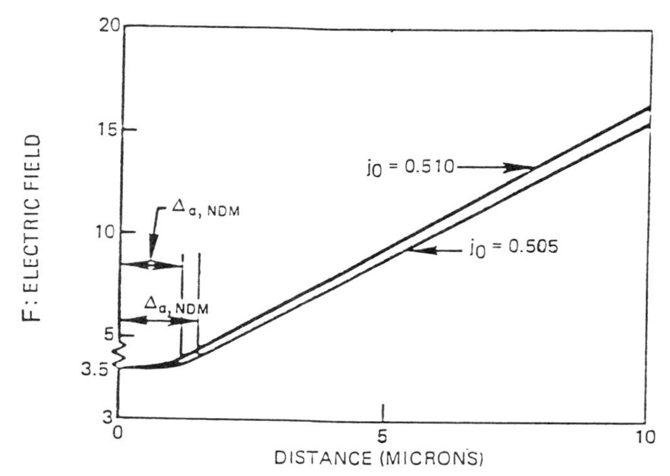

FIGURE 5-9. Normalized electric field versus distance (in micrometers) for a 10-μm-long element with $N_0 = 10^{15}\ \text{cm}^{-3}$, $\mu = -0.2$, $F/F_p = 3.5$ ($v_c = 0.5$), and $j_0 = 0.510$ and 0.505, (after Grubin and Kaul, 1975).

where

$$F_a = F_v + (j_0 - v_s)(l' - \Delta_{a,\text{NDM}})/v_s \tag{5-16}$$

The plots of Fig. 5-6 and 5-8 show that the effect of a cathode-adjacent depletion layer is to lead to sublinearity in the dc current versus average electric field relation. The contribution of a stable dc accumulation layer is different. A plot of current density versus average electric field that includes low bias depletion layer contributions and high bias accumulation layer profiles is shown in Fig. 5-10. This calculation is for a 10-μm-long NDM element doped to $10^{15}\ \text{cm}^{-3}$.

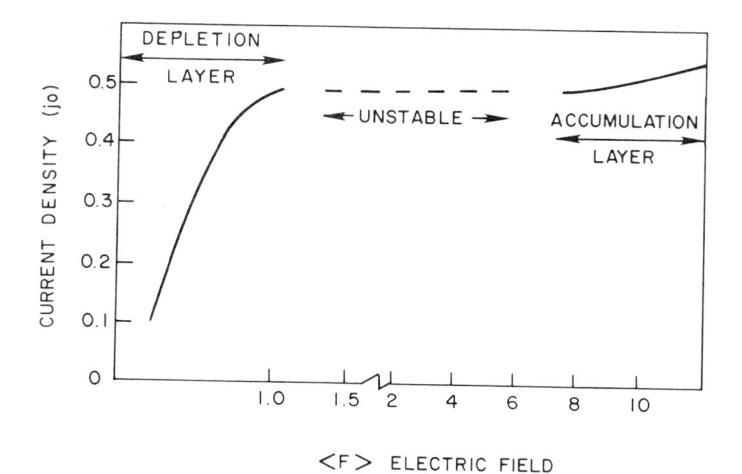

FIGURE 5-10. Current density versus average electric field for an NDM element with the parameters of Fig. 5-9. The range of voltages for depletion, unstable, and accumulation layer profiles are indicated. (After Grubin and Kaul, 1975.)

Note the dashed line separating the two branches. Within this region the profiles are unstable, a point discussed by Shaw et al. (1979) with reference to the topic of NDM amplification.

5.3. STABILITY

5.3.1. Formulation of the Problem

The previous chapters and section provided us with the means of discussing the stability of NDM elements. However, we must first define the type of instability that we are interested in; there are several classes of these.

In the case of Gunn's early experiments (1964), the NDM semiconductor was part of a mostly resistive circuit for which the NDM element was stable for a range of low bias values. At higher values of bias the appearance of periodic coherent current oscillations was regarded as an unstable state.

In another case we can imagine an ac signal superimposed on a dc stable configuration. When the amplitude of the ac source is sufficiently high, large-signal transit-time microwave oscillations may result. These oscillations are also regarded as an instability.

In our stability considerations we attempt a small-signal replication of Gunn's experiment, that is, we include the NDM element in the resistive circuit of Fig. 5-11 and eliminate all external ac sources. We then seek the conditions under which a steady time-dependent current circulates throughout the circuit. We do this by examining one Fourier component of the circulation current. The important condition is that the net ac voltage across the NDM element and load resistor is zero. Expressed in terms of impedances, we have (normalized to the low field resistance of the device, R_0)

$$\delta\Phi(t) = [\mathrm{Re}\, Z(\omega) + R_L/R_0]a(\omega) \sin \omega t + \mathrm{Im}\, Z(\omega)a(\omega) \cos \omega t = 0 \tag{5-17}$$

Thus

$$a(\omega)[\mathrm{Re}\, Z(\omega) + R_L/R_0] = a(\omega)\, \mathrm{Im}\, Z(\omega) = 0 \tag{5-18}$$

Since we are interested in the condition when a time-dependent current is circulating throughout the circuit of Fig. 5-11, we seek the criteria for which

$$\mathrm{Re}\, Z(\omega) + R_L/R_0 = 0 \tag{5-19}$$

and

$$\mathrm{Im}\, Z(\omega) = 0 \tag{5-20}$$

are satisfied and note that the usual circumstance under which Eq. (5-18) is satisfied is $a(\omega) = 0$.

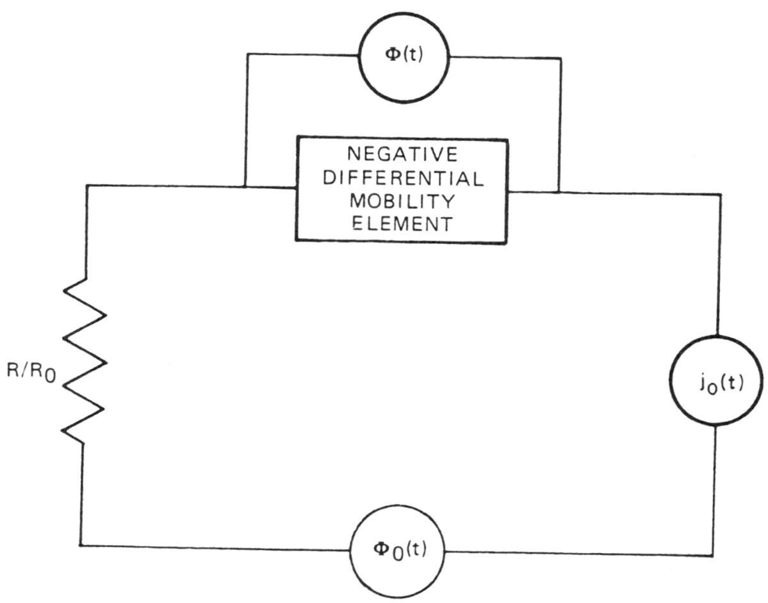

FIGURE 5-11. Circuit used in the small signal calculations. Parameters are in normalized units. (After Shaw et al., 1979.)

The conditions of Eqs. (5-19) and (5-20) refer to an electrically unstable condition, in the small-signal sense. The question is: What does the small-signal instability have to do with the large-signal instability associated with Gunn's original experiment? Small-signal behavior is not the same as large-signal behavior, and it is conceptually possible that a small-signal instability occurring at one bias level could be damped by reducing the bias level. A similar reduction in bias level for a propagating domain would not necessarily result in domain quenching until the domain reached the anode. Furthermore, the linearized equations of this chapter do not permit a rigorous discussion of the transition from the small- to the large-signal region. Rather, numerical computations are required in which the dynamic equations are subject to small but gradual increases in an applied ac source. But a system that does not exhibit a small-signal instability will not exhibit a large-signal instability. A small-signal analysis can therefore provide a delineation of the stable and unstable regions. Experience has indicated that the delineation of small-signal stable and unstable regions is approximately the same as that of the large-signal case, indicating that the small-signal instability is a precursor to the large-signal instability. With this in mind, we evaluate Eqs. (5-19) and (5-20) for two cases:

1. The stability of uniform $F(x)$ and nonuniform $\delta F(x, t)$ profiles (this case has received the most attention in the literature);
2. The stability of nonuniform depletion layer profiles (this case is relevant to common cathode originated instabilities).

5.3.2. Stability of Uniform F(x) and Nonuniform δF(x, t) Profiles

From Eqs. (5-19) and (5-20), we examine solutions of the equation

$$\frac{l'}{j_0}\frac{\exp(-\lambda) + \lambda - 1}{\lambda^2} - \frac{1}{\sigma_c}\frac{\exp(-\lambda) - 1}{\lambda} + \frac{R_L}{R_0} = 0 \tag{5-21}$$

where $\lambda \equiv t(0, l')(\mu + i\omega)$. The most frequently discussed case, and the one considered next, is that of McCumber and Chynoweth (1966) with $\sigma_c \to \infty$ and $R_L/R_0 = 0$. In this limit the zeros of the impedance lead to the conditions

$$\exp[|\mu|\, t(0, l')] \sin\theta = \theta \tag{5-22a}$$

$$\exp[|\mu|\, t(0, l')] \cos\theta = 1 + |\mu|\, t(0, l') \tag{5-22b}$$

where $\theta = \omega t(0, l')$. McCumber and Chynoweth (1966) have shown that the lowest nonzero root of the above equation is

$$|\mu|\, t(0, l') = 2.09 \tag{5-23a}$$

$$\omega t(0, l') = \pm 7.46 \tag{5-23b}$$

Equation (5-23a) translates, in ordinary units (see the Appendix), into the instability condition

$$N_0 l = 2.09 \frac{\epsilon}{e} \frac{v_p}{|dv/dF|} \frac{J_0}{N_0 e v_p} \text{ cm}^{-2} \tag{5-24}$$

Thus, the $N_0 l$ product depends upon the magnitude of the differential mobility of the NDM region and on the current density. For the values given in Table 5-1, $J_0 \cong \frac{1}{2} N_0 e v_p$ and $dv/dF = 0.2\mu_0$, we find

$$N_0 l \cong 10^{11} \text{ cm}^{-2} \tag{5-25}$$

as a condition for a small-signal instability. Thus, we see that a condition can be found that allows current to circulate in the circuit of Fig. 5-11, and that this condition depends on the sample's length and doping level. The circulation current does not require the presence of an external ac source. This result also appears to be qualitatively applicable when a finite load is included in the calculation.

We next ask, what type of picture does the instability criterion provide? The answer must be given in terms of the assumption of a uniform field profile. We thus imagine that the electric field profile within the NDM element is uniform and that at low bias levels is within the ohmic portion of the dc $v(F)$ curve. For this case, Re $Z(\omega) + R_L/R_0$ is always positive and no ac current circulates within the circuit. Now consider the situation at elevated bias levels. According to the criteria of Eq. (5-23), entering the NDM region does not guarantee the presence of a circulating current. For this to occur the differential mobility times

TABLE 5-1. GaAs Parameters Used in the Calculations

Parameters	Γ(000)	L(111)	Common
Number of equivalent valleys	1	4	
Effective mass (m_e)	0.067	0.222	
Γ–L separation (eV)			0.33
Polar optical scattering			
Static dielectric constant			12.90
High-frequency dielectric constant			10.92
LO phonon (eV)			0.0354
Γ–L scattering			
Coupling constant (eV/cm)			0.800×10^9
Phonon energy (eV)			0.0278
L–L scattering			
Coupling constant (eV/cm)		2.0×10^9	
Phonon energy (eV)		0.0354	
Acoustic scattering			
Deformation potential (eV)	7.0	9.2	
Nonpolar scattering (L)			
Coupling constant (eV/cm)		0.300×10^9	
Phonon energy (eV)		0.0343	

transit-time product must be sufficiently high [see Eq. (5-23a)]. Thus, if we had the situation where the shape of the $v(F)$ curve within the NDM region was a function of electric field and the criteria of Eq. (5-23) were satisfied only when the field was well into the NDM region (or its modification due to a finite load), we would expect an instability to be delayed until the electric field was in excess of the NDM threshold field. For the situation of a delayed instability we might also expect to see prior to a current instability:

1. A dc negative differential resistance; and
2. ac power delivered to the load by the NDM element when an external ac source is present.

With regard to the dc negative resistance, the analysis indicates that the electric field within the NDM element rearranges itself to prevent the existence of dc negative differential resistance under isothermal conditions.

The experimental situation with regard to the N_0l product criterion was discussed in the introductory sections of this chapter. But more specifically, Thim and Barber (1966) reported the observation of dc stable negative rf resistance in samples with N_0l products below $5 \times 10^{11}\ \text{cm}^{-2}$. Samples with N_0l products exceeding this value always broke out into uncontrolled oscillations. Thim and Barber also reported that all samples that exhibited rf negative resistance in the microwave range showed a positive differential resistance at dc. They also reported results in which increases in applied bias resulted in increases in negative conductance until the latter exceeded the conductance of the load. At this point oscillations broke out.

In summary, the McCumber and Chynoweth study provided a broad explanation for much of the NDM device behavior. Difficulties in matching some of their conclusions with experiments can for the most part be traced to the assumption of uniform fields within the device. This assumption has since been relaxed by a number of workers. In particular, Kroemer (1968) analyzed nonuniform field effects by separating the NDM semiconductor into an NDM region and a lossy ohmic region. He was able to show that the N_0l product criterion was, in addition, dependent on the length of the lossy region of the NDM element. Shaw et al. (1969) and Grubin et al. (1973) examined the stability of the depletion layer profile, determining the role of the bias level and the influence of the quiescent part of the NDM element on the instability. They were also able to establish a set of threshold conditions for the current instability (Shaw et al., 1979). This is discussed in the next section.

5.3.3. Stability of Nonuniform Depletion Layer Profiles

The bias dependence of $Z(\omega) + R_L/R_0$ and the conditions for an instability are illustrated in Fig. 5-12 for a depletion layer profile. The figure will be discussed at the end of this section. First, we discuss an evaluation of $Z(\omega) + R_L/R_0$ in the limit where nonlocal effects dominate (Shaw et al., 1979). In this case the zeros of $Z(\omega) + R_L/R_0$ require that

$$\sin(\theta_c + 2\Psi_2 + \Psi_1) = 0 \tag{5-26a}$$

$$\cos(\theta_c + 2\Psi_2 + \Psi_1) < 0 \tag{5-26b}$$

$$\theta_c + 2\psi_2 + \psi_1 = (2m - 1)\pi \tag{5-26c}$$

where m is a positive integer. When the conditions of Eq. (5-17) are satisfied, we obtain from Eq. (5-21)

$$\frac{j_0 \exp[|\mu|\, t(0, \Delta)](1 + |\mu|)}{l'(\mu^2 + \omega^2)(1 + \omega^2)^{1/2}} = \frac{R_L}{R_0} \tag{5-27}$$

Interpreting the above equations broadly, Eq. (5-26) determines the frequency–transit-time combination for an instability. Equation (5-27) determines the length of the depletion layer necessary for an instability. Generally, the larger R_L/R_0 is, the longer the depletion length must be.

For purposes of evaluation we now rearrange Eqs. (5-26) and (5-27). Noting that $\tan\omega$ is of the order of magnitude of ω, for ω varying from zero to unity, and therefore that $\arctan\omega$ is of the order of ω for this frequency range, and that $\theta_c = \omega t(0, \Delta) > \omega$, we find

$$2\Psi_2 = \arctan[2\mu\omega/(\mu^2 - \omega^2)] \cong (2m - 1)\pi - \theta_c \tag{5-28a}$$

and

$$\tan\theta_c = 2\mu\omega/(\omega^2 - \mu^2) \tag{5-28b}$$

which can be arranged to read

$$\tan(\theta_c/2) = -|\mu|/\omega \tag{5-29}$$

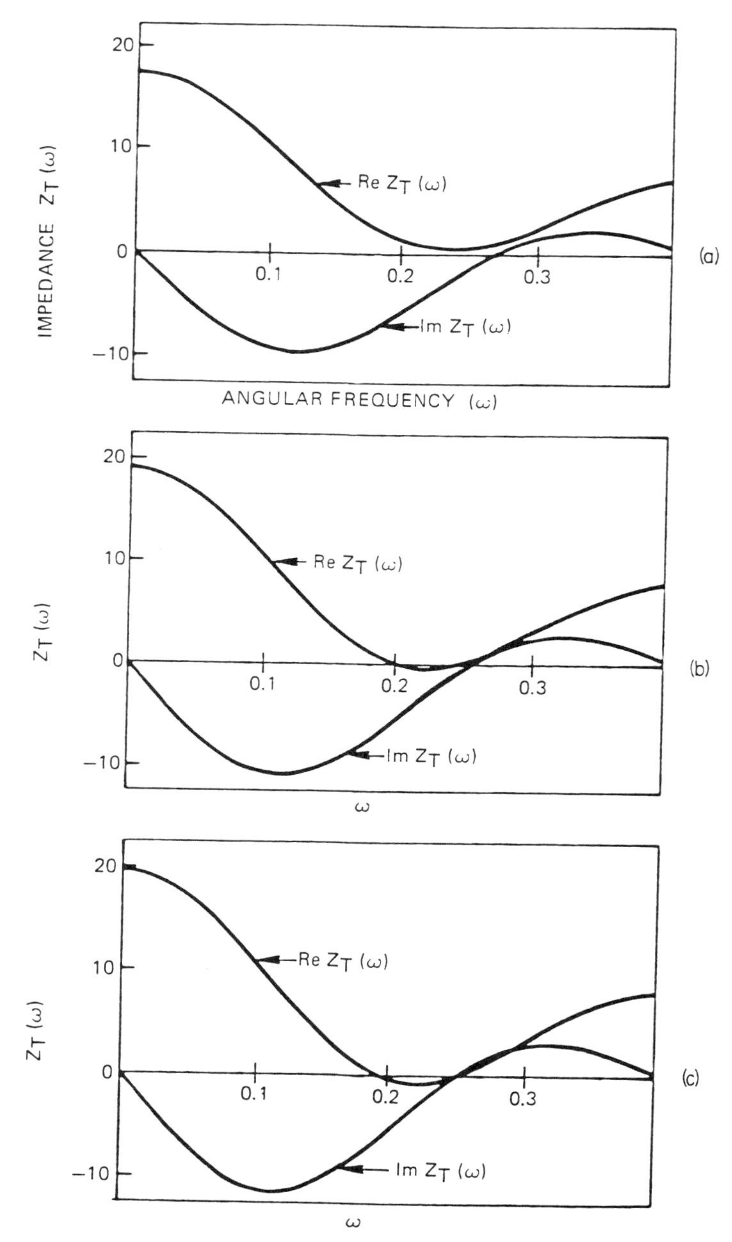

FIGURE 5-12. Approach to an instability for a depletion layer within a 10-μm-long NDM element with the parameters of Fig. 5-9. Here $Z_T(\omega) = Z(\omega)5$. (a) $j_o = 0.488$, (b) $j_o = 0.489$, (c) $j_o = 0.490$. (After Grubin and Kaul, 1975.)

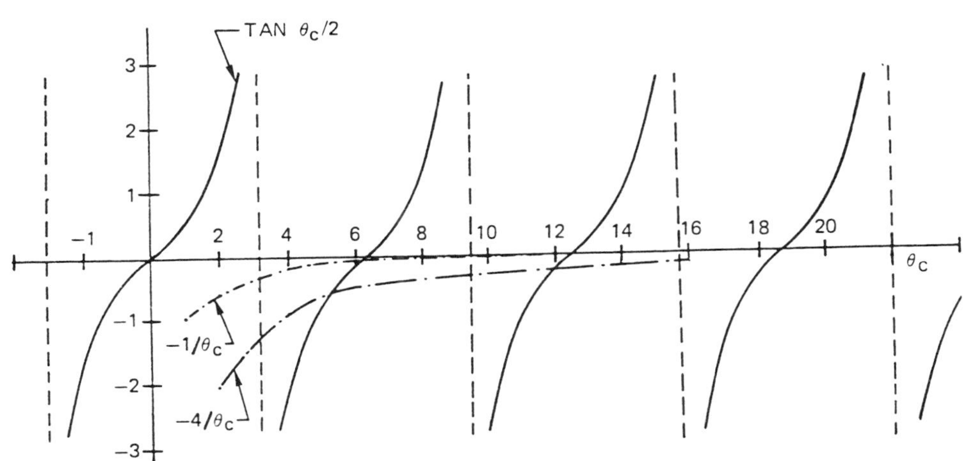

FIGURE 5-13. Superimposed plots of tan $\theta_c/2$ and $-|\mu|\,t(0, l)/\theta_c$ used to illustrate the solutions to Eq. (5-18) (after Shaw et al., 1979).

Equation (5-27) can also be arranged as

$$\frac{-\sin\theta_c \exp|\mu|\,t(0, \Delta)}{2\theta_c} = \frac{R_L}{R_0}\frac{l'}{j_0 t(0, \Delta)}\frac{|\mu|}{1 + |\mu|}(1 + \omega^2)^{1/2} \equiv \frac{R_L}{R_0} \tag{5-30}$$

With regard to Eq. (5-29), Fig. 5-13 displays superimposed plots of tan $(\theta_c/2)$ versus θ_c, and $-|\mu|\,t(0, \Delta)/\theta_c$ versus θ_c for different values of $t(0, \Delta)$. The solutions of interest fall in the range $\pi < \theta_c < 2\pi$ and occur for all values of $t(0, \theta)$. Generally, the paired values of ω and $t(0, \theta)$ have a reciprocal relation, i.e., the larger the value of $t(0, \Delta)$, the smaller the frequency ω.

Equations (5-29) and (5-30) yield the following information. For a given value of R_L/R_0 there is a bias level j_0 large enough for different values of $|\mu|\,t(0, \Delta)$ to have a value for Eq. (5-30) to be satisfied. For this value of $|\mu|\,t(0, \Delta)$, Eq. (5-29) provides the instability frequency. Two particular solutions discussed by Kroemer (1968) illustrate this point.

Case 1. In this case a solution exists when $R_L/R_0 = 11.7$ and

$$\theta_c = |\mu|\,t(0, \Delta) = 3\pi/2 \tag{5-31a}$$

Note:

$$\exp[|\mu|\,t(0, \Delta)]/3\pi \cong 11.7 \tag{5-31b}$$

With $|\mu| = 0.2$, $N_0 = 10^{15}\ \text{cm}^{-3}$, and $F_c = 3.5$ (in dimensionless units), we have

$$\omega = 0.2 \qquad (f = 32\ \text{GHz in real units})$$

and for $j_0 \cong 0.5$

$$\Delta_{d,\text{NDM}} = 14.3 \qquad [\text{for a depletion length of } 3.2\ \mu\text{m; see Eq. (5-5)}]$$

Case 2. For the second case, we increase R_L/R_0 to 146. A solution exists for

$$\theta_c = 4\pi/3 \quad \text{and} \quad |\mu|\,t(0, \Delta) = 4\pi/\sqrt{3} \tag{5-32a}$$

Note:

$$\frac{3 \exp |\mu|\, t(0, \Delta)}{16\pi/\sqrt{3}} \cong 146$$

For the same material parameters as in Case 1,

$$\omega = 0.115 \qquad (f = 18.5\,\text{GHz in real units}) \tag{5-32b}$$

and

$$\Delta_{d,\text{NDM}} = 20.6 \qquad (\text{for a depletion length of } 4.5\,\mu\text{m}).$$

Comparing the results of the two cases with that of Eq. (5-23) for uniform electric fields, we are immediately impressed with the instability requirement on $|\mu|\, t(0, \Delta)$. In this section it is shown to exceed that of McCumber and Chynoweth (1966). This is not surprising in view of the fact that we are including the effects of the load. Also, the greater the value of load resistance, the greater the value of $|\mu|\, t(0, \Delta)$ and the lower the value of the instability frequency. In principle there is nothing in the preceding equations to prevent an instability from occurring for any value of load resistance.

It is useful at this time to recall several earlier points. First, from Eq. (5-5) we see that for the values of $|\mu|\, t(0, \Delta)$ given previously

$$\begin{aligned} v_c - j_0 &\cong (1 - j_0)0.009 \qquad (\text{for Case 1}) \\ v_c - j_0 &\cong (1 - j_0)0.0007 \qquad (\text{for Case 2}) \end{aligned} \tag{5-33}$$

Thus, a condition of approximate cathode boundary charge neutrality, with $v_c \cong j_0$, is required for depletion layer instabilities. Secondly, the instability equations do not require that the cathode velocity v_c be equal to v_p (unity in dimensionless units). Thus, the results of this section reiterate an earlier important statement—the field within the NDM element need not everywhere be within the NDM region for an instability to occur. (For the examples of this section only the cathode region is within the NDM region.)

At this point is is useful once again to rewrite Eq. (5-30) to obtain an analytical form for $|\mu|\, t(0, \Delta)$. Using (Eq. 5-29), we find

$$|\mu|\, t(0, \Delta) \cong \log\left(\frac{R_L}{R_0}\frac{l'}{v_c}\frac{\mu^2}{1 + |\mu|}\right) + \log\left[\frac{(1 + \omega^2)^{1/2}}{\sin^2(\theta_c/2)}\right] \tag{5-34}$$

where we have replaced j_0 by v_c. Generally, it is possible to ignore the last term $(1 + \omega)^{1/2} \cong 1$, and for increasing values of $|\mu|\, t(0, \Delta)$, $\theta_c \to \pi$. thus

$$|\mu|\, t(0, \Delta) \cong \log\left(\frac{R_L}{R_l}\frac{l'}{v_c}\frac{\mu^2}{1 + |\mu|}\right) \tag{5-35}$$

is the instability condition.

Another point worth noting is that in the approximations leading to Eq. (5-27) we neglected the lossy contribution of the NDM element in comparison with the contributions of the NDM region and the transition region. For long NDM elements these losses may be significant. If we neglect the capacitive effects associated with this region, we can approximate these losses in the stability criterion by adding a term $(l' - \Delta)/l'$ to the right-hand side of Eq. (5-27). We can also make the corresponding change in Eq. (5-35) by replacing R_L/R_0 by $R_L/R_0 + (l' - \Delta)/l'$. Then, in the limit $R_L/R_0 < 1$ and $l' > \Delta$, Eq. (5-35) reads

$$|\mu|\, t(0, \Delta) \cong \log \frac{l' \mu^2}{v_c(1 + |\mu|)} \tag{5-36}$$

as the stability criterion for long NDM elements (Grubin et al., 1973). From Eqs. (5-35) and (5-36) and the stability arguments discussed in the preceding section, we can conclude that for a dc current level below that required to satisfy these equations, $|\mu|\, t(0, \Delta)$ will be too small and the system is stable (although in the presence of an external ac source gain may be possible). For current levels that exceed the requirements of Eqs. (5-35) and (5-36), the system is unstable. We emphasize that the stability criterion depends upon (1) the load, (2) the sample length, and (3) the cathode boundary field as expressed through the value of v_c.

The approach to an instability for depletion layer profiles with a finite load resistance is displayed in Fig. 5-12. The important feature of the figure is that the instability occurs as the bias current level increases to the value $j \cong v_c$.

We summarize the principal results of this section by writing down the stability criterion for long NDM elements in ordinary units. We find the NDM element to be unstable when the following inequality holds [from Eq. (5-36)]:

$$T_2 > \tau_2 \log \frac{l\tau}{v(F_c)\tau_2^2(1 + \tau/\tau_2)} \tag{5-37}$$

where $T_2[=\tau t(0, \Delta)]$ is the time of flight across the NDM region and $\tau_2[=\tau/|\mu|]$ is the magnitude of the differential dielectric relaxation time of the NDM region. The above inequality is satisfied for NDM elements with values of current density approximately equal to

$$J_c = N_0 e v(F_c) \tag{5-38}$$

5.3.4. Stability of Nonuniform Accumulation Layer Profiles

The stability problems already discussed have exclusively involved depletion layer profiles. While these were the earliest to be considered, the discussion was expanded soon after the observations of Perlman et al. (1970) were reported. An explanation of the results required the presence of stable accumulation layer profiles in samples whose $N_0 l$ products exceeded the critical value. These accumulation layer profiles are of the type displayed in Fig. 5-9.

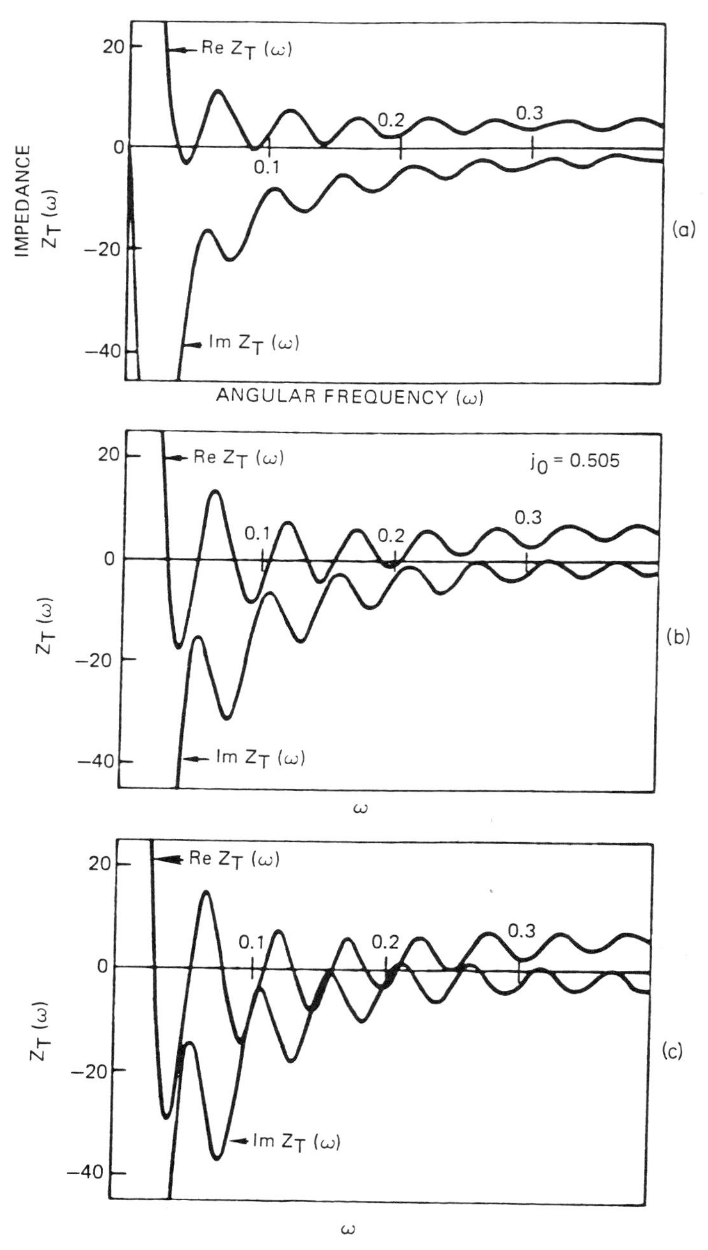

FIGURE 5-14. As in Fig. 5-12, but for accumulation layer profiles. (a) $j_0 = 0.510$, (b) $j_0 = 0.505$, (c) $j_0 = 0.504$. (After Grubin and Kaul, 1975.)

A small-signal stability analysis of accumulation layer profiles presents conceptual difficulties in that the stable layer represents a postinstability state. Small-signal analyses cannot go from a large-signal instability to a small-signal stable state. Large-signal numerical simulation is usually required for this analysis. But we can turn the problem around and proceed in the opposite direction—that is, from a stable accumulation layer to an unstable one. We can then place bounds on the accumulation layer stability conditions. But here too there are interpretive difficulties, because identification of the region that dominates the instability frequency is less clear than for the depletion layer case. For accumulation layers, both the NDM region, with a periodicity dependent on $t(0, \Delta)$, and the SDV region, with a periodicity dependent on $t(\Delta, l)$, are for dominance. The result is that the instability is not necessarily initiated at a frequency characteristic of either region. Analytical examination of the problem is difficult. We therefore use numerical techniques to compute the impedance (Shaw et al., 1979). The results are displayed in Fig. 5-14, where we show the approach to an instability for accumulation layer profiles with a finite load resistance. As in the preceding discussion, the important conclusion of the figure is that the instability occurs as the bias current j_0 decreases to the value $j_0 \cong v_c$, thereby emphasizing the wide applicability of the instability criteria expressed by Eq. (5-38).

5.4. GaAs DEVICE SIMULATION FROM THE BOLTZMANN TRANSPORT EQUATION

5.4.1. Introduction

Mobility models are expected to have limited usefulness in submicron and high-frequency devices. The extent to which this is true is material dependent and is most likely to be revealed by device simulations. Complete one- and two-dimensional simulations involving solutions of the Boltzmann transport equation (BTE) are only sparsely available. Rather, only pieces of the problem have been treated. Here Monte Carlo, iteration, and moment methods are used; in the following we shall illustrate some results, drawing on GaAs, since the nonlocal spatial and temporal contributions are enhanced dramatically by the transferred electron effect. Our approach will emphasize the displaced Maxwellian moment equations, primarily because of intuitive advantages and for the relative ease with which space-charge contributions can be handled.

The device-momentum equations (Blotekjaer, 1970) for a displaced Maxwellian of the form given by

$$f_j(k) = C_0 \exp[-\hbar^2(k - k_{di})^2/2m_j k_B T_{ei}] \tag{5-39}$$

where $\hbar k_{di} = m_j v_{di}$ is the average drift momentum of the electron gas, C_0 a normalization constant, and T_{ei} the electron temperature, with the subscript i

referring to the ith valley, are

$$(\partial n_i/\partial t) + \nabla(v_i n_i) = (\partial n_i/\partial t)_c \tag{5-40}$$

$$(\partial P_i/\partial t) + \nabla(v_i P_i) = -en_i F - \nabla(n_i k_B T_i) + (\partial P_i/\partial t)_c \tag{5-41}$$

$$(\partial W_i/\partial t) + \nabla(v_i W_i) = -en_i v_i F - \nabla(v_i n_i k_B T_i) - \nabla q_i + (\partial W_i/\partial t)_c \tag{5-42}$$

Here

$$P_i = m_i n_i v_i \tag{5-43}$$

and

$$W_i = \tfrac{3}{2} n_i k_B T_i + \tfrac{1}{2} m_i n_i v_i^2 \tag{5-44}$$

W_i is regarded as the average total kinetic energy density.

These equations have simple physical interpretations. For Eq. (5-40) the increase of electron density plus the outflow of electrons equals the increase of density due to collisions (conservation of particles). For Eq. (5-41) the left-hand side is the rate of change plus the outflow of momentum density of the ith valley. The right-hand side represents the forces exerted by the electric field and by the electron pressure $n_i k_B T_i$ and the rate of momentum density gained in collisions (conservation of momentum). For Eq. (5-42) the left-hand side contains the rate of change plus outflow of total kinetic energy density. The right-hand side represents the energy supplied by the electric field, the work performed by the electron pressure, the divergence of the heat flow q_i, and the rate of change of total kinetic energy density due to collisions (conservation of energy).

When the transport equations are derived by integration of the distributions (or any symmetric distributions), the heat flow vanishes. In spite of this, following Blotekjaer (1970), we may allow for heat conduction by assuming

$$q_i = -\kappa_i \nabla T_i \tag{5-45}$$

where K_i is the heat conductivity of the electron gas in the ith valley. This term is believed to represent the most important effect of a non-Maxwellian distribution function (Blotekjaer, 1970).

The $(\partial/\partial t)_c$ terms in Eqs. (5-40)–(5-42) represent scattering integrals. They may be given in approximate form, for example, for polar optical scattering, or in exact form for the displaced Maxwellian. In the following simulations we choose the latter and use the integrals summarized by Butcher (1967) (see also Blotekjaer and Lunde, 1969). The scattering integrals are then represented as

$$(\partial n_i/\partial t)_c = (-n_i/\tau_{n_{ij}}) + (n_j/\tau_{n_{ji}}) \tag{5-46}$$

$$(\partial P_i/\partial t)_c = -P_i/\tau_{P_i} \tag{5-47}$$

$$(\partial W_i/\partial t)_c = -\tfrac{3}{2}(n_i k_B T_i/\tau_{E_{ij}}) + \tfrac{3}{2}(n_j k_B T_j/\tau_{E_{ji}}) \tag{5-48}$$

Early calculations using Eqs. (5-46)–(5-48) were performed for the set of scattering curves for a $\Gamma-X$ orientation shown in Fig. 5-15. The parameters used in the calculations are given in Table 5-2. These results should be compared to those of Bosch and Thim (1974). We retain the $\Gamma-X$ orientation because most of the early moment equations were evaluated for this ordering. Changing to a $\Gamma-L$ orientation (see, e.g., Littlejohn et al., 1978) would offer only quantitative differences (see Table 5-1) and might obscure some of the discussion.

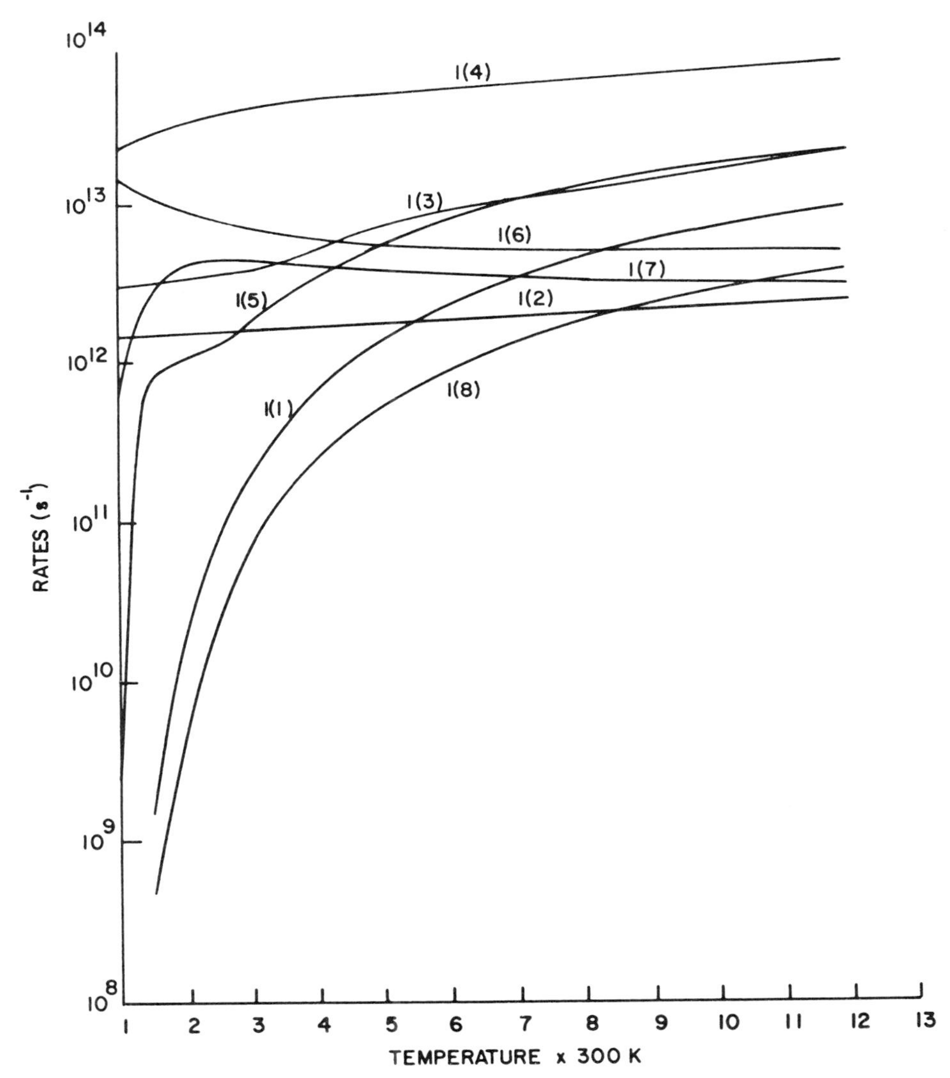

FIGURE 5-15. Scattering rates for the parameters of Table 5-2. Integrals are from Butcher (1967). Curves shown are I(1), CV particle; I(2), SV particle; I(3), CV momentum; I(4), SV momentum; I(5), CV energy, I(6), SV/CV energy; I(7), SV energy; I(8), CV/SV energy. C(S)V ≡ central (satellite) valley. (After Grubin et al., 1982.)

TABLE 5-2. GaAs Parameters Used in the Calculations

Parameters	Γ(000)	$X(100)$	Common
Number of equivalent valleys	1	3	
Effective mass (m_e)	0.067	0.40	
Γ–X separation (eV)			0.36
Lattice constant (Å)			5.64
Density (g/cm^3)			5.37
Polar optical scattering			
Static dielectric constant			12.53
High-frequency dielectric constant			10.82
LO phonon (eV)			0.0354
Γ–X Scattering			
Coupling constant (eV/cm)			0.621×10^9
Phonon energy (eV)			0.0300
X–X Scattering			
Coupling constant (eV/cm)		1.064×10^9	
Phonon energy (eV)		0.0300	
Acoustic scattering			
Deformation potential (eV)	7.0	7.0	
Acoustic velocity (cm/s)			5.22×10^5

5.4.2. Recovery of the Semiconductor Equations; Comparison to Nonlocal Equations

In analyzing transport from the nonlocal balance equation we are sometimes interested in recovering the ordinary semiconductor equations and, hence, mobility concepts. For a single parabolic conduction band we can obtain the usual semiconductor equations through judicious neglect of select space and time derivatives. For example, in the case of continuity, neglect of avalanching allows us to set $(\partial n/\partial t)_c = 0$. We then obtain current continuity for one valley:

$$(\partial n/\partial t) + \nabla(vn) = 0 \tag{5-49}$$

For the momentum term, we set $(\partial P_i/\partial t)_c = -P_i/\tau_p$, neglect the terms $\partial P_i/\partial t$, and the contribution from $\nabla(vP)$ and obtain

$$nv = P/m = -n\mu F - (\mu/e)\nabla(nk_BT) \tag{5-50}$$

where $\mu = e\tau_p/m$. Then using the Einstein relation $D(F) = \mu k_B T/e$, we obtain

$$nv = -n\mu F - \nabla(nD) \tag{5-51}$$

We see that within the framework of the Boltzmann picture even the use of the "semiconductor" Eq. (5-51) is suspicious. Further complications arise in multi-valley calculations where the mobility is taken as a weighted average

$$\langle\mu(F)\rangle = \frac{n_c\mu_c(F) + n_s\mu_s(F)}{n_c + n_s} \tag{5-52a}$$

where c and s refer to central and satellite valley, respectively, and the diffusion coefficient is given, with limited validity, by

$$D(F) = \frac{k_B}{e}\frac{n_c\mu_c(F)T_c + n_s\mu_s(F)T_s}{(n_c + n_s)} \tag{5-52b}$$

Since we have a more general set of transport equations than the semiconductor equations, we are in a position to isolate differences between the two approaches, even on such a relatively direct topic such as domain propagation. Cheung and Hearn (1972) examined this question by considering the mobility and scattering rates to be unique functions of electron temperature rather than of field. They are unique functions of field only in steady state. In their study the particle current flux is given by a multivalley version of Eq. (5-50):

$$J_i = -n_i\mu_i(T_i)\left[F + \frac{k_B}{en_i}\frac{\partial}{\partial x}(n_iT_i)\right] \tag{5-53}$$

$$\frac{\partial n_i}{\partial t} = -\frac{\partial J_i}{\partial x} + \frac{n_j}{\tau_{n_{ji}}}(\tau_{n_{ji}}) - \frac{n_i}{\tau_{n_{ij}}}(\tau_{n_{ij}}) \tag{5-54}$$

with

$$\frac{\partial}{\partial t}\left(\frac{3}{2}n_ik_BT_i\right) = -eJ_iF + \frac{3}{2}\frac{-n_ik_BT_i}{\tau_{E_{ij}}} + \frac{n_jk_BT_j}{\tau_{E_{ji}}} - \frac{\partial}{\partial x}\left(\frac{5}{2}J_ik_BT_i\right) \tag{5-55}$$

Note that Eq. (5-53) neglects inertial terms and hence "overshoot" effects, which we shall discuss shortly. Figure 5-16 shows a comparison of domain size using the semiconductor equations and the BTE. We can see significant differences.

We consider some of these inertial effects more closely by solving the full set of equations, (5-40)–(5-42), for a one-dimensional 5000-Å GaAs element with a donor distribution as shown in Fig. 5-17. The element is part of a resistive circuit. The carrier dynamics are examined at two instants of time. (Note: Serious objections can be raised to the use of a "jellium" distribution insofar as any combination of decreased donor density or size reduction will necessarily introduce effects due to the discrete nature of the donors. This will be ignored in the following discussion.)

As the bias is turned on there is an increase in potential across the device and a corresponding increase in current and field. The field is computed self-consistently and its slope reflects any incomplete screening of inhomogeneities by the mobile carriers. For the device in the schematic configuration of Fig. 5-17, as the field rises energy relaxation is incomplete and velocity overshoot contributions are dramatic (see Figs. 5-18 and 5-19). Here velocity is computed from the equation

$$\langle v(x, t)\rangle = \frac{P_c/m_c + P_s/m_s}{n_T(x, t)} \tag{5-56}$$

where n_T is the total number of mobile carriers.

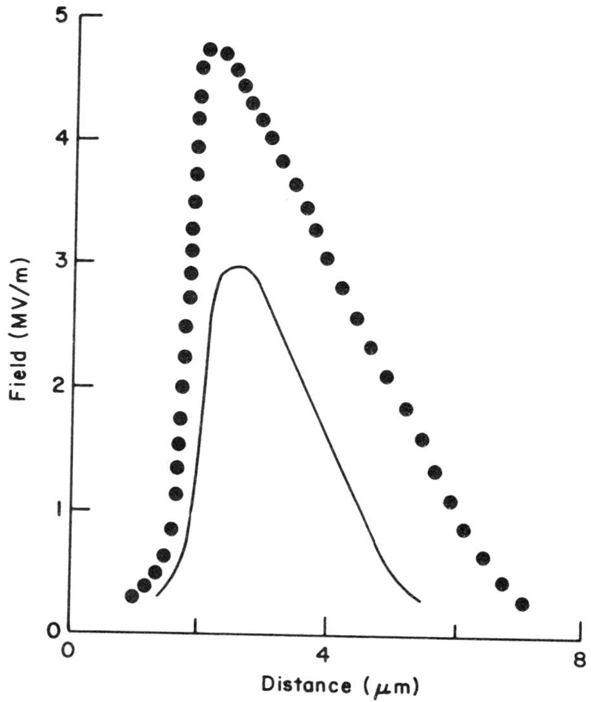

FIGURE 5-16. Field profiles for a high-field-propagating domain. Dotted curve is obtained from the mobility equations. Solid curve is obtained using Eqs. (5-53)–(5-55). (After Cheung and Hearn, 1972.)

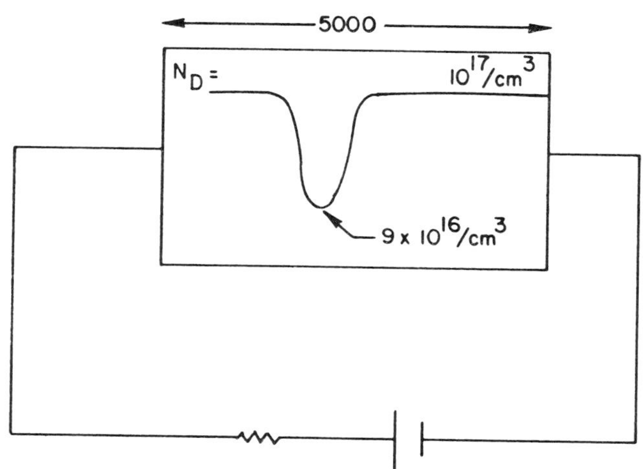

FIGURE 5-17. Schematic representation of device and circuit configuration for submicron homogeneous field profiles. The inhomogeneous doping profile is treated as a "jellium" distribution. The beginning and end of this inhomogeneity are displayed in Fig. 5-19c. The circuit equation for this calculation is $\Phi_B = \Phi + IR$. We set R the to low-field resistance of the element, $F_0 = 4.3\,\text{kV cm}^{-1}$ and $l = 5000$ Å. Generally, Φ_B is turned on at a finite rate. For the calculations discussed, $\Phi_B = 3.0$. (After Grubin et al., 1982.)

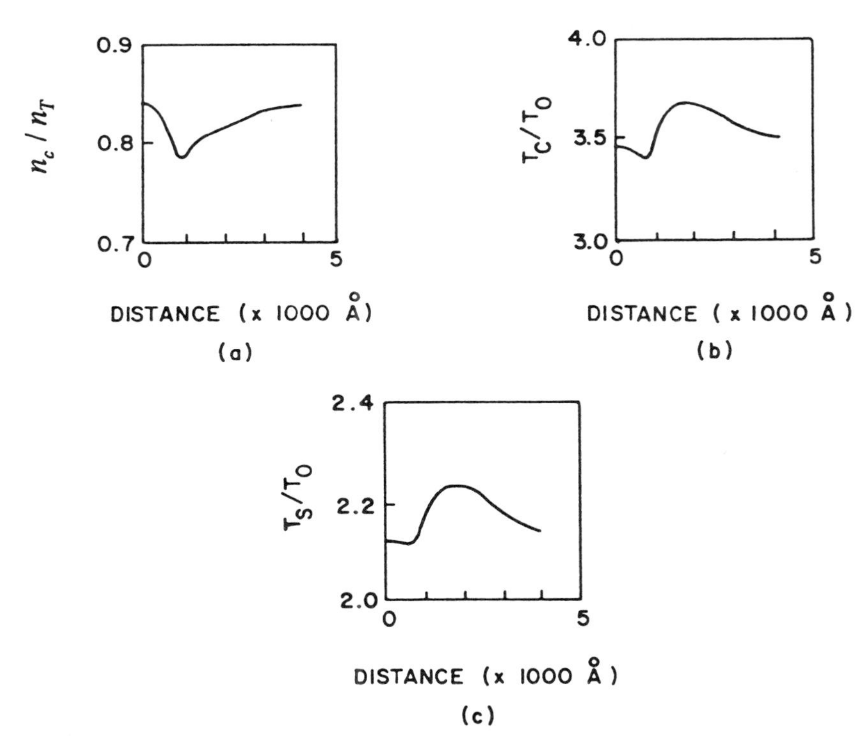

FIGURE 5-18. (a) Fractional central valley population versus distance; (b) central valley temperature versus distance; (c) satellite valley temperature versus distance. Computation occurs at time $t/\tau_0 = 4$, where τ_0 is the LO phonon intervalley scattering time and equals 0.32 ps [see Fig. 5-15, curve I(3)] and T_0 is room temperature. The gradient of n is zero at the boundaries. All nonuniformities are due to the notch. (After Grubin et al., 1982.)

An important point to make here is that when spatial gradients occur the carrier velocity can overshoot its equilibrium value even though all time derivatives ($\partial/\partial t$) are zero. This was emphasized by Kroemer (1978), who estimated substantial overshoot when $\partial F/\partial x \gtrsim F_{\text{th}}$/mean free path, where the subscript "th" designates the NDM threshold field. Thus, spatial overshoot should be most pronounced in devices biased near the NDM threshold. The results shown in Figs. 5-20 and 5-21 are for bias fields substantially higher than the NDM threshold. In these figures the fields are high enough to accommodate almost complete transfer. Here the carrier temperature is reduced, the energy relaxation rates are shorter, and there is virtually no overshoot. The results are essentially the same as we would obtain using the steady state curves.

Nonuniform field calculations are the clear order of business in future numerical simulations (see e.g., Wu and Shaw, 1989), but a catalog of results that are not moot have yet to be produced. Until this is done we are forced to rely heavily on uniform field analyses.

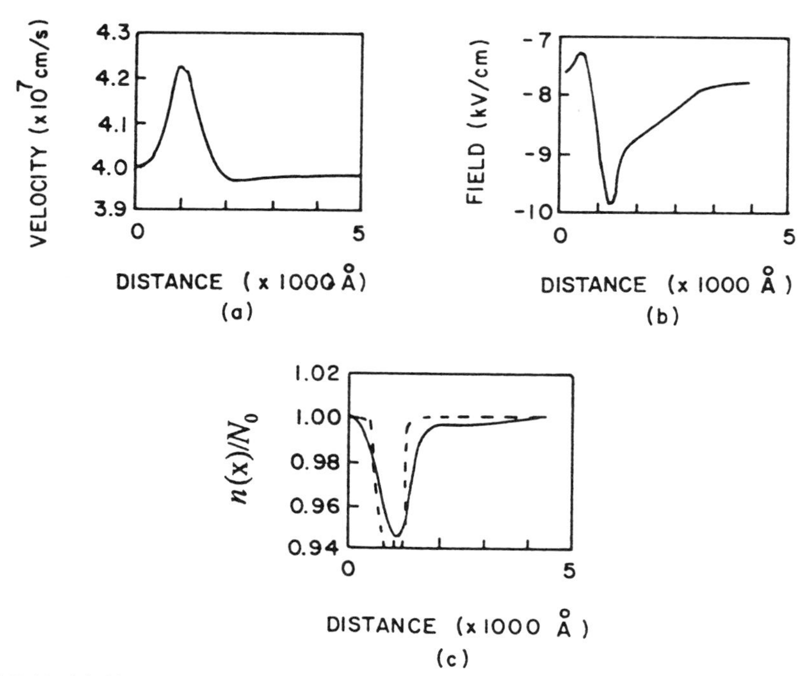

FIGURE 5-19. (a) Mean velocity versus distance; (b) field versus distance; (c) free carrier density versus distance at time $t/\tau_0 = 4$. Dashed curve denotes background density. (After Grubin et al., 1982.)

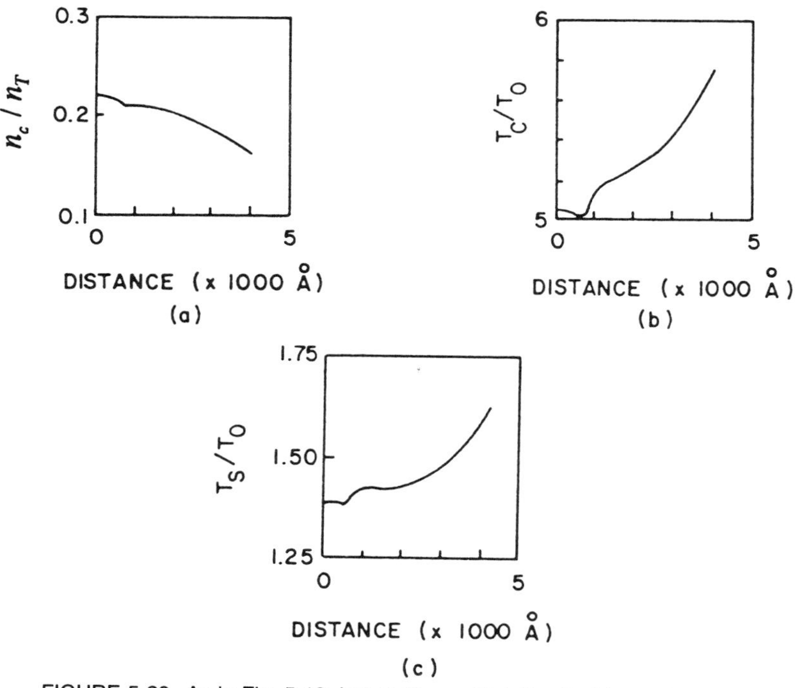

FIGURE 5-20. As in Fig. 5-18, but at $t/\tau_0 = 16$ (after Grubin et al., 1982).

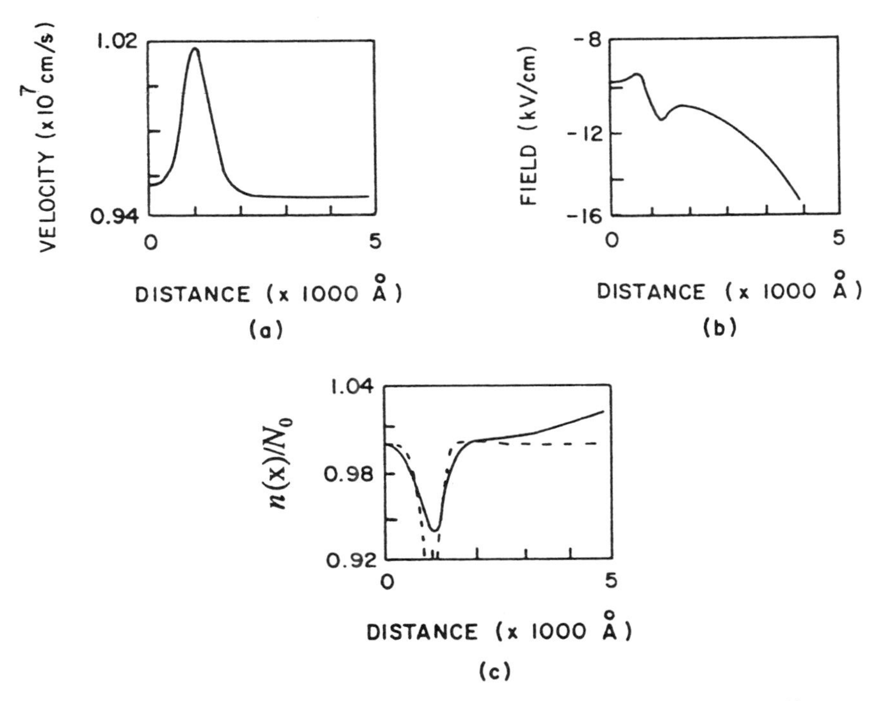

FIGURE 5-21. As in Fig. 5-19, but at $t/\tau_0 = 16$ (after Grubin et al., 1982).

5.4.3. Uniform Field Transients

For uniform fields, the transport equations simplify considerably:

$$\frac{a_i\, dn_i}{dt} = \frac{-n_i}{\tau_{n_{ij}}} + \frac{n_j}{\tau_{n_{ji}}} \tag{5-57}$$

where a_i denotes the number of equivalent ith valleys. For $\mathbf{p}_i = \mathbf{P}_i/n_i$, we have

$$\frac{d}{dt} n_i \mathbf{p}_i = -en_i\mathbf{F} - \frac{n_i \mathbf{p}_i}{\tau_{\mathbf{p}_i}} \tag{5-58}$$

$$\frac{d}{dt}\left(\frac{n_i p_i^2}{2m_i} + \frac{3}{2} n_i k_B T_i\right) = -\frac{en_i \mathbf{F}\mathbf{p}_i}{m_i} - \frac{3n_i k_B T_i}{2\tau_{E_{ij}}} + \frac{3n_j k_B T_j}{2\tau_{E_{ji}}} \tag{5-59}$$

It is instructive at this time to rewrite the momentum-balance equation as

$$\frac{dp_i}{dt} + p_i\left(\frac{1}{\tau_{p_i}} + \frac{d \log n_i}{dt}\right) = -eF \tag{5-60}$$

In this form the effect of intervalley transfer enters as an additional transient momentum scattering term (Grubin et al., 1979).

Dynamic overshoot effects are the consequence of differences in momentum and energy relaxation times. In multivalley semiconductors the overshoot contributions therefore appear in the momentum, as well as the velocity computations. We shall illustrate this for both the central and satellite valleys and for situations where electrons starting with zero drift velocity are subjected to a sudden change in electric field.

For the central valley and at low values of bias, the electron temperature is approximately equal to room temperature and ordinary time-dependent dynamic

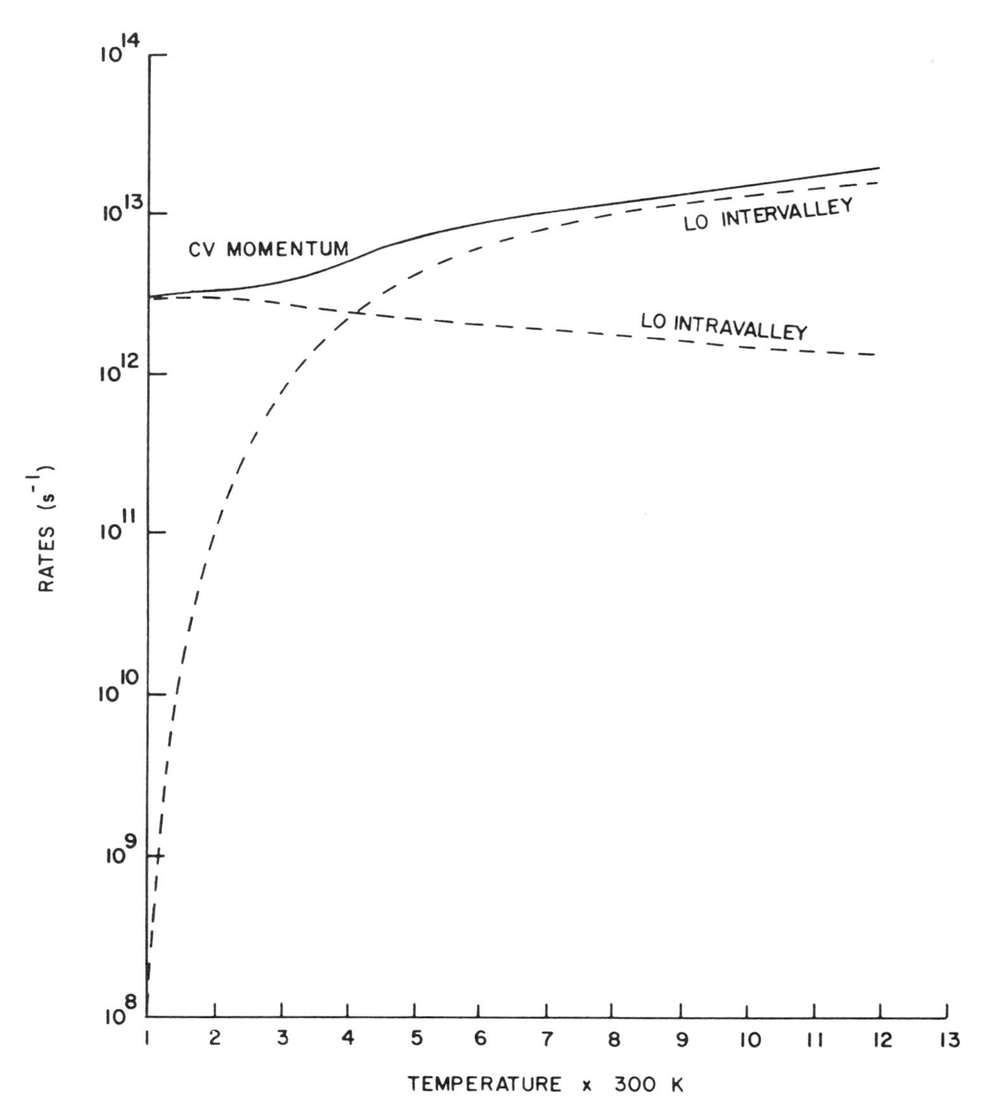

FIGURE 5-22. Central valley momentum scattering rates versus electron temperature (after Grubin et al., 1982).

behavior occurs. At elevated bias levels the electron temperature is substantially increased and the momentum-relaxation time, owing to strong intervalley coupling, decreases with increasing temperature (Fig. 5-22). Thus, we see overshoot, in that the final momentum is below the peak momentum (Fig. 5-23a). (We point out that above moderate increases in electron temperature, LO phonon intravalley and ionized impurity scattering do not provide a momentum-relaxation time that decreases with energy; intervalley phonons are required. Indeed, for ionized impurity scattering, the relaxation time increases with energy (see, for example, Smith, 1978). During this same time interval the increasing central valley temperature (Fig. 5-23b) results in electron transfer and the momentum density $n_c p_c$ shows an even greater overshoot (Fig. 5-23c).

We now examine the contribution of the term $d(\log n)/dt$ appearing in Eq. (5-60). For the central valley, where at $t = 0$, $n_s \cong 0$, and $p_c = p_s = 0$, this term is approximately zero. However, when we consider the transient behavior of the satellite valley, the time derivative of $\log n_s$ is important because the change in the satellite population, relative to the original number present, is quite large. In addition, note that the satellite valley momentum-relaxation contribution is

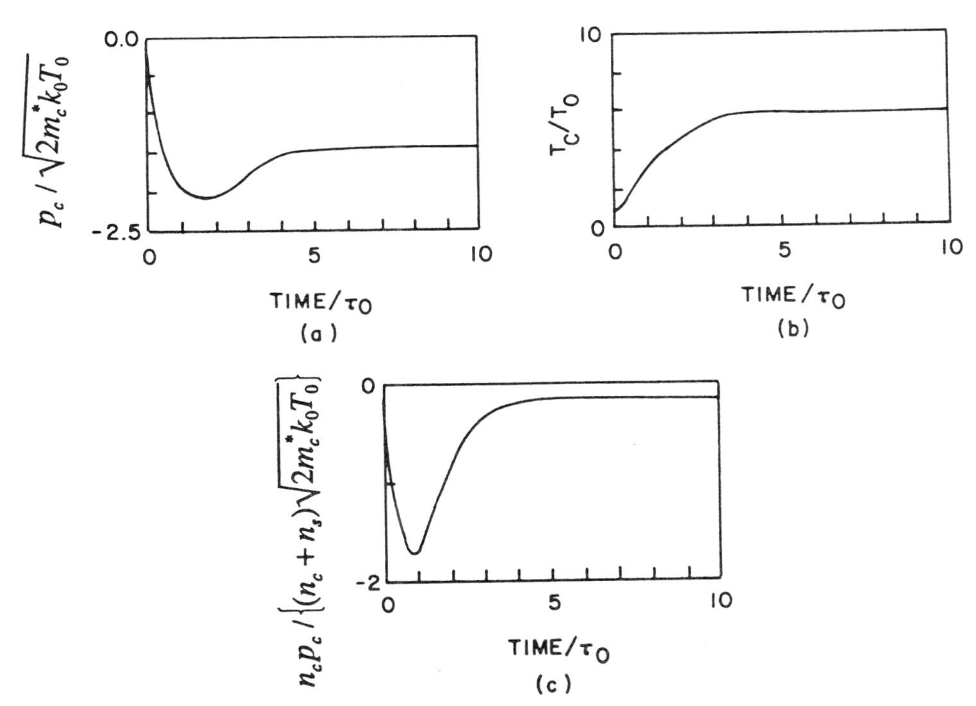

FIGURE 5-23. Transient central valley behavior: (a) momentum versus time; (b) temperature versus time; (c) momentum density versus time. The circuit in this calculation is the same as that of Fig. 5-17. Here, because the field is uniform, we write the circuit equation as $F_B = F_D + R/R_0 i$, where $i = I/N_0 e v_p A$ with $v_p = (2m_c^{-1} k_B T_0)^{1/2} = 3.7 \times 10^7$ cm sec^{-1}. For this calculation N_0 is 10^{17} cm^{-3} and we have included geometrical capacitance. Also, the normalized bias F_B is turned on suddenly to the value $F_B = 5$, which corresponds to an average field of 21.5 kV cm^{-1} ($\tau_0 = 0.32$ ps). (After Grubin et al., 1982.)

almost an order of magnitude larger than that of the central valley. Thus, in a time considerably shorter than that associated with the central valley, the satellite valley momentum reaches the value

$$p_s = -eF(t)\left(\frac{1}{\tau_{ps}} + \frac{d \log n_s}{dt}\right)^{-1}$$

where, because the satellite temperature remains close to room temperature for large changes in field (Bott and Fawcett, 1968), the scattering rates may often be taken as approximately constant. Combining both scattering contributions, we see some "overshoot" due to differential repopulation but none as dramatic as that associated with the central valley. Figure 5-24 illustrates this point. For part (a): 1 corresponds to the increasing momentum prior to any significant electron transfer; 2 represents the scattering rate due to repopulation; 3 is the steady-state value. In Fig. 5-24c, we show the product of $n_s' p_s/(n_c + 3n_s')$. Here the prime indicates population of a single satellite valley ($n_s = 3n_s'$). We see that as far as the contribution to velocity overshoot is concerned, the electron transfer tends to wipe it out.

The average drift velocity versus time is shown in Fig. 5-25 and is computed

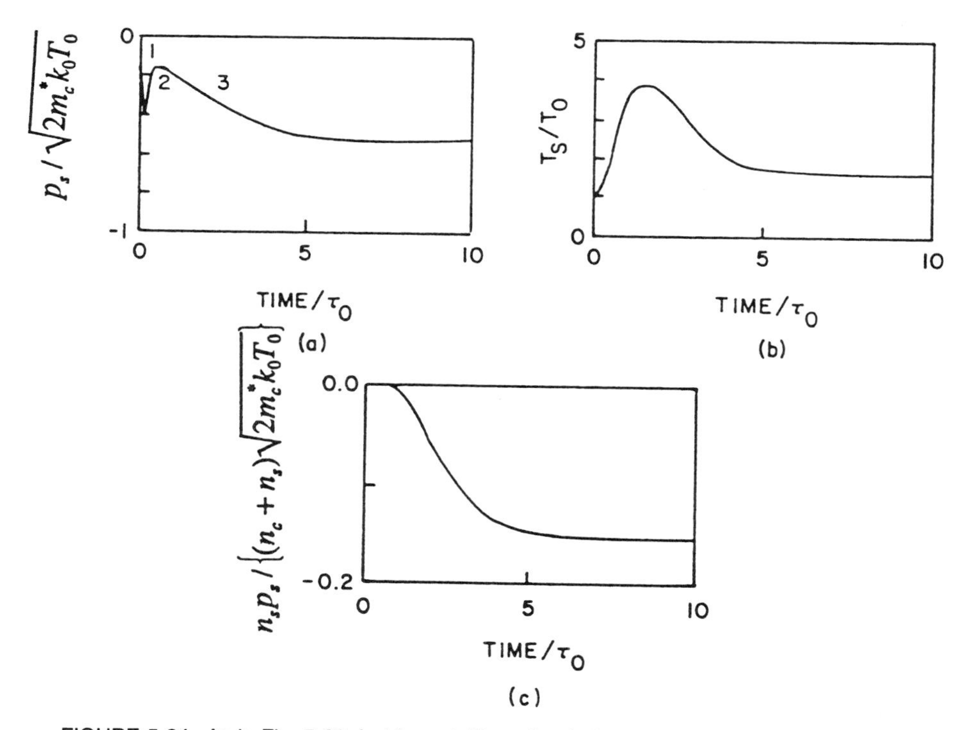

FIGURE 5-24. As in Fig. 5-23, but for satellite valley behavior (after Grubin et al., 1982).

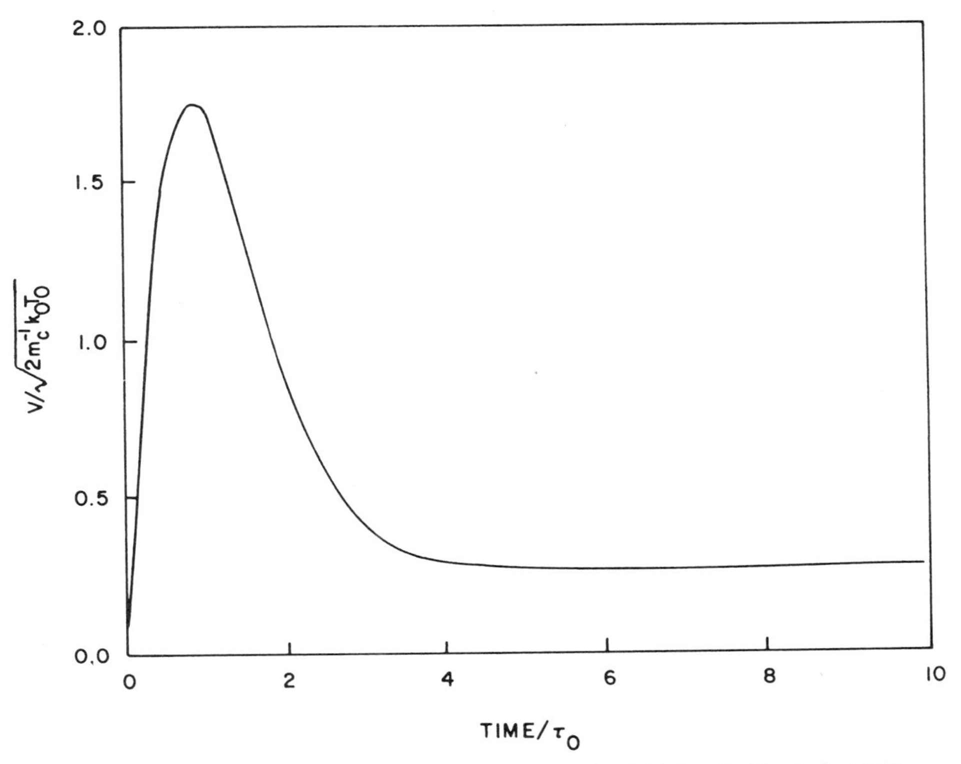

FIGURE 5-25. Mean carrier velocity, from Figs 5-23 and 5-24 (after Grubin et al., 1982).

from

$$v = \frac{(n_c p_c / m_c) + (n_s p_s / m_s)}{n_c + n_s} \tag{5-61}$$

Perhaps the most remarkable aspect of this result is the very large peak velocity prior to steady state (see, e.g., Ruch, 1972). This result has been one among many that has led some to suppose that narrow-channel devices will yield higher carrier velocities. To some extent high overshoot velocities are illusory, as they are very sensitive to rise time. We shall illustrate this now for a sequence of trapezoidal bias pulses, each with a varying rise time [see Fig. 5-26 (Grubin et al., 1976)].

The first set of results is for a relatively slow rise time and the dynamic curves come very close to the steady-state curves (see Fig. 5-27). A more significant departure from steady state occurs for the somewhat steeper rise time. In Fig. 5-28 we see some asymmetry in the time dependence of the central valley population and temperature and an increase in the peak velocity. In the final sequence we show results for a very short rise time. We see a dramatic increase in the peak velocity and clear asymmetry in the carrier dynamics (Fig. 5-29). Indeed, the final point of approximately zero field and velocity is not an

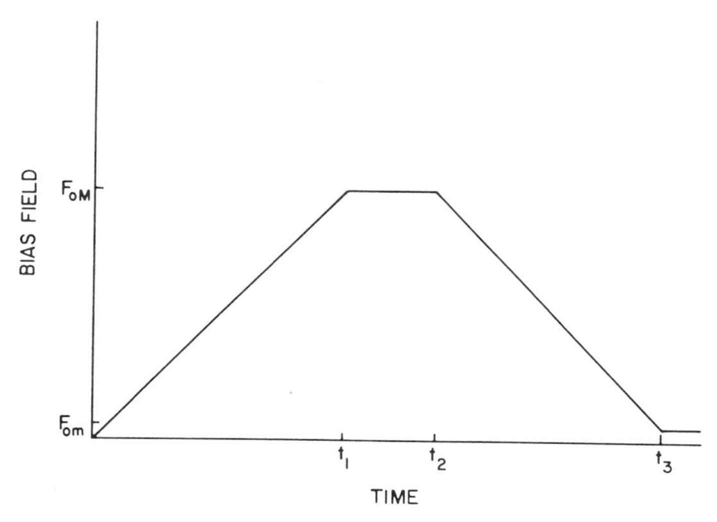

FIGURE 5-26. Time-dependent bias field used to show the rise-time dependence of the transient velocity. For all of these calculations, $F_{OM} = 5$ (21.5 kV cm^{-1}) and $F_{Om} = 0.1$ (0.43 kV cm^{-1}). (After Grubin et al., 1982.)

equilibrium state. Rather, we have a dramatic example of velocity undershoot. A longer time is needed for the electron temperature to approach equilibrium. There are strong implications here for upper-frequency limits of device operation.

5.4.4. *Determination of the Maximum Frequency for Small-Signal, Large-Signal, and Self-Excited Oscillations*

Perhaps the earliest attempt to examine the upper-frequency limit for large-signal oscillations was that of Butcher and Hearn (1968). Using a set of displaced Maxwellian electron distributions for each valley, Butcher and Hearn solved the set of differential equations [see Eqs. (5-57)–(5-59) for the time-dependent electron temperatures, drift velocities, and valley populations] for a dc bias field plus rf field. The results of their study are shown in Fig. 5-30.

In Fig. 5-30, the mean drift velocity and satellite population are shown as functions of a total field consisting of a dc field of 15 kV cm^{-1} and a 60-GHz rf field with an amplitude of 13.1 kV cm^{-1}. The arrows indicate the direction of increasing time and the dashed lines represent the static relationships. Near the maximum field of 28.1 kV cm^{-1} the satellite population (Fig. 5-30a) approximates the static value quite well because the field is stationary and the population variation is nearly saturated. The curve is qualitatively similar to those of Fig. 5-28. We again see the higher satellite population on the downswing and the absence of NDM. We point out that these curves very likely constitute the earliest attempt at including overshoot contributions. For dc bias levels of around 10 to 20 kV cm^{-1} and ac levels of 13.1 to 18.6 kV cm^{-1}, they obtained an upper-frequency limit of 100 GHz.

From a device physics point of view a driven oscillator probably lies

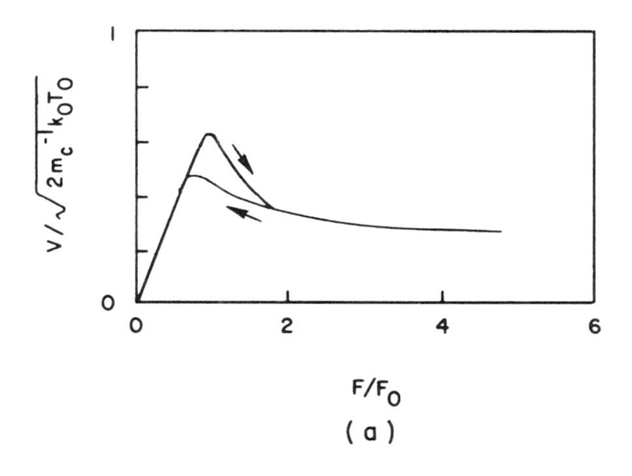

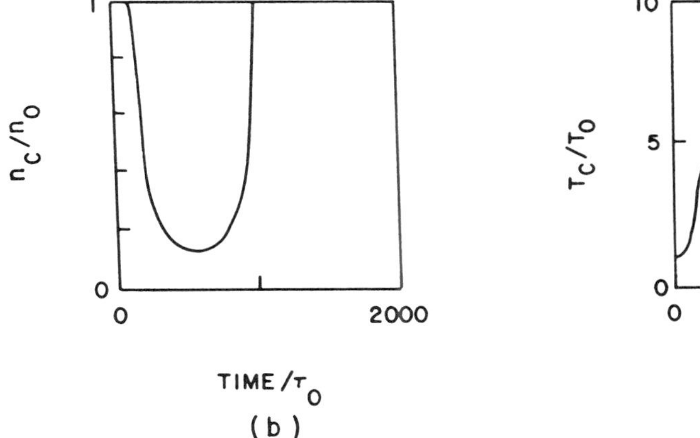

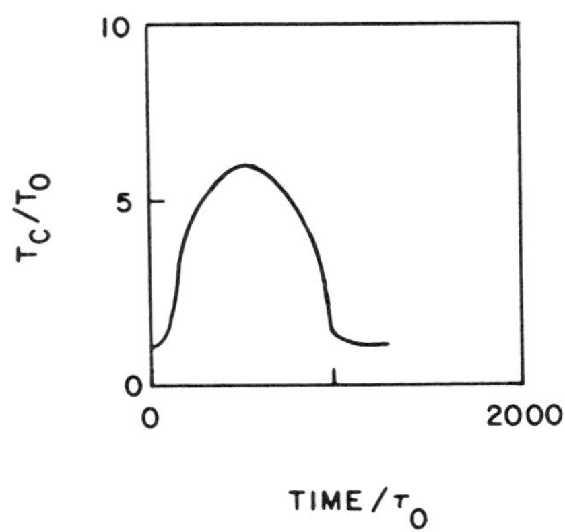

FIGURE 5-27. For this pulse, $t_1 = 500/\tau_0$ (see Fig. 5-26), $t_2 = 600/\tau_0$, and $t_3 = 1200/\tau_0$. (a) Mean velocity versus field ($F_0 = 4.3\,\mathrm{kV\,cm^{-1}}$); (b) central valley populations versus time; (c) central valley temperature versus time. (After Grubin et al., 1982.)

somewhere between a small-signal and a large-signal self-excited oscillator. The self-excited oscillator is perhaps the most interesting of the three because it highlights the tenacious balance between electron transfer and sustained oscillations. It is extremely sensitive to contact, space-charge, and circuit conditions (Shaw et al., 1979). We examine the upper frequency limit for the device in a circuit with reactive elements. Figure 5-31 shows the oscillation.

The circuit differential equation for this oscillation is given by Eq. (2-176), repeated here for convenience:

$$\Phi_B = \frac{1}{\omega_0^2}\frac{d^2\Phi}{dt^2} + \frac{Z_0}{\omega_0}\frac{dI_c}{d\Phi}\frac{d\Phi}{dt} + RC\frac{d\Phi}{dt} + \Phi + RI_c \tag{5-62}$$

$$\omega_0 = (LC)^{-1/2}, \qquad Z_0 = (L/C)^{1/2}$$

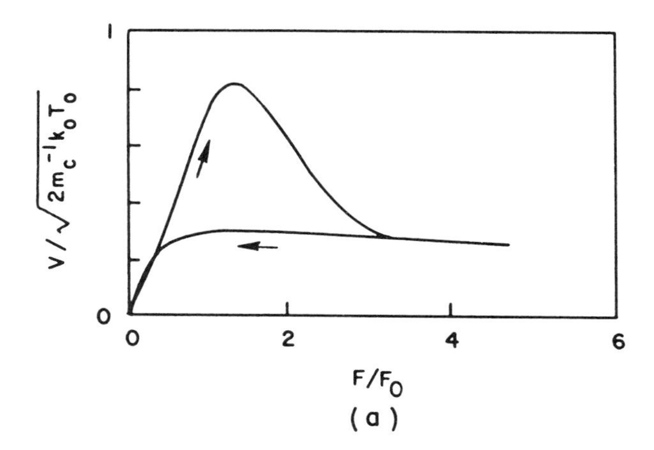

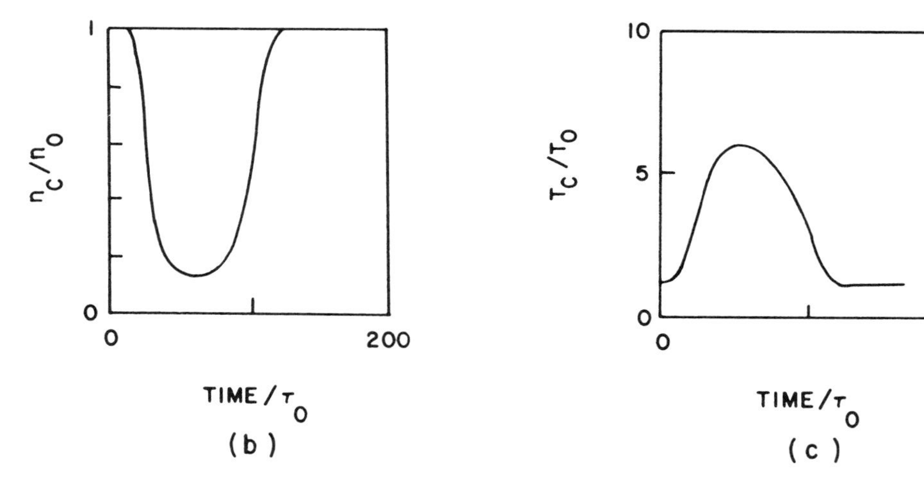

FIGURE 5-28. As in Fig. 5-27, but here $t_1 = 50/\tau_0$, $t_2 = 60/\tau_0$, and $t_3 \cong 110/\tau_0$ (after Grubin et al., 1982).

where $dI_c/d\Phi$ represents resistance. When $dI_c/d\Phi > 0$, Eq. (5-62) yields damped oscillations. When $dI_c/d\Phi < 0$, the oscillations grow in amplitude. As $dI_c/d\Phi$ changes sign during each cycle, the correct set of circuit, bias, and device parameters can yield sustained circuit-controlled oscillations (Shaw et al., 1979).

Several aspects of a self-excited oscillation are displayed in Fig. 5-31. The current through the load resistor is displayed in part (c); the dynamic voltage and I vs. Φ, obtained by eliminating time between current and voltage, is shown in part (a). We also display the mean velocity (dynamic conduction current) in part (b). The details of the oscillation will be discussed next.

As the field across the device increases and exceeds threshold, intervalley

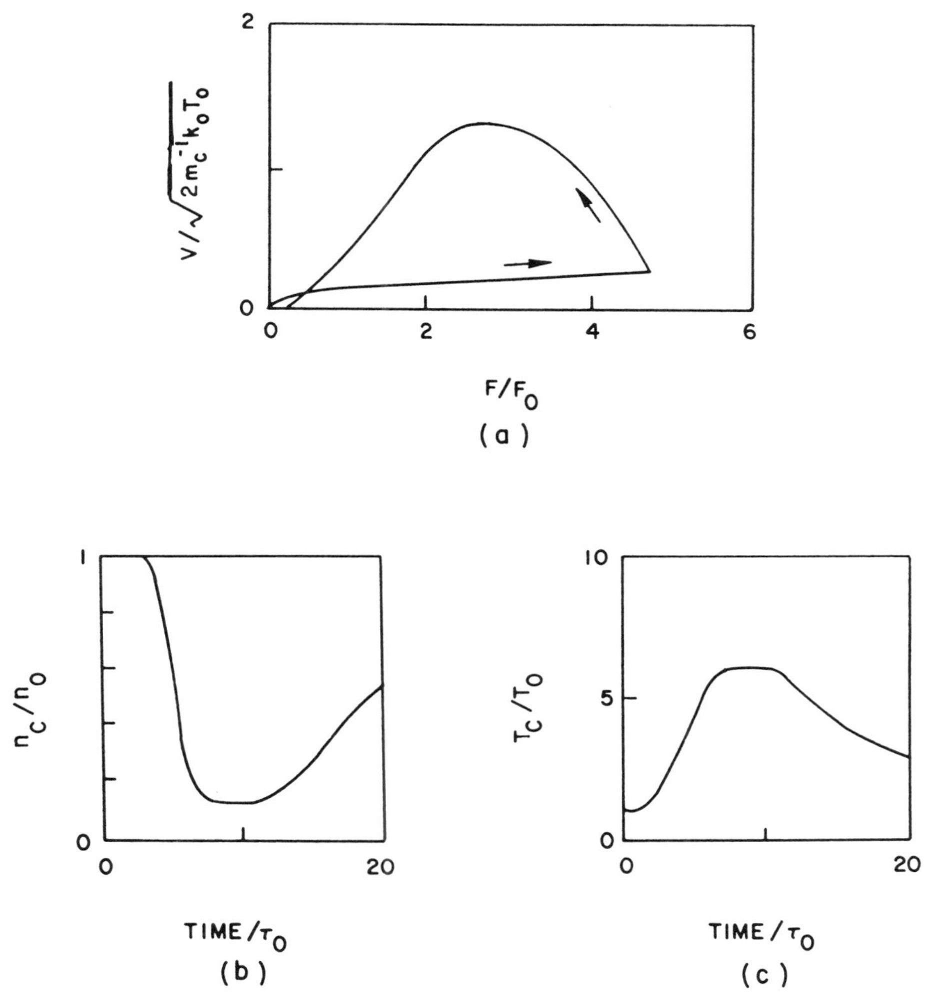

FIGURE 5-29. As in Fig. 5-27, but here $t_1 = 5/\tau_0$, $t_2 = 10/\tau_0$, and $t_3 = 15/\tau_0$ (after Grubin et al., 1982).

transfer begins to occur. However, because the field changes more rapidly than the electron temperature, more carriers are retained in the central valley, with higher momenta than steady state would dictate. This effect is responsible for the higher peak conduction current. But if the increasing electric field sustains high fields for a sufficient duration, enough carriers will transfer to produce NDM, which must be of sufficient magnitude for sustained self-excited oscillations.

The field on the downswing again changes more rapidly than the electron temperature and more carriers are retained in the satellite valley; i.e., we achieve transient undershoot. Now, although NDM is not necessary on the downswing

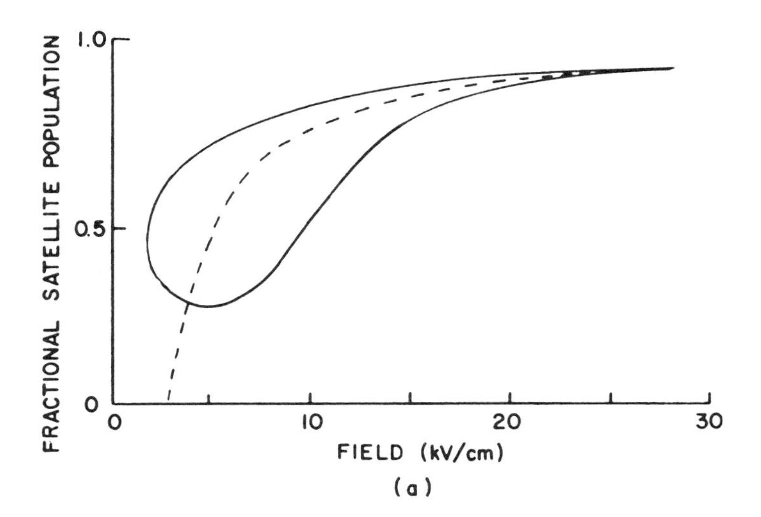

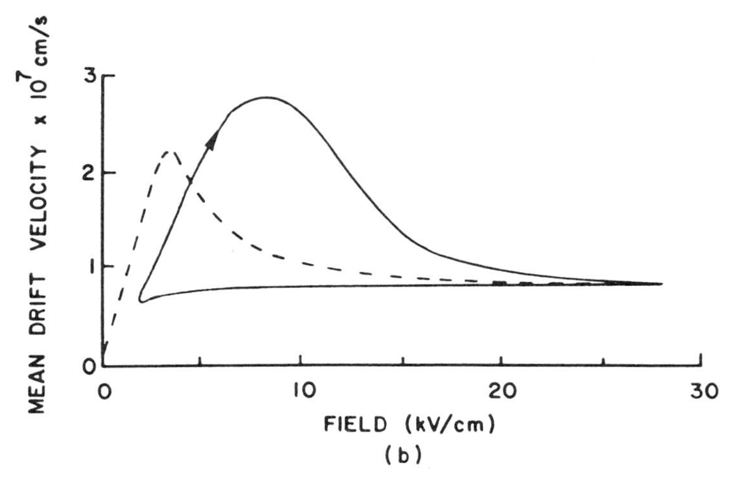

FIGURE 5-30. Plots against field: (a) fractional satellite population; (b) mean drift velocity for a 60-GHz rf field of 13.1 kV cm^{-1} superimposed on a dc field of 15 kV cm^{-1}. The dashed curves give the static values. (After Butcher and Hearn, 1968.)

(Shaw et al., 1979), enough carriers must be returned to the high-mobility valley for both transfer on the up-swing and NDM to occur. This means that the field must change slowly enough to allow the relaxation of carriers from the satellite to the central valley. If the field changes too rapidly, too many carriers are retained in the satellite valley and the NDM is too weak to sustain steady state oscillations. In Fig. 5-32 we plot the maximum frequency of the self-excited oscillations as a function of dc bias.

The large variations in field and carrier temperature for both self-excited and large-signal-driven oscillations should result in upper-frequency limits

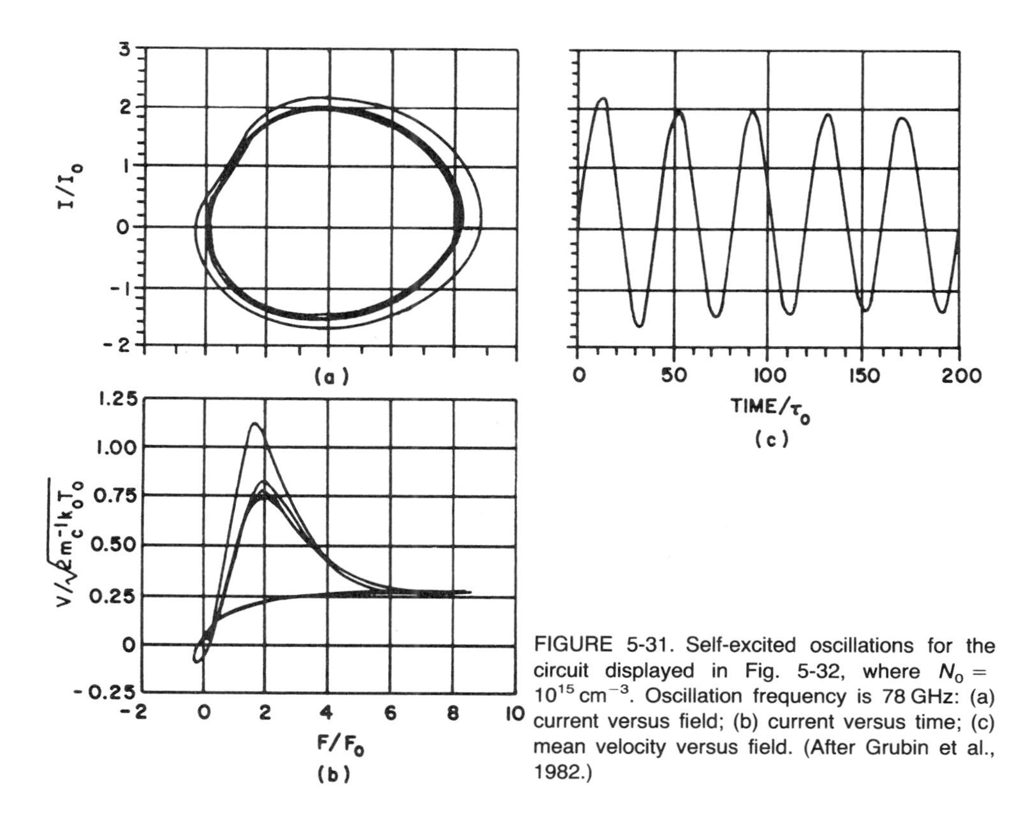

FIGURE 5-31. Self-excited oscillations for the circuit displayed in Fig. 5-32, where $N_0 = 10^{15}\ \text{cm}^{-3}$. Oscillation frequency is 78 GHz: (a) current versus field; (b) current versus time; (c) mean velocity versus field. (After Grubin et al., 1982.)

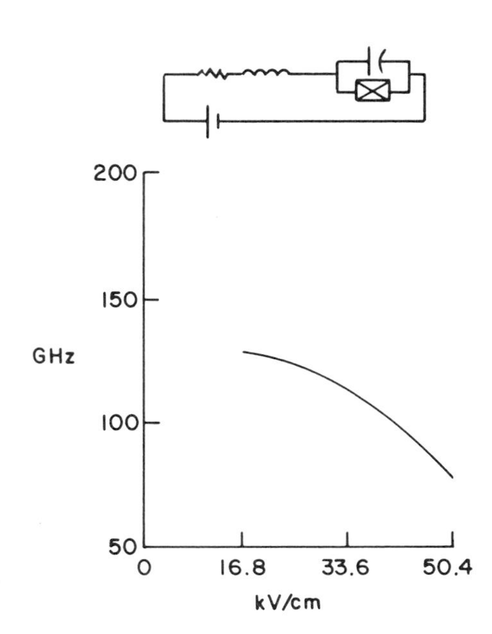

FIGURE 5-32. Maximum frequency for self-excited oscillations (after Grubin et al., 1982).

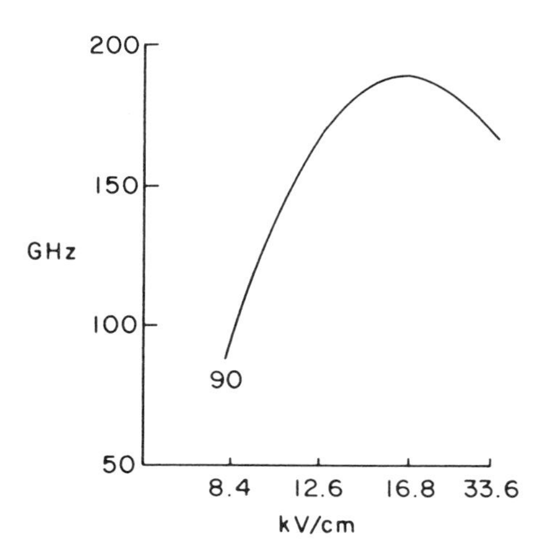

FIGURE 5-33. Maximum frequency for small-signal oscillations (after Grubin et al., 1982).

quantitatively different from that obtained for small-signal oscillations. We have examined the latter by perturbing a nonlinear element in steady state. The perturbation is a square wave voltage pulse and the resulting response is then Fourier-analyzed. The results are shown in Fig. 5-33, where we plot the maximum frequency of the small-signal negative conductance as a function of dc bias. The most significant feature here is that f_{max} for small-signal operation is significantly greater than that for large-signal operation. We note that when similar calculations with the nonlinear element driven by an ac source of controlled amplitude and frequency are performed the results bridge the small- and large-signal oscillation calculations.

The explanation for differences in the large- and small-signal results lies in the energy-scattering rates (see Fig. 5-15). For large-signal oscillations the transient temperature in both the central and satellite valleys oscillates over a larger range than that for the small-signal oscillations and samples lower scattering rates. This gives rise to the reduced maximum frequency for the large-signal oscillation.

Detailed analysis of the frequency limitation of the transferred electron effect in GaAs using the SETEM approach was discussed by Rolland, et al. (1979). They were able to show that reasonable efficiencies could be obtained with frequencies up to 150 GHz and offered the uniform field mode as an alternative means for circumventing the drastic size restrictions usually associated with millimeter wave devices.

5.4.5. *Length Dependence of NDM*

We have been discussing the upper-frequency limit of transferred electron devices from the circuit viewpoint and the transfer and return of electrons between the central and subsidiary valleys. In the analysis we have not explicitly considered the following problem: When an electron enters the active region of a device, it accelerates in the presence of an applied electric field. If the initial drift velocity of the carrier is low, is the transit length sufficient to cause electron

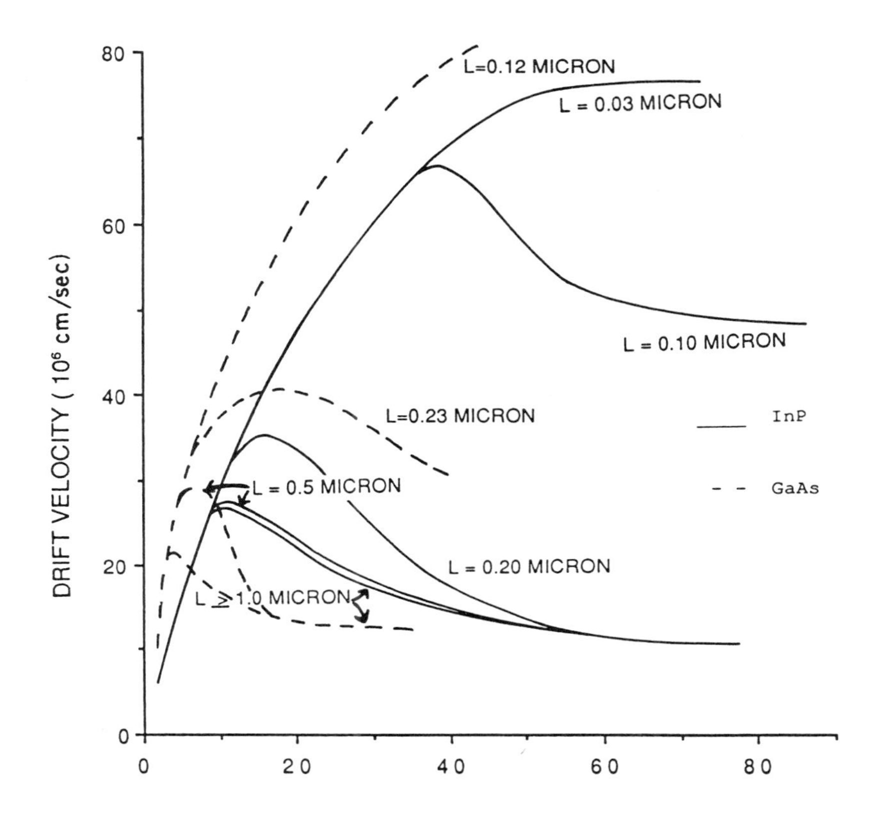

FIGURE 5-34. Length dependence of NDM in GaAs (dashed) and InP (solid) (after Wu et al., 1991).

transfer and negative differential mobility? The answer lies in earlier calculations. If the carrier experiences a sudden change in field, the mean initial transient ($t \ll LO$ phonon scattering time) will increase approximately linearly with time, followed by a region where v will approach $-e\tau F/m^*$ for a single valley. If the transit time is short enough to prevent significant transfer, the NDM will be weak, if it exists at all. Figure 5-34 summarizes the results for GaAs (Grubin et al., 1979; Wang, 1979) and InP (Wu et al., 1991), where for uniform fields the device length is a derived quantity. Here we have

$$L = \int_0 v(t)\, dt \tag{5-63}$$

The velocity versus field curves and the velocity versus time curves of the type discussed provide an indication of why there is interest in submicron devices. The possibility exists that very high velocities over very short distances can be obtained. But, again, a word of caution: The calculations of Fig. 5-34 are for carriers subjected to sudden changes in field. As we have seen, a finite rise time dramatically reduces this peak; the experimental results will therefore yield lower velocities than anticipated by the zero rise time calculations.

5.5. THE INFLUENCE OF BOUNDARY CONDITIONS

5.5.1. Introduction

It is now generally accepted that electrical instabilities in bulk semiconductors are controlled by the details of the boundary as well as the details of the interior regions. By boundary we mean the metal–semiconductor interface, the n^+–n^- interface, the semiconductor–vacuum interface, etc. Because of their technological importance, we will emphasize contacts to near- and submicrometer-sized devices in this chapter. The situation with submicrometer devices is such that, by virtue of the thin interior region, the interface is expected to exercise principal control over transport within the semiconductor and devices constructed thereof.

Transport within any device, particularly with regard to boundaries, is three dimensional. The distribution function within the device mirrors scattering events at the boundaries, particle confinement, and a host of detailed surface properties. Difficulties arise simply in describing the role of the boundary theoretically and identifying its influence experimentally. Because of this, the discussion here will be confined to transport normal to the interface. Particular emphasis will be placed upon the identification of the role of the boundary in controlling transport in near- and submicrometer-length devices.

In examining the role of the boundary, cognizance is taken of the work of Hess and Iafrate (1985) on the dependence of transport on the energy and velocity distribution of electrons entering a uniform field region. In the discussion below, however, emphasis is on spatially dependent transport in which both the space-charge and the field distributions within the device are nonuniform. The reason for including nonuniformities in the discussion is that they are consequences of the presence of contacts and/or the existence of nonuniformities in the doping profile. The significance of including them in the study lies in the fact that, as we have already pointed out, transient effects in the presence of spatial inhomogeneities are both qualitatively and quantitatively different from those calculated under uniform field conditions. Several examples illustrate these differences. First, under uniform field conditions, long-time steady state velocities show the presence of a dc negative differential conductivity in gallium arsenide arising from electron transfer. Under nonuniform field conditions, where current rather than velocity is the relevant quantity, calculations for devices with injecting, partially blocking contacts and highly nonuniform n^+-n^--n^+ structures show highly nonlinear current–voltage relationships. These $I(\Phi)$ characteristics do not, however, display negative differential conductance. (The absence of dc NDC from the calculation reported later does not imply the universal absence of dc NDC from transferred electron semiconductors. It is possible to envision the mathematical possibility of a boundary with a region of NDC, which when coupled to a transferred electron semiconductor, will yield dc NDC.) Another point of importance involves the character of the transient. For uniform fields the signature of velocity overshoot lies in an initial high peak velocity followed by electron transfer and a rapid settling toward steady state. Under nonuniform field conditions the initial transient is dependent upon the structure of the device. For n^+-n^--n^+ regions the initial transient sustains major position-dependent displacement current contributions. These displacement current contributions arise from the internal rearrangement of electric fields and have the effect of increasing the lapsed time before the field reaches its steady state value. This results in a decreased velocity overshoot transient but not a decreased spatial overshoot, as discussed below.

It must also be recognized that the role of metal boundaries and/or properly designed heterostructure interfaces is significantly different from the role of the n^+-n or n-n^+ interface on device operation. The key element here, even for transport normal to the interface, may be carrier confinement. A 0.25-μm structure with carriers confined to this region will differ in behavior from a 1-μm-long n^+-n^--n^+ element in which the n^- region is only 0.25 μm in length. For the n^+-n^--n^+ structure at sufficiently high fields, enough of the potential can fall across the downstream n^+ region to cause it to maintain high current densities and electron transfer within it.

A key element in the study of these devices lies in the description of the interface and how it is modeled. There are several philosophical approaches that we may take. In one-dimensional descriptions the metal–semiconductor interface may be treated as a mathematical boundary, with the variables chosen to represent the boundary dictated by the form of differential equations chosen to describe transport within the semiconductor. For example, in the drift and diffusion formulation of transport, the equation for total current is often expressed in terms of a second-order partial differential equation in field. Thus, the boundary conditions involve specifying the field at the cathode and anode. In one study (Shaw et al., 1969) the electric field was specified as a time-independent value and the resulting dc current voltage characteristic and time-dependent behavior, when it occurred, was shown to be a sensitive function of the chosen boundary value. More general discussions have included a time-dependent cathode field (Grubin, 1978).

Another point of view may tend to ignore the mathematical boundary as an appropriate representation of the interface effect. Instead, at a position far removed from the boundary, an effective field may be introduced to account for the consequences of, e.g., a dipole layer, or indeed the dipole layer may be introduced (Zur et al., 1983). The region must then be coupled to a set of time-dependent rate equations that account for either thermionic emission or field-assisted tunneling through the generated barrier (DeGroot et al., 1984).

Independent of the point of view taken to model the effect of the interface in the presence of an applied field, the carriers will enter the semiconductor with a well-defined distribution of energies that are likely to be significantly different from those far from the interface. A case in point is gallium arsenide, where the following question may be asked: When the distribution of carriers, velocity, and energy in the Γ, L, and X valleys are known at both the up and downstream interfaces, then through solution to the governing interior equations it may be expected that the current–voltage relation and transient behavior of the structure in principle is predictable. Given this, can the obverse be seen? Namely, can we extract from a given set of electrical measurements, on near- and submicrometer structures, a family of interfacial characteristics with which material variations lead to predictive device behavior? This approach is clearly iterative and has been attempted. It may also be necessary if we have any hopes of engineering structures for high-speed applications. Indeed, there are already indications that this approach may be successful. The evidence lies in the success of the boundary field models to explain, on one level, the broad range of electrical behavior of gallium arsenide and indium phosphide (Shaw et al., 1979) and the apparent relationship of these boundary field models to the energy, momentum, and carrier distribution of the entering electrons.

The preceding discussion expresses the construction and viewpoint of this section: device boundaries and interfaces dictate that transport must reflect their presence. The purpose of this section is to illustrate this. The discussion is separated into two distinct parts, with the first part dealing with the equations governing near- and submicrometer transport. The description of transport is through moments of the Boltzmann transport equation. The second part of the discussion deals with boundary- and length-dependent transport. Initially, several uniform field transient calculations are included to introduce the language of transient transport and to form a basis for comparison with the nonuniform field results.

The nonuniform field results are discussed in Sec. 5.5.3. Here, two distinct classes of devices are considered. The first consists of a uniformly doped structure in which all space charge nonuniformities arise from variations in the upstream boundary (cathode) conditions. The second device structure is the n^+-n^--n^+ structure, in which nonuniformities in the space charge arise primarily from the n^+-n^- and n^--n^+ interfaces. Transient calculations with both structures show distinct local displacement current contributions, which will camouflage, in many cases, the presence of transient overshoot.

A brief summary of the basic findings of the study is contained in Sec. 5.5.4.

5.5.2. Transport through Moments of the Boltzmann Transport Equation

Spatial and temporal transients are determined through solution of a set of coupled equations. These include Poisson's equation

$$\nabla^2\Phi = +(e/\epsilon)(n - N_0) \tag{5-64}$$

where N_0 is a prespecified background concentration and n denotes the free carrier contribution arising from various portions of the conduction band. For the discussion below, only two sections of the conduction band are considered, Γ and L. Thus, we have

$$n = n_1 + n_2 \tag{5-65}$$

where n_1 designates the population of the Γ valley and n_2 the population of the L valley.

Poisson's equation is coupled to the first three moments of the Boltzmann transport equation, the first set of which involves continuity. For the Γ valley we have

$$\frac{\partial n_1}{\partial t} = -\frac{\partial}{\partial x_j}\left(\frac{n_1 \hbar k_1^j}{m_1}\right) - n_1\Gamma_1 + (n - n_1)\Gamma_2 \tag{5-66}$$

where Γ_1 denotes the rate at which carriers are scattered from the Γ valley to all sections of the L valley and Γ_2 denotes return scattering, and summation over equal indices is assumed. For parabolic bands we assume

$$\hbar k_1^j = m_1 v_1^j \tag{5-67}$$

An equation similar to Eq. (5-66) describes transient population changes in the L valley. When the two are combined, a global continuity equation results:

$$\frac{\partial n}{\partial t} = -\frac{\partial}{\partial x_j}\left[n_1 \frac{\hbar k_1^j}{m_1} + (n - n_1)\frac{\hbar k_2^j}{m_2}\right] \tag{5-68}$$

The quantity

$$n_1(\hbar k_1^j/m_1) + (n - n_1)(\hbar k_2^j/m_2) \equiv C^j \tag{5-69}$$

is the velocity flux density of the system. It is convenient to relate this term to a mean spatially dependent drift velocity

$$v^j = C^j/n \tag{5-70}$$

Note that the total current density,

$$J^j = -eC^j + \epsilon(\partial F^j/\partial t) \tag{5-71}$$

is conserved; i.e.,

$$\frac{\partial J^j}{\partial x_j} = 0 \tag{5-72}$$

The second pair of moment equations is that of momentum balance. For the Γ-valley carrier we have

$$\frac{\partial}{\partial t} n_1 \hbar k_1^j = -\frac{\partial}{\partial x_i}\left(\frac{\hbar k_1^i}{m_1} n_1 \hbar k_1^j\right) + e n_1 \frac{\partial \Phi}{\partial x_j} - \frac{\partial}{\partial x_i}\psi_1^{ij} - n_1 \hbar k_1^j \Gamma_3 \tag{5-73}$$

Here Γ_3 represents the net rate of momentum scattering and ψ^{ij} represents the components of the pressure tensor

$$\psi_1^{ij} = \frac{1}{4\pi^3}\frac{\hbar^2}{m_1}\int (k - k_1)_i (k - k_1)_j f\, d^3k \tag{5-74}$$

For the situation in which f represents a displaced Maxwellian

$$\psi_1^{ij} = n_1 k_B T_1 \delta_{ij} \tag{5-75}$$

where T_1 is the electron temperature of the Γ-valley carriers. For the situation in which there are nonspherical contributions to f, additional diagonal as well as off-diagonal components of the pressure tensor arise. For the following calculations the distribution function has been generalized from the displaced Maxwellian

$$f_0 = A \exp[-\hbar^2(k - k_1)^2/2m_1 k_B T_1] \tag{5-76}$$

into the form (see, for example, Sommerfeld, 1956)

$$f = \left(1 + a_i \frac{\partial}{\partial k_i} + a_{ij} \frac{\partial^2}{\partial k_i\, \partial k_j} + a_{ijk} \frac{\partial^3}{\partial k_i\, \partial k_j\, \partial k_k} + \cdots\right) f_0 \tag{5-77}$$

subject to the conditions

$$\frac{1}{4\pi^3} \int f\, dk = \frac{1}{4\pi^3} \int f_0\, d^3k \tag{5-78}$$

$$\frac{1}{4\pi^3} \frac{\hbar^2}{m_1} \int f(k - k_1)^2\, dk = \frac{1}{4\pi^3} \frac{\hbar^2}{m_1} \int f_0(k - k_1)^2\, d^3k \tag{5-79}$$

The nonspherical nature of the distribution function suggests the separation

$$\psi_1^{ij} = \psi_{01}^{ij} + \hat{\psi}_1^{ij} \tag{5-80}$$

where ψ_{01}^{ij} is given by Eq. (5-75) and $\hat{\psi}_1^{ij}$ represents the additional contribution. The nonspherical contributions are not calculated from first principles. Instead, the treatments of fluid dynamics are followed with

$$\hat{\psi}_1^{ij} = -\mu_1\left(\frac{\partial v_1^i}{\partial x_j} + \frac{\partial v_1^j}{\partial x_i} - \frac{2}{3} \frac{\partial v_1^k}{\partial x_k} \delta_{ij}\right) \tag{5-81}$$

where it is noted that

$$\sum_{i=1}^{3} \hat{\psi}_1^{ii} = 0 \tag{5-82}$$

In one dimension (along x),

$$\hat{\psi}_1^{ij} = -\frac{4}{3} \mu_1 \frac{\partial v_1^x}{\partial x} \tag{5-83}$$

In two dimensions the derivative of the stress tensor is

$$\frac{\partial}{\partial x_i}\hat{\psi}_1^{ij} = i\left[(-\mu_1)\left(\frac{4}{3}\frac{\partial^2 v^x}{\partial x^2} + \frac{\partial^2 v^y}{\partial x^2} + \frac{\partial^2 v^x}{\partial x\,\partial y} - \frac{2}{3}\frac{\partial^2 v^y}{\partial x\,\partial y}\right)\right]$$
$$+ j\left[(-\mu_1)\left(\frac{4}{3}\frac{\partial^2 v^y}{\partial y^2} + \frac{\partial^2 v^x}{\partial y^2} + \frac{\partial^2 v^y}{\partial x\,\partial y} - \frac{2}{3}\frac{\partial^2 v^x}{\partial x\,\partial y}\right)\right] \tag{5-84}$$

In the discussion below, an even simpler version of Eq. (5-84) is assumed:

$$\frac{\partial}{\partial x_i}\hat{\psi}_1^{ij} = -\hat{\mu}_1\frac{\partial^2 v_1^j}{\partial x_i^2} \tag{5-85}$$

with the constraint Eq. (5-78) only approximately satisfied. Thus, the relevant equation for momentum balance is

$$\frac{\partial}{\partial t}n_1\hbar k_1^j = -\frac{\partial}{\partial x_i}\left(\frac{\hbar k_1^i}{m_1}n_1\hbar k_1^j\right) + en_1\frac{\partial\Phi}{\partial x_j} - \frac{\partial}{\partial x_j}n_1 k_B T_1 + \frac{\mu_1\hbar}{m_1}\frac{\partial^2 k_1^j}{\partial x_i^2} - n_1\hbar k_1^j\Gamma_3 \tag{5-86}$$

For the L valley the relevant momentum balance equation is

$$\frac{\partial}{\partial t}(n-n_1)\hbar k_2^j = -\frac{\partial}{\partial x_i}\frac{\hbar k_2^j}{m_2}(n-n_1)\hbar k_2^j + e(n-n_1)\frac{\partial\Phi}{\partial x_j}$$
$$- \frac{\partial}{\partial x_j}(n-n_1)k_B T_2 + \mu_2\frac{\hbar}{m_2}\frac{\partial^2 k_2^j}{\partial x_i^2}$$
$$- (n-n_1)\hbar k_2^j\Gamma_4 \tag{5-87}$$

The third and final pair of balance equations is that associated with energy transport. Straightforward application of the moment equations yields

$$\frac{\partial}{\partial t}W_1 = -\frac{\partial}{\partial x_j}\frac{\hbar k_1^j}{m_1}W_1 + n_1 e\frac{\hbar k_1^j}{m_1}\frac{\partial\Phi}{\partial x_j} - \frac{\partial}{\partial x_j}\frac{\hbar k_1^i}{m_1}\psi_1^{ij} - \frac{\partial}{\partial x_j}Q_1^j$$
$$- n_1 U_1\Gamma_5 + (n-n_1)U_2\Gamma_6 \tag{5-88}$$

where

$$U_1 = \tfrac{3}{2}k_B T_1 \qquad U_2 = \tfrac{3}{2}k_B T_2 \tag{5-89}$$

$$W_1 = n_1\left(\frac{\hbar^2 k_1^2}{2m_1} + U_1\right) \tag{5-90}$$

and

$$Q_1^j = \frac{\hbar^3}{8\pi^3 m_1^2}\int (k-k_1)_j(k-k_1)_i^2 f\,d^3k \tag{5-91}$$

where the summation convention over i is assumed. For spherically symmetric distribution functions, Q_1^j is zero. For nonspherical situations it represents a flow of heat and is treated phenomenologically through analogy to Fourier's law

$$Q_1^j = -\kappa_1 \frac{\partial}{\partial x_j} T_1 \tag{5-92}$$

It is important to note at this point that the relationships given by Eqs. (5-81) and (5-92) are not fundamental. Rather, they are expressions of our ignorance of the detailed role of the distribution on transport, particularly near boundaries.

In the analysis that follows, Eq. (5-88) is not solved. Rather, it is combined with Eqs. (5-66) and (5-87) to yield

$$\begin{aligned}\frac{\partial}{\partial t} n_1 U_1 = &-\frac{\partial}{\partial x_j}\frac{\hbar k_1^j}{m_1} n_1 U_1 - \frac{2}{3} n_1 U_1 \frac{\partial}{\partial x_j}\frac{\hbar k_1^j}{m_1} + \kappa_1 \frac{\partial^2}{\partial x_j^2} T_1 \\ &+ \frac{\hbar^2 k_1^2}{2m_1}[2n_1\Gamma_3 - n_1\Gamma_1 + (n - n_1)\Gamma_2] - n_1 U_1 \Gamma_5 \\ &+ (n - n_1)U_2\Gamma_6\end{aligned} \tag{5-93}$$

In Eq. (5-93) the nonspherical contributions of the stress tensor, [Eq. (5-73)] are ignored. For the second species of carriers

$$\begin{aligned}\frac{\partial}{\partial t}(n - n_1)U_2 = &-\frac{\partial}{\partial x_j}\frac{\hbar k_2^j}{m_2}(n - n_1)U_2 - \frac{2}{3}(n - n_1)U_2 \frac{\partial}{\partial x_j}\frac{\hbar k_2^j}{m_2} + \kappa_2 \frac{\partial^2}{\partial x_j^2} T_2 \\ &+ \frac{\hbar^2 k_2^2}{2m_2}[2(n - n_1)\Gamma_4 + n_1\Gamma_1 - (n - n_1)\Gamma_2] \\ &- (n - n_1)U_2\Gamma_7 + n_1 U_1 \Gamma_8\end{aligned} \tag{5-94}$$

Equations (5-64), (5-66), (5-68), (5-86), (5-87), (5-93), and (5-94) are the equations governing transport in the systems considered in this chapter. The equations are more general than others in that nonspherical contributions to the (BTE) moments have been included. The scattering integrals Γ_1–Γ_8 and the form they take have been discussed in the past where these evaluations have been in terms of the displaced Maxwellian only. These integrals have not been generalized to include nonspherical contributions.

The governing equations are expressed in dimensionless form prior to transformation into difference equations. The dimensionless equations are discussed in the Appendix to this chapter. Solution of the governing equation requires imposition of boundary conditions. These represent a crucial aspect of the study and are discussed as they are needed. The band structure parameters used in the study for two-level transfer are also discussed in the Appendix.

5.5.3. Solutions of the Governing Equations

5.5.3.1. Uniform Fields

Calculations for uniform fields were discussed in Sec. 5.4.3. But since they offer an important starting point for examining transients under nonuniform field conditions, we briefly repeat and extend some of our prior arguments. Uniform fields result from assuming a donor level N_0 that is spatially constant at the boundary and specifying that

$$n_x = n_{1x} = V_{1x} = V_{2x} = T_{1x} = T_{2x} = 0 \tag{5-95}$$

at both the cathode and anode boundaries. The subscript x in Eq. (5-95) denotes a first derivative. Figure 5-35 displays the velocity transient for a 1-μm-long element with a doping level of $5.0 \times 10^{15}\ \text{cm}^{-3}$. The parameters involved are listed in Table 5-1. The length specification is artificial. For each calculation the bias was raised in one time step from 0.01 V to the value indicated in the figure. Note the high carrier velocity occurring at approximately 0.5 ps and the long-term asymptotic lower steady-state value. Also apparent in the figure is the presence of

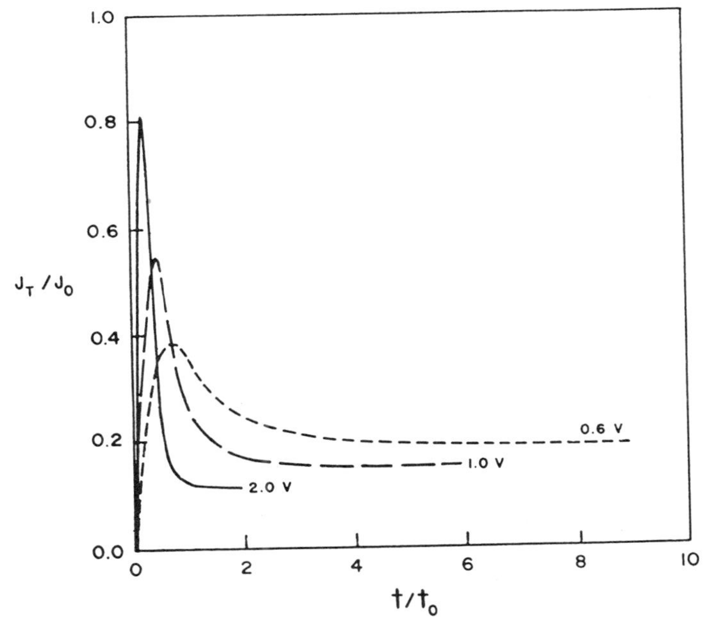

FIGURE 5-35. Magnitude of the current transient [Eq. (5-8)] following application of a sudden change in bias. Parameters for this calculation are listed in Table 5-1. The results of the calculation are qualitatively similar to those obtained in many studies, the first for GaAs being Ruch (1972). The terminus of each calculation reflects the physical time required for steady state. The longest time duration is that associated with the lowest bias level. For this calculation 2.0 V corresponds to an average field of $20\ \text{kV cm}^{-1}$; 1.0 V yields $10\ \text{kV cm}^{-1}$; etc. $J_0 = 8 \times 10^4\ \text{A cm}^{-2}$; $t_0 = 1$ ps. (After Grubin et al., 1985.)

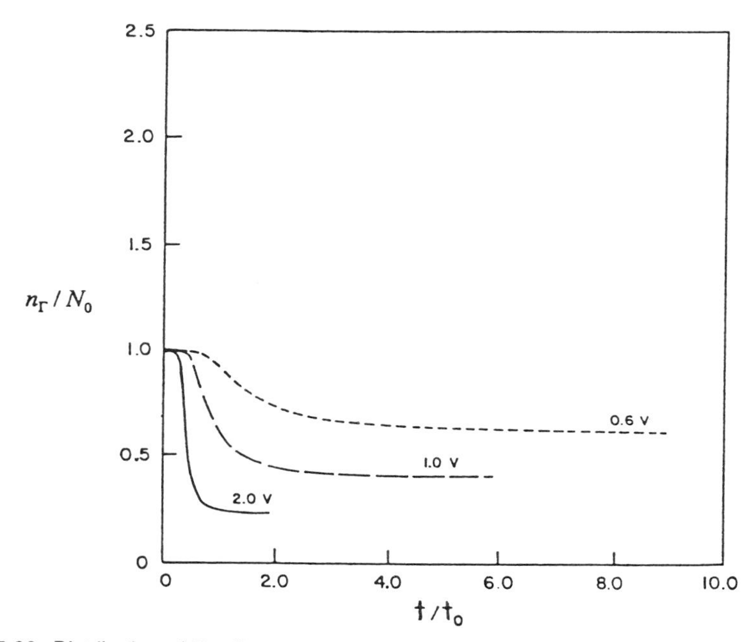

FIGURE 5-36. Distribution of Γ-valley carriers as a function of time for the parameters of Fig. 5-35. Note the delay in electron transfer, which is shortest for the highest bias level. $N_0 = 5 \times 10^{15}\,\text{cm}^{-3}$; $t_0 = 1$ ps. (After Grubin et al., 1985.)

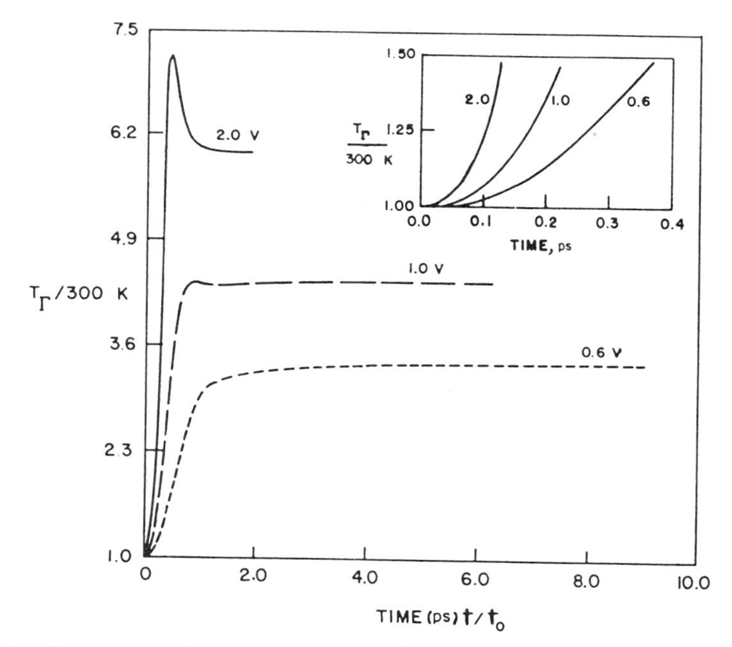

FIGURE 5-37. Transient distribution of temperature following application of a sudden change in bias for the parameters of Fig. 5-35. The presence of a temperature overshoot is noted, a feature resulting from the enhanced scattering at elevated temperature. The inset displays the temperature during the first 0.4 ps and demonstrates the onset of scattering. $t_0 = 1$ ps. (After Grubin et al., 1985.)

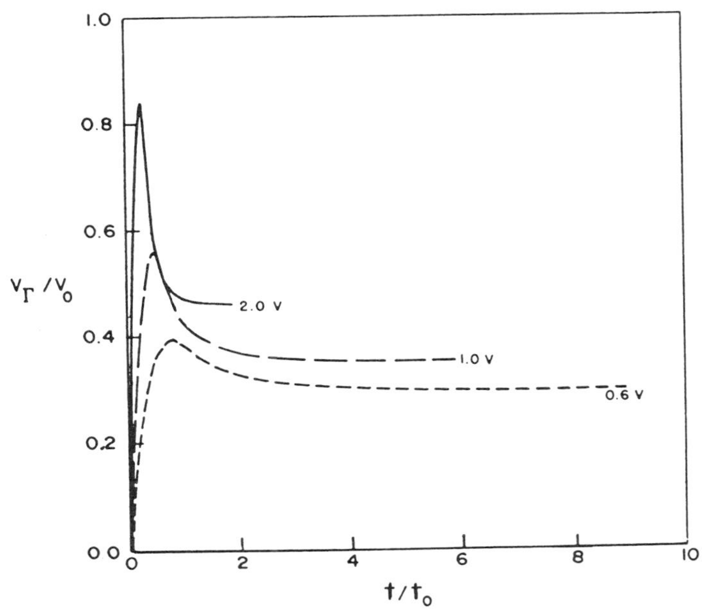

FIGURE 5-38. Transient Γ-valley velocity distribution for the parameters of Fig. 5-35. The initial velocity peak corresponds closely in value to the peak current transient prior to electron transfer. The decreased velocity represents enhanced scattering at elevated temperatures. $v_0 = 10^8\ \text{cm s}^{-1}$; t_0 = ps. (After Grubin et al., 1985.)

a region of NDM. Figure 5-36 displays the time rate of change of carriers in the Γ valley. Electron transfer is apparent for times after the peak velocity is reached. Figure 5-37 displays the time dependence of the electron temperature following application of the voltage pulse. The feature to be noted from this figure and Eq. (5-93) is that for uniform fields any time dependence in T_1 is due entirely to scattering events and is thus a measure of when ballistic transport may be ignored. Another point of interest is that under uniform field conditions the population of carriers in either the central or satellite valley is governed by the scattering rates, which are in turn governed by the value of the carrier temperature. This will be featured prominently later when contact effects are considered. The time dependence of the Γ-valley velocity is displayed in Fig. 5-38. (The long-time asymptotic values do not and should not display NDM. The mean steady state distribution of velocities as well as that within the Γ valley is shown in Fig. 5-39. Nonparabolic effects are not included here.)

5.5.3.2. Nonuniform Fields and Uniform Doping

The origin of nonuniform fields and space-charge layers in uniformly doped structures lies in the conditions imposed at the upstream and downstream boundaries. Under conditions where current flows through the structure, the upstream boundary conditions manifest themselves as cathode boundary current–field relationships (see, e.g., Czekaj et al., 1985, 1988; Wu and Shaw, 1989; Wu

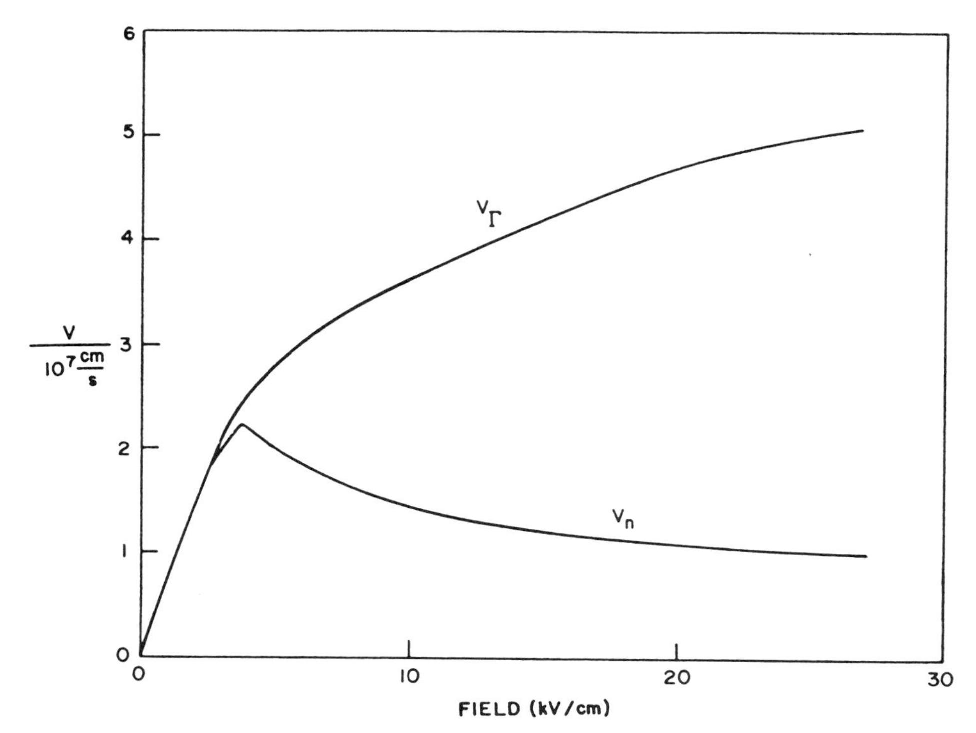

FIGURE 5-39. Steady state field-dependent velocity for electrons in the Γ-valley of GaAs for the parameters of Table 5-1. Also, steady state mean field-dependent electron velocity v_n. [see Eq. (5-70).] (After Grubin et al., 1985.)

et al., 1991). It is the influence of the cathode boundary that will dominate the following discussion. To develop the concept of boundary controlled transport, several qualitative features of the mathematics governing transport are considered.

Under time-independent steady state conditions, the velocity flux density

$$C = n_1(x)V_1(x) + [n(x) - n_1(x)]V_2(x) \equiv n(x)V(x) \tag{5-96}$$

is a constant independent of position. Denoting through the subscript c the carrier density and the mean velocity at the first computed point within the semiconductor, the following exercise is performed:

$$C - N_0V_c = (n_c - N_0)V_c \tag{5-97}$$

where N_0 is the uniform background doping level. For the purpose of specificity V_c is assigned to be a monotonically increasing function of field and to have the form represented by the curve N_0V_c in Fig. 5-40. Note that for uniform field conditions V_c would necessarily be the same as the bulk field-dependent velocity and exhibit NDM. Also included in Fig. 5-40 is a sketch of one possible variation of n_cV_c. The field dependence of n_c is thereby defined implicitly in Fig. 5-40. It is

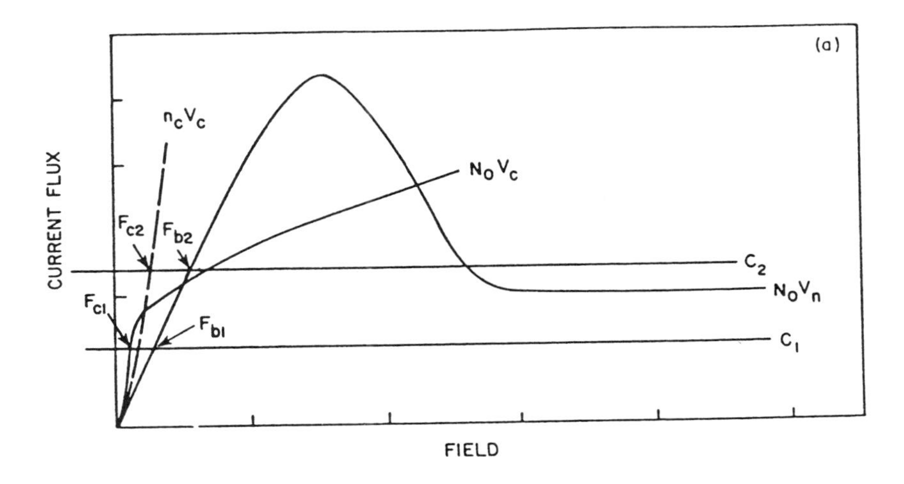

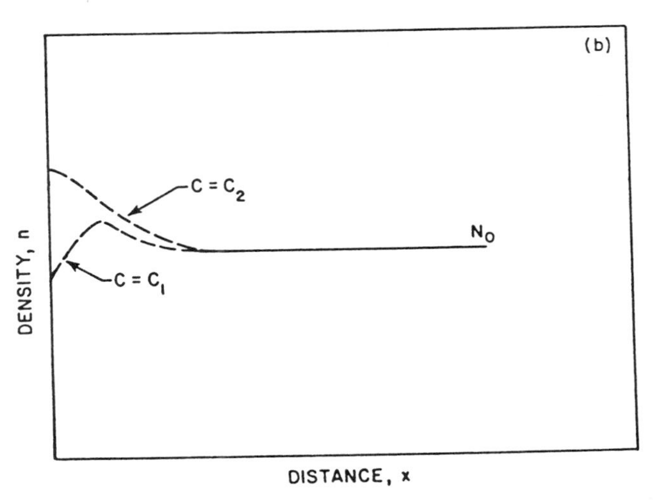

FIGURE 5-40. (a) Schematic representation of a current–field relationship within the interior of the semiconductor, N_0v_n. (C_1 and C_2 represent constant current levels in the device.) The cathode field for the low current level case is denoted by F_{c1}. The neutral interior field is represented by F_{b1}. Similar remarks apply to the higher current level. (It is important to note that studies using the drift and diffusion equations indcate that for $\Phi_x \approx 0$ at the boundary, the transition from cathode depletion to cathode accumulation requires all three characteristics, N_0v_c, n_cv_c, and N_0v_n to intersect at the same point (Kroemer, 1986b). (b) Schematic of possible cathode adjacent depletion and accumulation, followed by broad depletion, for the two bias current levels of Fig. 5-40a. (After Grubin et al., 1985.)

also noted that N_0V_c and n_cV_c are chosen to intersect, although there is no a priori reason to assume any universality to this property. Figure 5-40 also includes a schematic of the velocity flux density, N_0V_n (assuming NDM), associated with uniform fields and two horizontal lines representing two different values of the current flux density within the device.

Figure 5-40a takes on significance when the intersection of the line of constant current C and the neutral field characteristic N_0V_n is taken to represent uniform field region values within the interior of the semiconductor; the intersection of C with the cathode characteristic n_cV_c is taken to represent field values at the boundary of the semiconductor (Kroemer, 1968).

First consider the low-current case C_1. Here, the assumed current–field relationships are such that for

$$F_c > F_{b1} \tag{5-98}$$

$N_0 > n_c$. For a specific distance between the upstream boundary and the interior of the structure, a range of charge depletion forms, as sketched in Fig. 5-40b. Next consider the higher-current case C_2. For this situation

$$N_0V_c(F_c) < n_cV_c(F_c) \tag{5-99}$$

and a region of local charge accumulation forms at the upstream boundary. Because the field dependence of the mean carrier velocity exhibits a region of NDM the downstream interior field is either greater than or less than the cathode field and either a range of charge accumulation forms within the interior of the structure or a range of charge depletion forms within the interior. The latter is illustrated in Fig. 5-40b. The stability of this profile has been discussed in Sec. 5-3.

Consider Fig. 5-41a with a different set of upstream boundary characteristics. For the low-current case we have

$$F_c < F_{b1} \tag{5-100}$$

and a region of charge accumulation layer forms over a specific distance between the upstream boundary and the interior of the structure. However, at the upstream boundary

$$N_0V_c(F_c) > n_cV_c(F_c) \tag{5-101}$$

indicating that a region of local charge depletion forms at the upstream boundary. A sketch of a possible space charge profile is shown in Fig. 5-41b. For the high-current case, both within the interior and at the upstream boundary, regions of charge accumulation form. A sketch of this layer is also shown in Fig. 5-41b, and a boundary condition capable of generating this condition is shown in Fig. 5-42.

The preceding discussion indicates that the interplay between the boundary and the interior of the semiconductor introduces a rich variation in the space-charge distribution. A situation evoking considerable interest with respect to this interplay is one that may be regarded as a singular solution. This occurs when the current flux density C_2 intersects the neutral characteristic at two points and the curves n_cV_c, N_0V_c, and N_0V_n intersect at the same field value (thus, $V_n = V_c$ and $N_0 = n_c$). The general description and consequences of the approach

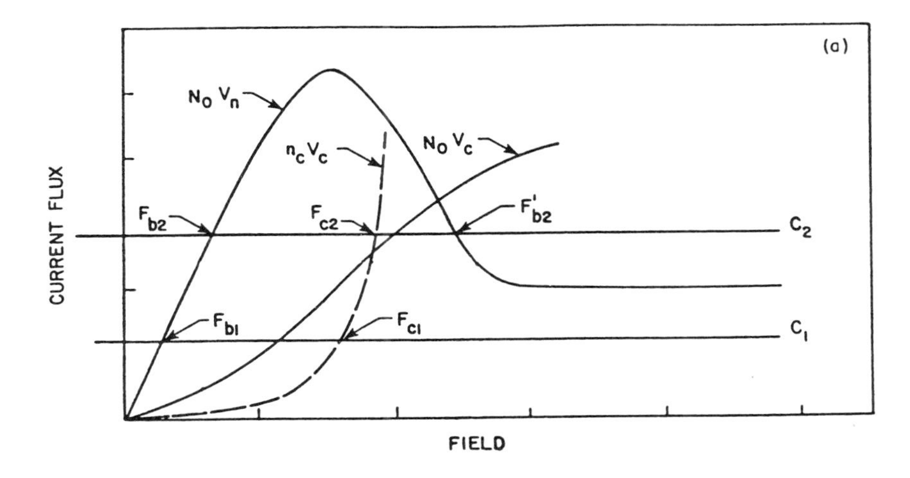

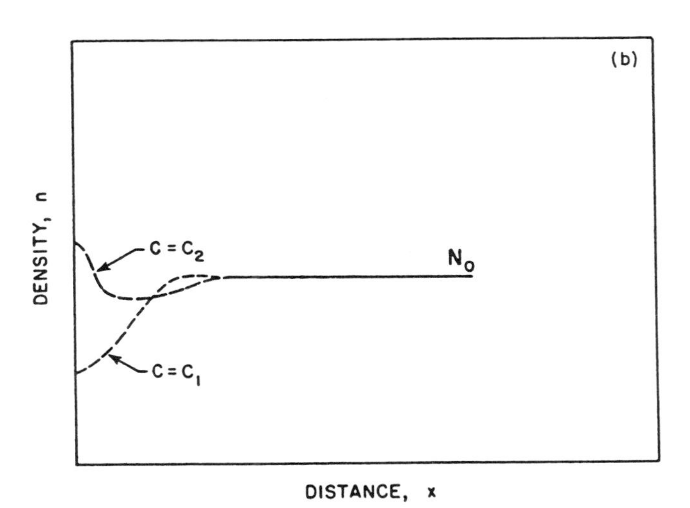

FIGURE 5-41. (a) Current–field relationship as in Fig. 5-40a but for a different set of N_0v_c and n_cv_c curves. (b) Schematic of possible cathode adjacent depletion and accumulation followed by broad accumulation for the two bias levels of Fig. 5-41a. (After Grubin et al., 1985.)

to this event in long samples, as a precursor for nucleation of high electric field traveling dipole layers, has been broadly delineated in a variety of publications (see, e.g., Shaw et al., 1969, 1979; Kroemer 1968) and discussed in Sec. 5.3. The consequence of this in terms of solutions to the BTE moments is discussed in Sec. 5.5.3.5.

It should be apparent from the preceding discussion that the detailed description of the influence of the boundary requires a description of the field dependence of the mean entrance velocity and carrier distributions. In the following calculations, in which transport is described through solutions to moments of the BTE, these field dependencies are constrained by the boundary

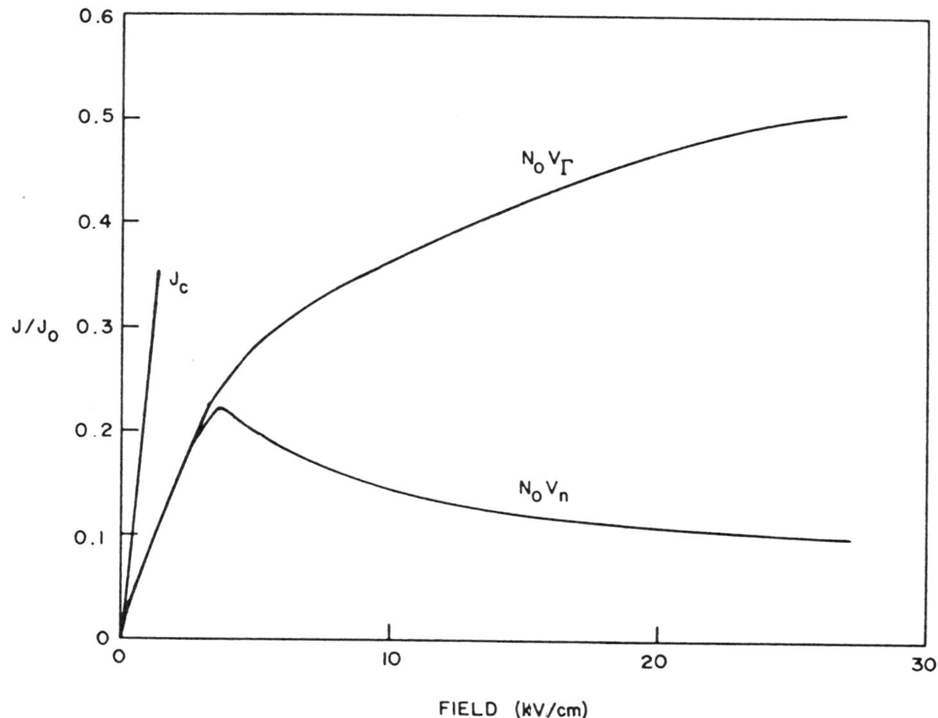

FIGURE 5-42. Data of Fig. 5-39 plus cathode current field relation, J_c, for the accumulation layer boundary. $J_0 = 8 \times 10^4$ A cm^{-2} (After Grubin et al., 1985.)

conditions to the governing equation and are expressed as solutions of

$$C = n_c V_c(F_c) \rightarrow F_c = g(C) \tag{5-102}$$

When F_c is a double-valued function of C a regional approach is taken.

5.5.3.3. Steady State and Transient Behavior for an Injecting Cathode ($L = 1.0\ \mu$m)

The preceding discussion is independent of the detailed description provided by the governing equations chosen to represent device transport. The governing equations and their associated boundary conditions provide a mechanism by which a set of contact descriptors can be extracted. For example, it is expected that the specific properties of the physical contact or boundary will influence the distribution of carriers within the valence and conduction bands of the semiconductor in the vicinity of the boundary. (One of the earlier studies involving the role of the boundaries on transient transport through solutions of the BTE was that of Gray, et al., 1975.) Furthermore, under conditions of finite bias in which current is transported through the device the influence of the contact is expected to affect the entrance velocities (Shaw et al., 1969). In the discussion that follows a very simple set of boundary conditions is imposed to represent the effects of the physical boundary. The importance of these boundary conditions is to create nonuniform fields. As will be seen, the boundary conditions chosen are not the result of an exhaustive study. Rather, they are an initial effort. For example, in the following discussion, the initial sharing of carriers between the Γ and L portions of the conduction band is controlled by specifying a value for the electron temperature at the cathode boundary. In addition, a representation of the entrance velocity is through a cathode contact mobility. This is identified in the calculation beginning with Fig. 5-43.

Figures 5-43–5-46 are calculations performed for a gallium arsenide structure with the same material parameters as that of the uniform field calculations. Here, however, the boundary conditions are different. At the cathode

$$\begin{gathered} n_x = n_{1xx} = 0, \qquad V_1 = -15\,625F, \qquad V_{2x} = 0 \\ T_1 = 300\,K, \qquad T_{2x} = 0 \end{gathered} \tag{5-103}$$

and at the anode

$$n_{xx} = n_{1xx} = V_{1xx} = V_{2xx} = T_{1xx} = T_{2xx} = 0 \tag{5-104}$$

where the double x subscript denotes a second derivative. The consequences of this set of boundary conditions is that the Γ-valley electrons enter the structure with a velocity in excess of the steady state uniform field value. Specification of the Γ-valley temperature at 300 K ensures that the relative cathode carrier contribution of the L valley is negligible. Furthermore, the fact that the mean velocity of the L-valley carriers is significantly below that of the Γ-valley carriers provides a demonstration that the cathode current field relation is dominated by the Γ valley carriers:

$$J_c - e[nV_{1c} + (n - n_1)V_{2c}] \cong ne\mu_c F \tag{5-105}$$

While Eq. (5-105) is significant in providing a description of the dominating carrier at the cathode, it alone will not determine whether the cathode is carrier depleted, neutral, or accumulated. The moment equations coupled to Poisson's equation must be solved. Qualitative information, however, can be obtained for the specific set of boundary conditions given by Eq. (5-103) through use of the mobility approximation. Because of the inherent limitations of the mobility approximation the consequences of its use must be regarded as relevant only if insight is provided in the interpretation of the exact solution.

The qualitative information is obtained through a calculation of the transit time of a carrier within the vicinity of the cathode. Because the transit time is necessarily a positive quantity, inequalities arise which express cathode depletion, neutrality, and accumulation. The transit time between the cathode and an interior point x is

$$t(x) = \int_0^x \frac{dX'}{V(X')} \tag{5-106}$$

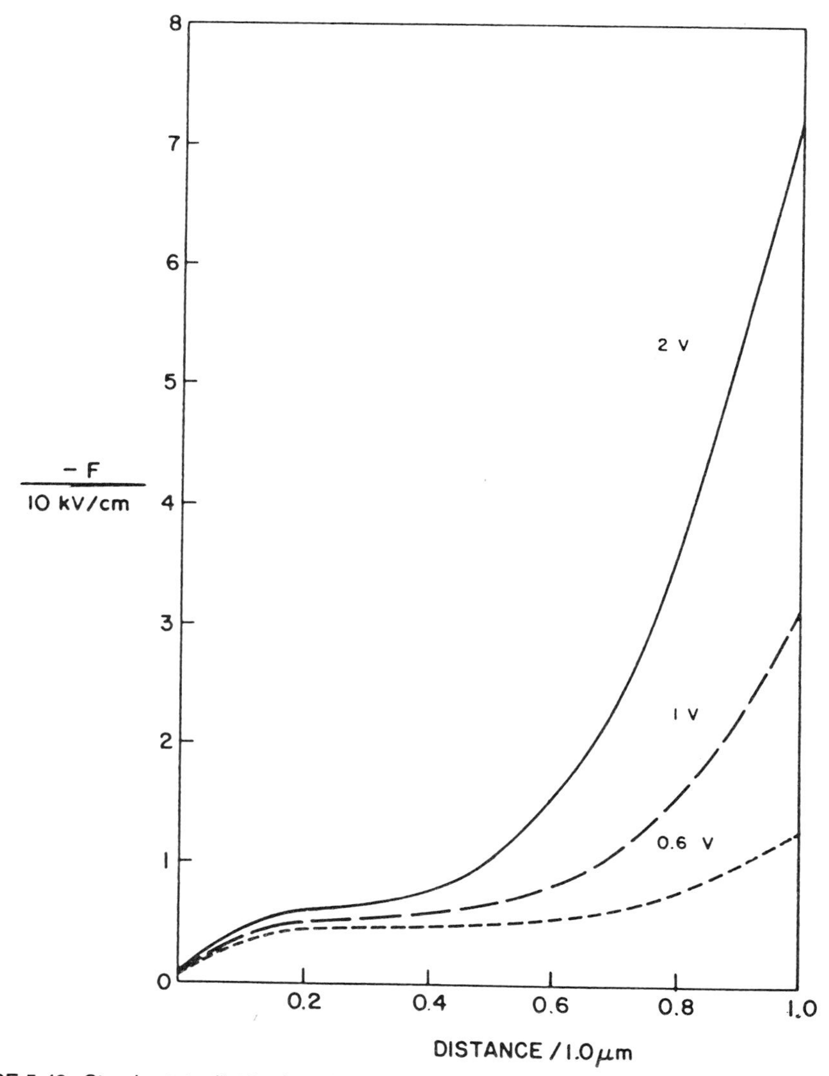

FIGURE 5-43. Steady state distribution of field within a 1-μm-long GaAs element at three bias levels; $T = 300$ K. (See appendix of this chapter for boundary conditions.) Electron transfer occurs downstream from the cathode resulting in a downstream accumulation of carriers. (After Grubin et al., 1985.)

Assuming a constant mobility for the Γ-valley carriers and the significant approximation

$$nV \cong n_1 V_1 \tag{5-107}$$

then for carriers within the vicinity of the cathode the arguments leading to Eq. (5-105) imply that $V \cong -\mu_1 F$. This last statement, when coupled to Poisson's

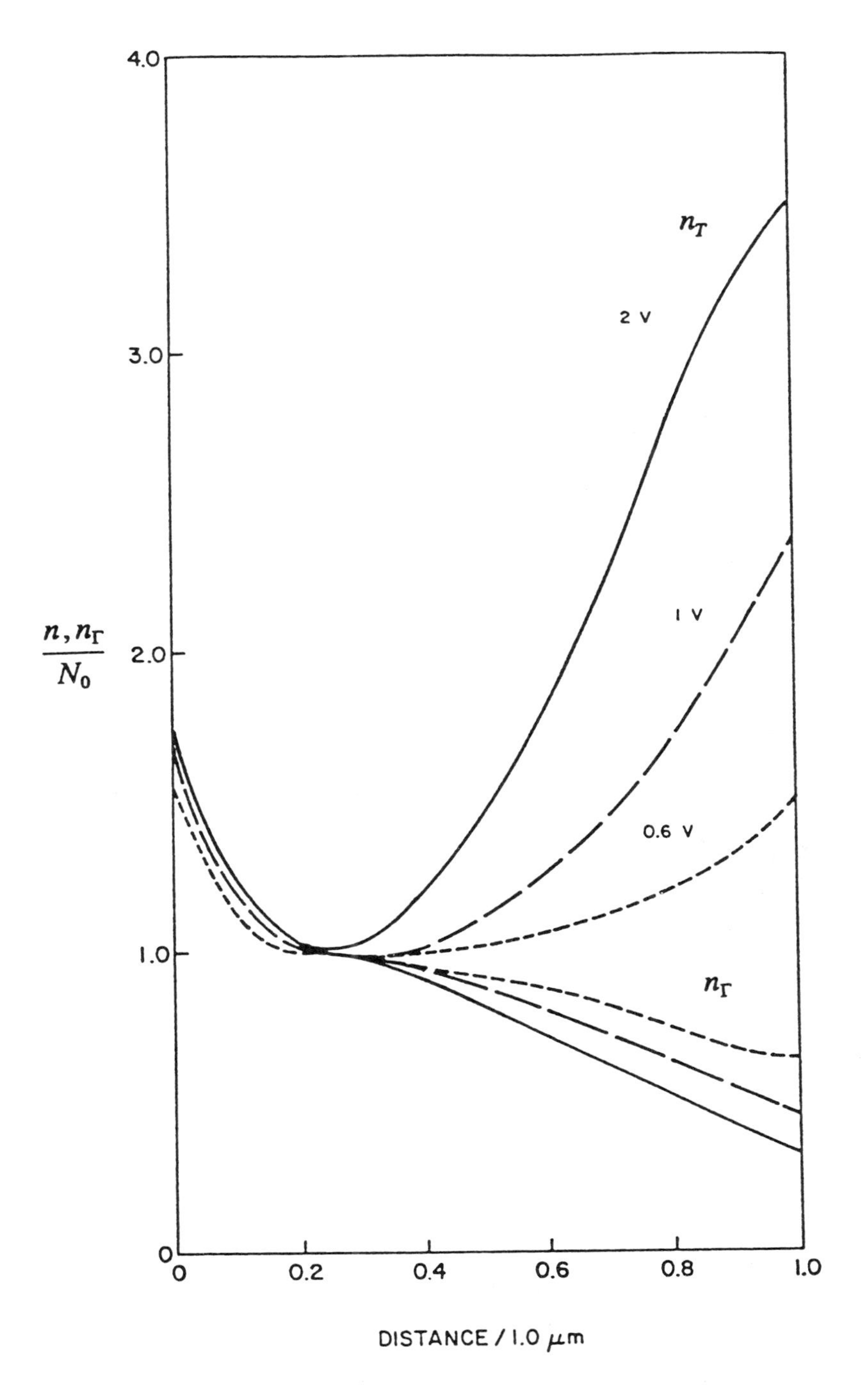

FIGURE 5-44. Distrubtion of total and Γ-valley carrier density for the parameters of Fig. 5-43. Electron transfer begins within 0.2 μm downstream from the cathode. By comparing Figs. 5-36 and 5-44 it is noted that electron transfer at 6,10, and 20 kV cm^{-1} significantly lags behind the uniform field value. (After Grubin et al., 1985.)

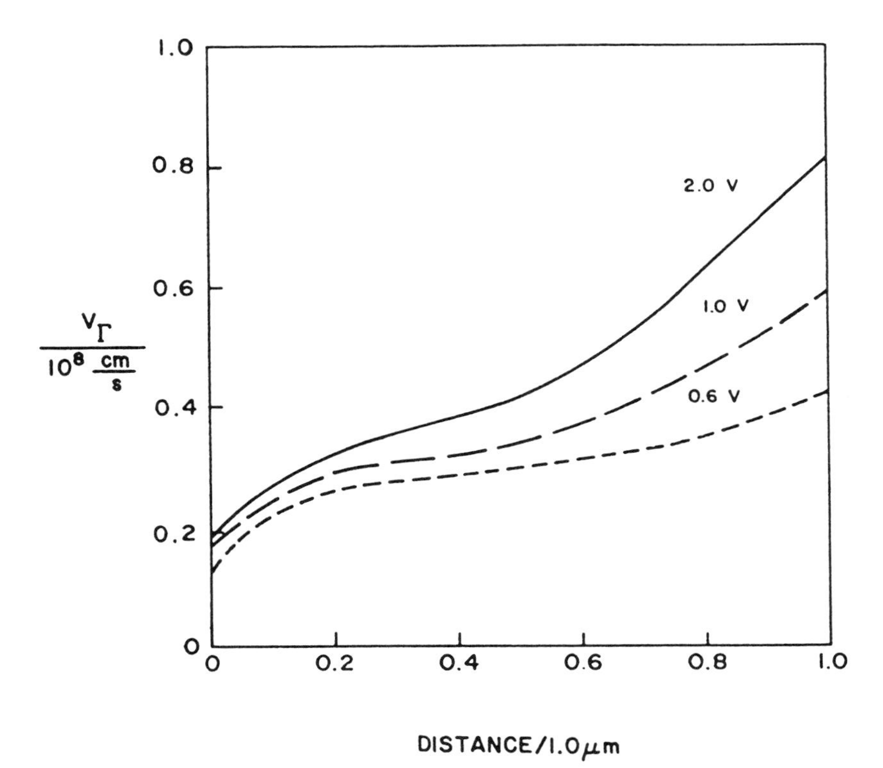

FIGURE 5-45. Distribution of Γ-valley velocity for the parameters of Fig. 5-43. At a bias of 2 V and a field of 20 kV cm^{-1} the Γ valley velocity is slightly in excess of the uniform field calculation. At a bias of 1 V and a field of 10 kV cm^{-1}, the difference between the nonuniform and uniform field velocity is even greater. This excess is a consequence of a lower value of electron temperature at these given field values. (After Grubin et al., 1985.)

equation, yields

$$t(x) = \tau_0 \log\left[\frac{J - N_0 e\mu_1 F_c}{J - N_0 e\mu_1 F(x)}\right] \tag{5-108}$$

where

$$\tau_0 = \epsilon / N_0 e\mu_1 \tag{5-109}$$

is the dielectric relaxation time of the Γ-valley carriers. For a cathode boundary condition consistent with $J = n_c e\mu_c F_c$ we have

$$t(x) = \tau_0 \log\left\{\left(1 - \frac{N_0\mu_1}{n_c\mu_c}\right) \Big/ \left[1 - \frac{N_0\mu_1 F(x)}{n_c\mu_c F_c}\right]\right\} \tag{5-110}$$

Since the requirement that the transit time be positive must be met, two

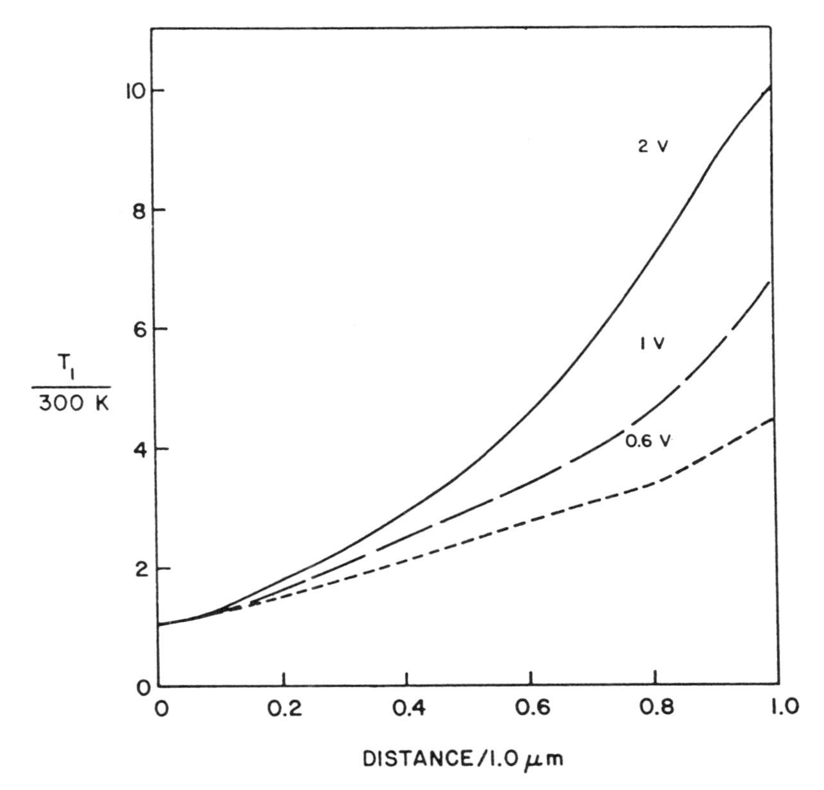

FIGURE 5-46. Temperature distribution within the Γ valley for the parameters of Fig. 5-43. See comments associated with Fig. 5-45. (After Grubin et al., 1985.)

inequalities emerge

$$n_c\mu_c > N_0\mu_1, \qquad F(x) < F_c \tag{5-111a}$$

$$n_c\mu_c < N_0\mu_1, \qquad F(x) > F_c \tag{5-111b}$$

For Eq. (5-111a), local charge accumulation is present at the cathode. Note that the condition $F(x) < F_c$ is stronger than necessary and requires that $n_c > N_0$. In Eq. (5-111b) cathode depletion occurs with $n_c < N_0\mu_1/\mu_c$. The results of the following discussion of the simulation are consistent with Eq. (5-111b), as demonstrated in Fig. 5-42. For Eq. (5-111b) reference is made to the discussion of Grubin and Kreskovsky (1983). Figure 5-42 is a plot of the computed dc current versus field relationship at the cathode boundary. It is approximately linear with only a marginal variation in field. The cathode field is effectively pinned. For reference purposes the current–field relationship for the uniform field structure is also shown.

The characteristics of the uniform field curve and the cathode current–field relationship are different; for a constant current through the semiconductor at least two different field values result at intersection. The cathode boundary field is lower than that of the neutral field intersection, a result that is consistent with cathode accumulation.

The steady state time-independent distributions of electric field, carrier density, Γ-valley velocity, and electron temperature are displayed in Figs. 5-43–5-46 for various bias levels. While the calculation displays the excess carrier velocity at elevated bias levels, there is also an enhanced electron transfer and the dc current shows saturation. The clear consequence of the transfer is that the current does not scale with velocity. This latter feature is reflected in the current–voltage relationship shown in Fig. 5-47.

With regard to the current–voltage characteristic, while the current does not scale with velocity, and thus does not fully reflect overshoot contributions, its high bias level is above that associated with the equilibrium steady state velocity–field relationship, while below that associated with the Γ-valley velocity. The excess above V_n is due predominantly to the cathode boundary condition that allows for a high level of injected charge. The depression below V_n is due to electron transfer. It is also noted that there is virtually no electron transfer near the cathode. Most of it occurs near the anode, and the effect of electron transfer leads to saturation in the current density. Another feature of the nonuniform field calculation lies in the clear absence of NDC, a phenomenon present in uniform field calculations.

The significant qualitative differences between the steady state uniform field characteristics and those associated with nonuniform fields suggest some differences in the transient characteristics. This is indeed the case, as we shall discuss in what follows.

Figure 5-48 displays the current transient following application of a voltage pulse. The first point we emphasize is that the plot is that of current rather than velocity. The second point is that the current transient is ostensibly similar to that associated with velocity overshoot. There is, however, a fundamental difference

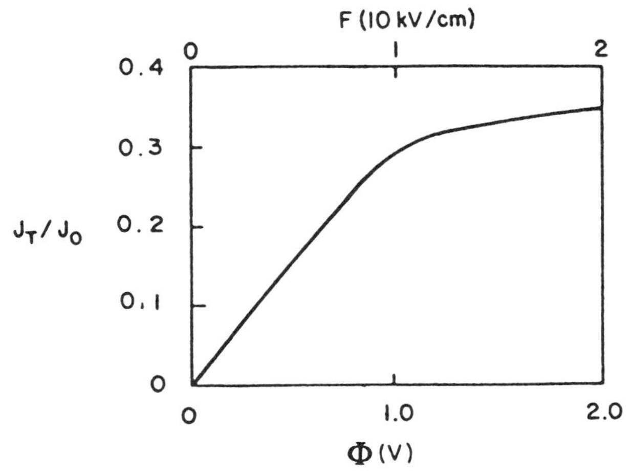

FIGURE 5-47. Steady state current density versus applied voltage and average field for the 1.0-cm-long structure with the parameters of Fig. 5-43. $J_0 = 8 \times 10^4$ A cm^{-2}. (After Grubin et al., 1985.)

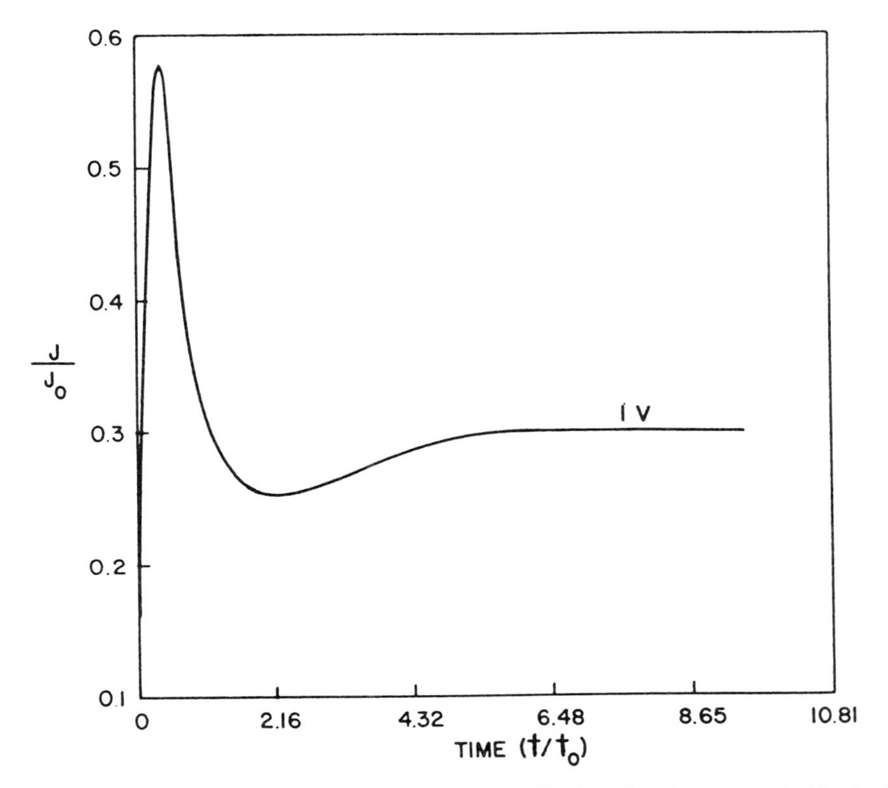

FIGURE 5-48. Magnitude of current transient following application of a step change in bias to 1.0 V for a 1.0 μm-long device at 300 K with the parameters of Fig. 5-43. Current peak is similar to that of Fig. 5-35. Steady state velocity is above that of the uniform field case. It is noted that the time required for steady state is approximately 50% longer than that associated with the steady state calculation of Fig. 5-35. $J_0 = 8 \times 10^4$ A cm^{-2}; $t_0 = 1$ ps.(After Grubin et al., 1985.)

between the two. For both uniform and nonuniform fields, during the first time step the field throughout the structure is increased by an amount equal to the change in applied voltage divided by device length. This introduces a one-time-step displacement current whose magnitude is computationally dependent and therefore nonphysical. For uniform fields, all displacement current contributions cease after the initial time step. For nonuniform fields all the time-dependent field evolution is accurately calculated following the initial time step. Here, with the cathode boundary introducing a cathode adjacent accumulation layer, the time dependence introduces a layer that propagates toward the anode boundary. This propagation is accompanied by field rearrangement and internal point-by-point displacement current contributions.

Figure 5-49 shows the space- and time-dependent evolution of the electric field within the device, and Fig. 5-50 the current density distribution. The effect of the boundary condition is to introduce a propagating accumulation layer originating at the cathode, while downstream from the anode the field is approximately uniform during the first 0.5 ps, becoming highly nonuniform as

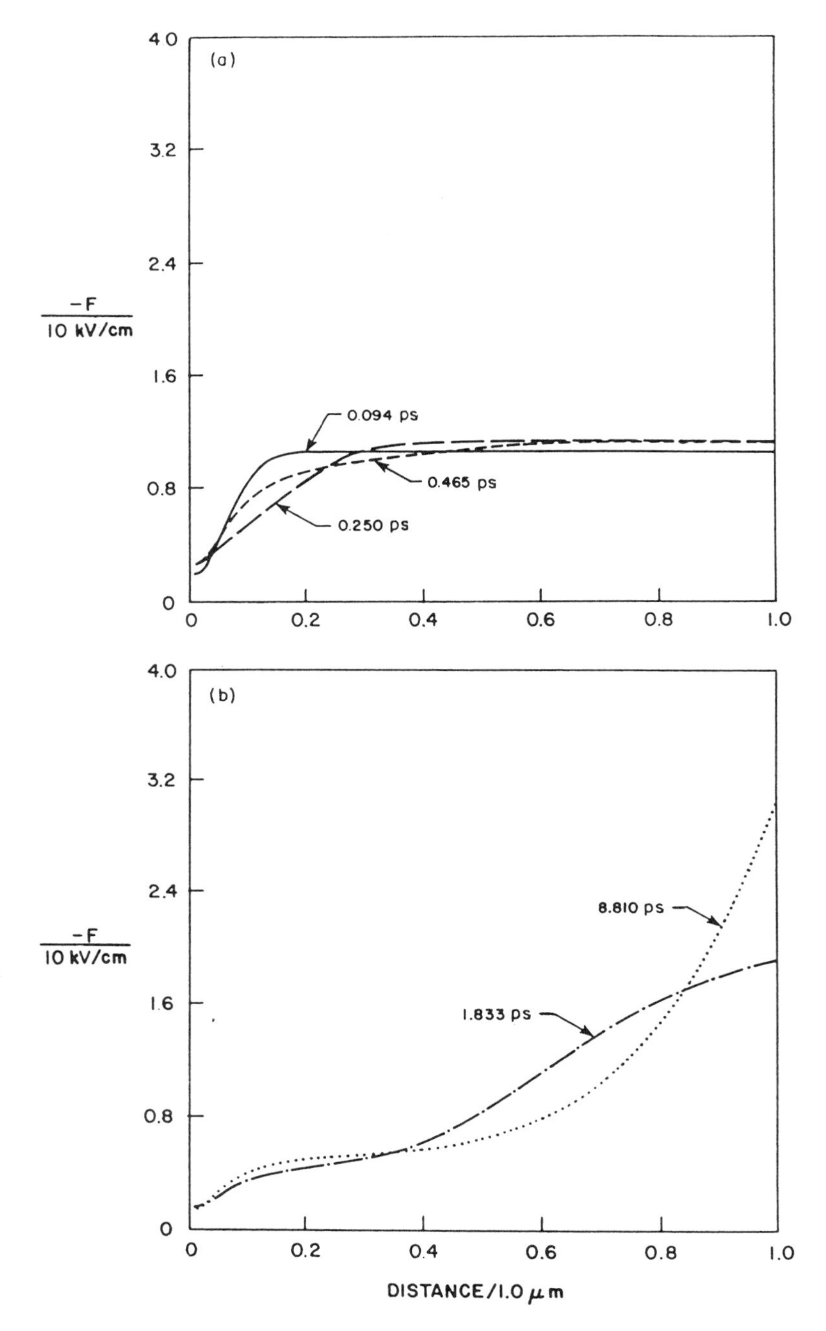

FIGURE 5-49. Distribution of electric field at successive instants of time following applications of a step change in voltage for the parameters of Fig. 5-48. During the first time step, the field increases from its steady state value at a bias of 0.01 V by an amount equal to 9.99 kV cm^{-1} [(1−0.01) V μm^{-1}]. Subsequent time dependence shows a space charge layer propagating toward the anode. (a) During the first 0.5 ps the field downstream from the propagating accumulation layer is spatially uniform. Within this region transients are governed by the uniform field velocity overshoot transient. (b) During the long-time transient, electron transfer occurs and relaxation differs from that of the uniform field transient. (After Grubin et al., 1985.)

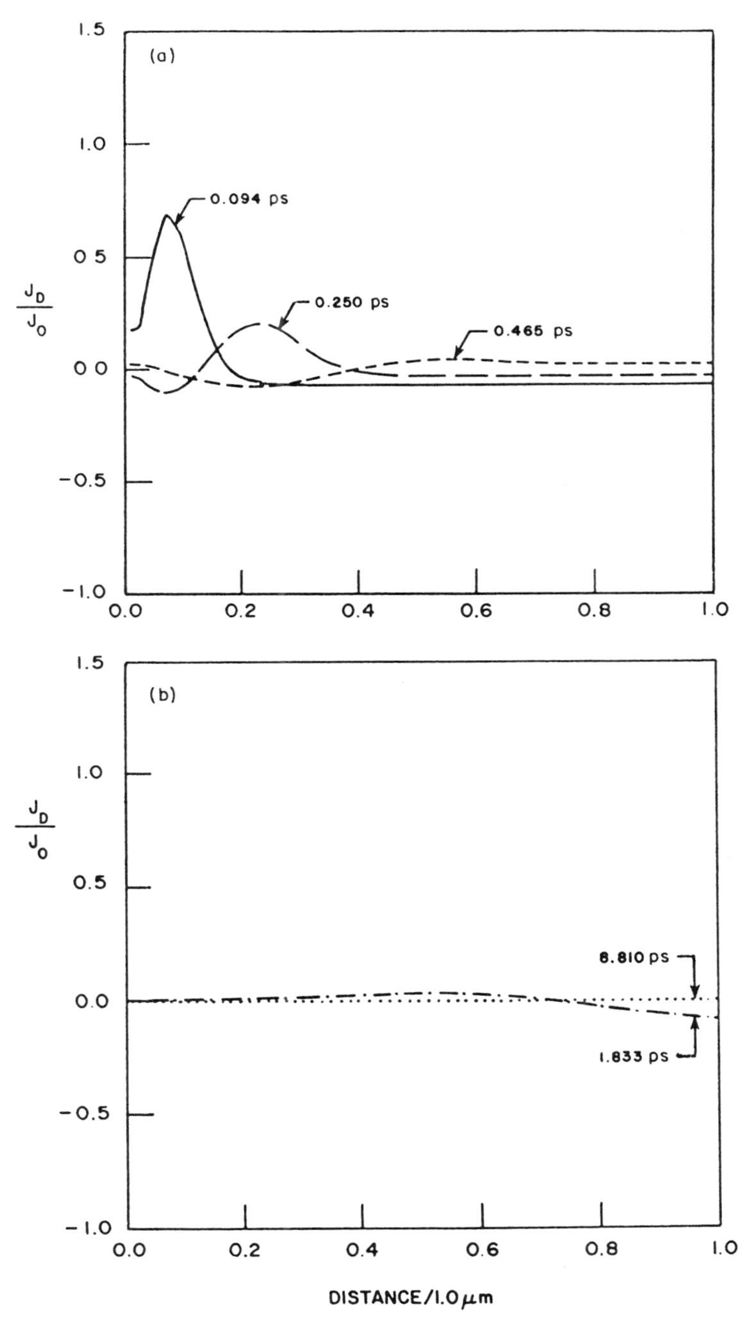

FIGURE 5-50. (a) and (b) Displacement current at 5 instants of time for the parameters of Fig. 5-48. Initial displacement currents are strong and accompany the moving accumulation layer. $J_0 = 8 \times 10^4\ \mathrm{A\ cm^{-2}}$. (After Grubin et al., 1985.)

steady state is approached. The early time transients dictate that the displacement current contributions will be significant within the vicinity of the propagating accumulation layer. In the latter regions the familiar velocity transients obtained from the uniform field calculations arise. At later times propagation continues, but is accompanied by electron transfer. The long-time transient differs from that of uniform fields. [We note from Figs. 5-51 and 5-52 the absence, for $t < 0.5$ ps, of any significant transfer downstream from the moving space charge layer. The carrier velocity (Fig. 5-53) downstream from the moving space charge layer sustains high values common to overshoot.]

There are three dominating features of the preceding calculations. The first two are the boundary conditions on the Γ-valley temperature and mean carrier velocity. The third is the length of the structure. As discussed earlier, the specification of the Γ-valley electron temperature provides dominant control in the calculations of the relative population of the Γ-valley carriers at the cathode. For the calculations of Figs. 5-44–5-53, specifying T_1 at 300 K resulted in virtually the entire sea of cathode carriers as Γ-valley carriers. In a study performed earlier (Grubin and Kreskovsky, 1983) in which the device length was 2.0 μm and the Γ-valley velocity was subject to a mobility boundary condition as in Eq. (5-103), the results were qualitatively similar for $T_1 = 300$ K and a Γ-valley boundary mobility greater than that of the low-field steady state mobility of the Γ-valley carriers. In those calculations the total set of boundary conditions was somewhat different from those employed in the discussion of Figs. 5-44–5-53 but there were several definite trends. For example, by retaining a suitably high cathode mobility and by elevating the electron temperature, space-charge accumulation at the cathode was retained, but the relative proportion of Γ-valley carriers at the cathode decreased. Again, obversely, retaining a cathode temperature of $T_1 = 300$ K but reducing the boundary mobility of the Γ-valley carrier in steady state results in a partial depletion of carriers at the cathode and concomitant increase in the cathode field to values in excess of that within neutral regions interior to the device. Each of these results is consistent with the qualitative arguments contained in Eqs. (5-106)–(5-111).

The immediate conclusion that can be drawn from the set of referenced results is that the presence of space charge accumulation or depletion at the cathode is dominated by the field dependence of the entering carrier velocity, vis-à-vis that within the interior of the device. The conclusion is now applied to a problem of high visibility: transit-time Gunn domain instabilities in GaAs.

5.5.3.4. Steady State and Transient Behavior for a Partially Blocking Cathode (L = 5.00 μm)—Dipole Domain Oscillations

The structure under consideration is "long" with respect to submicrometer dimensions—the device length is 5.00 μm. The boundary conditions here are different from those used for Figs 5-44–5-53. In this case those of Eq. (5-93) are repeated, with two critical variations:

$$V_1 = -4000\,F \quad \text{and} \quad T_1 = 1200\,\text{K} \tag{5-112}$$

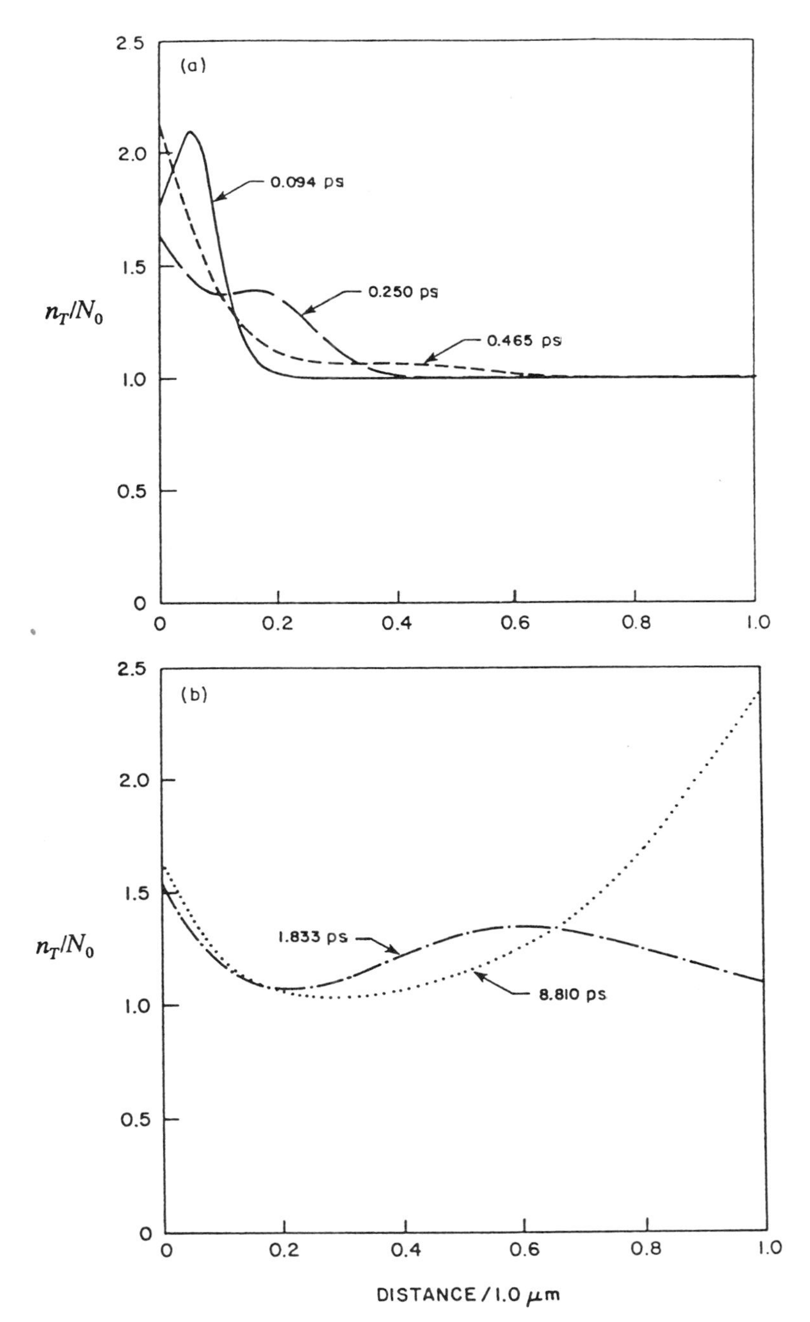

FIGURE 5-51. (a) and (b) Transient distribution of total charge following application of a step change in potential for the parameters of Fig. 5-48. Note that downstream from the propagating accumulation layer the charge distribution is flat as reflected additionally in the flat field profile of Fig. 5-49a. Space charge accumulation occurs during the longer time interval. (After Grubin et al., 1985.)

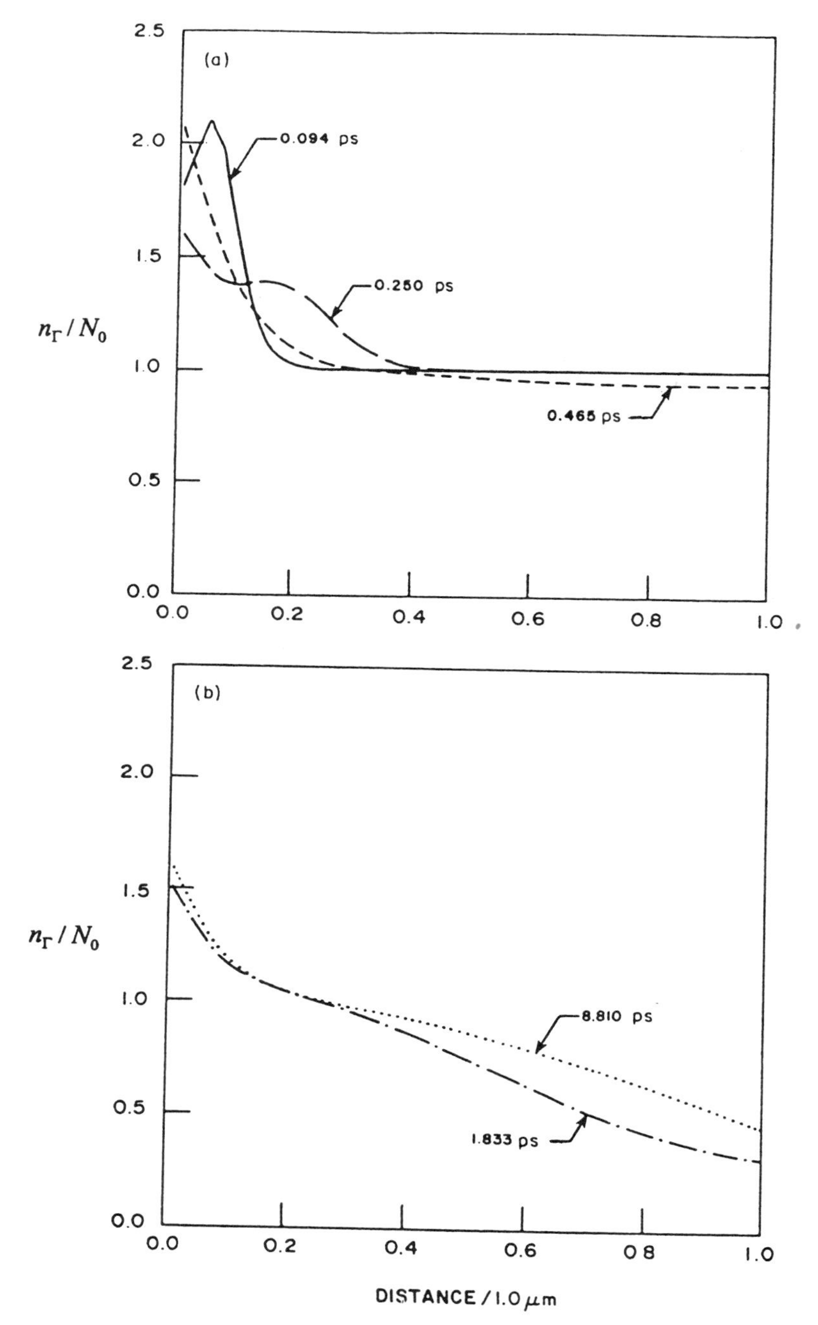

FIGURE 5-52. (a) and (b) Transient distribution of Γ-valley carrier density for the parameters of Fig. 5-48. Note that within the first 0.5 ps, very little transfer occurs. (After Grubin et al., 1985.)

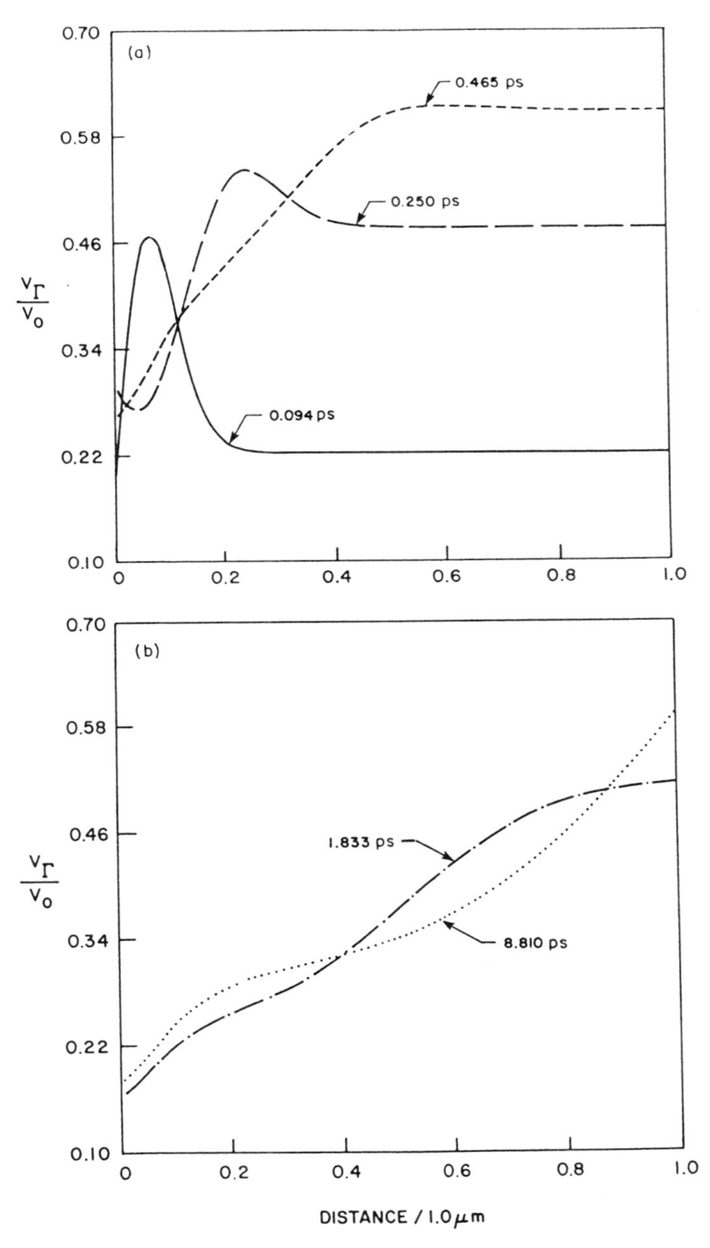

FIGURE 5-53. (a) and (b) Transient distribution of the Γ-valley velocity following application of a voltage pulse for the parameters of Fig. 5-48. The velocity layer propagates and shows a tendency to lead to transient changes in the Γ-valley carrier density. Downstream, the velocity transient is relatively uniform for $t < 0.5$ ps and tends to follow the uniform field transient of Fig.5-38. Differences from the uniform field calculations occur during the long-time transient. $v_0 = 10^8$ cm s^{-1}. (After Grubin et al., 1985.)

As in the case of the accumulated cathode, the situation represented by Eq. (5-112) can be described qualitatively by Eqs. (5-106)–(5-110) with the modifications

$$nV \approx n_1 V_1 \cong \beta n V_1 \tag{5-113}$$

when β represents an average of the fraction of Γ-valley to total carriers within the cathode region. With this change

$$t(x) = \tau_0 \log\left\{\left(1 - \frac{N_0\mu_1}{\beta n_c \mu_c}\right) \Big/ \left[1 - \frac{N_0 \mu_1 F(x)}{\beta n_c \mu_c F_c}\right]\right\} \tag{5-114}$$

The requirement that the transit time be positive leads to the inequality

$$\beta n_c \mu_c < N_0 \mu_1, \qquad F(x) > F_c \tag{5-115}$$

for a restricted range of field values. Equation (5-115) is discussed in detail below; it produces cathode depletion.

Figure 5-54 displays the field distribution, whose most obvious characteristic is that of a broad depletion region adjacent to the cathode. The characteristics of this depletion region are that with increasing bias the depletion zone broadens, the cathode field increases, and the downstream field begins to approach a constant value. This latter feature manifests itself as hard saturation in current versus voltage.

Figure 5-55 is a display of the carrier density in the Γ-valley as well as total carrier density. It is first noted that for all of the bias values chosen the Γ-valley carrier density displays partial depletion in the vicinity of the cathode boundary. It may be anticipated that this will manifest itself as an excess carrier velocity at the cathode (Fig. 5-56). Note also that as the bias level increases the total charge at the cathode shows a diminished depletion and there is a weak region of charge accumulation downstream. With regard to the Γ-valley velocity, this follows the pattern dictated by current continuity and cathode adjacent charge depletion. The carrier velocity at the cathode sustains values in excess of that within the neutral interior regions of the semiconductor.

Figure 5-57 displays the dc current–voltage relationship for the structure. Several points are noteworthy. The first is the absence of NDC even though the neutral interior region is characterized by a region of NDM. The second point to note is that current saturation occurs at values below that associated with the 1-μm-long device.

The cathode current–field relationship is displayed against the neutral field characteristic in Fig. 5-58. In addition, the cathode boundary neutral field characteristic is also shown. The curves display an apparent tendency to intersect within the region of NDM, resulting in two approximately neutral regions sustaining different values of field and velocity. Under a well-defined set of conditions this configuration is electrically unstable and leads to the nucleation and propagation of high-field domains. For the configuration under consideration an increase in bias level from 2.0 to 3.0 V results in transient local cathode

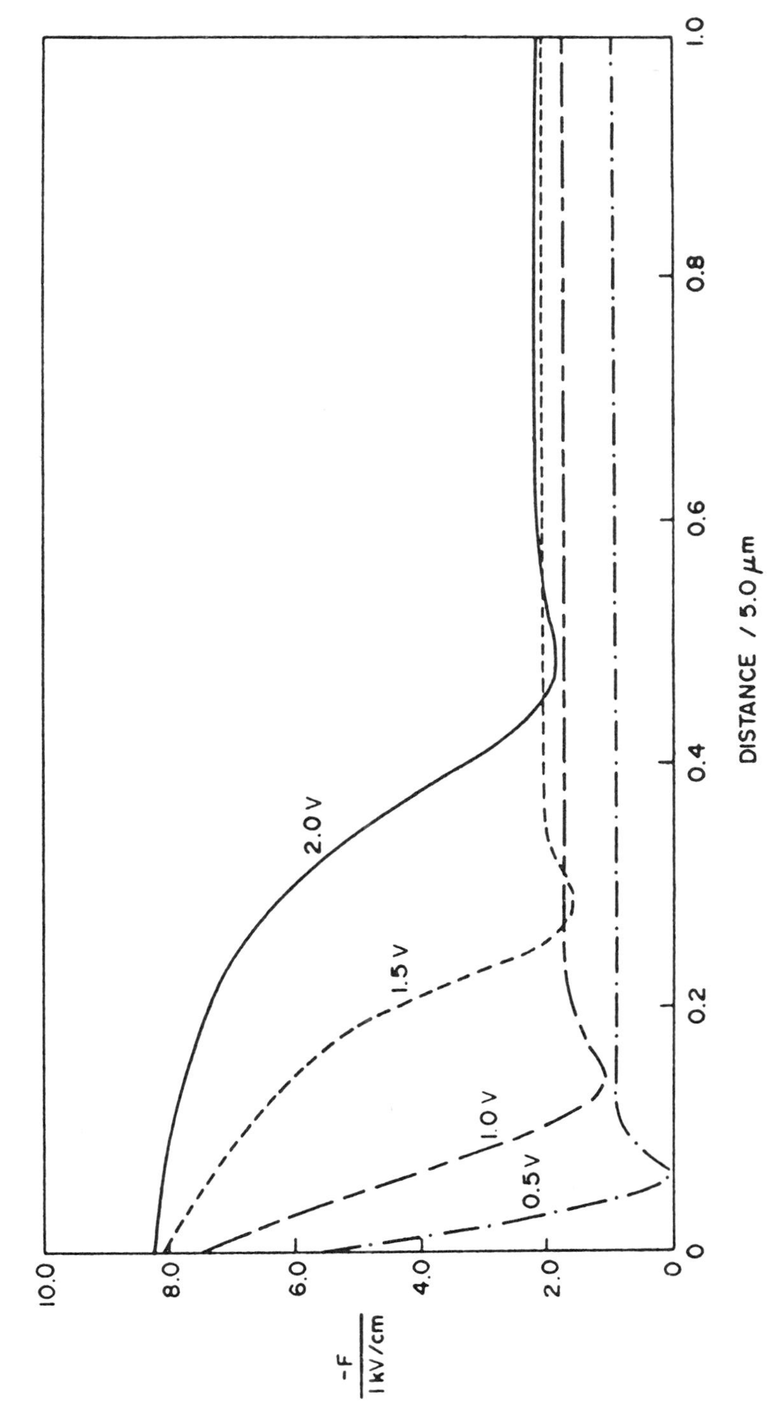

FIGURE 5-54. Steady state distribution of electric field within the interior of a 5.0-μm-long uniformly doped GaAs structure at various bias levels. Parameters are given in the appendix of this chapter. It is noted that the field at the cathode, in response to the boundary conditions, is qualitatively different from that associated with Figs. 5-43–5-55. Here, the field decreases from the cathode to the anode. In the vicinity of the anode the field is uniform. The net decrease in field is consistent with a cathode region partially depleted of carriers. Note that prior to reaching the downstream portion of the structure, the field displays a miminum followed by a change in slope. This change in slope represents the presence of a region of local charge accumulation. (After Grubin et al., 1985.)

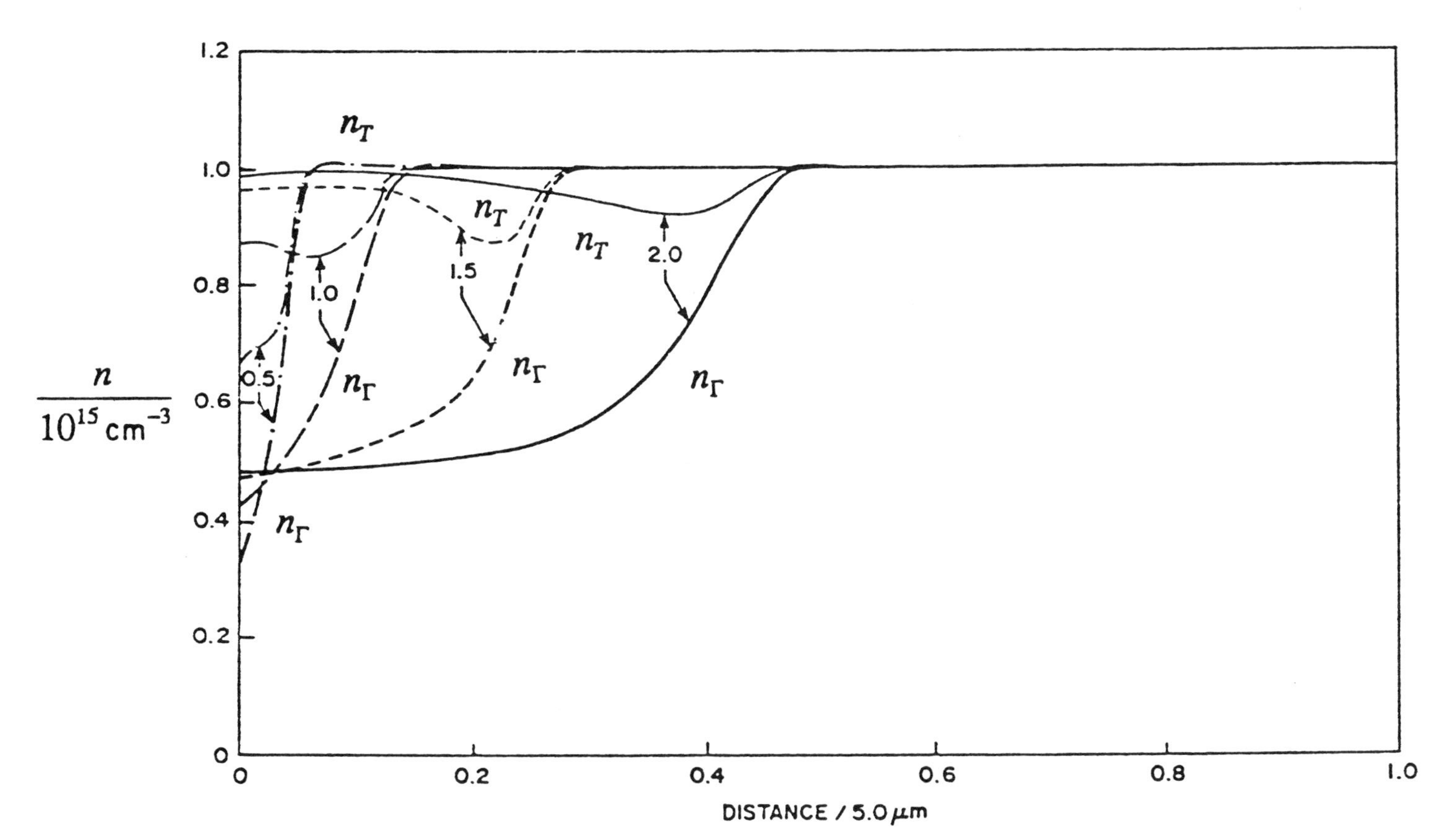

FIGURE 5-55. Distribution of total (n_T) and Γ-valley (n_Γ) carriers within the device for the parameters of Fig. 5-54. Note that there is a net depletion of total charge within the vicinity of the cathode and that this depletion is reduced as the bias level is raised. For all bias levels the Γ-valley population is below that of the total carrier density. (After Grubin et al., 1985.)

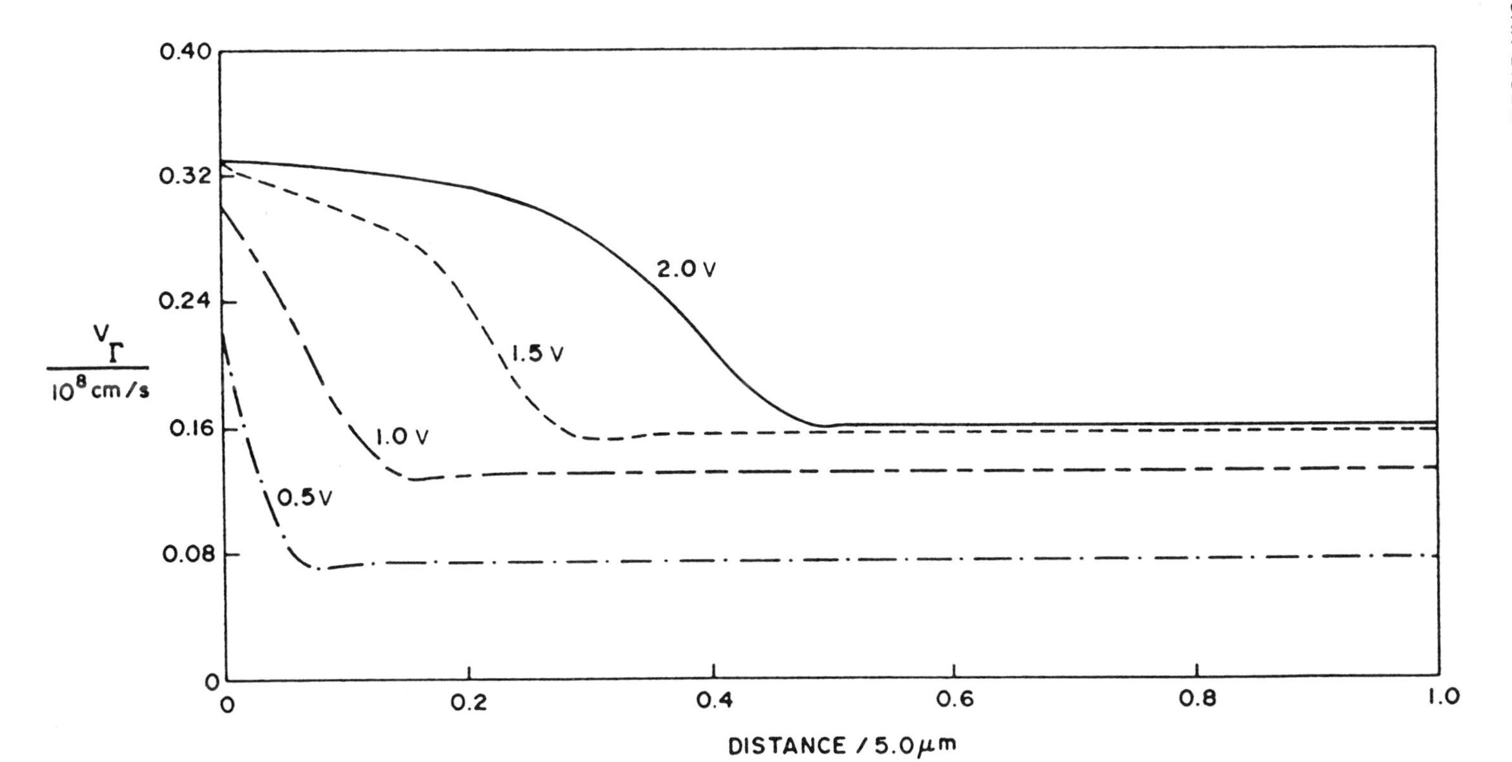

FIGURE 5-56. Steady state velocity distribution of Γ-valley carriers for the parameters of Fig. 5-54. Note that unlike the velocity distribution of the Γ-valley electrons of an injecting contact where the carrier velocity is greatest at the anode, for this length structure the Γ-valley velocity is greatest at the cathode. Note further that the change in cathode velocity with increased bias is very small at high bias levels and reflects the presence of current saturation. (After Grubin et al., 1985.)

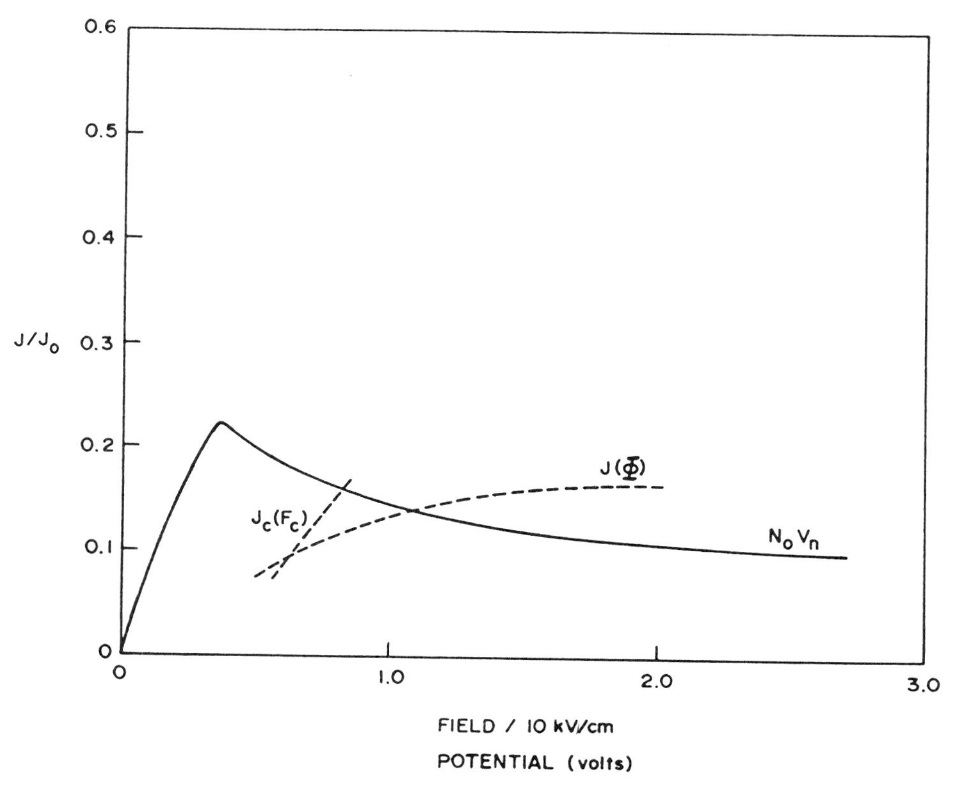

FIGURE 5-57. Steady state current–voltage characteristics $J(\Phi)$ for the partially depleted cathode structure with a length of 5.00 μm. Parameters are those of Fig. 5-54. Of significance here is the fact that saturation in current occurs at an average field significantly below that of the 1.0-μm-long device. Also shown are the cathode field relation $J_c(F_c)$ and the neutral field characteristic $N_0 v_n$; $J_0 = 8 \times 10^4$ A cm^{-2}. (After Grubin et al., 1985.)

adjacent accumulation and subsequent dipolar propagation, as displayed in Fig. 5-59. The details of Figs. 5-59 and 5-60 show the transient transformation of the space-charge layer (as reflected in the electric field distribution) from a depletion layer to a propagating dipole layer. The dipole layer is quenched at the anode boundary and repeated transit-time oscillations occur. The time-dependent oscillations are displayed in Fig. 5-61 and occur after an initial transient that is qualitatively similar in structure to that associated with the accumulated cathode and the uniform field transients. Indeed, the peak current is a reflection of both overshoot and the influence of the cathode boundary condition, which reduces its value to a level below that of the uniform field transient. It is important to note that while the development of a set of conditions for initiating a propagating domain is of clear technological significance, it plays a secondary role to the major thrust of this section—the conditions at the cathode are the single most pervasive influence on the behavior of near- and submicrometer-length semiconductor devices, much as they are for longer devices (Shaw et al., 1979).

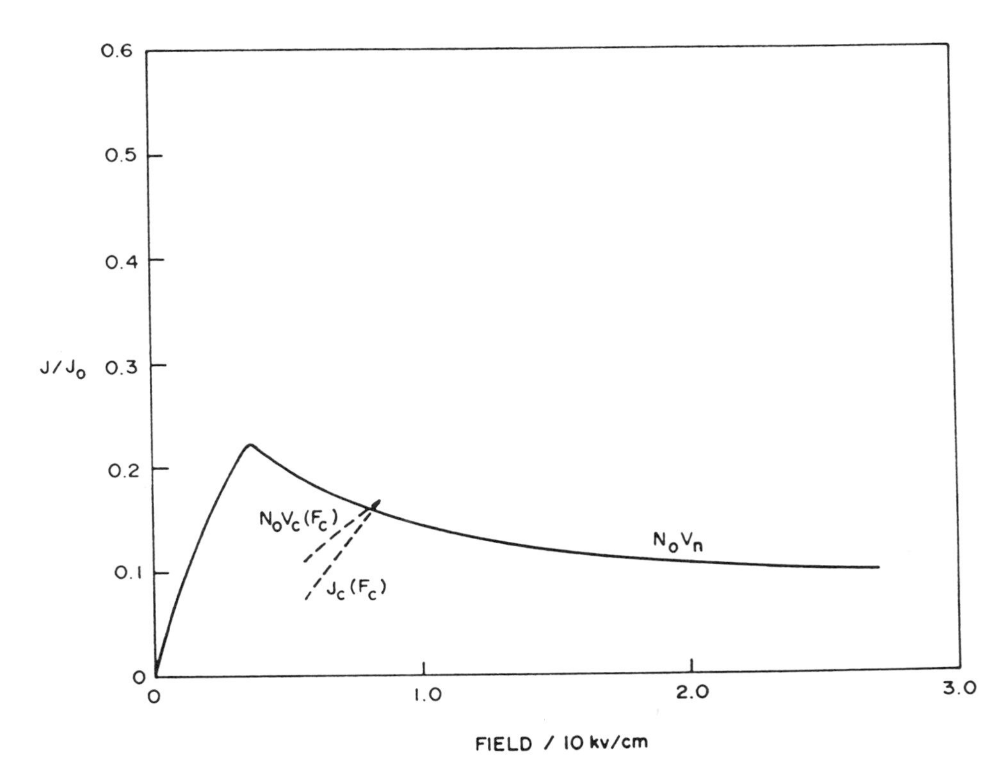

FIGURE 5-58. Neutral current density–field relationship for gallium arsenide N_0v_n, the cathode current field relation $J_c = n_cv_c$, and the neutral cathode field relation N_0v_c. Parameters are those of Fig. 5-54; $J_0 = 8 \times 10^4$ A cm^{-2}. (After Grubin et al., 1985.)

5.5.3.5. Nonuniform Fields and Length Scaling

While the calculations in Secs. 5.5.3.3. and 5.5.3.4 were for structures of different lengths, the emphasis was on the effects of the boundary. However, the effects of length scaling, insofar as they affect the velocity field relationship (vis-à-vis Fig. 5-42) will influence the electrical transient and the steady state field profiles. This is illustrated for two situations. The first situation is for a uniformly doped structure with the same boundary conditions as given by Eq. (5-103) but with a length of 0.25 μm. The second structure considered is that of a n^+-n^--n^+ device with a 1-μm cathode-to-anode spacing but with a variable-length n^- region.

The calculations for the 0.25-μm-long device are displayed in Figs. 5-62–5-70. The steady state electric field distribution is displayed in Fig. 5-62 for the indicated bias levels. Note that although the average fields for the 0.25-μm device and the 1.0-μm device are the same, the field distributions are quantitatively different. The difference lies in the fact that at the lower bias levels only a marginal amount of electron transfer occurs within the shorter structure. Note also that the electric field at the cathode is low, as for the 1-μm-long device.

Figure 5-63 displays the steady state population of the Γ-valley as well as the total carrier density. The first point to note here is that the density of carriers for

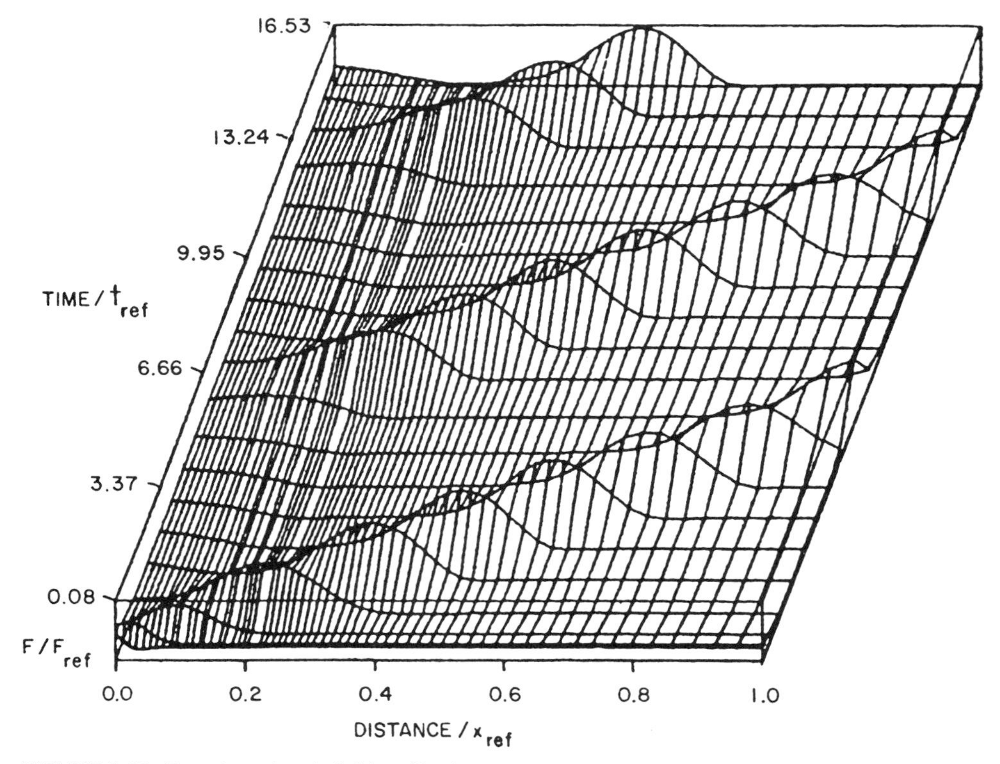

FIGURE 5-59. Transient electric field profile showing nucleation and propagation of high field domain. Note that propagation is accompanied by low downstream field values and residual cathode adjacent depletion. Parameters are those of Fig. 5-54; $F_{ref} = 2\,kV\,cm^{-1}$; $t_{ref} = 5$ ps, $x_{ref} = 5\,\mu m$, and $\Phi_{ref} = 3$ V. (After Grubin et al., 1985.)

the given bias level exceeds that for the 1-μm-long element. The second point is that considerably less electron transfer occurs downstream from the cathode. There is, however, a far more significant aspect to the quantitative differences between the results of the 0.25- and 1.0-μm devices. The carrier and velocity distributions for the two structures are different. These differences are, in part, a result of the fact that conditions at the upstream boundary are sensitively dependent upon the proximity of the collecting contact. Further evidence for this is provided by the velocity distribution displayed in Fig. 5-64, which shows higher entrance velocities, but lower exit velocities.

Figure 5-65 is a plot of current versus voltage for the 0.25-μm-long device. Again, two points are emphasized. The first point shows the absence of NDC. The second point is that the presence of increased levels of charge injection yield an increase in the drive current over that of the 1-μm-long device.

The transient characteristics at 0.25 μm are displayed in Figs. 5-66–5-70. The results are quantitatively different from that associated with the 1-μm device. The first difference is displayed in the current transient (Fig. 5-66), which shows a higher peak current and a smaller current dropback. As revealed in the time-dependent distributions of fields (Fig. 5-67), the higher peak current (greater

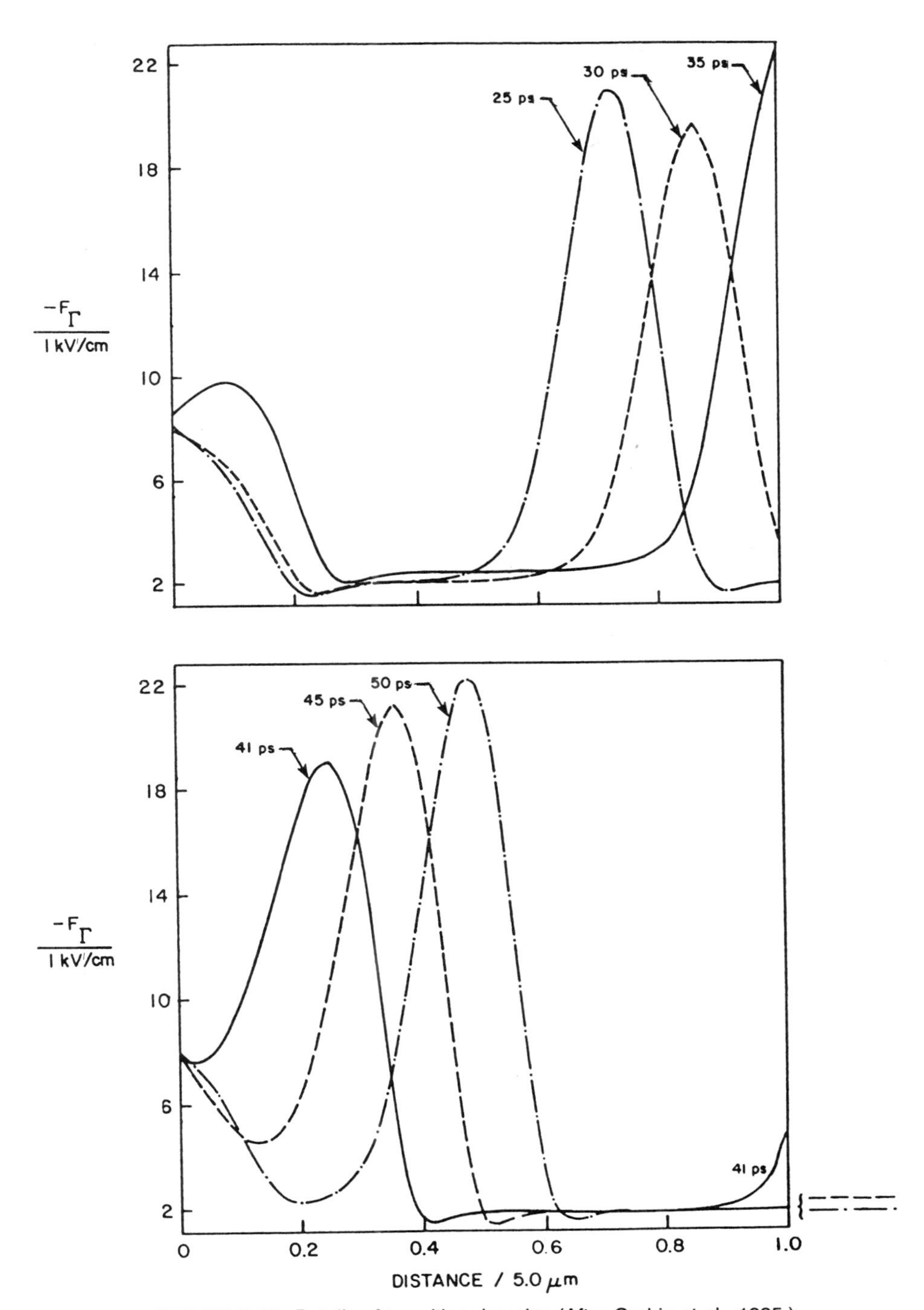

FIGURE 5-60. Details of transiting domains (After Grubin et al., 1985.)

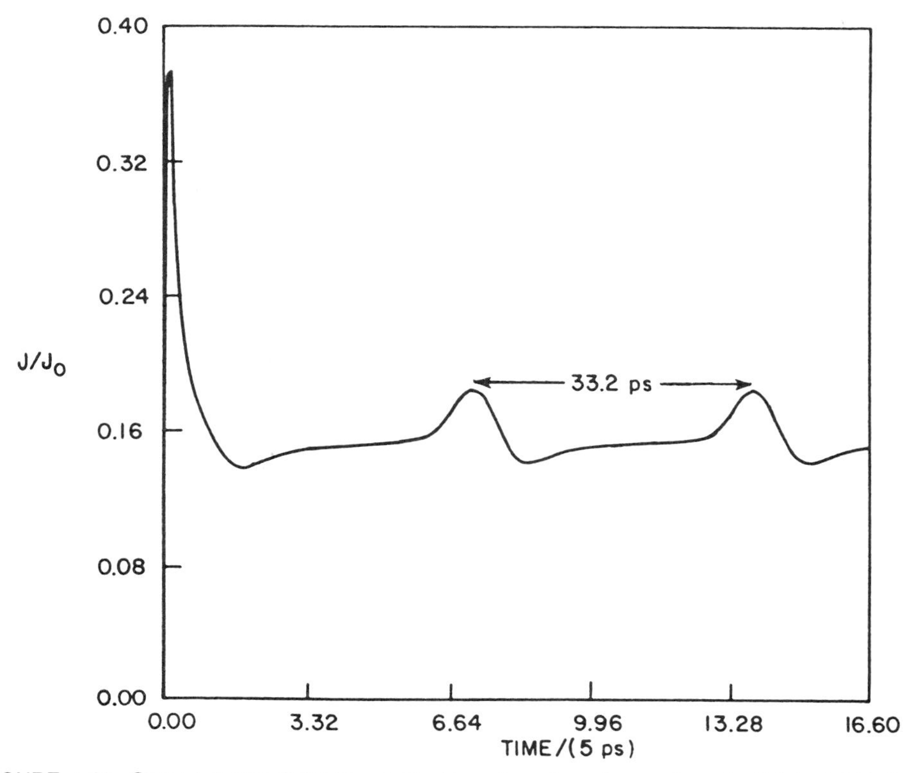

FIGURE 5-61. Current transit following application of a step change in potential to 3.0 V. Initial transient is a reflection of nonequilibrium transient transport. The steady long-time transient reflects the nucleation, propagation, and quenching of a propagating high-field dipole layer. Parameters are those of Fig. 5-54; $J_0 = 8 \times 10^4$ A cm^{-2}; $\Phi = 3$ V. (After Grubin et al., 1985.)

than 25%) is in large part due to displacement current contributions. The higher long-time steady state current level reflects the increased injection level (Figs. 5-68 and 5-69) over that of the 1.0-μm calculation. In this regard it is again pointed out that the exit velocity for the 0.25-μm structure is below that for 1.0 μm (Fig. 5-70). The final point of interest involves the time to reach steady state: This time is shorter for the 0.25-μm structure, but not a factor of 4 shorter. The time scales involved in the approach to steady state involve nontransit-time contributions.

5.5.3.6. Transients in n^+-n^--n^+ Structures and Length Scaling

The final two-terminal structure considered is the n^+-n^--n^+ device, and there are several key features to note. The first is that the dominant interfaces for this structure, the n^+-n^- and n^--n^+ interfaces, are not the physical boundaries of the device and are thus likely to have a different effect on its electrical behavior. The second feature of importance lies in the fact that the electric field profile is highly nonuniform in steady state; this may dominate the transient and

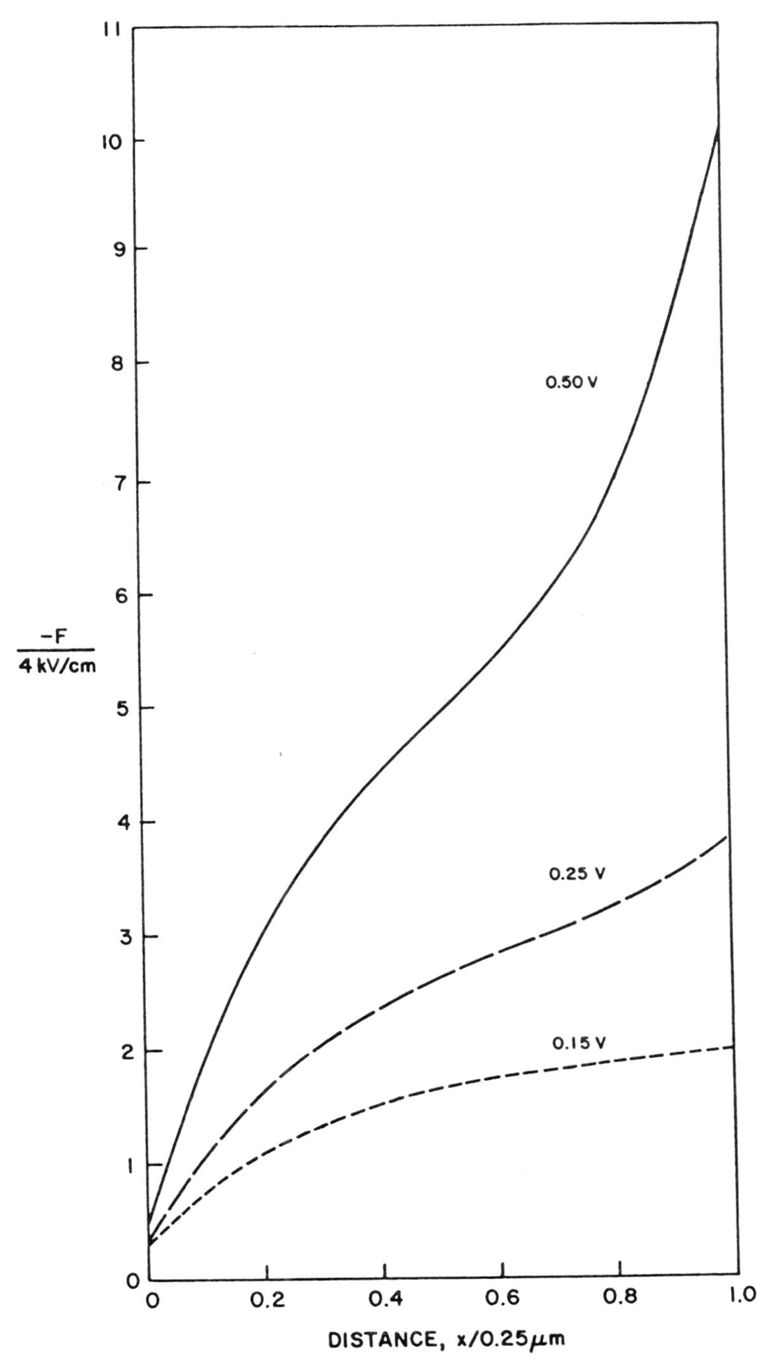

FIGURE 5-62. Steady state distribution of electric field for 0.25 μm uniform structure with injecting cathode contacts. Average field across the structure is the same as that of Fig. 5-43 for a 1.0-μm-long device. Parameters for this calculation are listed in the Appendix of this chapter. (After Grubin et al., 1985.)

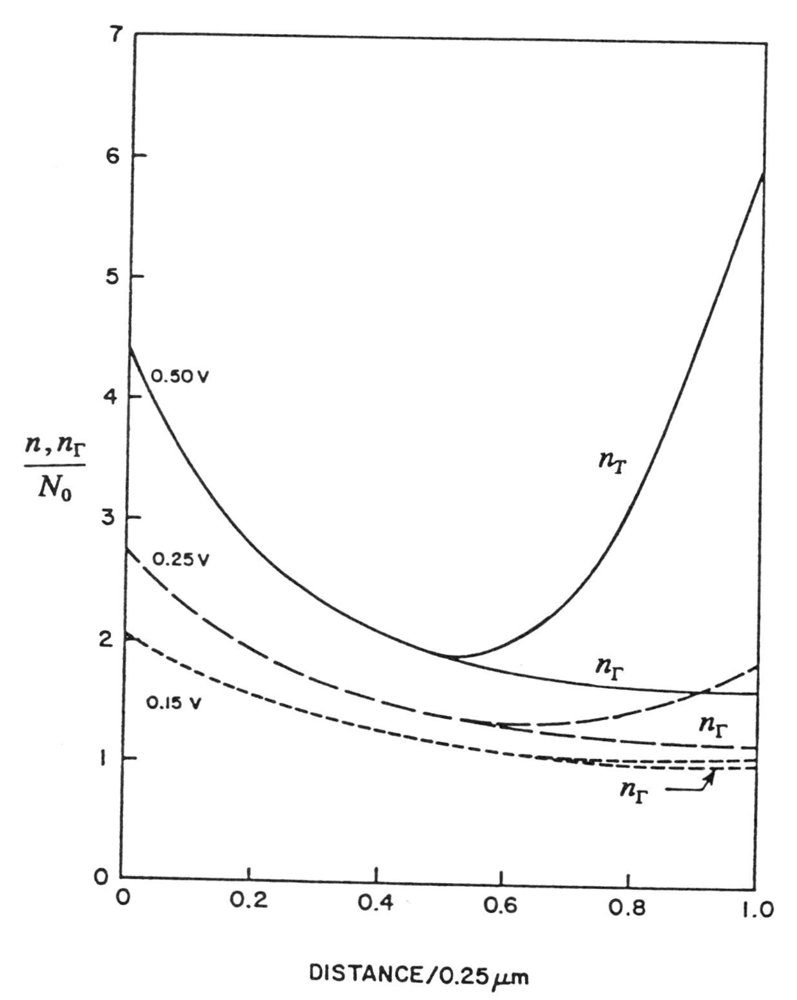

FIGURE 5-63. Distribution of total (n_T) and Γ-valley (n_Γ) carrier density. Only marginal transfer occurs for the lowest bias level. Substantial transfer occurs at the higher bias levels. At all bias levels, the injection level is extremely high, and n_Γ exceeds N_0. ($N_0 = 5 \times 10^{15}\,\text{cm}^{-3}$.) (After Grubin et al., 1985.)

completely camouflage all submicrometer effects. Third, for sufficiently small n^- regions the influence of the n^+-n^- and n^--n^+ interfaces for carrier confinement may be less prominent. Thus, this last two-terminal structure offers the most serious example of the influence of the interface and the length of the critical submicrometer region on the electrical characteristics of submicrometer structures. To avoid conflict with the influence of the true metal confining contacts, the physical boundary conditions at the cathode are taken as

$$n = N_0, \qquad N_1 = n_{1\text{eq}}, \qquad V_{1x} = 0, \qquad V_{2x} = 0$$
$$T_1 = 300\ \text{K}, \qquad T_{2x} = 0 \tag{5-116}$$

At the anode all second derivatives are set to zero.

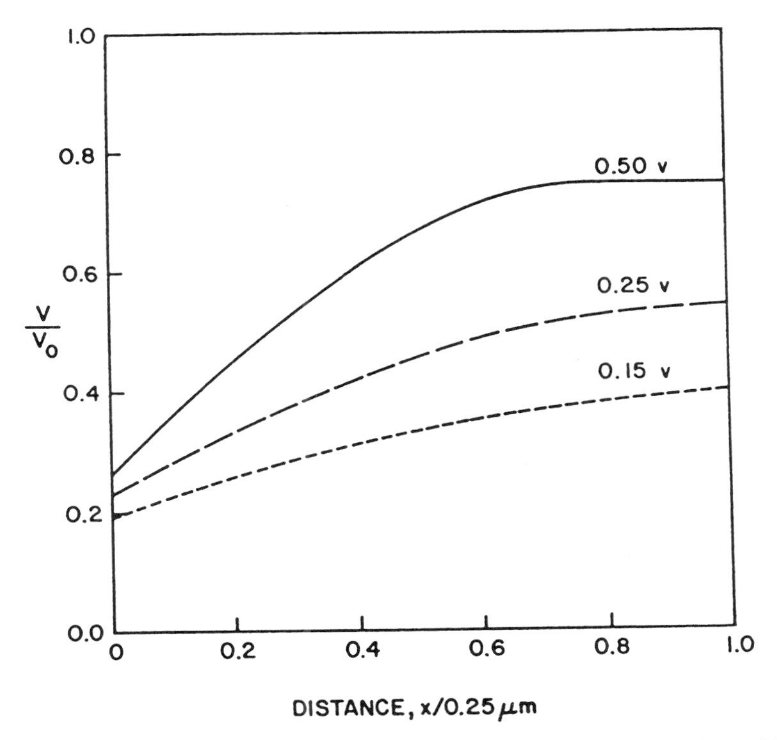

FIGURE 5-64. Steady state distribution of the Γ-valley carrier velocity at three values of bias. Note that the velocity increases from cathode to anode, corresponding to a decrease in the Γ-valley carrier density. Parameters are those of Fig. 5-62; $v_0 = 10^8\ \mathrm{cm\ s^{-1}}$. (After Grubin et al., 1985.)

The n^+-n^--n^+ calculations performed are for the one-dimensional structure of Fig. 5-71, in which the n^- region is assigned a nominal doping level of $10^{15}\ \mathrm{cm}^{-3}$ and the n^+ region is at $10^{17}\ \mathrm{cm}^{-3}$. The length of the n^- region is specified at the doping level of $10^{16}\ \mathrm{cm}^{-3}$, and varied from 0.416 to 0.116 μm. The entire structure is fixed at a length of 1.0 μm. The design of the structure dictates that nonuniform fields and charge densities form within it. Thus, the relevant experimental quantity is again the current density, rather than the velocity. The first set of results is shown in Fig. 5-72, which displays the total current flowing through the device following application of a voltage pulse of magnitude 1.0 V.

As for the uniform N_0 studies, the calculation is performed in two stages. The first involves obtaining a steady state solution at 0.01 V. For the second, using this as an initial condition, the bias is raised in one time step to 1.0 V. Application of the bias in one time step replicates the procedure of most of the uniform field calculations.

As seen in Fig. 5-72, the current displays an initial peak at approximately 0.15 ps, followed by a drop in current and a subsequent rise toward a steady state value. For uniform field calculations in which the voltage is increased in one time step, as discussed earlier, there is an initial displacement current whose magnitude is determined entirely by the computational time step. Thereafter, all displacement currents are zero and all transients are particle current transients.

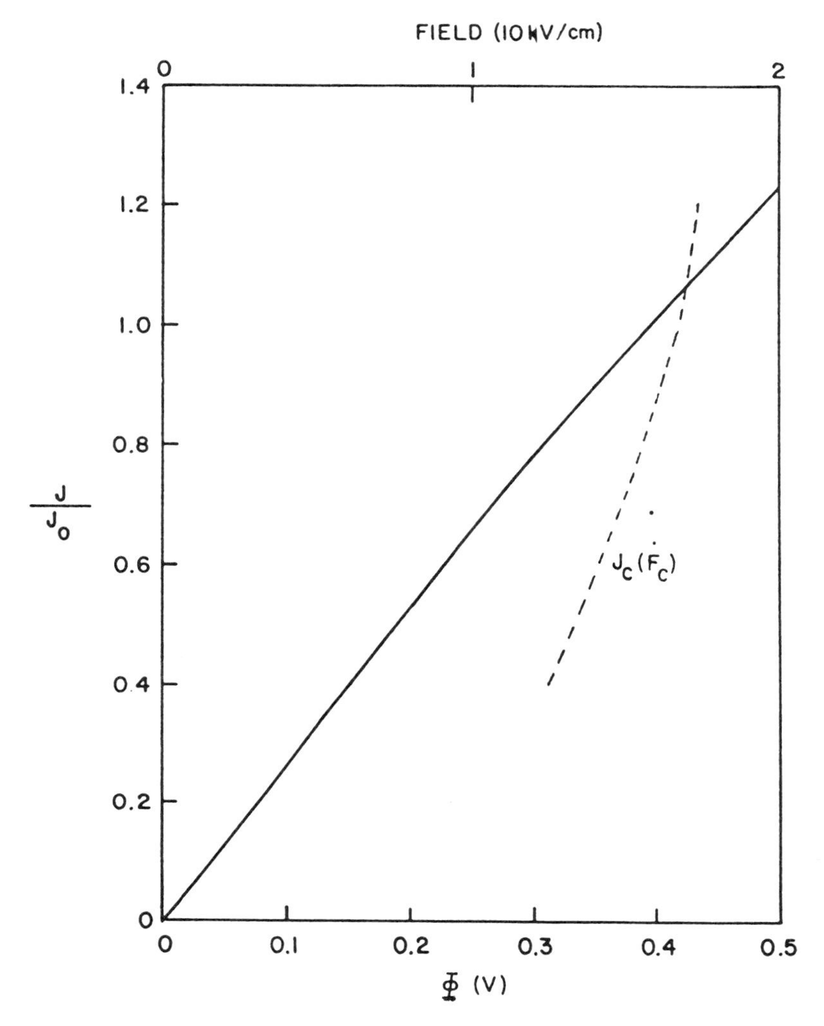

FIGURE 5-65. Steady state current density versus applied voltage and average field for a 0.25-μm-long structure with injecting contacts. Excessively high current levels are primarily due to high levels of space charge injection. Also shown is the cathode current field relationship $J_c(F_c)$. Parameters are those of Fig. 5-62; $J_0 = 8 \times 10^4\ \mathrm{A\ cm^{-2}}$. (After Grubin et al., 1985.)

(Note that with a finite load displacement currents would exist.) The situation with the nonuniform field calculation and displacement current contribution is different. Figure 5-73 displays the particle current through the device at selected instants of time. A comparison of the magnitude of the particle and total current indicates that within certain key regions of the device (particularly near the n^+-n^- and n^--n^+ interface regions) the displacement current dominates the current level. The general conclusion of this calculation is that since the initial transient is strongly influenced by displacement current contributions, it would be inap-

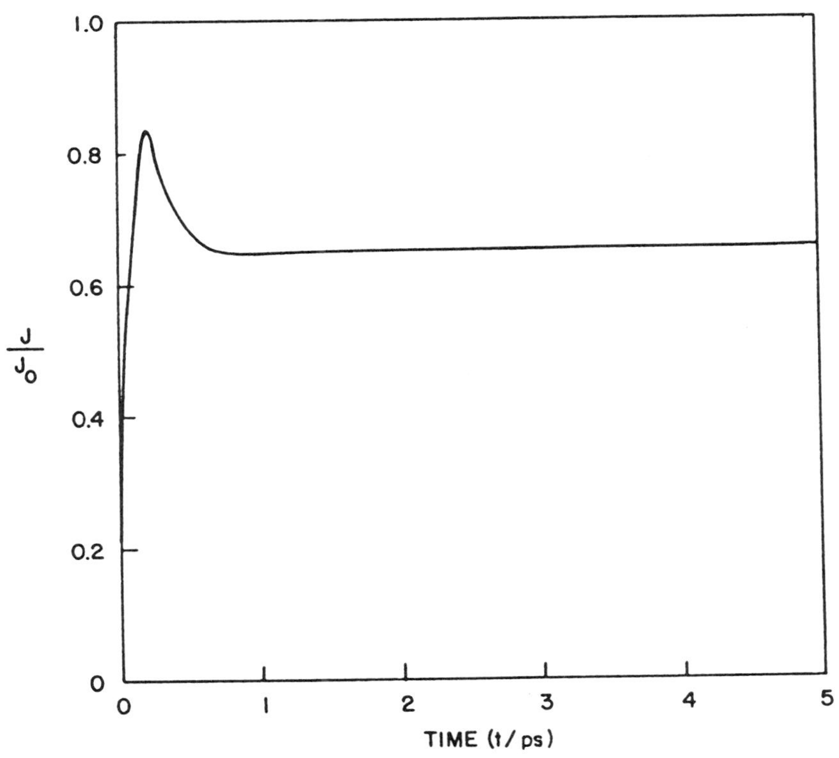

FIGURE 5-66. Magnitude of current transient following application of a step change in potential for a 0.25-μm-long device with injecting contacts. The current peak exceeds that of both the uniform field structure and the 1.0-μm device with injecting contacts. The steady state current level is above that of the 1.0-μm device with injecting contacts. The time of relaxation is 4.9 ps, which is approximately 40% less than that of the 1.0-μm device with the same average field. Given the fact that the structure is 0.25-μm in length, this result provides evidence that the relaxation effects are influenced by nontransit time effects. Parameters are those of Fig. 5-62; $J_0 = 8 \times 10^4 \text{ A cm}^{-2}$. (After Grubin et al., 1985.)

propriate to assume that the initial current transient is a measure of velocity overshoot.

The details of the transient, specifically as it relates to displacement current contributions, are reflected in the time dependence of the electric field and potential profiles (Figs. 5-74 and 5-75) and the spatially dependent charge density and electron temperature profiles (Figs. 5-76 through 5-79). Note, however, that as in the uniform field calculations, immediately following the voltage step the electric field increases everywhere by the ratio of the applied bias to the length of the structure (in this case 10 kV cm^{-1}). This initial increase introduces a displacement current whose magnitude does not correctly represent the physical transient, but rather the impulsive change in the applied potential over a single small, but finite, time step. Physically accurate calculations follow the initial time step and are discussed below for the cases illustrated in Fig. 5-80.

Prior to the application of the step potential a retarding field is formed at the

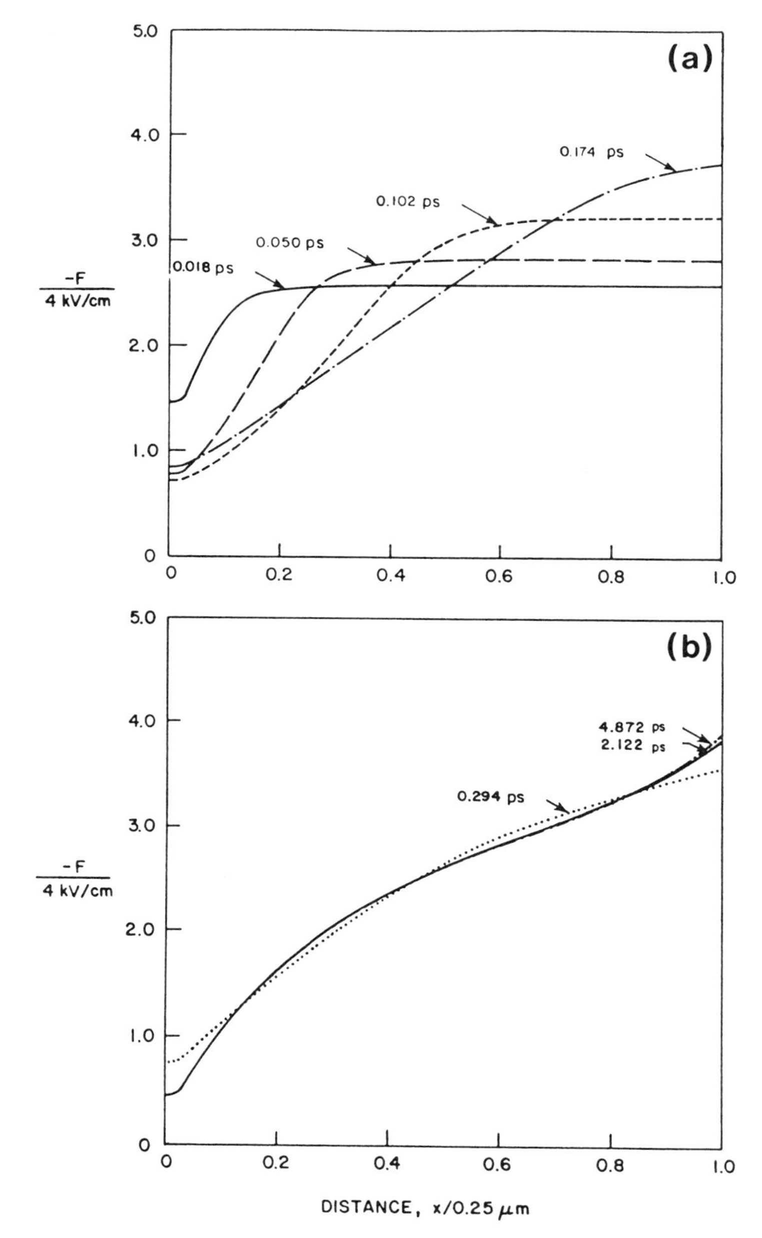

FIGURE 5-67. (a), (b) Time-dependent evolution of electric field distribution for the 0.25-μm device subject to a step change in bias of 0.25 V. During the first 0.1 ps, the field propagates downstream from the cathode, indicating a propagating accumulation layer. the field downstream from the cathode is relatively uniform. To satisfy the constraints of constant voltage across the device there are displacement current distributions at the bottom half of the structure that account for much of the difference in the peak currents associated with the 0.25-μm and 1.0-μm devices. Parameters are those of Fig. 5-66. (After Grubin et al., 1985.)

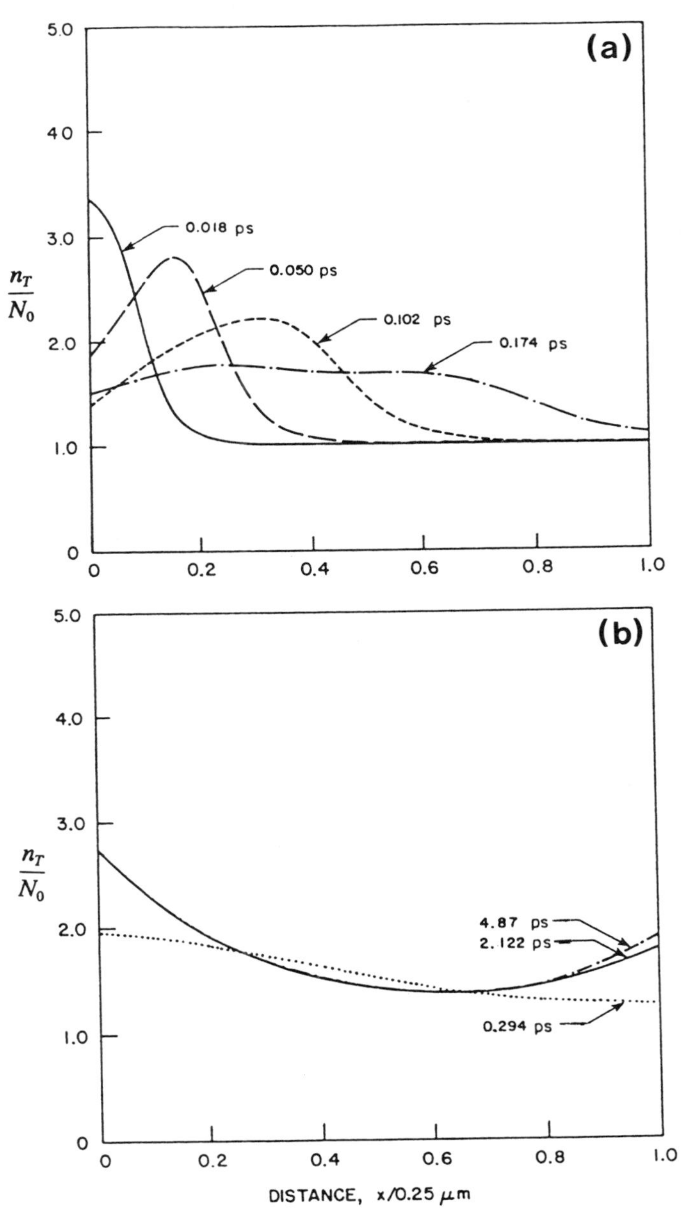

FIGURE 5-68. (a), (b) Time-dependent evolution of total carrier density within the 0.25-μm device. Initial propagation characteristics are similar to those of the 1.0-μm devices. Proximity effects are introduced after 0.1 ps and differences in the 0.25-μm and 1.0-μm calculations arise. Parameters are those of Fig. 5-66; $N_0 = 5 \times 10^{15}$ cm^{-3}. (After Grubin et al., 1985.)

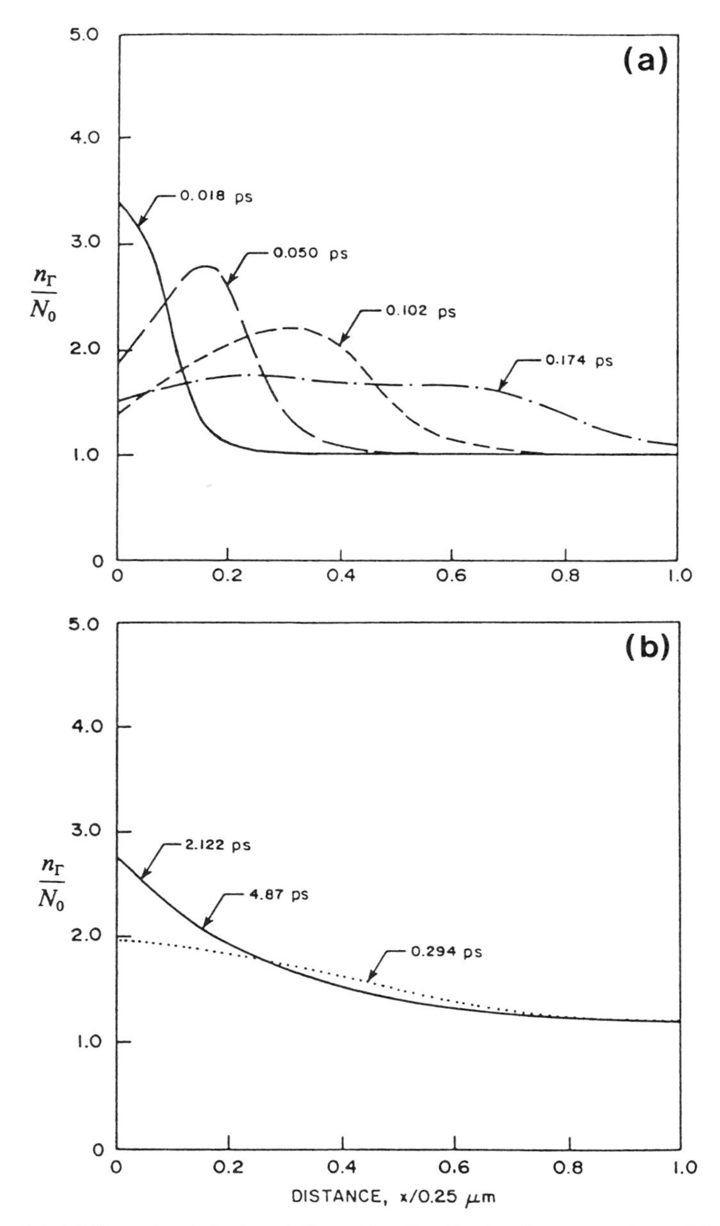

FIGURE 5-69. (a), (b) Time-dependent evolution of the Γ-valley for the parameters of Fig. 5-66. Note the electron transfer after 2 ps. (After Grubin et al., 1985.)

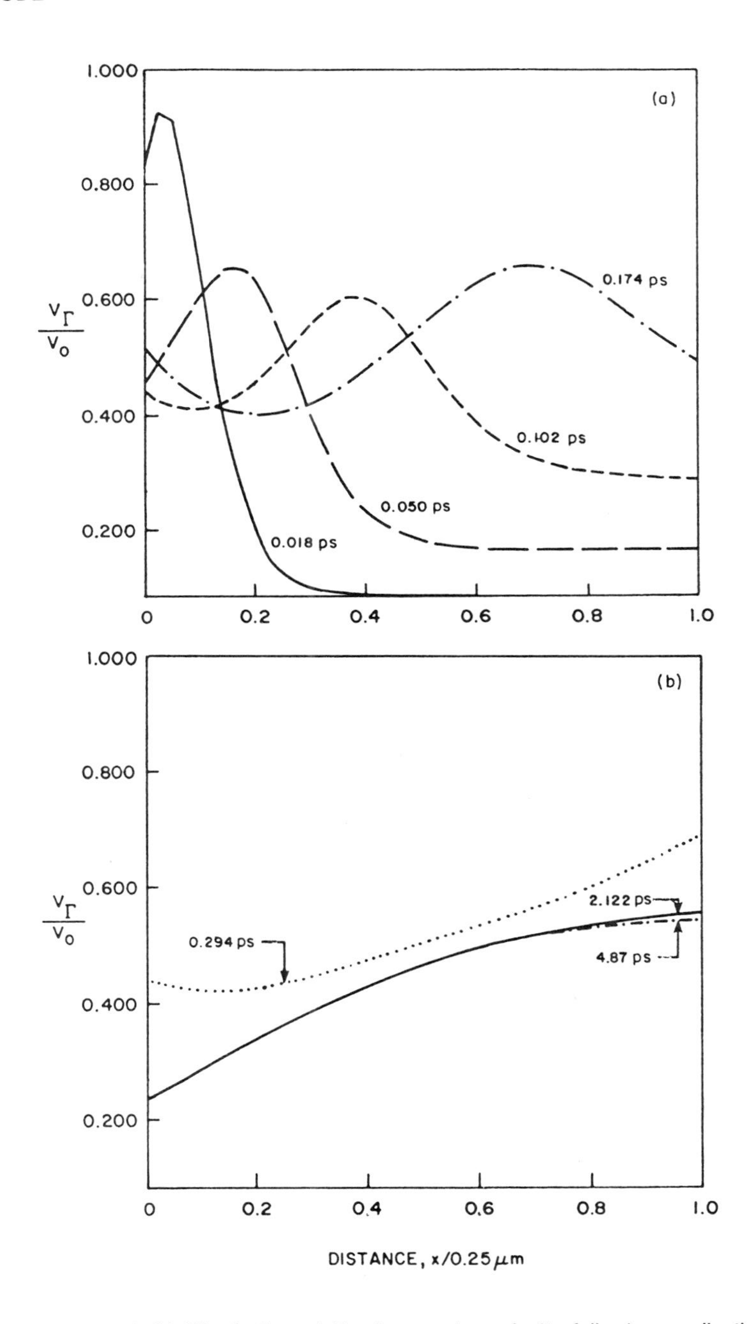

FIGURE 5-70. (a) and (b) Distribution of Γ-valley carrier velocity following application of a step change in bias of 0.25 V. The initial velocity distribution is similar to that found in the 1.0-μm transient study. Downstream velocity values during the first 0.1 ps are higher than that of the 1.0-μm calculation, corresponding in part to the presence of slightly higher downstream fields. Parameters are those of Fig. 5-66; $v_0 = 10^8\ \mathrm{cm\ s^{-1}}$. (After Grubin et al., 1985.)

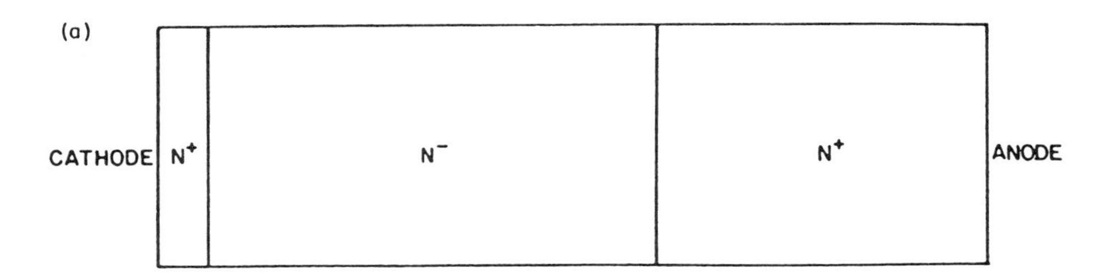

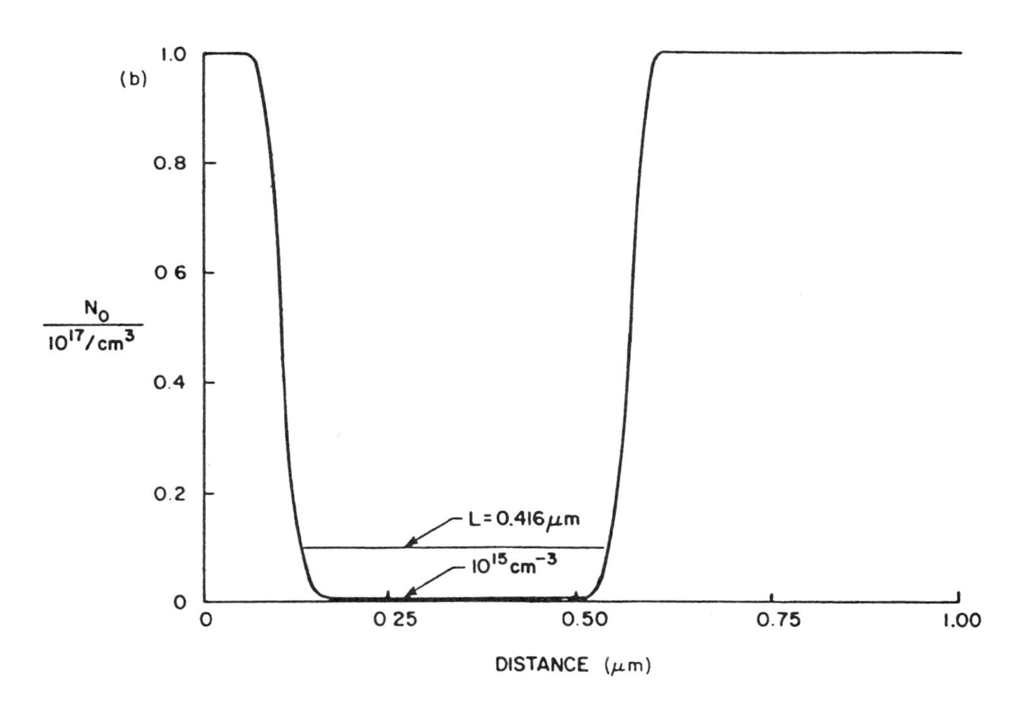

FIGURE 5-71. (a) One-dimensional structure used for calculations. (b) Donor distribution of the n^+-n^--n^+ structure used in the study. In the calculations, the width of the n^--region (defined at a donor level of $10^{16}\ cm^{-3}$) varied from 0.416 to 0.166 μm. In all calculations the width of the upstream n^+ region was unchanged. (After Grubin et al., 1985.)

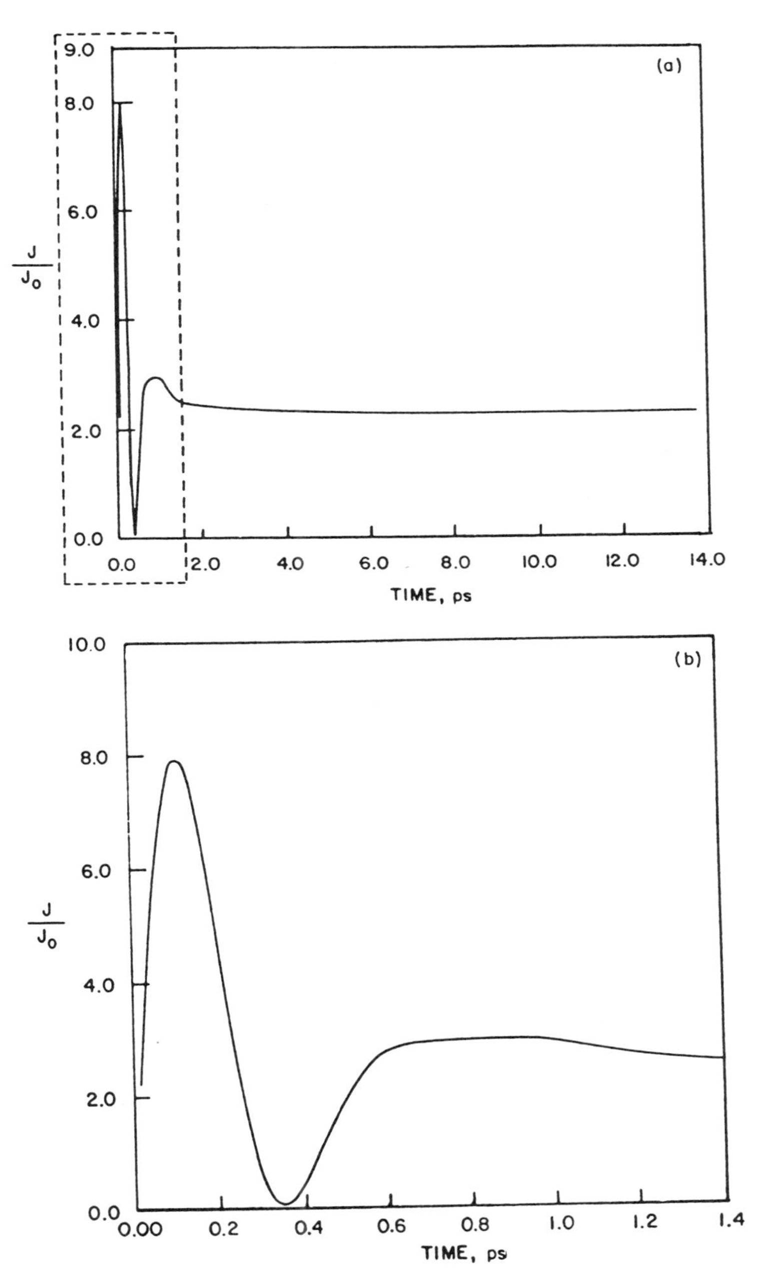

FIGURE 5-72. (a) Time-dependent current following application of a step change in bias to 1.0 V for the n^+-n^--n^+ structure with an n^--region of 0.416 μm. The structure of the current profile displays significant quantitative differences from that of the uniform donor calculations. Firstly, the peak in the current occurs within 0.10 ps, which is below that of the uniform donor calculations. Secondly, there is a strong current minimum, followed by relaxation. Steady state is achieved after approximately 15 ps. Parameters of the calculation are listed in the Appendix of this chapter. (b) Magnification of dashed area. (After Grubin et al., 1985.)

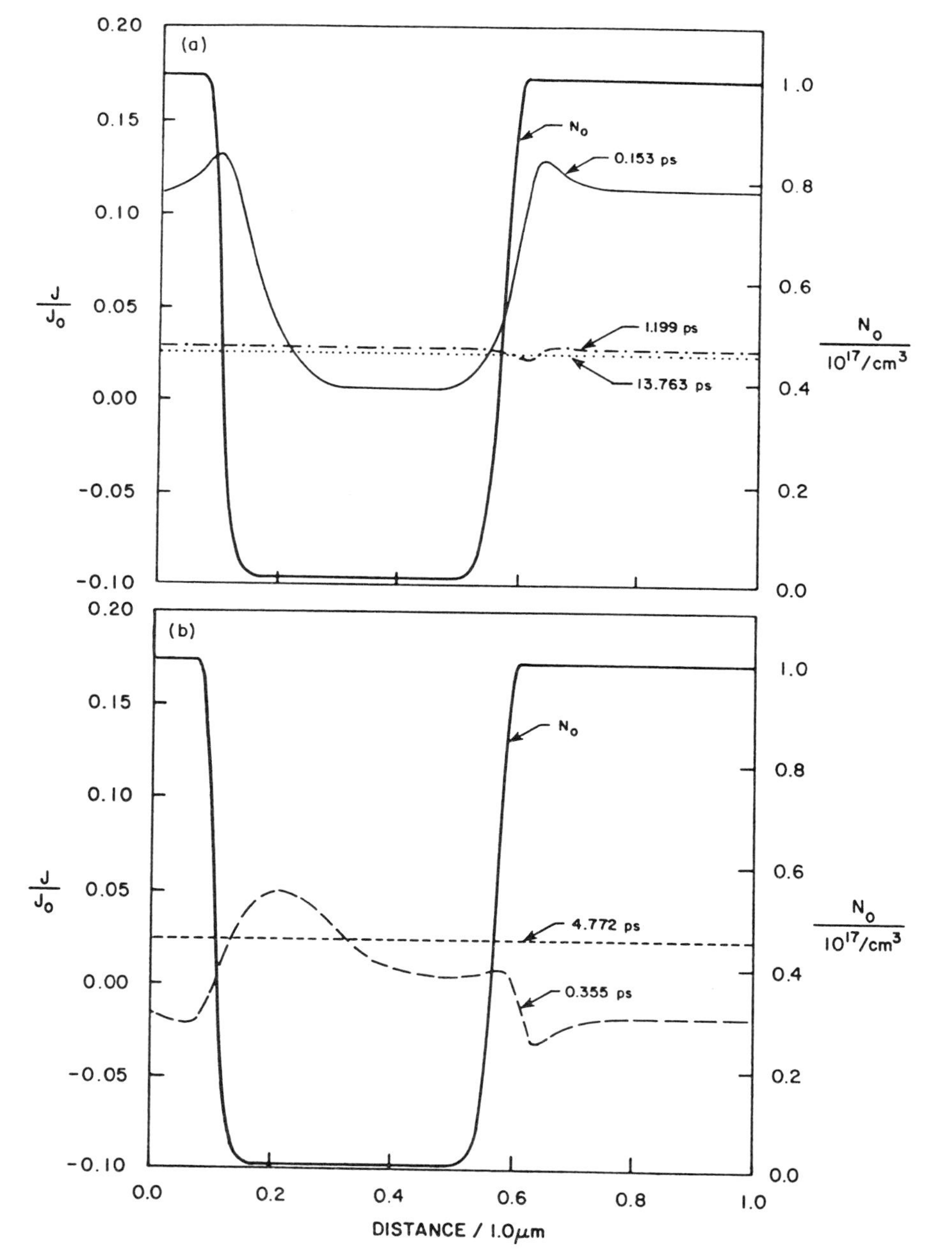

FIGURE 5-73. (a), (b) Spatial distribution of particle current at different instants of time for the parameters of Fig. 5-72. Also shown is the donor distribution N_0. The largest spatial variation in particle current occurs near the interfacial boundaries. (After Grubin et al., 1985.)

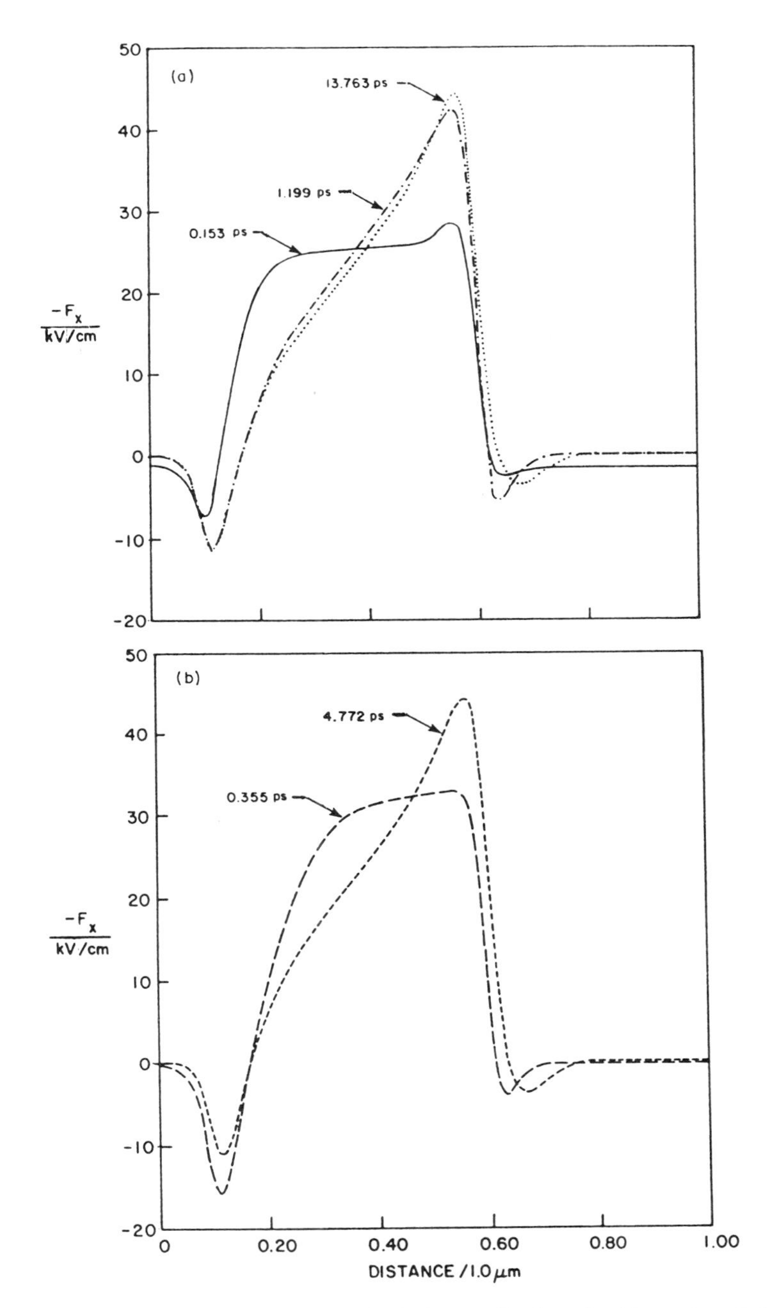

FIGURE 5-74. (a), (b) Distribution of electric field after application of a bias pulse. Note the strong temporal variation in field at the upstream interface within the n^- region. The propagation characteristics associated with the electric field distribution under uniform donor conditions are camouflaged here by the spatial rearrangements within the interface region. In addition, note the presence of the strong retarding field, one that is characteristic of n^+-n^--n^+ structures (see also Cook and Frey, 1981). Parameters are as in Fig. 5-72. (After Grubin et al., 1985.)

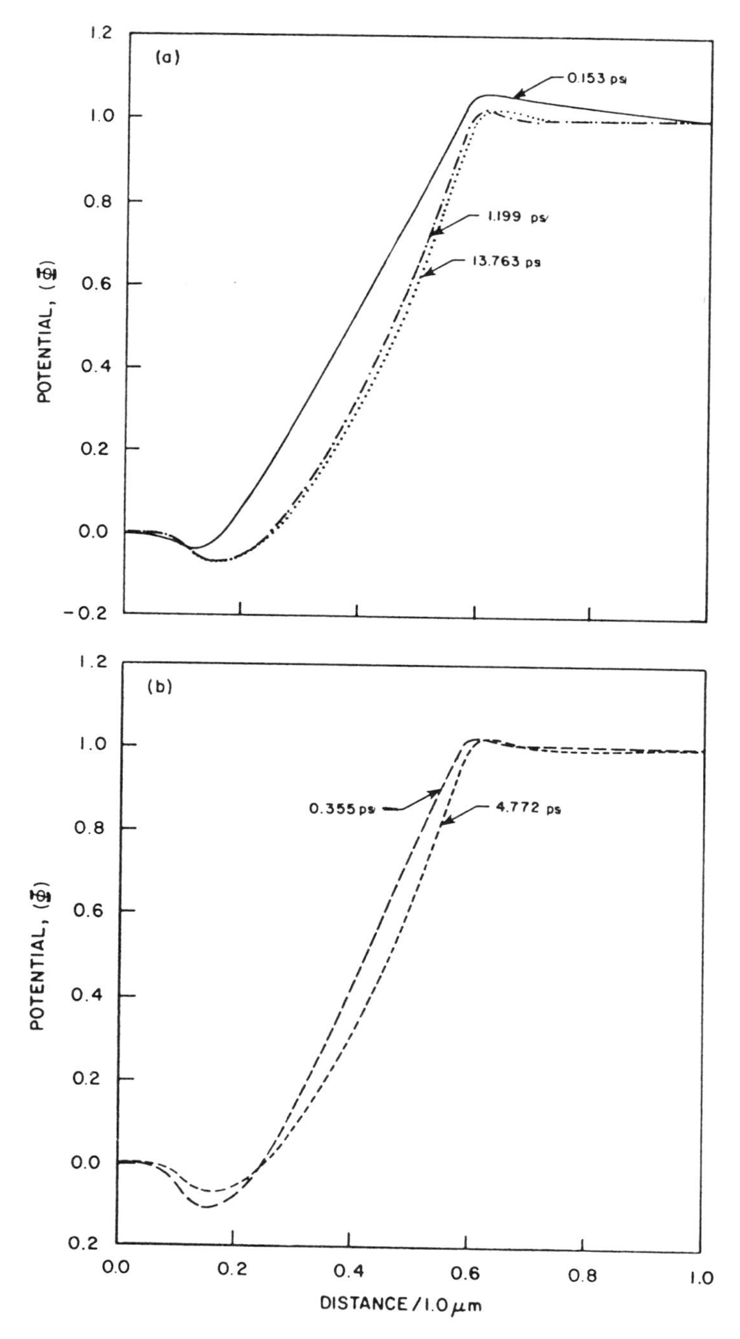

FIGURE 5-75. (a), (b) Spatial distribution of potential within the n^+-n^--n^+ structure at different instants of time. Note that in the steady state approximately 1.0 V falls across the 0.41-μm-long region. It may be anticipated that this will lead to large-scale injection in the n^- region. Parameters are as in Fig. 5-72; l = 0.416 μm. (After Grubin et al., 1985.)

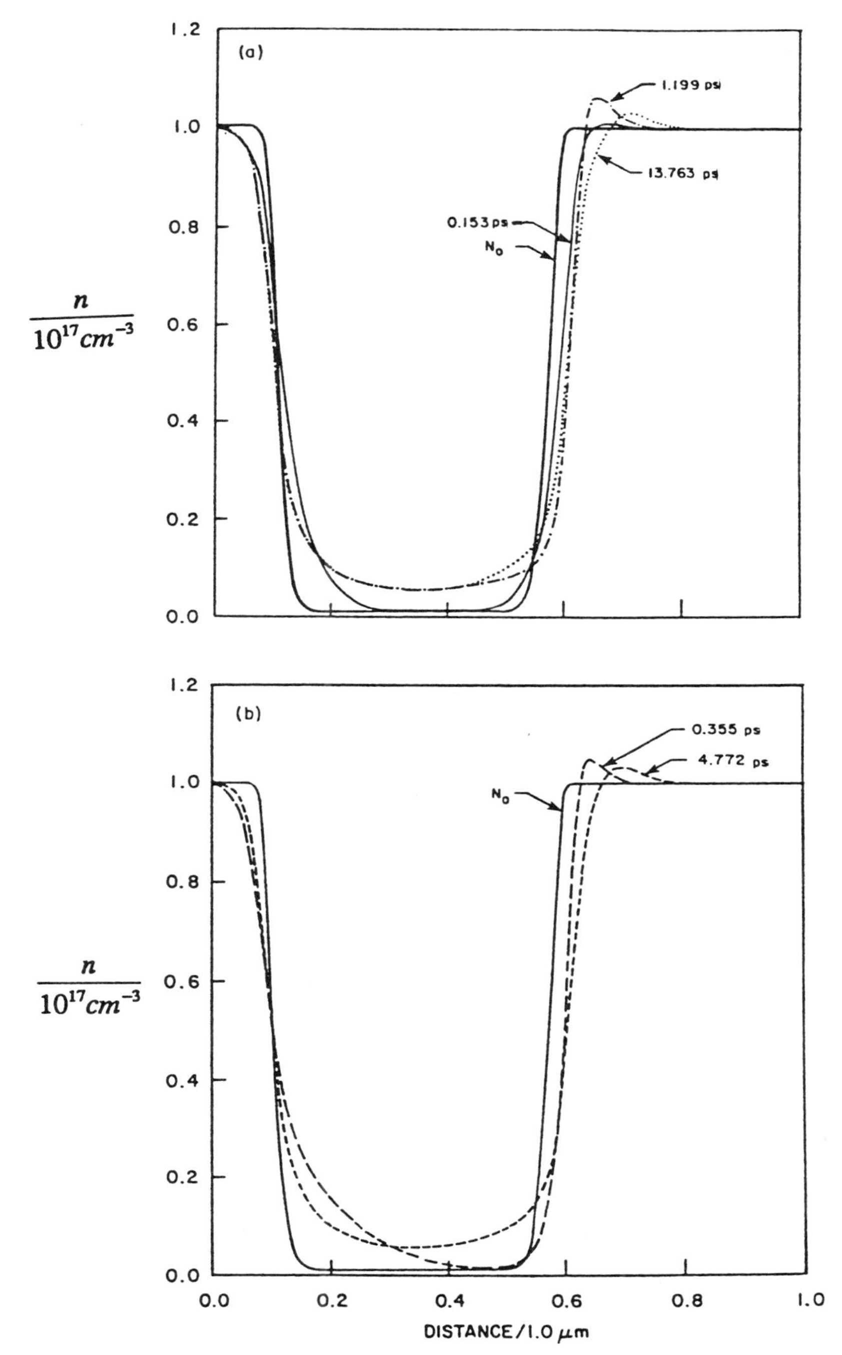

FIGURE 5-76. (a), (b) Time-dependent evolution of the total carrier concentration within the n^+-n^--n^+ structure. During the early transient, the space charge layer displays a characteristic propagation downstream from the cathode, as seen in the uniform donor calculations. Insofar as the form of the electric field profile is controlled by differences between n and N_0, these propagation characteristics lose the distinction that emerges from the uniform donor calculations. The injection level is almost an order of magnitude higher than the n^- donor level. Parameters are as in Fig. 5-72. (After Grubin et al., 1985.)

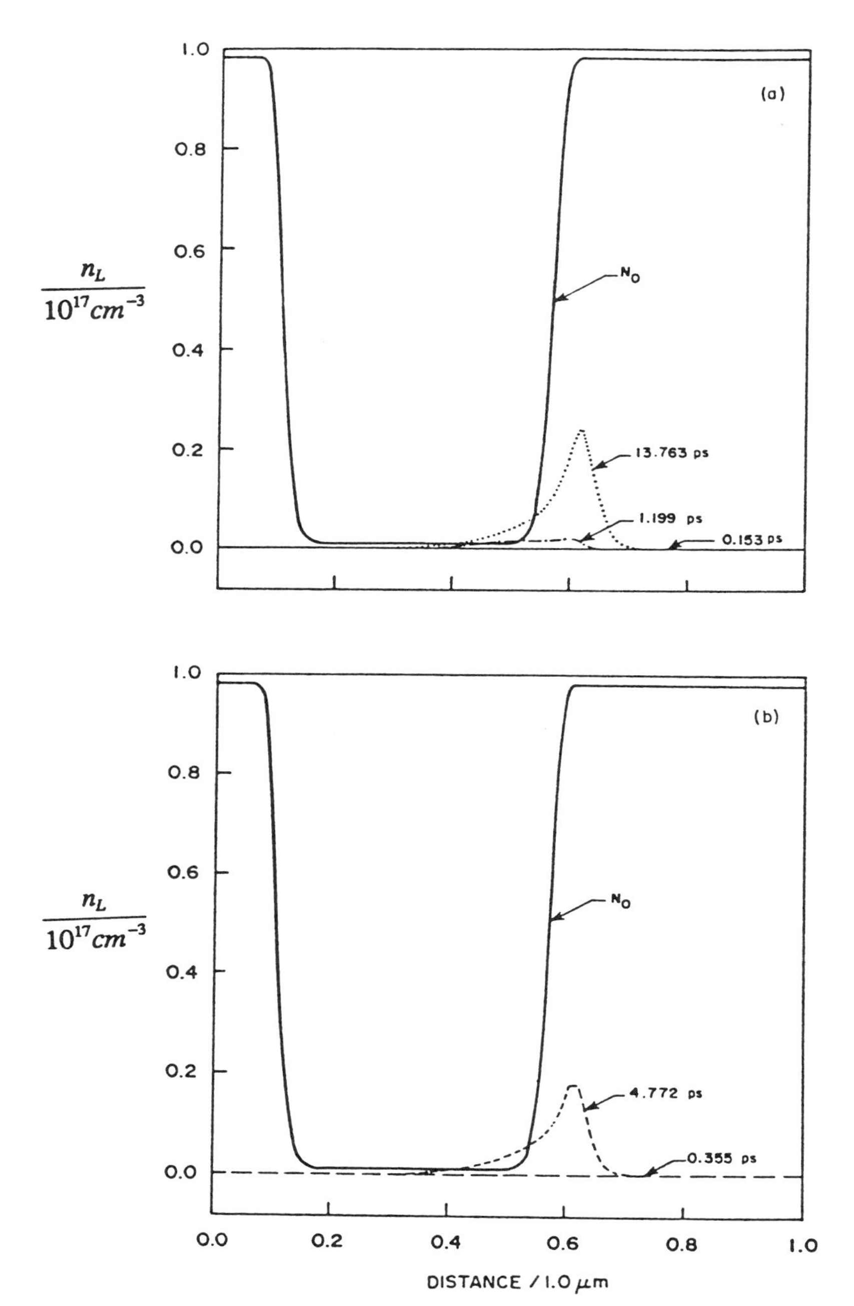

FIGURE 5-77. (a), (b) Time-dependent evolution of the distribution of satellite valley carriers. An inconsequential number of carriers are scattered into the satellite valley during the early transient ($t < 0.45$ ps). Electron transfer is apparent in the steady state. Parameters are as in Fig. 5-72 (After Grubin et al., 1985.)

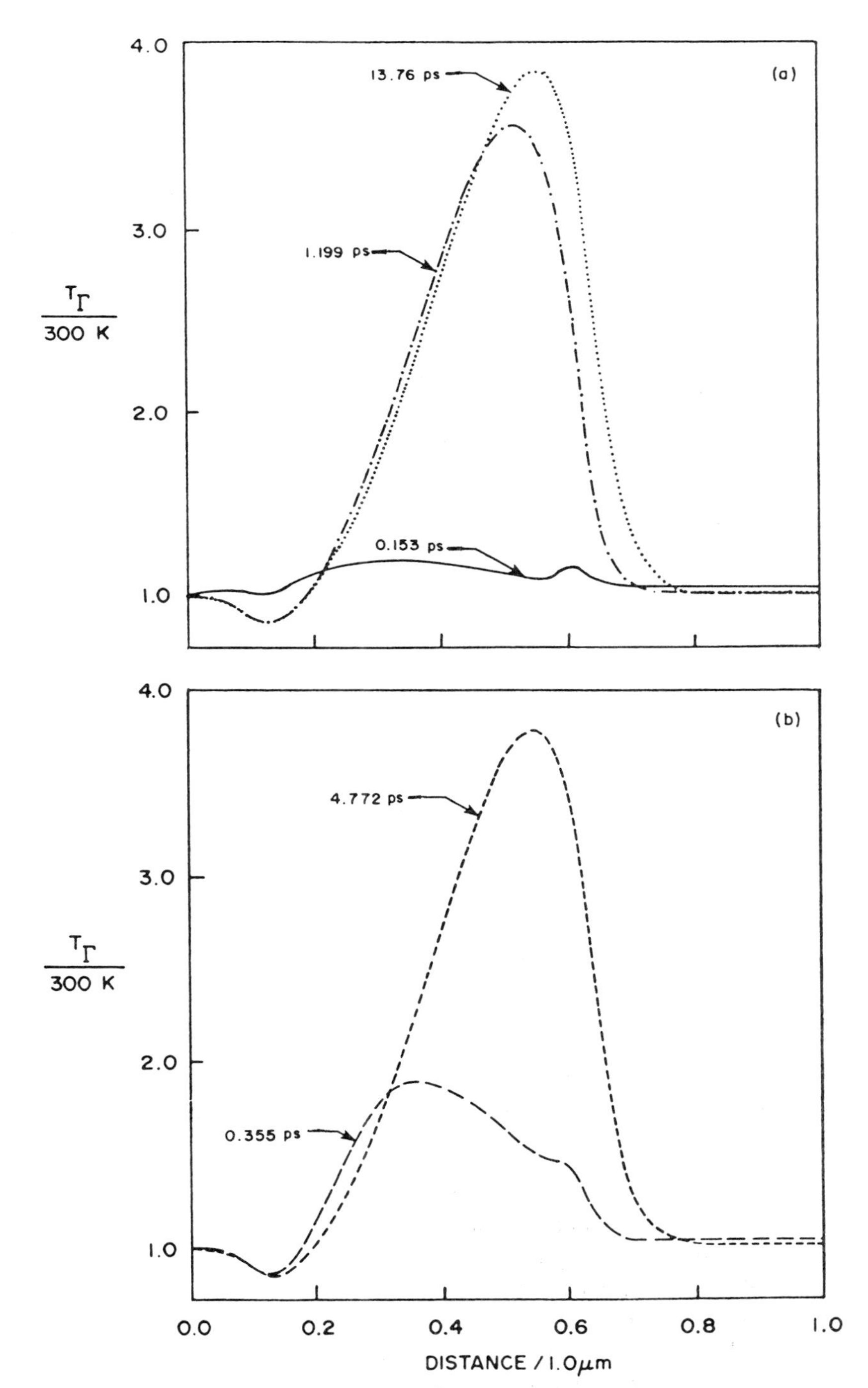

FIGURE 5-78. (a), (b) Time-dependent evolution of the spatial distribution of Γ-valley temperature within the n^+-n^--n^+ structure. Note that in steady state an apparent cooling of the carriers occurs within the region near the n^+-n^- interface where a large excess of carriers is present (see also Cook and Frey, 1981). Parameters are as in Fig. 5-72. (After Grubin et al., 1985.)

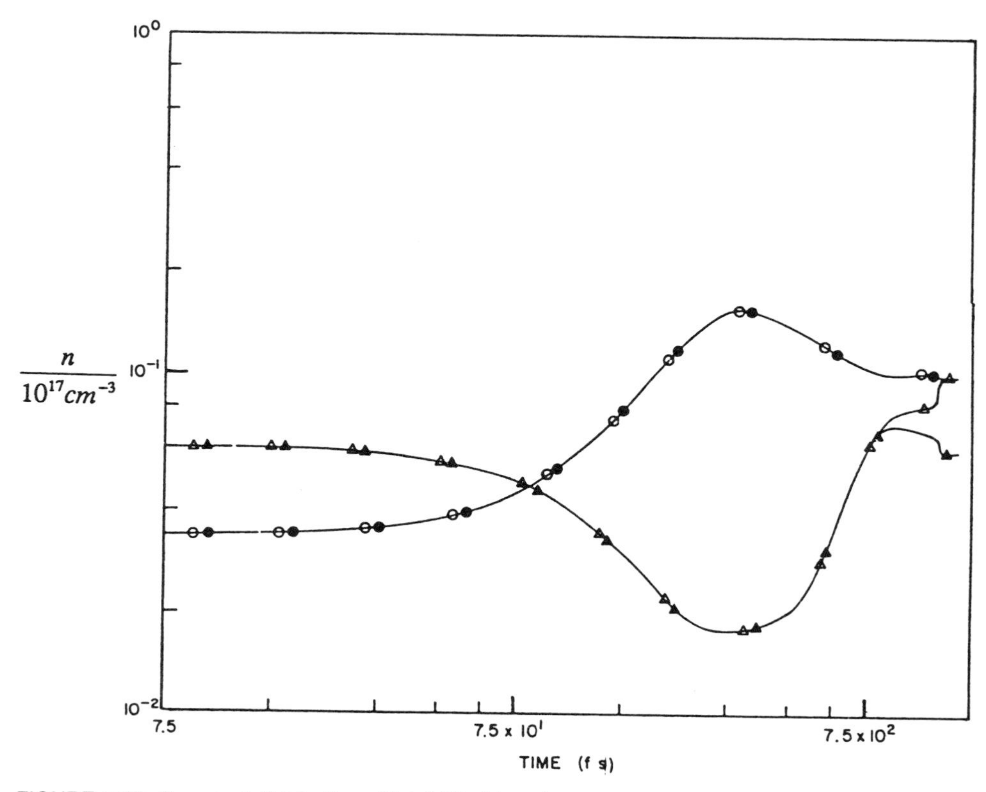

FIGURE 5-79. Transient distribution of total (○, △) and Γ-valley (●, ▲) carrier density at two points within the n^+-n^--n^+ structure. At 0.2 μm (○, ●), there is no electron transfer of significance. At 0.5 μm (△, ▲), electron transfer occurs at the end of the transient. Parameters are as Fig. 5-72. (After Grubin et al., 1985.)

upstream n^+-n^- interface, limiting further injection of space charge into the n^- region. This retarding field, which at its maximum is positive in sign, is significantly reduced following application of the step potential; carrier injection into the n^- region is thereby resumed. Two events accompany this enhanced injection. First, to accommodate the increased charge within the n^- region, Gauss's law dictates that the electric field within the region must become increasingly negative. Second, the space charge injection is self-limiting in that as the process of injection proceeds, the retarding field begins to arise from the excess charge injected into the n^- region. This point was also made by East and Blakey (1984), who also examined the dependence of current and voltage on the n^- region length. A second point of importance here concerns determining which portion of the structure dominates its transport. It may be intuitively expected that for the structure considered, it is the n^- region that dominates. This appears to be the case in the preceding discussion. But we may expect that for a sufficiently small n^- region, no single region dominates. In the calculations reported here the absence of a single dominating region becomes apparent at higher voltage levels and for the case when $L_{n^-} = 0.116\ \mu$m. These results are

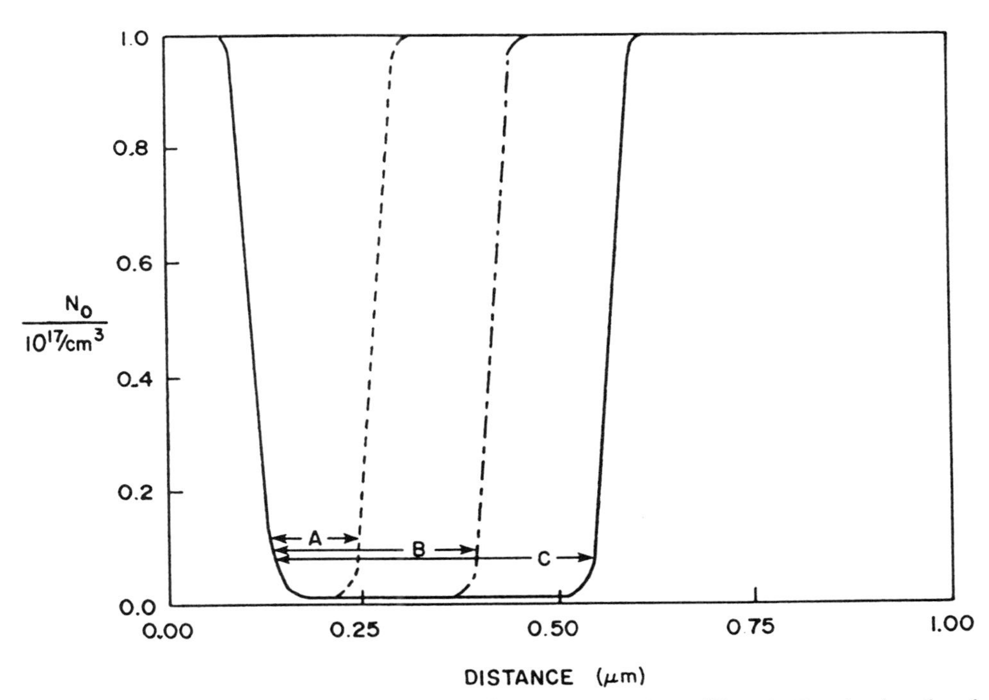

FIGURE 5-80. Donor concentration for n^+-n^--n^+ structure with three different n^- region lengths. *A*, $L_{n^-} = 0.116\ \mu m$; *B*, $L_{n^-} = 0.266\ \mu m$; *C*, $L_{n^-} = 0.416\ \mu m$. (After Grubin et al., 1985.)

illustrated in Figs. 5-80–5-86, with particular attention paid to voltage sharing and electron transfer in the n^+ region as the n^- region is reduced in size. Figure 5-80 is a sketch of the background doping level associated with the variable n^- region. Within these regions and at a bias of 1 V, the potential is calculated self-consistently and is displayed in Fig. 5-81. It is noted that for n^- regions of length 0.266 and 0.416 μm, most of the potential drop is across the n^- region. For the smallest region a substantial potential drop falls across the n^+ region. The origins of this enhanced potential drop may be found in examining the self-consistently computed charge distribution (Fig. 5-82), which shows the presence of excess charge accumulation at the downstream n^--n^+ interface, resulting in a change in sign of the curvature of the potential. The distribution of Γ-valley carriers is displayed in Fig. 5-83, where the presence of substantial electron transfer in the n^+ region is noted. The carrier velocity (Fig. 5-84) and electron temperature (Fig. 5-85) within the Γ-valley display the expected increases for the shorter n^- region. The electric field distribution, shown in Fig. 5-86, displays higher field values within the n^+ region.

The significance of the preceding result is that while variations in the total charge density tend to screen variations in the doping profile of the structure, the potential drop across the downstream n^- region may be small enough to allow a substantial drop across the downstream n^+ regions, thereby permitting electron transfer to occur away from the n^- region. This, of course, is not unexpected. It

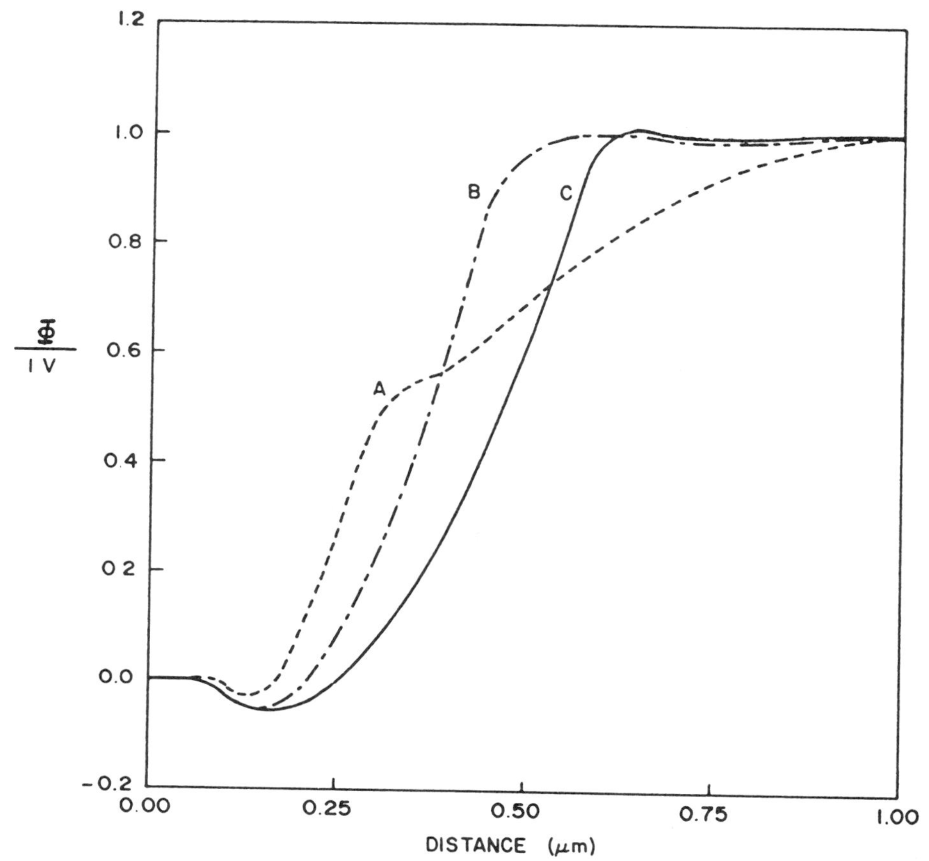

FIGURE 5-81. Steady state distribution of potential for structures *A*, *B*, and *C* subject to a bias of 1.0 V. For structures *B* and *C* the potential drop is confined mainly to the n^- region. For structure *A* a significant fraction of potential falls across the n^+ region. (After Grubin et al., 1985.)

is implicit in the design of Gunn oscillators with doping variations designed to act as domain nucleation sites. The current–voltage characteristics, therefore, are expected to reflect a complex set of electrical phenomena.

Figure 5-87 displays a series of current–voltage curves for n^+-n^--n^+ structures with the indicated n^- region length. Each curve displays J/J_{ref} versus Φ/Φ_{ref}. J_{ref} is the computed value of current at $\Phi_{\text{ref}} = 0.25$ V. The value of J_{ref} is indicated in the figure caption. Because of the intuitive relationship between the space-charge injection properties of the submicrometer n^+-n^--n^+ structure and those associated with Child's law, a power law, $J \propto \Phi^y$, was extracted. Note that J_{ref} increases as the n^- region decreases in length. At low bias levels the current–voltage relationship appears to follow a power relationship that is slightly less than $J/J_{\text{ref}} = (\Phi/\Phi_{\text{ref}})^y$ with $y = 1.7$ (as compared to a Child's law relationship where $y = 1.5$). At higher values of bias there is enhanced sublinearity in the current–voltage relationship due, in part, to electron transfer to the satellite valleys.

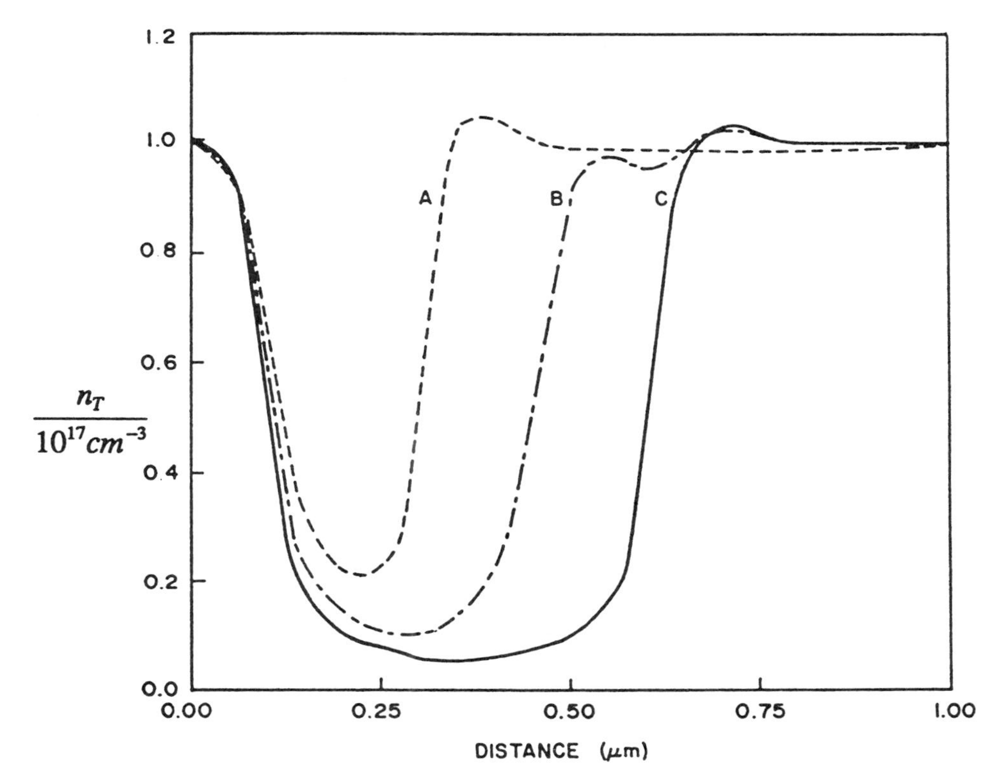

FIGURE 5-82. Steady state distribution of total carrier concentration for structures *A, B,* and *C.* Note that, for all three structures, the free carrier concentration closely traces the donor variation. (After Grubin et al., 1985.)

As indicated above, a considerable amount of electron transfer occurs in the downstream portion of the n^+ region when the n^- region is decreasing in length. Indeed, the detailed calculations indicate that the relative amount of electron transfer increases as the n^- region decreases in length. At first glance this result appears to contradict all that has been discussed about transport in submicrometer devices. But it is not unusual when it is realized that as the n^- region decreases in length a greater fraction of the voltage drop falls across the n^+ regions of the device. It is this latter feature that is responsible for the enhanced transfer. To place this in different terms, the active region length of the device increases as the n^- region becomes insignificantly small.

5.6. *InP SUBMICROMETER DEVICE SIMULATION FROM THE BOLTZMANN TRANSPORT EQUATION*

Submicrometer InP TEDs have been analyzed recently with an emphasis on length effects (Wu and Shaw, 1989; Wu et al., 1991). The static and dynamic characteristics of such devices also deviate substantially from those of longer

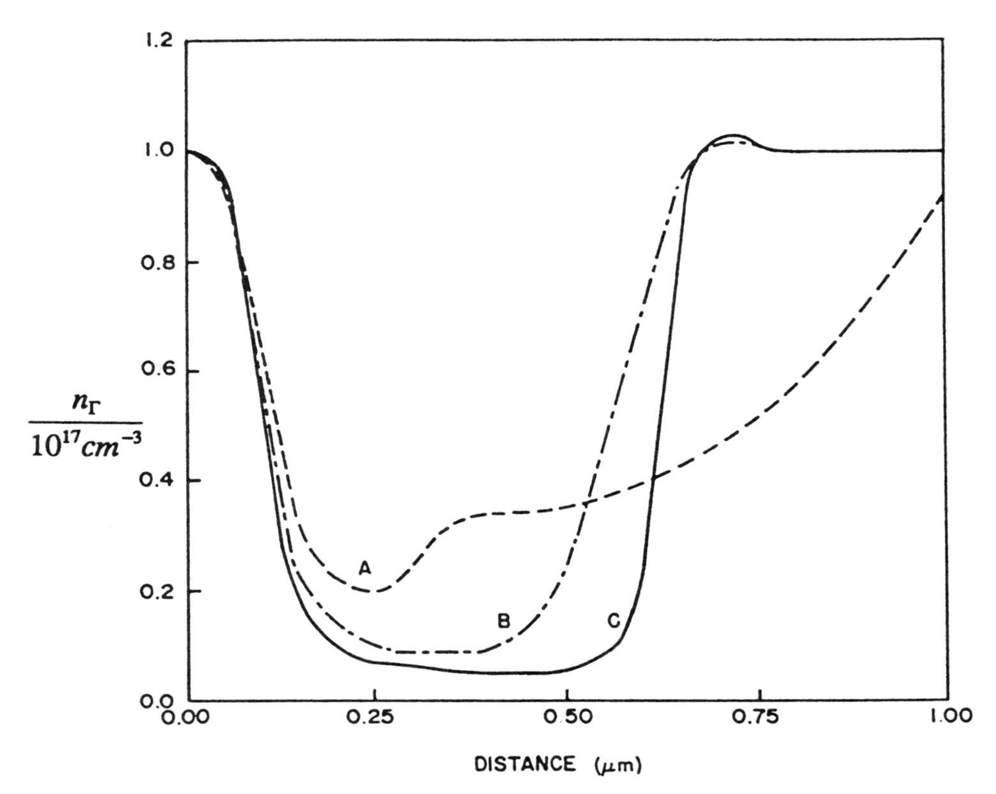

FIGURE 5-83. Distribution of Γ-valley carriers. The least amount of electron transfer occurs for the widest structure, *C*. For structure *A*, transfer continues to the anode contact and is a consequence of a large potential drop across the downstream n^+ region. (After Grubin et al., 1985.)

devices. The upper limit on the transit time frequency for InP TED was found to be about 230 GHz for a 0.2 μm device with a dc to ac conversion efficiency of about 0.3%. The same length effect was found in the $v(F)$ curves of InP as was found in GaAs (see Fig. 5-34). InP loses its negative slope characteristic at a shorter length (~0.05 μm) than GaAs (~0.1 μm). Design rules for determining the optimum frequency and efficiency via the interrelated parameters were also determined. There is, however, reason to be concerned about the InP results. For the nonuniform field analysis we should retain all the spatial and temporal terms in the moment equations. However, in order to maintain numerical stability, Rosencher's (1981) stabilization procedure was used, which employs the cold plasma approximation (this assumes $dP/dx = 0$, where $P = nkT + \rho(dJ/dx)$ is the carrier pressure and ρ is the carrier viscosity) and the adiabatic approximation (which assumes $dQ/dx = 0$, where $Q = -\kappa_T(dT/dx)$ is the thermal conductance of the carriers and κ_T is the thermal conductivity of the carriers). These are the higher moments involved in the lower moment equations that often cause a lack of convergence (numerical instability) in the mathematical model. Wu et al. (1991) adopted a physical condition in their attempt to validate the use of the cold plasma approximation: The velocity of propagation of collective disturbances

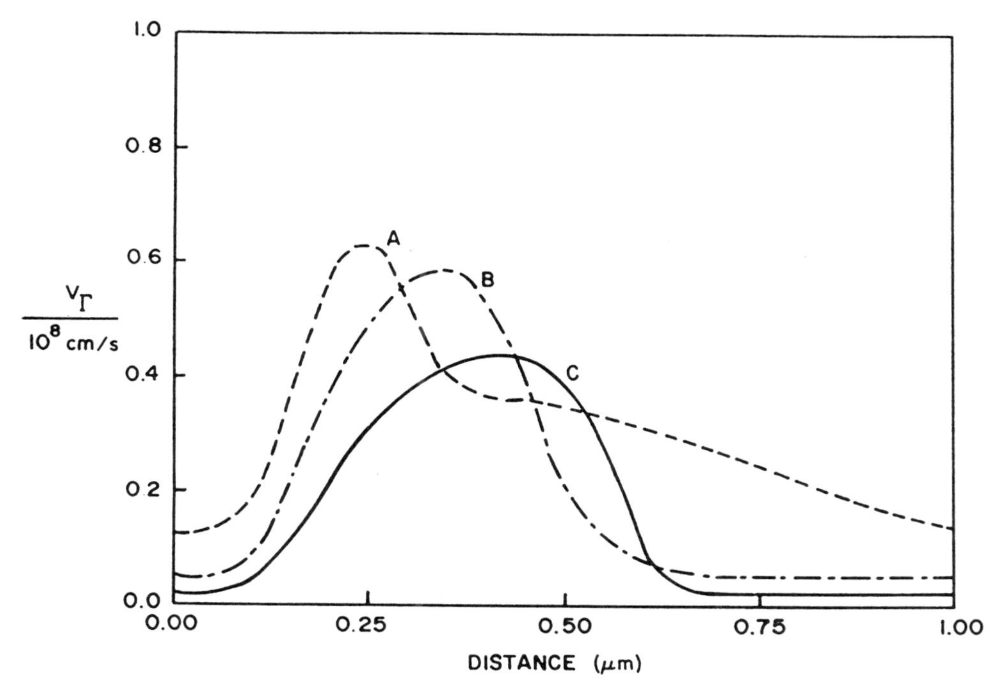

FIGURE 5-84. Distribution of Γ-valley carrier velocity for structures *A, B,* and *C.* The peak velocity gradually increases as the n^- region decreases in length. Additionally, the upstream and downstream carrier velocities increase as the n^- region decreases in length. (After Grubin et al., 1985.)

should be large compared to the thermal velocities of the particles (electrons). However, this violates the equilibrium zero current condition and could invalidate the use of the cold plasma approximation. Under low-field conditions this argues against the exclusion of the pressure term. Nevertheless, at high fields (hot electrons), the influence of the pressure term in the moment equations may perhaps be reduced. To investigate the validity of the cold plasma approximation, Wu et al. (1991) studied the retention of the nk_BT term in P, but with omission of the viscous force term when solving the conservation equations. The solutions exhibited a dramatic deviation from what was expected: the transit frequencies become extremely high, which contradicted their theoretical predictions (an equal areas rule for long devices and a different areas rule for short devices), and because of the spatial gradients of n and T, the dipole domain turns into an accumulation layer that propagates through the device with a velocity near the peak of the $v(F)$ curve. In comparing the results with the dynamic domain transport characteristics of GaAs TEDs that have been reported by Grubin and Kreskovski (1985), who solved the conservation equations by choosing constant values of P and κ_T in order to retain all spatial terms, they concluded that much of the diffusive effects from the $n\kappa_BT$ term are balanced out by the viscous force term. The validity of using the cold plasma approximation was also indicated by comparing their results with the results of a near-micrometer analysis and experiments (Czekaj et al., 1988). The same conclusion for the adiabatic

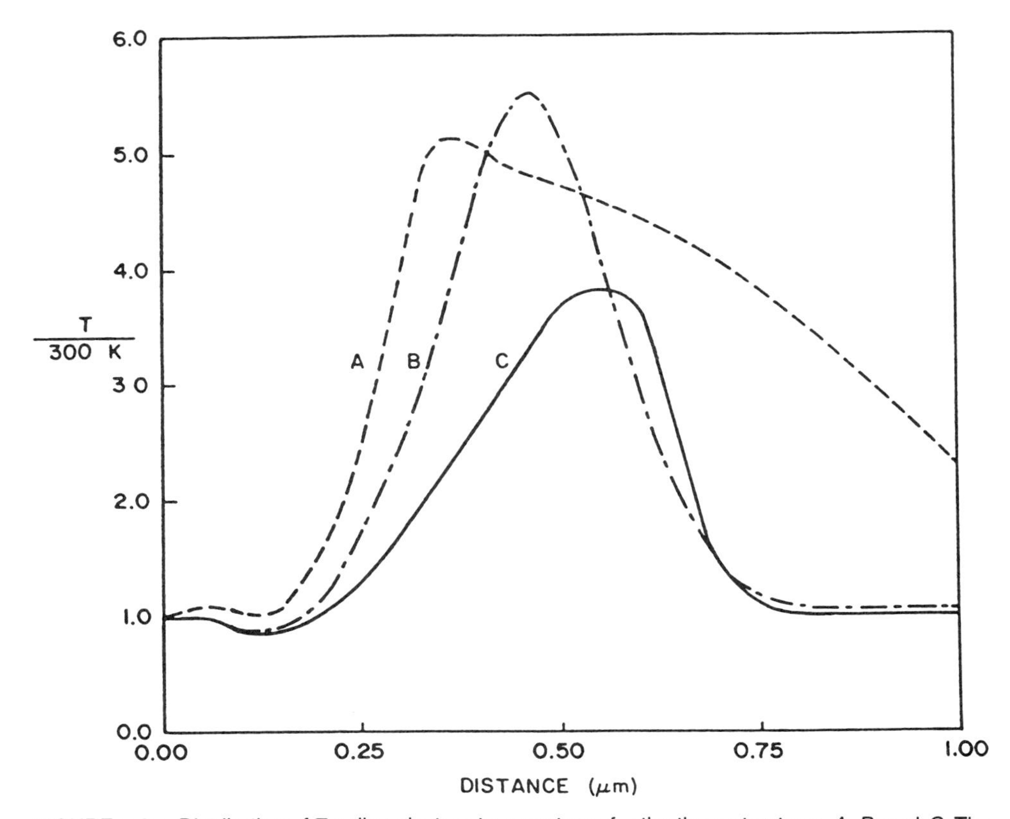

FIGURE 5-85. Distribution of Γ-valley electron temperatures for the three structures *A*, *B*, and *C*. The electron temperature distribution is qualitatively different for structure *A*. A longer downsteam n^+ region is needed before the temperature approaches 300 K. (After Grubin et al., 1985.)

approximation in the thermal term has been cited by Bosch and Thim (1974). Insofar as the results of Wu et al. (1991) agree with the above trends, there is some confidence in their analysis. However, we are presently not aware of any published experimental data for submicrometer InP TED. Until sufficient experimental results are available to make comparison with theory profitable, we are unable to state precisely what their approximation does to the results, and thus to the ultimate utility of the analysis. The realization of actual submicrometer InP TED samples becomes one of the future studies that will enable us to correlate theoretical predictions with experimental data.

The cathode mobility model has been shown to be useful in simulating low-resistance linear contacts (Grubin and Kreskovski, 1985; Wu and Shaw, 1989; Czekaj et al., 1988). Although the model can be shown to be valid over the very short region of cathode "contact," it still fails to completely reveal the actual hot electron configuration in the presence of a field gradient. Furthermore, it is difficult to relate this model to an actual contact structure. For the device structure considered in the analysis, the metal–semiconductor junction at the cathode contact region can be fabricated and stabilized by a deposition and

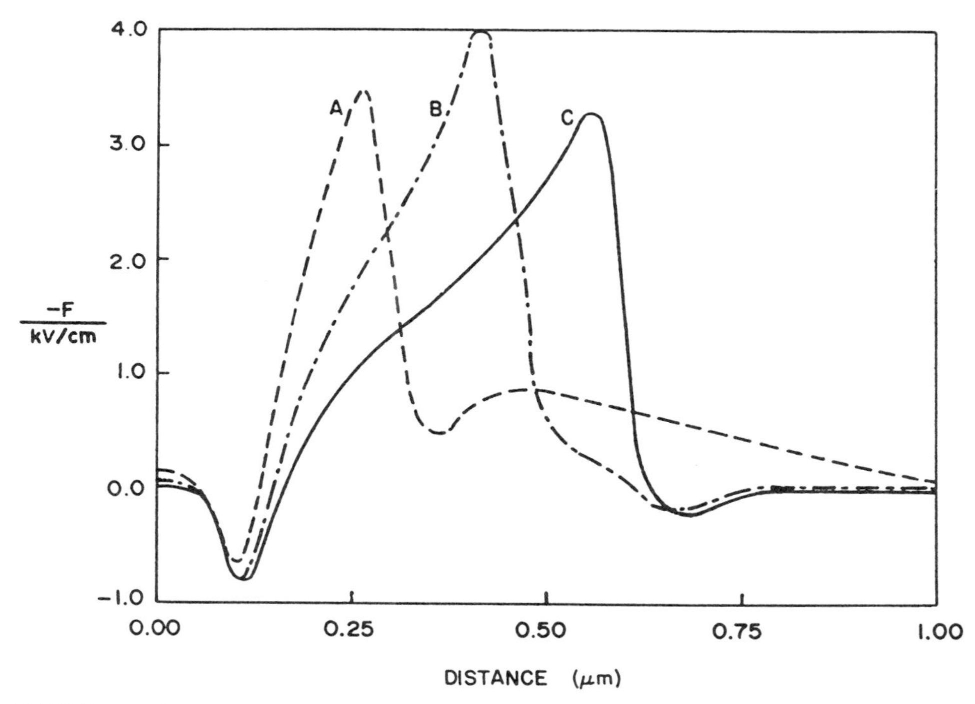

FIGURE 5-86. Steady state distribution of electric field for the three structures *A*, *B*, and *C*. Note that for structure *A*, a large residual field remains across the downstream n^+ layer. (After Grubin et al., 1985.)

annealing process that involves various combinations of temperature and time. The difficulty of relating the cathode mobility model to these annealing parameters makes the effort of replacing the abstract model by a more sophisticated cathode scattering model attractive. For example, in order to impose real electron–phonon kinetics in the contact region with proper scattering mechanisms, the boundary conditions can be set by segregating the scattering parameters of the contact area from the bulk region in the sense of making it easier to match the bulk BTE solution. In order to correlate the physical model to the fabrication process, the scattering parameters at the contact region that involve the phonon energy and deformation potential are the ones that most likely can be related to physical effects that occur during the annealing process. This study will further enhance the completeness of the numerical analysis, and also reduce the gap between theory and experiment.

5.7. *SUMMARY AND CONCLUSIONS*

The major technological interest in transient transport arises from the predictions of unusually high mean carrier velocities. The initial discussion of these high velocity values was for uniform space charge distributions, but the

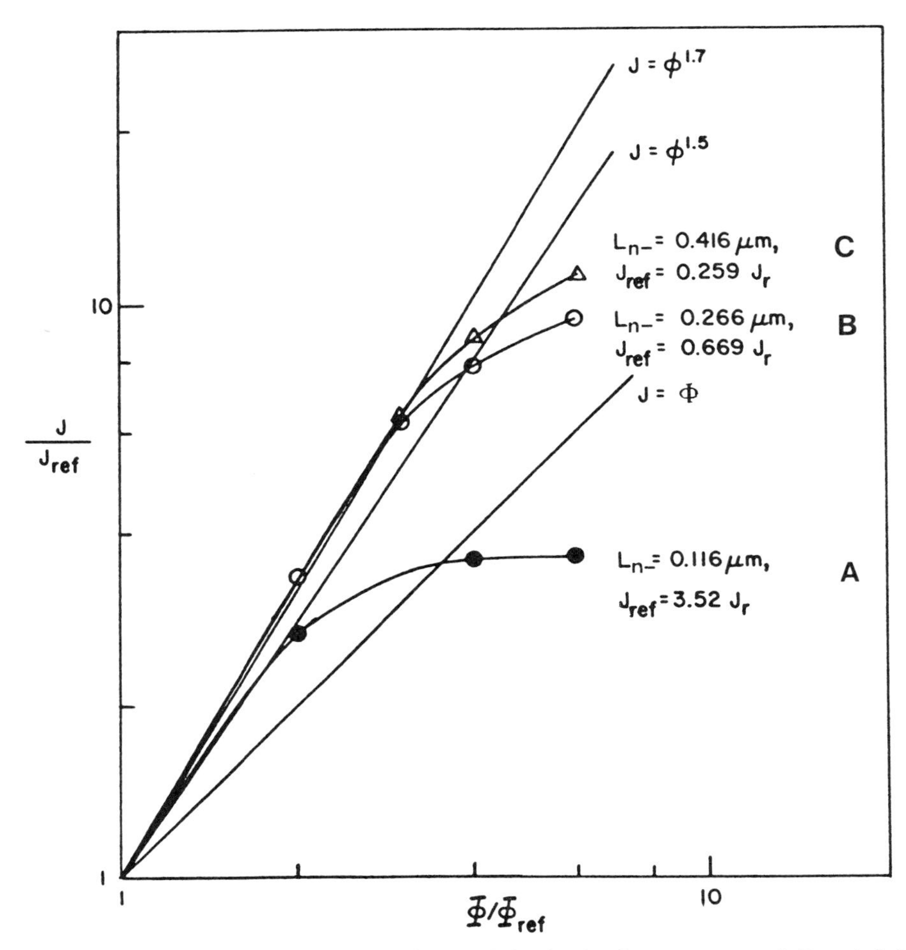

FIGURE 5-87. Steady state current–voltage characteristics for the three structures *A*, *B*, and *C*. The current level for structure *A* is higher than that of *B*, which in turn is higher than *C*. Note that the low-field resistance of structure *A* is the lowest of the three. Also included for reference are the Child's law $J = \Phi^{1.7}$, $J = \Phi^{1.5}$, and $J = \Phi$. $J_r = 1.6 \times 10^4\ \mathrm{A\,cm^{-2}}$; $\Phi_{ref} = 0.25\ \mathrm{V}$. (After Grubin et al., 1985.)

results were thought to be relevant for those situations where the mean carrier energy was insufficient to lead to substantial electron transfer in gallium arsenide. Thus, the trend developed toward submicrometer-scale devices. The complication that arises in submicrometer devices is that the boundary conditions will be the determinant as to whether high velocities will be attained. Additionally, the constraints of current continuity dictate whether high velocities will be accompanied by high carrier densities. For example, in the case of injecting contacts the velocity of the entering carriers was significantly below that within the interior of the semiconductor. The situation was reversed for the case of partially blocking contact conditions.

Several critical results emerged from the discussion: (1) Transient overshoot in submicrometer structures reflects the presence of velocity overshoot and displacement current effects. It is not possible, in a simple way, to separate the two, with the result that transient measurements of overshoot require extreme care in interpretation. (2) Relaxation times to steady state are dominated by the dominating boundary; e.g., either the metal contact or the critical interface. Relaxation times do not scale linearly with device length. The relaxation time scales monotonically with length. (3) Transient overshoot effects are dependent upon rise times and the time for relevant field rearrangement within the structure to occur. (4) The maximum frequency of operation in a conventional fundamental transit time is near 100 GHz for GaAs and 200 GHz for InP. Other materials, particularly the ternary compounds, could possibly produce higher upper frequency limits.

In this chapter we emphasized stability calculations for uniform and nonuniform field profiles, and presented results for both GaAs and InP devices, focusing on numerical simulations and the role of the cathode boundary conditions.

APPENDIX: DIMENSIONLESS EQUATIONS USED IN THE NUMERICAL SIMULATIONS

The continuity equations in dimensionless form are as follows:
(i) Equation (5-66):

$$\frac{\partial n_1^*}{\partial t^*} = \frac{\partial n_1^* V_1^{*j}}{\partial x_j^*} - n_1^* f_1 + (n^* - n_1^*) f_2$$

(ii) Equation (5-68):

$$\frac{\partial n^*}{\partial t^*} = -\frac{\partial}{\partial x_j^*} (n_1^* V_1^{*j} + (n^* - n_1^*) V_2^{*j})$$

The dimensional terms are identified in the following tabulation:

$$n_1^* = n_1/n_{\text{ref}} \qquad V_1^* = v_1/V_{\text{ref}}$$

$$n^* = n/n_{\text{ref}} \qquad V_2^* = v_2/V_{\text{ref}}$$

$$f_1 = \Gamma_1/\Gamma_{\text{ref}} \qquad f_2 = \Gamma_2/\Gamma_{\text{ref}}$$

$$x^* = x/x_{\text{ref}} \qquad t^* = t/t_{\text{ref}}$$

with

$$t_{\text{ref}} = x_{\text{ref}}/V_{\text{ref}} \qquad \Gamma_{\text{ref}} = 1/t_{\text{ref}}$$

The momentum balance equations in dimensionless form are as follows:
(i) Equation (5-86):

$$\frac{\partial n_1^* V_1^{*i}}{\partial t^*} = -\frac{\partial n_1^* V_1^{*j} V_1^{*i}}{\partial x_j^*} + \mathrm{Pf}\frac{n_1^*}{m_1^*}\frac{\partial \Phi^*}{\partial x_i^*} - \frac{1}{yM^2}\frac{\partial}{\partial x_i^*} n_1^* R_1^* T_1^* + \frac{\mu_1^*}{\mathrm{Re}\cdot m_1^*}\frac{\partial^2 V_1^{*i}}{\partial x_j^{*2}} - n_1^* V_1^{*i} f_3$$

(ii) Equation (5-87):

$$\frac{\partial (n^* - n_1) V_2^{*i}}{\partial t^*} = \frac{\partial}{\partial x_j^*}(n^* - n_1^*) V_2^{*i} V_2^{*j} + \mathrm{Pf}\frac{(n^* - n_1^*)}{m_2^*}\frac{\partial \Phi^*}{\partial x_i^*} - \frac{1}{yM^2}\frac{\partial}{\partial x_i^*}(n^* - n_1^*) R_2^* T_2^* + \frac{\mu_2^*}{\mathrm{Re}\cdot m_2^*}\frac{\partial^2 V_2^{*i}}{\partial x_j^{*2}} - (n^* - n_1^*) V_2^{*i} f_4$$

The dimensionless terms and parameters are identified in the following tabulation:

$$\Phi^* = \Phi/\Phi_{\mathrm{ref}}$$
$$m_1^* = m_1/m_{\mathrm{ref}}, \qquad m_2^* = m_2/m_{\mathrm{ref}}$$
$$R_1 = k_B/m_1, \qquad R_2 = k_B/m_2$$
$$R_{\mathrm{ref}} = k_B/m_{\mathrm{ref}}$$
$$R_1^* = R_1/R_{\mathrm{ref}} = \frac{1}{m_1^*}, \qquad R_2^* = R_2/R_{\mathrm{ref}} = \frac{1}{m_2^*}$$
$$f_3 = \Gamma_3/\Gamma_{\mathrm{ref}}, \qquad f_4 = \Gamma_4/\Gamma_{\mathrm{ref}}$$
$$\mu_1^* = \mu_1/\mu_{\mathrm{ref}}, \qquad \mu_2^* = \mu_2/\mu_{\mathrm{ref}}$$

with

$$\mathrm{Pf} = e\Phi_{\mathrm{ref}}/m_{\mathrm{ref}}V_{\mathrm{ref}}^2, \qquad M = V_{\mathrm{ref}}/V_a$$
$$V_a = (\tfrac{5}{3}R_{\mathrm{ref}}T_{\mathrm{ref}})^{1/2}, \qquad \gamma = \tfrac{5}{3}$$
$$\mathrm{Re} = x_{\mathrm{ref}}V_{\mathrm{ref}}n_{\mathrm{ref}}m_{\mathrm{ref}}/\mu_{\mathrm{ref}}$$

The energy balance equations in dimensionless form are as follows:
(i) Equations (5-93):

$$\frac{\partial n_1^* T_1^*}{\partial t^*} = -\frac{\partial n_1^* T_1^* V_1^{*j}}{\partial x_j^*} - (\gamma - 1) n_1^* T_1^* \frac{\partial V_1^{*j}}{\partial x_j^*} + \frac{1}{\mathrm{Re}\cdot\mathrm{Pr}}\frac{1}{m_1^* C_{v_1}^*} \times \frac{\partial}{\partial x_j^*}\left(\kappa_1^*\frac{\partial T_1^*}{\partial x_j^*}\right) - n_1^* T_1^* f_5 + (n^* - n_1^*) T_2^* f_6 + \gamma(\gamma - 1) M^2 \frac{V_1^{*j} V_1^{*j}}{2C_{v_1}^*}[2n_1^* f_3 - n_1^* f_1 + (n^* - n_1^*) f_2]$$

(ii) Equation (5-94):

$$\frac{\partial (n^* - n_1^*)T_2^*}{\partial t^*} = -\frac{\partial (n^* - n_1^*)T_2 V_2^{*j}}{\partial x_j^*} - (\gamma - 1)(n^* - n_1^*)T_2^* \frac{\partial V_2^{*j}}{\partial x_j^*}$$
$$+ \frac{1}{\mathrm{Re}\cdot\mathrm{Pr}}\frac{1}{m_2^* C_{v_2}^*}\frac{\partial}{\partial x_j^*}\left(\kappa_2^* \frac{\partial T_2^*}{\partial x_j^*}\right)\gamma(\gamma - 1)M^2 \frac{V_2^{*j}V_2^{*j}}{2C_{v_2}^*}$$
$$\times [2(n^* - n_1^*)f_4 + n_1^* f_1 - (n^* - n_1^*)f_2]$$
$$- (n^* - n_1^*)T_2^* f_7 + n_1^* T_1^* f_8$$

The dimensionless terms and parameters are identified in the following tabulation:

$$f_5 = \Gamma_5/\Gamma_{\mathrm{ref}}, \qquad f_6 = \Gamma_6/\Gamma_{\mathrm{ref}}$$
$$f_7 = \Gamma_7/\Gamma_{\mathrm{ref}}, \qquad f_8 = \Gamma_8/\Gamma_{\mathrm{ref}}$$
$$\kappa_1^* = \kappa_1/\kappa_{\mathrm{ref}}, \qquad \kappa_2^* = \kappa_2/\kappa_{\mathrm{ref}}$$
$$C_{v1} = \tfrac{3}{2}R_1, \qquad C_{v2} = \tfrac{3}{2}R_2, \qquad C_{v_{\mathrm{ref}}} = \tfrac{3}{2}R_{\mathrm{ref}}$$
$$C_{v_1}^* = C_{v_1}/C_{v_{\mathrm{ref}}}, \qquad C_{v_2}^* = C_{v_2}/C_{v_{\mathrm{ref}}}$$
$$P_r = C_{v_{\mathrm{ref}}}\mu_{\mathrm{ref}}/\kappa_{\mathrm{ref}}$$

Poisson's equation in dimensionlesss form is as follows:
(i) Equation (5-64):

$$\frac{\partial^2 \Phi}{\partial x_j^{*2}} = S_n[n_1^* + (n^* - n_1^*) - N_0^*]$$

The dimensionless terms and parameters are

$$S_n = \frac{x_{\mathrm{ref}}^2 e n_{\mathrm{ref}}}{\Phi_{\mathrm{ref}}\epsilon}, \qquad N_0^* = \frac{N_0}{n_{\mathrm{ref}}}$$

The boundary conditions in dimensionless form are as follows:
(i) Equation (5-95):

$$\frac{\partial n^*}{\partial x^*} = \frac{\partial n_1^*}{\partial x^*} = \frac{\partial V_1^*}{\partial x^*} = \frac{\partial V_2^*}{\partial x^*} = \frac{\partial T_1^*}{\partial x^*} = \frac{\partial T_2^*}{\partial x^*} = 0$$

at

$$x^* = 0, \qquad \Phi^* = 0$$
$$x^* = L, \qquad \Phi^* = \Phi_a$$

(ii) Equation (5-103):

$$\frac{\partial^2 n^*}{\partial x^{*2}} = \frac{\partial^2 n_1^*}{\partial x^{*2}} = 0, \qquad V_1^* = \mu_c^* \frac{\partial \Phi^*}{\partial x^*}, \qquad \mu_c^* = \frac{\mu_c \Phi_{\mathrm{ref}}}{x_{\mathrm{ref}} V_{\mathrm{ref}}}$$

TABLE 5-A1. Parameters

	Dimensionless reference quantities Figure number				
Parameters	1–4	9–19	21–27	28–36	37–53
Device length (μm)	1.0	1.0	5.0	0.25	1.0
x_{ref} (μm)	1.0	1.0	5.0	0.25	0.75
n_{ref} (cm^{-3})	5.0×10^{15}	5.0×10^{15}	5.0×10^{15}	5.0×10^{15}	1.0×10^{17}
t_{ref} (ps)	1.0	1.0	5.0	0.25	0.75
Γ_{ref} (10^{12}/s)	1.0	1.0	0.2	4.0	1.33
κ_{ref} (J/K cm s)	2.0×10^{-6}	2.0×10^{-6}	2.0×10^{-6}	2.0×10^{-6}	4.14×10^{-5}
μ_{ref} (gm/cm s)	5.74×10^{-11}	5.7×10^{-11}	5.7×10^{-11}	5.7×10^{-11}	1.15×10^{-9}
Re	53.17	53.17	265.86	13.29	39.88
Sn	8.16	8.16	204.00	0.51	91.84
Φ^*	0.6 1.0 2.0	0.6 1.0 2.0	0.5 1.0 1.5 2.0 3.0	0.15 0.25 0.50	1.0
T_c^*	—	1.0	4.0	1.0	1.0
μ_c^*	—	1.56	0.08	6.25	—

Common parameters		
$V_{ref} = 10^8$ cm/s	$m_1^* = 1$	$\kappa_1^* = 1.0$
$\Phi_{ref} = 1.0$ V	$m_2^* = 3.1$	$\kappa_2^* = 1.0$
$m_{ref} = 6.10 \times 10^{-29}$ g	$R_1^* = 1.0$	Pr = 9.40
$k_B = 1.38 \times 10^{-23}$ J/K	$R_2^* = 1.0$	Pf = 2.62
$R_{ref} = 2.26 \times 10^5$ J/K g	$\mu_1^* = 1.0$	$M = 2.97$
$T = 300$ K	$\mu_2^* = 1.0$	
$V_a = 3.36 \times 10^7$ cm/s		

(iii) Equation (5-104):

$$\frac{\partial^2 n^*}{\partial x^{*2}} = \frac{\partial n_1^*}{\partial x^{*2}} = \frac{\partial^2 V_1^*}{\partial x^{*2}} = \frac{\partial^2 V_2^*}{\partial x^{*2}} = \frac{\partial^2 T_1^*}{\partial x^{*2}} = \frac{\partial^2 T_2^*}{\partial x^{*2}} = 0, \qquad \Phi^* = 0$$

(iv) Equation (5-116):

$$n^* = N_0^*, \qquad n_1^* = \frac{f_2}{f_1 + f_2} N_0^*, \qquad \frac{\partial V_1^*}{\partial x^*} = \frac{\partial V_2^*}{\partial x^*} = 0$$

$$T_1^* = T_c^*, \qquad \frac{\partial T_2^*}{\partial x^*} = 0, \qquad \Phi^* = 0$$

where f_1 and f_2 are evaluated at $T = T_{ref}$. Common parameters and dimensionless reference quantities are given in Table 5-A1.

6

Superconducting Junctions

6.1. INTRODUCTION

Although there are a few superconducting structures that exhibit S or NNDC characteristics, one class of devices, Josephson junctions, which are *not* readily classified as NDC elements, (1) have a potentially important role in the future technology; (2) exhibit $I(\Phi)$ curves that are readily analyzable in the mode of this text, so that, for completeness, we will include superconducting phenomena in our treatment and describe Josephson Junctions as pseudo-NNDC-type switching elements.

Superconducting electronics primarily involve fabrication of Josephson junctions, either the important tunnel junction or the "weak-link" junction. The tunnel junction requires a thin insulating barrier between superconductors and is well developed theoretically. Weak-link junctions have the superconductors separated by a bridge of normal metal, semiconductor or small area superconductor. The weak-link junctions behave in a manner that depends critically on the geometry and type of material employed. Their understanding is less well developed than the tunnel junction; we will emphasize the tunnel junction in this chapter. But bear in mind that the Josephson effect (Josephson, 1962, 1965) is a general property of *any* two weakly-coupled superconductors.

Before discussing the phenomenon of superconductivity it is important to realize that the fabrication of simple tunnel junctions has challenging aspects. Uniform insulating barriers 20–50 Å thick must be made. Further, we shall see that capacitance must be minimized, areas must be kept small and thermal problems avoided (see Chap. 8). Material problems also abound; the most successful junctions are presently made from Pb–In alloys such as $Pb_{0.84}In_{0.12}Au_{0.04}$ and $Pb_{0.71}Bi_{0.29}$, Nb and Nb–Pb alloys (Beasley and Kircher, 1981).

In order to understand how superconducting tunnel junction devices operate, we must first understand tunneling processes in normal semiconductors and metals. These have already been treated in Chap. 3. It then remains to first explain tunneling between either a normal metal or a semiconductor and a superconductor, and the coupling between two superconductors separated by a nonsuperconducting region. An excellent general reference for the following discussions is the text by Van Duzer and Turner (1981); we will borrow substantially from their treatment. Further, a special issue on Josephson junction devices has been published by the *IEEE Transactions on Electron Devices* (Oct. 1980, Vol. ED-27), Matisoo (1980a,b) has contributed two useful articles on Josephson logic and memory, and Beasley and Kircher (1981) have provided an excellent review article on materials and fabrication techniques. The physics and applications of the Josephson effect have been discussed by Barone and Paterno (1982) and Likharev (1986). Nonlinear effects and chaos in Josephson junctions and devices have been reviewed by Kautz and Monaco (1985) and Kurkijärvi (1991). Again, we shall emphasize the switching properties of superconducting tunnel junctions in the chapter, with the understanding that a wide spectrum of other potentially useful structures exist that are based on the coupling of superconductors put in close proximity to one another (both tunneling and weak-link systems: see Van Duzer and Turner, 1981, Chap. 5; the special issue of IEEE-ED mentioned above).

A presentation of the details of the microscopic theory of superconductivity is not what is intended in this chapter. Rather, we aim to describe some of the properties of the superconducting state that are applicable to our needs.

One major difference between a superconductor and a normal metal is that a superconductor exhibits a small temperature-dependent energy gap centered at the Fermi energy, E_F. The gap vanishes at the critical temperature T_C, above which the superconductor becomes a normal metal. The existence of the gap was predicted by the first successful microscopic theory of superconductivity, that of Bardeen, Cooper, and Schrieffer (1957), the BCS theory. It was shown by Cooper (1956) that a normal Fermi gas is unstable under conditions where the net electron–electron interaction is positive, i.e., attractive. If a positive interaction could be induced the ground state energy would be lowered. Indeed, by understanding the role played by the phonon coupling, so that the interaction between electrons is mediated by phonons, a positive interaction was shown to be present at sufficiently low temperatures and provided the "pairing" framework which proved to be the underlying mechanism for the most common form of superconductivity.

It is not easy to visualize the difference between the normal and superconducting states. However, a simple conceptual understanding can be appreciated from the following considerations (Schwartz and Frota-Passoa, 1982). In the current-carrying normal state individual electrons can be thought of as runners on a playing field moving with different velocities and having a nonzero average drift velocity. Scattering centers such as phonons and impurities are represented by rocks. A runner can hit a rock, change his momentum, and lose energy. This phenomenon represents a finite resistance to the flow of electrons. In the superconducting state the electrons form pairs, with every pair having the same net momentum. The current-carrying superconducting state is represented by a block of paired runners, with all having the same momentum. Each pair has other pairs directly in front and behind having the same momentum. If a pair runs into a rock, the pairs in close proximity prevent the stumbling pair from changing its momentum. The presence of the rocks is neutralized by the uniform cadence of pairs moving with the same velocity. This is a cooperative effect. Indeed, the essence of the superconducting state, that which produces the energy gap, is twofold: electron pairing; identical momenta for all pairs, i.e., a macroscopic occupation of a quantum state. This indicates that the wave function describing the superconducing state will be representative of pairs of electrons having the same center of mass momentum, **k**, and, therefore, has a macroscopic significance. It is usually written as $\psi_S = |\psi_S| e^{i\theta}$, where $|\psi_S|$ is the amplitude and θ the phase given by $\mathbf{k} \cdot \mathbf{r}$. The macroscopic pair density in the superconducting state is $|\psi_S|^2$. When $\mathbf{k} = 0$ (non-current-carrying state) $\psi_S = |\psi_S|$ and the phases are locked together to minimize the energy. If there is a net **k**, a gradient of phase develops.

In order to show the existence of a superconducting energy gap the ground state energy spectrum must be determined. To do this the BCS ground state at $T = 0$ K is written as the product of the occupation operators for all pair states:

$$|\psi\rangle = \prod_{\mathbf{k}} [u_{\mathbf{k}} + v_{\mathbf{k}} b_{\mathbf{k}}^*] |0\rangle \tag{6-1}$$

where **k** represents the state of a pair of electrons, one having wave vector **k** and spin up, the other having wave vector $-\mathbf{k}$ and spin down, $v_{\mathbf{k}}^2$ is the probability that a pair state is occupied, $u_{\mathbf{k}}^2 = 1 - v_{\mathbf{k}}^2$ is the probability that a pair state is empty, and $|0\rangle$ is the vacuum state. $b_{\mathbf{k}}^*$ is the pair creation operator. Using this state function we find the energy of the superconducting ground state, W, relative to the normal ground state as the expectation value of the reduced Hamiltonian (Van Duzer and Turner, 1981, Chap. 2):

$$W = 2 \sum_{\mathbf{k} > \mathbf{k}_F} \epsilon_{\mathbf{k}} v_{\mathbf{k}}^2 + 2 \sum_{\mathbf{k} < \mathbf{k}_F} |\epsilon_{\mathbf{k}}| u_{\mathbf{k}}^2 + \sum_{\mathbf{k},\mathbf{k}'} \Phi_{\mathbf{k}\mathbf{k}'} u_{\mathbf{k}} v_{\mathbf{k}} u_{\mathbf{k}'} v_{\mathbf{k}'} \tag{6-2}$$

where $\mathbf{k}_F$ is the Fermi wave vector, $\epsilon_{\mathbf{k}}$ is the Bloch state energy relative to the Fermi energy, and $\Phi_{\mathbf{k}\mathbf{k}'}$ is the scattering amplitude between states **k** and **k**′. To find the equilibrium state Eq. (6-2) is minimized with respect to $v_{\mathbf{k}}^2$ to yield for the occupation probability

$$v_{\mathbf{k}}^2 = \tfrac{1}{2}[1 - \epsilon_{\mathbf{k}}/(\Delta_{\mathbf{k}}^2 + \epsilon_{\mathbf{k}}^2)^{1/2}] \tag{6-3}$$

where the *gap parameter* $\Delta_{\mathbf{k}}$ is defined by

$$\Delta_{\mathbf{k}} = -\sum_{\mathbf{k}'} \Phi_{\mathbf{k}\mathbf{k}'} v_{\mathbf{k}'} u_{\mathbf{k}'} \tag{6-4}$$

Further, by defining

$$E_{\mathbf{k}} \equiv (\Delta_{\mathbf{k}}^2 + \epsilon_{\mathbf{k}}^2)^{1/2} \tag{6-5}$$

we can write Eq. (6-3) as

$$v_{\mathbf{k}}^2 = \tfrac{1}{2}(1 - \epsilon_{\mathbf{k}}/E_{\mathbf{k}}) \tag{6-6}$$

and Eq. (6-4) as

$$\Delta_{\mathbf{k}} = -\sum_{\mathbf{k}'} \Phi_{\mathbf{k}\mathbf{k}'} \frac{\Delta_{\mathbf{k}'}}{2E_{\mathbf{k}}} \tag{6-7}$$

In order to solve Eq. (6-7) some assumptions are now made. First, it is assumed that $\Phi_{\mathbf{kk}'} = -\Phi$, a constant, if both $|\epsilon_{\mathbf{k}}|$ and $|\epsilon_{\mathbf{k}'}|$ are less than the Debye energy $\hbar\omega_D$. If not, $\Phi_{\mathbf{kk}'}$ is taken as zero. Therefore, $\Delta_{\mathbf{k}} = \Delta(0)$, a constant, for $|\epsilon_{\mathbf{k}}| < \hbar\omega_D$ and is zero otherwise. Hence Eq. (6-5) becomes

$$E_{\mathbf{k}} = [\Delta^2(0) + \epsilon_{\mathbf{k}}^2]^{1/2} \tag{6-8}$$

With these assumptions, by converting the sums to an integral in Eq. (6-7), we obtain for $N_n(0)\Phi \ll 1$, where $N_n(0)$ is the normal density of states at E_F,

$$\Delta(0) \cong 2\hbar\omega_D e^{-1/N_n(0)\Phi} \tag{6-9}$$

This is the gap parameter at $T = 0\,\text{K}$, typically on the order of 1 meV.

If we perform a similar calculation at finite temperatures, the relation between the temperature-dependent gap parameters and temperature is found to be (see Van Duzer and Turner, 1981, Chap. 2)

$$\frac{1}{N_n(0)\Phi} = \int_{-\hbar\omega_D}^{\hbar\omega_D} \frac{\tanh\{[\epsilon^2 + \Delta^2(T)]^{1/2}/2k_BT\}}{2[\epsilon^2 + \Delta^2(T)]^{1/2}}\, d\epsilon \tag{6-10}$$

where we have assumed a constant density of states, and k_B is the Boltzmann constant.

Since the gap must vanish at T_C, Eq. (6-10) yields in the limit $N_n(0)\Phi \ll 1$

$$k_BT_C = 1.13\hbar\omega_D e^{-1/N_n(0)\Phi} \tag{6-11}$$

Comparison with Eq. (6-9) shows us that

$$2\Delta(0) = 3.52k_BT_C \tag{6-12}$$

which is a relation between the superconducting energy gap at $T = 0\,\text{K}$ and the critical temperature.

Before discussing tunneling effects in a superconductor it is important for us to understand why an energy gap appears in the superconducting density of states versus energy curve and what form the density of states curve actually has. To do this we will make use of Fig. 6-1. But first we must appreciate that the ground state, comprising electron pairs, is different from the excited states, which occur when pairs are broken. The excited states, with energies outside the range of the Fermi energy plus the gap parameter, can be either holelike or electronlike (quasiparticles), and it is knowledge of their density that is actually required. In a normal metal any excited Bloch state above E_F can be occupied by an electron and any excited Bloch state below E_F can be occupied by a hole. The excitations occupy Bloch states and their density is identical to the Bloch state density. This is not the case for a superconductor, for there can be no excited states in the forbidden range centered at the Fermi energy, as shown in Fig. 6-1b. In the superconductor, as in the normal metal, for every value of k we have a corresponding excitation energy E given by Eq. (6-8). Denoting the Bloch state

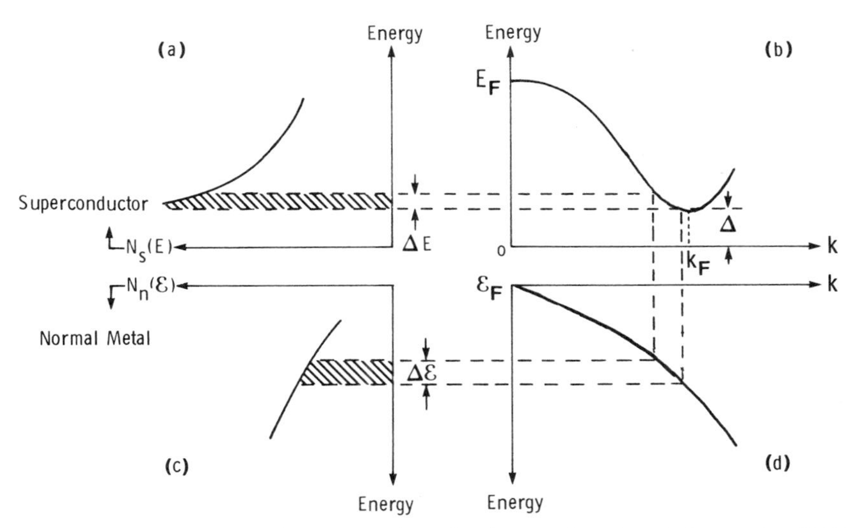

FIGURE 6-1. Diagram showing by construction why the density of excitation states $N_s(E)$ has a singularity at $E = \Delta$. The two cross-hatched zones have the same area—for each Bloch state, there is an excitation state. (a) and (b) represent the superconducting density of states $N_s(E)$ and dispersion relations $E(k)$, respectively, and (c) and (d) the same $N_n(\epsilon)$ and $\epsilon(k)$ for a normal metal. (After Van Duzer and Turner, 1981.)

energies by ϵ, states in the range dk will fall in a range of excitation energies dE and in a *corresponding* range of Bloch state energies $d\epsilon$. The number of states in dE will be by definition equal to the number in $d\epsilon$. Now consider the range of Bloch state energies $\Delta\epsilon$ shown in Figs. 6-1c and 6-1d. There will be a corresponding number of states shown cross-hatched in Fig. 6-1c. We then project this range of ϵ onto the $\epsilon(k)$ curve for the normal metal (Fig. 6-1d). Next, projecting up to Fig. 6-1b, we obtain the smaller range of E for the superconductor. Finally, projecting to Fig. 6-1a, we see that in order to preserve the same number of states as in the normal state, we require a larger density of superconducting excitation states $N_S(E)$. Clearly, as k approaches k_F, $N_S(E)$ approaches infinity. Now, since

$$N_S(E)\,dE = N_n(\epsilon)\,d\epsilon \tag{6-13}$$

then

$$N_S(E) = N_n(\epsilon)\,d\epsilon/dE \tag{6-14}$$

Equation (6-8) then yields

$$N_S(E) = \frac{N_n(\epsilon)E}{[E^2 - \Delta^2(0)]^{1/2}} \tag{6-15}$$

But, since $N_n(\epsilon)$ is slowly varying and almost equal to $N_n(0)$ over the energy

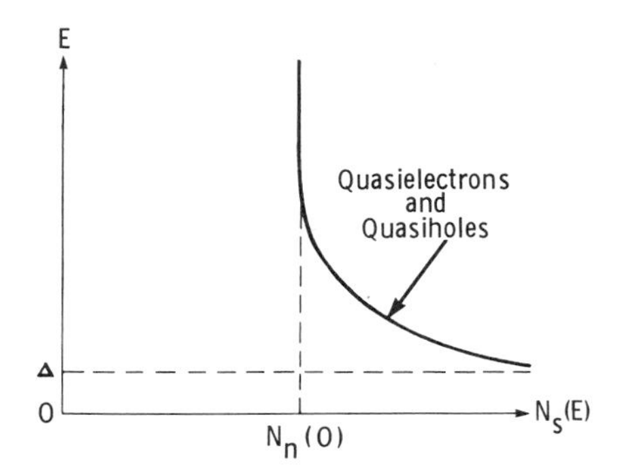

FIGURE 6-2. Density of states diagram for a superconductor. An expanded scale is used for the region around k_F and the density of Bloch states is assumed to be constant over the very small range of excitation energies considered. (After Van Duzer and Turner, 1981.)

range of interest (typically less than 0.1 eV) we can write

$$N_S(E) \cong \frac{N_n(0)E}{[E^2 - \Delta^2(0)]^{1/2}} \tag{6-16}$$

which we show in Fig. 6-2. No states exist in the range $0 \leq E < \Delta$, and for E larger than a few times Δ, $N_S(E) \cong N_n(\epsilon)$. It is convenient to draw Fig. 6-2 in "semiconductorlike" fashion, and this is done in Fig. 6-3. Here the excitation energy is increased positively outward from the gap center. As we have shown in Chap. 3, it is this type of diagram that is useful for understanding tunneling calculations.

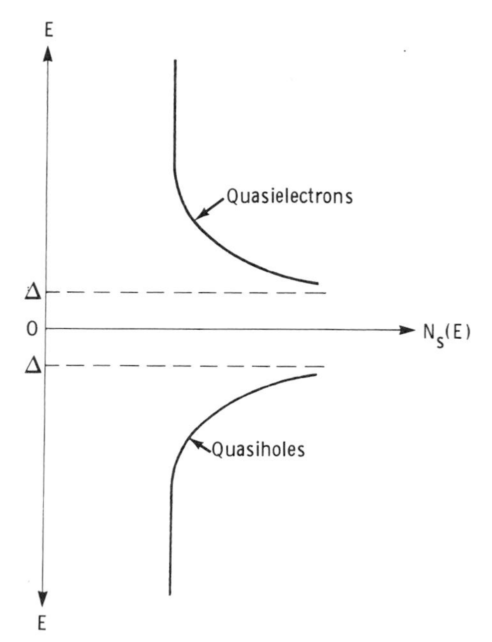

FIGURE 6-3. Density of states versus excitation energy in the "semiconductorlike" representation. In this diagram, excitation energy is measured positively in both directions. (After Van Duzer and Turner, 1981.)

6.2. QUASIPARTICLE TUNNELING

In Chap. 3 we discussed the tunnel diode and the analytical techniques employed to predict its $I(\Phi)$ characteristics. Here we apply this understanding to the density of states structure of a superconductor and treat both normal–insulator–superconductor (NIS) and superconductor–insulator–superconductor (SIS) tunneling events that occur via the motion of excitations through a thin insulating region. For the superconducting density of states we use Eq. (6-15) for $|E| \geq \Delta$ and zero for $|E| < \Delta$. Figure 6-4 shows the NIS structure at $T = 0$, where we have used the approximation that all pair states are occupied below k_F and empty above k_F (Van Duzer and Turner, 1981, Chap. 2). The shaded areas are those energies available for hole occupancy; unshaded areas can accommodate electrons. When $T > 0$ a "Boltzmann Tail" develops at E_F. In the normal metal electron excitations appears above E_F and hole excitations below E_F. In the superconductor quasielectron excitations appear above Δ in the unshaded region and quasihole excitations appear below Δ in the shaded region. That is, some of the states both above and below k_F become partially occupied by excitations. To calculate the tunneling current we introduce the dummy variable E' shown in Fig. 6-4. Following the same argument developed in Chap. 3 for the tunnel diode, we can immediately write down the tunneling current. For tunneling transitions from the normal metal to the superconductor we have

$$I_{\text{sn}}(T = 0) = \frac{2\pi e A}{\hbar} \int_{\Delta}^{e\Phi} |T|^2 N_n(E' - e\Phi) N_S(E')\, dE' \tag{6-17}$$

for the oppositely directed case $I_{ns} = 0$.

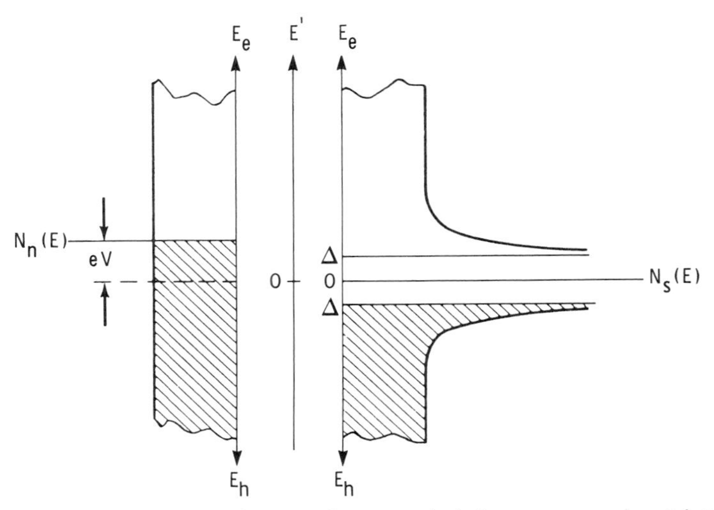

FIGURE 6-4. Diagram for determining tunneling currents between a normal metal and a superconductor for $T = 0$ K. In this kind of diagram, all the states above the gap are assumed empty and those below, full, for $T = 0$ K. (After Van Duzer and Turner, 1981.)

Using Eq. (6-15) we obtain

$$I_{sn}(T=0) = \frac{G_n}{e}\int_{\Delta}^{e\Phi} \frac{E'\,dE'}{(E'^2-\Delta^2)^{1/2}} = \begin{cases} (G_n/e)[(e\Phi)^2-\Delta^2]^{1/2} & \text{for } e\Phi \geq \Delta \\ 0, & e\Phi < \Delta \end{cases} \tag{6-18}$$

where G_n is the normal tunneling conductance (NIN) given by

$$G_n = (2\pi e^2 A/\hbar)\,|T|^2\,N_l(0)N_r(0) \tag{6-19}$$

where l and r denote left and right, respectively. When $T > 0$ the tunneling integral becomes

$$I_{sn} = \frac{G_n}{e}\int_{-\infty}^{\infty} \frac{E'}{(E'^2-\Delta^2)^{1/2}}[f(E'-e\Phi)-f(E')]\,dE' \tag{6-20}$$

where the Fermi function, f, takes into account the presence of excitations. Figure 6-5 shows the resulting $I(\Phi)$ characteristics. Note from Eq. (6-18) that

$$\left(\frac{dI_{sn}}{d\Phi}\right)_{T=0} = \begin{cases} G_n\,|\Phi|/[\Phi^2-(\Delta/e)^2], & |e\Phi| \geq \Delta \\ 0, & |e\Phi| < \Delta \end{cases} \tag{6-21}$$

A measurement of the conductance as $T \to 0$ yields the form of the superconducting density of states [Eq. (6-15)].

The situation for quasiparticle tunneling in the SIS structure for $T > 0$ is shown in Fig. 6-6. Here we have two types of processes that must be considered: a pair can break, creating a local hole excitation while injecting an electron excitation across the insulator; an excited electron can tunnel across the insulator. The right-to-left tunneling current is given by

$$I_{ss} = \frac{G_n}{e}\int_{-\infty}^{\infty} \frac{|E'-e\Phi|}{[(e\Phi-E')^2-\Delta_1^2]^{1/2}}\frac{|E'|}{[E'^2-\Delta_2^2]^{1/2}} \times [f(E'-e\Phi)-f(E')]\,dE' \tag{6-22}$$

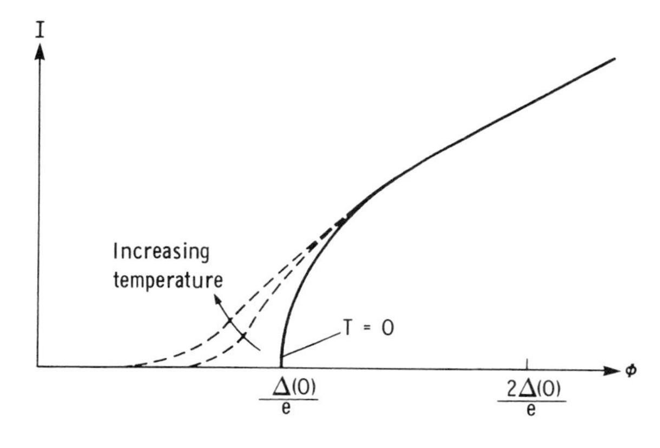

FIGURE 6-5. Quasiparticle tunneling current between a normal metal (left) and a superconductor (right) versus voltage Φ for various temperatures. Note that as the temperature increases both the excitation occupancy and the gap change. Here, as elsewhere, $2\Delta(0)$ is the energy gap at $T = 0$ K. (After Van Duzer and Turner, 1981.)

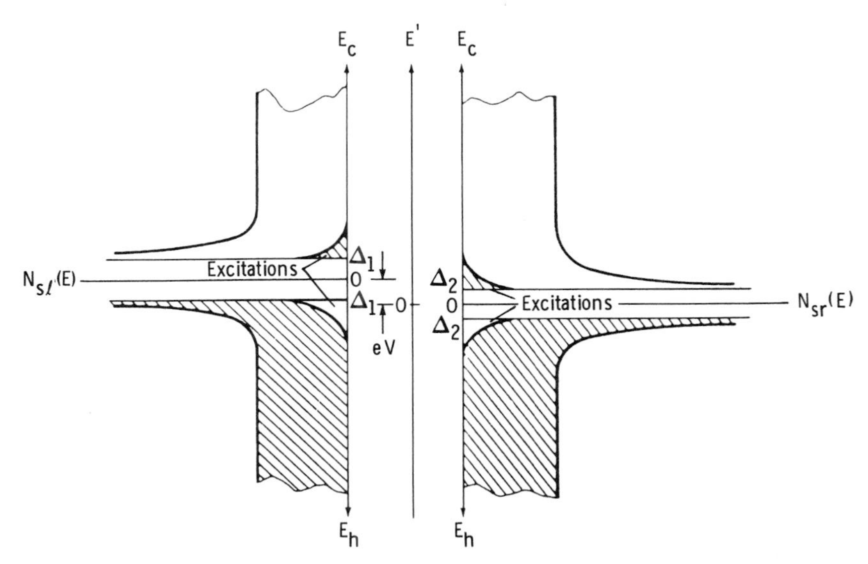

FIGURE 6-6. Diagram for determination of tunneling currents between two different superconductors with $T \neq 0$ K and an applied voltage Φ (after Van Duzer and Turner, 1981).

At $T = 0$ this has the form shown in Fig. 6-7. The more interesting situation from our viewpoint, however, is the $T > 0$ case shown in Fig. 6-8. At low temperature we observe a region of NNDC between $|\Delta_1 - \Delta_2|/e$ and $(\Delta_1 + \Delta_2)/e$. The reason for this can be readily seen from Fig. 6-6. At low voltage, excited electrons on the left see a high density of states to tunnel into. As the voltage is increased, the density of states decreases and so does the tunneling current, as in the tunnel diode when the semiconducting band gap is reached. When the voltage reaches $(\Delta_1 + \Delta_2)/e$ pairs on the left break and the tunneling current starts to increase again. Hence, an SIS tunnel junction provides a real NNDC region, as does the tunnel diode. Experimentally, studies of NNDC regions in superconducting junctions were made by Shapiro et al. (1962). Good quantitative agreement was obtained using a simple one-dimensional model based on the BCS theory.

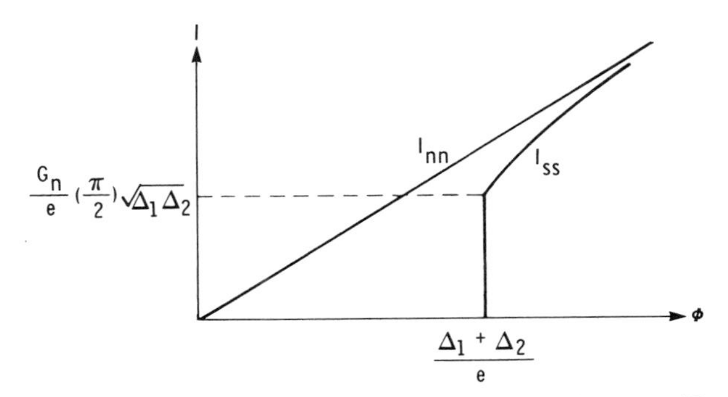

FIGURE 6-7. Quasiparticle tunneling current I_{ss} between superconductors at $T = 0$ K compared with the tunneling current I_{nn} between the same materials in the normal state versus voltage Φ (after Van Duzer and Turner, 1981).

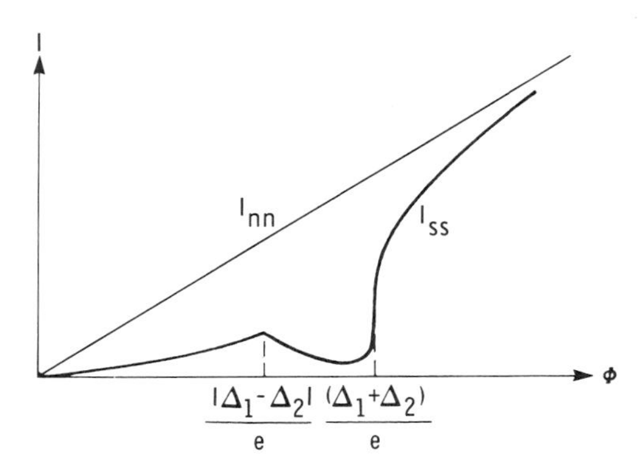

FIGURE 6-8. Quasiparticle tunneling current I_{ss} between superconductors for $T \neq 0$ K (after Van Duzer and Turner, 1981).

6.3. PAIR TUNNELING—THE JOSEPHSON EFFECT

Besides those tunneling events due to excitations, Josephson (1962) showed that pair tunneling could also make a contribution to the current (Anderson and Rowell, 1963). Here, however, current would flow with no voltage drop across the junction. If we assume that pairs can tunnel through the insulator, then we expect that the wave functions on either side of the junction might be coupled in some manner, even though we also expect that their phases are different. Feynman et al. (1965) present an argument based on this coupling that leads to the well-known Josephson relations. The approach is to employ a pair of coupled Schrödinger equations, with each expressing the temporal evolution of the wave function on a particular side of the junction:

$$
\begin{aligned}
i\hbar \frac{\partial \psi_1}{\partial t} &= E_1 \psi_1 + K\psi_2 \\
i\hbar \frac{\partial \psi_2}{\partial t} &= E_2 \psi_2 + K\psi_1
\end{aligned}
\tag{6-23}
$$

Here E_1 and E_2 are the pair energies. In the absence of pair coupling across the junction the small coupling parameter, K, vanishes. When the junction is sufficiently thin such that pair coupling occurs, K is treated as a small perturbing term.

If we apply a voltage source, Φ, across the junction so that $E_2 - E_1 = 2e\Phi$ and take the zero of energy midway between them, we can write Eq. (6-23) as

$$
\begin{aligned}
i\hbar \frac{\partial \psi_1}{\partial t} &= -e\Phi \psi_1 + K\psi_2 \\
i\hbar \frac{\partial \psi_2}{\partial t} &= e\Phi \psi_2 + K\psi_1
\end{aligned}
\tag{6-24}
$$

We then write the pair wave function in terms of an amplitude and a phase, where the phase, θ, is the same for all pairs and, for normalization requirements, the amplitude is the square root of the pair density:

$$\psi_k = \sqrt{n_{sk}}e^{i\theta_k} \tag{6-25}$$

where k is 1 or 2. Noting that $e^{i\theta_k} = \cos\theta_k + i\sin\theta_k$, Eqs. (6-24) and (6-25) yield

$$\frac{\partial n_{s1}}{\partial t} = -\frac{2}{\hbar}K(n_{s1}n_{s2})^{1/2}\sin\Delta\theta \tag{6-26}$$

$$\frac{\partial n_{s2}}{\partial t} = +\frac{2}{\hbar}K(n_{s1}n_{s2})^{1/2}\sin\Delta\theta \tag{6-27}$$

$$\frac{\partial \theta_1}{\partial t} = -\frac{K}{\hbar}\left(\frac{n_{s2}}{n_{s1}}\right)^{1/2}\cos\Delta\theta + \frac{e\Phi}{\hbar} \tag{6-28}$$

$$\frac{\partial \theta_2}{\partial t} = -\frac{K}{\hbar}\left(\frac{n_{s1}}{n_{s2}}\right)^{1/2}\cos\Delta\theta - \frac{e\Phi}{\hbar} \tag{6-29}$$

where the phase difference across the junction $\Delta\theta = \theta_1 - \theta_2$. For two superconducting electrodes separated by a distance l the current density flowing from electrode number one to electrode number two is

$$J = -l\frac{\partial}{\partial t}(2en_{s1}) = l\frac{2}{\partial t}(2en_{s2}) \tag{6-30}$$

where we note that the rate of decrease of pair density n_s in one superconductor is the negative of that in the other. Hence, Eqs. (6-26), (6-27), and (6-30) yield

$$J = J_c\sin\Delta\theta \tag{6-31}$$

where J_c, the *critical current density* (the maximum current that can be carried solely by electron pairs) is given by

$$J_c = \frac{4elK}{\hbar}(n_{s1}n_{s2})^{1/2} \tag{6-32}$$

Since K is not known, J_c must be obtained from a microscopic theory, and this has been done by Ambegaokar and Baratoff (1963). Equation (6-31) is the first Josephson relation; the second is obtained by subtracting Eq. (6-29) from (6-28):

$$\frac{\partial\Delta\theta}{\partial t} = \frac{2e}{\hbar}\Phi \tag{6-33}$$

The Josephson current must be added to the quasiparticle tunneling current, so at $T = 0\,\text{K}$ the complete $I(\Phi)$ characteristic for identical superconductors is

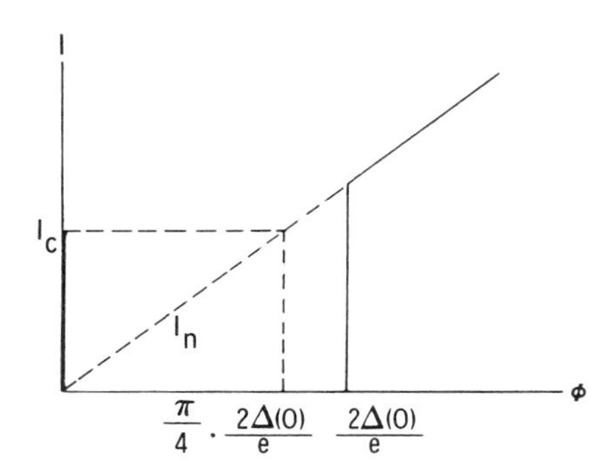

FIGURE 6-9. $I - \Phi$ characteristic for a Josephson junction at $T = 0$ K with pair tunneling. The maximum zero-voltage current is equal to the normal-state current at $\pi/4$ of the quasiparticle tunneling gap. (After Van Duzer and Turner, 1981.)

predicted to be as shown in Fig. 6-9. Note that pair current flows with no voltage drop across the junction. Indeed, for $\Phi = 0$, Eq. (6-33) yields $\Delta\theta = \Delta\theta_0 =$ const, and $J = J_c \sin \Delta\theta_0$ by (6-31). This is the dc Josephson effect.

If we drive the junction with a constant current source we see from Fig. 6-9 that once $I_c (= J_c A$, where A is the area of the junction) is exceeded, a voltage equal to $2\Delta(0)/e$ will appear across the junction for $\Phi \neq 0$. For a time-independent $\Phi \neq 0$, Eq. (6-33) can be integrated to give

$$\Delta\theta = \Delta\theta_0 + (2e/\hbar)\Phi t \tag{6-34}$$

and Eq. (6-31) yields

$$I = I_c \sin(\omega_J t + \Delta\theta_0) \tag{6-35}$$

where $\omega_j = 2e\Phi/\hbar$ and ac Josephson current flows at a frequency

$$f_J = \left(\frac{1}{2\pi}\right)\frac{2e}{\hbar}\Phi \tag{6-36}$$

Direct dc to ac conversion occurs. Since the coefficient of Φ in Eq. (6-36) equals 483.6×10^{12} Hz V^{-1}, the ac is produced at extremely high frequencies.

Figure 6-9 shows the effective $I_c(\Phi)$ curve; do not confuse the subscript c here with that for the critical Josephson current for a $T = 0$ K superconducting tunnel junction admitting a Josephson component in the absence of reactive components. Note that there is no current for $0 < \Phi < 2\Delta/e$; in our modeling we have always assumed that voltage or current gaps did not exist. Indeed, for $T > 0$ K and when realistic reactive components and effects are considered, representative $I_c(\Phi)$ curves do not have gaps. Further, as shown in Fig. 6-8, NNDC can exist for quasiparticle tunneling between different superconductors (Shapiro et al., 1962). It is also of significance that the low resistance state has essentially zero resistance and that a region of NDC does not appear for identical superconductors (although some circuit-controlled situations can be envisaged that may result in NNDC characteristics); that is why we use the term "pseudo-NNDC elements" to describe a Josephson junction.

When a magnetic field, **H**, is applied to a nonsuperconducting NNDC element it is often the case that the $I_c(\Phi)$ curve is affected in some manner other than that associated with simple magnetoresistance. For example, the $I_c(\Phi)$ curve for a transferred electron device (Chap. 5) with low boundary fields loses its NDC region entirely for sufficiently high fields (10–20 kG) (Shaw et al., 1979). The Josephson $I_c(\Phi)$ curve is perhaps the most sensitive to the effects of a magnetic field. Here the critical current, I_c, is a strong function of **H**, even at weak fields as low as a fraction of a Gauss. To understand how this occurs, consider a Josephson tunnel junction parallel to the yz plane, with Josephson (pair) and quasiparticle currents allowed to flow in the x direction. Apply a magnetic induction **B** in the y direction ($\mathbf{B} = \mu\mathbf{H}$, where μ is the permeability); the insulator is centered at $x = 0$. Neglect the magnetic field induced by currents in the x direction. The Meissner effect (Van Duzer and Turner, 1981, Chap. 1) only allows **B** to penetrate a short distance, λ, into the superconductors, which we will assume identical; $B = B_y^0 e^{-\lambda x}$, where the superscript denotes $x = 0$. Figure 6-10 shows the system we are treating (a) along with the distribution of shielding current J_{sh} required to exclude the flux from the bulk of the superconductor (b). In the presence of a magnetic field the phase difference between points Q_2 and Q_1 or P_2 and P_1 is given by

$$\Delta\theta = \theta_2 - \theta_1 + \frac{2e}{\hbar}\int_1^2 \mathbf{A}(x, t)\cdot d\mathbf{l} \tag{6-37}$$

where $\mathbf{B} = \operatorname{curl}\mathbf{A}$ (Van Duzer and Turner, 1981, Chap. 4). Choosing the London gauge, where $\nabla\theta$ is zero within each superconductor, we have for the difference of the phase differences at points P and Q

$$\Delta\theta(P) - \Delta\theta(Q) = \frac{2e}{\hbar}\left[\int_{P_1}^{P_2} \mathbf{A}(P, t)\cdot d\mathbf{l} - \int_{Q_1}^{Q_2} \mathbf{A}(Q, t)\cdot d\mathbf{l}\right] \tag{6-38}$$

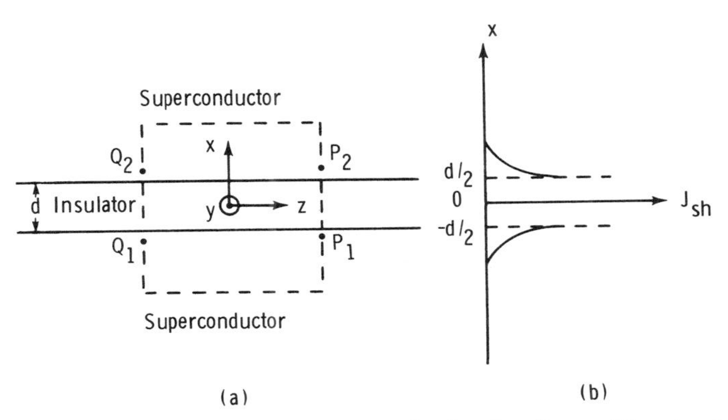

FIGURE 6-10. (a) Integration path (dotted) to relate flux in a junction and the phase difference along the junction. (b) Current density in the bounding superconductors. (After Van Duzer and Turner, 1981.)

But if the dotted area in Fig. 6-10 is chosen such that the horizontal path is sufficiently deep inside the superconductor, then the bracket on the right of Eq. (6-38) is just the total outward-directed magnetic flux, ϕ_y, contained inside the area, i.e.,

$$\Delta\theta(P) - \Delta\theta(Q) = \frac{2e}{\hbar}\phi_y \tag{6-39}$$

For an area of differential length dz, Eq. (6-39) becomes

$$\frac{\partial \Delta\theta}{\partial z} = \frac{2ed'}{\hbar} B_y^0 \tag{6-40}$$

where $d' = 2\lambda + d$. Integrating, we obtain

$$\Delta\theta(z) = \left(\frac{2ed'}{\hbar}\right) B_y^0 z + \Delta\theta(0) \tag{6-41}$$

Since $\Delta\theta$ is independent of y, we can first integrate Eq. (6-31) over y to obtain

$$J(z) = J_c(z)\sin\Delta\theta(z) \tag{6-42}$$

where $J_c(z) = \int J_c(y, z)\,dy$. Therefore

$$J(z) = J_c(z)\sin\left[\frac{2ed'}{\hbar} z + \Delta\theta(0)\right] \tag{6-43}$$

Integration over z for a rectangular junction of width L and depth W yields

$$I(B^0) = \left(\frac{kWJ_c}{ed'B^0}\right)\sin\Delta\theta(0)\sin\left(\frac{ed'B^0L}{\hbar}\right) \tag{6-44}$$

The Josephson current is a maximum when $\Delta\theta(0) = \pm\pi/2$, hence

$$I_c(B^0) = I_c(0)\left|\frac{\sin(e\phi'/\hbar)}{(e\phi'/\hbar)}\right| \tag{6-45}$$

where $I_c(0) = WLJ_c$ and the total magnetic flux through the junction $\phi = d'LB^0$. Figure 6-11 is a plot of the critical current as a function of magnetic flux. It is important to note that $\pi\hbar/e$ is the magnetic flux quantum, equal to 2.07×10^{-15} Wb. The reason that the $I_c(\Phi)$ curve has the form shown is that when an integer number of flux quanta are present in the junction the total Josephson current is reduced to zero since the spatial phase differences induced produce as much upward-flowing as downward-flowing currents; currents circulate in the junction due to the presence of n quantized fluxoids. As shown in Fig. 6-12, direct

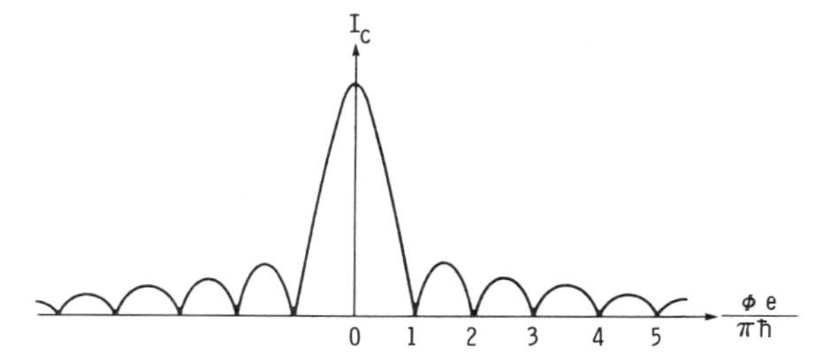

FIGURE 6-11. Dependence of the maximum zero-voltage current I_c in a junction with a uniform current density in a rectangular box as a function of the magnetic flux ϕ linking the junction. (Note that here we denote the magnetic flux as ϕ in deference to common usage. Do not confuse it with voltage Φ). (After Van Duzer and Turner, 1981.)

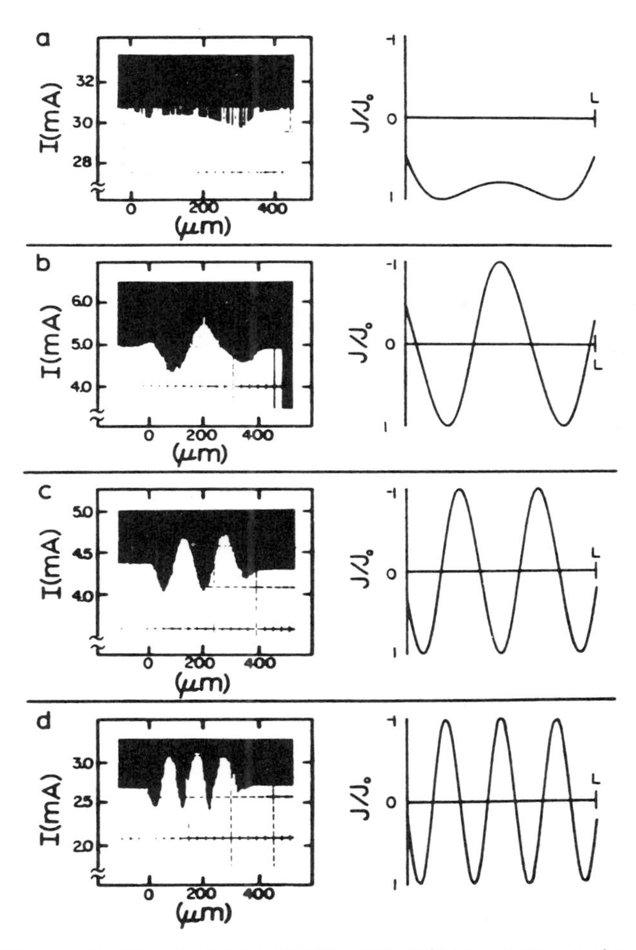

FIGURE 6-12. Figures on the left show maximum zero-voltage current as the beam is scanned across the length of the Pb–PbO_x–Pb junction at various magnetic fields. (a) $I_0 = 31.0$ mA, $B = 0$ G; (b) $I_0 = 5.0$ mA, $B = 1.6$ G; (c) $I_0 = 4.4$ mA, $B = 2.4$ G (d) $I_0 = 2.7$ mA, $B = 3.4$ G. The maximum zero-voltage current in the absence of the beam is visible at the edges of the photographs where the beam is not on the junction. (The junction extends from 0 to 370 μm.) Spikes in the figures are caused by temporary interruptions of the laser beam by bubbles in the liquid nitrogen. These could be partially eliminated by slowly pumping on the nitrogen bath. Figures on the right show the theoretical current distribution. (After Scheuermann et al., 1983.)

probing of such current distributions has been demonstrated (Scheuerman et al., 1983). They used a focused laser beam a few micrometers in diameter to produce local heating and measured the change in I_c while the laser beam was slowly scanned along the length of the junction. If the laser-induced reduction of the local current is proportional to the unperturbed current density of the spot, the change in I_c as a function of the laser beam position reflects the current distribution. Their experimental results agree well with the theoretical calculations of Owen and Scalapino (1967), who have also taken into account the effect of the magnetic field produced by the current in the films.

6.4. CIRCUITS AND DEVICES

The most general local lumped element circuit environment for any NDC element is shown in Fig. 2-36; Fig. 2-37 shows how the NDC element can be represented. When pair tunneling can occur we must add another path, $I_c \sin \Delta\theta$, parallel to the conductance current–voltage path previously denoted as $I_c(\Phi)$ in Fig. 2-38; we will now denote it as $I_c'(\Phi)$, and it represents quasiparticle tunneling. The new $I_c(\Phi)$ curve will now include effects due to pair tunneling. The added parallel current path will contain an element obtained from Eq. (6-31), $I_c \sin \Delta\theta$, and since there will be an inductance associated with pair tunneling too, an intrinsic inductive element L_{ip}. (In what follows we assume for the time being that there exists no coupling or interference between the parallel current paths.) The important circuit components are shown in Fig. 6-13, where L_{iqp} is the component of the intrinsic inductance associated with quasiparticles, L_l is the lead inductance, R_L is the sum of load, lead, and battery resistance, and $C = C_p + C_i$ is the sum of the package (shunt) and intrinsic capacitance. Here the system is driven by a dc voltage source Φ_B.

In Chap. 2 we showed that for NNDC elements where changes in voltage were large compared to changes in current during switching events or oscillatory phenomena, a reasonable approximation was to neglect (short out) the intrinsic inductance in the circuit. For Josephson junctions it is often the case that R_L is much larger than any resistance in the system; the dc driving system acts in this case like a constant current source. Under these conditions the effects of L_l will also be negligible. The most important approximate circuit to analyze, therefore, is shown in Fig. 6-14. In order to solve the circuit in a large signal analysis we require knowledge of the quasiparticle component $I_c'(\Phi)$, whose form for tunneling between identical superconductors and $T > 0\,\mathrm{K}$ will be similar to that

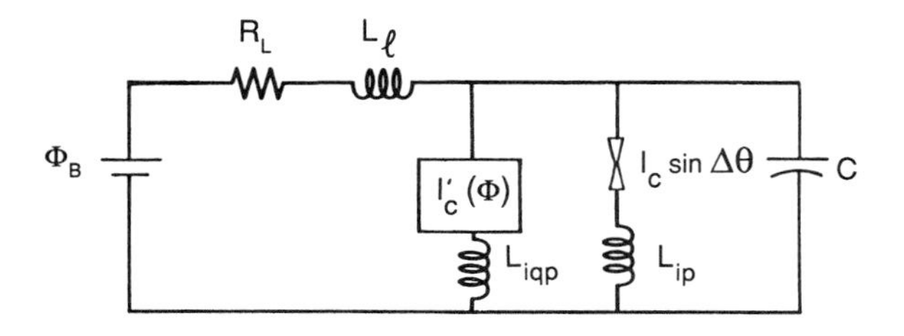

FIGURE 6-13. Circuit for a Josephson junction. All parameters are defined in the text.

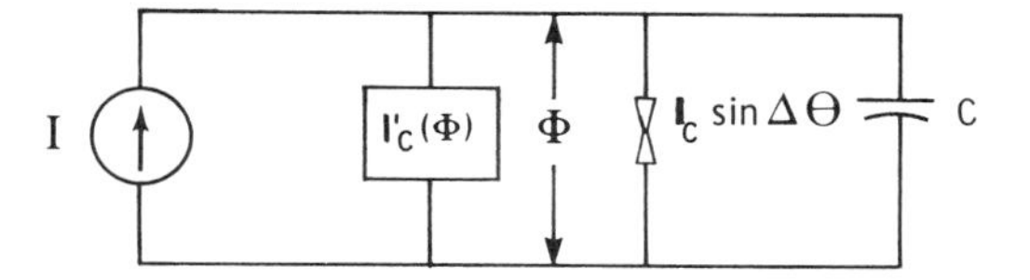

FIGURE 6-14. Circuit under analysis for a Josephson junction. The inductive effects are neglected in this approximation, and R_L is much larger than any other resistance. The constant current source approximates the latter condition.

shown in Fig. 6-8 but without an NNDC region. (Obviously, if the superconductors are not identical, an NNDC region might indeed be present.) As a first approximation, we analyze the circuit for the case where $I_c'(\Phi)$ is linear; i.e., we will treat the case where the quasiparticle contribution is modeled by a constant conductance G (this is a fairly good approximation for many weak-link systems). For a small junction where $\Delta\theta$ is independent of position, Kirchhoff's current law at the top node in Fig. 6-14 yields (Stewart, 1968; McCumber, 1968)

$$I = G\Phi + I_c \sin\Delta\theta + C\frac{d\Phi}{dt} \tag{6-46}$$

From the second Josephson relation [Eq. (6-33)] and Eq. (6-64) we obtain

$$I = \frac{\hbar C}{2e}\frac{d^2\Delta\theta}{dt^2} + \frac{\hbar G}{2e}\frac{d\Delta\theta}{dt} + I_c \sin\Delta\theta \tag{6-47}$$

which can be written

$$\frac{I}{I_c} = \beta_c \frac{d^2\Delta\theta}{d\alpha^2} + \frac{d\Delta\theta}{d\alpha} + \sin\Delta\theta \tag{6-48}$$

where

$$\alpha \equiv \omega_c t \equiv \left(\frac{2e}{\hbar}\right)\left(\frac{I_c}{G}\right)t \tag{6-49}$$

and

$$\beta_c \equiv \frac{\omega_c C}{G} = \left(\frac{2e}{\hbar}\right)\left(\frac{I_c}{G}\right)\frac{C}{G} \tag{6-50}$$

ω_c is the Josephson angular frequency corresponding to I_c and G. Equation (6-48) describes the "resistively shunted junction" (RSJ) model.

Equation (6-47) can be integrated directly only in the limit of vanishingly small capacitance; the "McCumber parameter" $\beta_c \to 0$. Under these conditions Eq. (6-48) is of the form

$$d\Delta\theta/(I/I_c - \sin\Delta\theta) = d\alpha \tag{6-51}$$

an expression whose integral is readily available in tables. In this limit we obtain for the average value of voltage $\Phi = \langle(\hbar/2e)(d\Delta\theta/dt)\rangle$

$$\begin{aligned} &\Phi = 0 && \text{for } I < I_c \quad \text{(dc Josephson effect)} \\ &\Phi = \left(\frac{I_c}{G}\right)\left[\left(\frac{I}{I_c}\right)^2 - 1\right]^{1/2} && \text{for } I > I_c \quad \text{(ac Josephson effect)} \end{aligned} \tag{6-52}$$

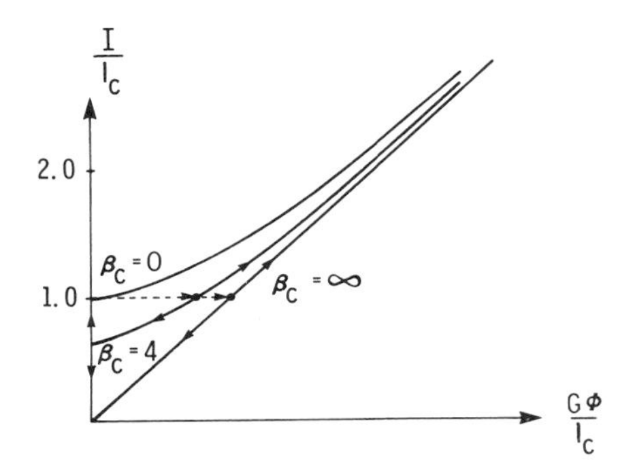

FIGURE 6-15. Normalized $I - \Phi$ characteristics for a Josephson junction described by the equivalent circuit of Fig. 6-14 for cases of negligible ($\beta_c = 0$) and dominating ($\beta_c = \infty$) capacitance (after Van Duzer and Turner, 1981).

Figure 6-15 shows the normalized $I(\Phi)$ curve for this case ($\beta_c = 0$). For values of $\beta_c > 0$ Eq. (6-48) must be integrated numerically; two representative cases are sketched in Fig. 6-15 ($\beta_c = 4$, $\beta_c = \infty$). When I_c is reached at constant current we switch along the load line of zero slope to the operating points at finite Φ shown on the $\beta_c = 4$ and $\beta_c = \infty$ curves. Upon reduction of the current below I_c, however, we do not switch back to the $\Phi = 0$ pair tunneling mode but rather develop a hysteresis and go smoothly back to a minimum current $(I/I_c)_{\min}$ dictated by the value of β_c [$(I/I_c)_{\min}$ is unity for $\beta_c = 0$ and zero for $\beta_c = \infty$].

The line $\beta_c = \infty$ corresponds to the case where only the conductance current path $G\Phi$ is the circuit. Under these conditions the ac current that flows must have an average value of zero. Hence, for finite values of β_c, at a given voltage the difference between the current at a particular value of β_c and the current for the $\beta_c = \infty$ case must be due to an ac component whose average value is greater than zero. For the case $\beta_c = 0$ we see, therefore, that at low voltages almost all the current is due to an ac component whose average value is slightly less than I_c. To understand why this phenomenon occurs note that as soon as a voltage develops across the junction the current develops an ac component. Since the source has ∞ resistance, this current flows partly into G and partly into C (depending on β_c). That component that flows into G must produce an ac voltage across G, which is thereby developed across the $I_c \sin \Delta\theta$ (pair) path. The voltage appearing across the Josephson junction has both dc and ac components; the rate of change of phase [Eq. (6-33)] is not constant any more and we obtain a nonsinusoidal temporal variation of current whose periodicity equals the Josephson frequency equivalent of the average voltage. Note, for example, that the case $\beta_c = \infty$ corresponds to $C = \infty$; the voltage across the junction is held constant, the rate of change of phase is constant, and the average value of ac current is zero.

The $I(\Phi)$ characteristics shown in Fig. 6-15 correspond to an idealized model where G is assumed constant, a reasonable approximation for some weak-link systems. A more representative situation for tunnel junctions models the quasiparticle contribution with an $I_c'(\Phi)$ curve shown in Fig. 6-8, with the NDC region absent when the superconductors are identical. For these more realistic situations a variety of unusual situations might possibly arise. For example, let us first consider the case of identical superconductors and a small, but nonzero, value of β_c. Depending on the exact form of $I_c'(\Phi)$ and the value of β_c, an NNDC region, induced by the capacitive effect ($\beta_c > 0$), could develop, as shown in Fig.

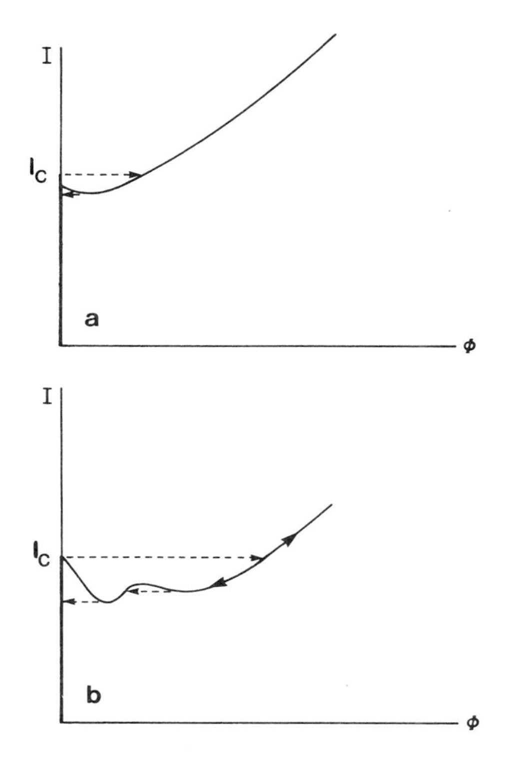

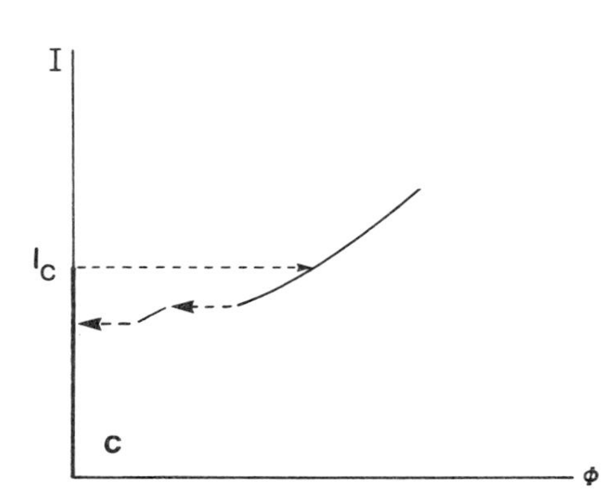

FIGURE 6-16. (a) NNDC region induced by the capacitative effect. (b) Double NNDC region caused by the capacitative effect in a tunnel junction composed of two different superconductors. (c) Experimentally observed $I(\Phi)$ curve for a state having two NNDC regions.

6-16a. Note that here, under constant current conditions, we would observe a switching event both *to* a $\Phi > 0$ state and back *from* a $\Phi > 0$ state. That is, a minimum finite voltage current state would be observed below which we would switch back to the $\Phi = 0$ state. For different superconductors sandwiching the tunnel junction, an $I(\Phi)$ characteristic like that shown in Fig. 6-16b might result. Here there are *two* NNDC regions, and under constant current conditions the observed $I(\Phi)$ curve, Fig. 6-16c, would exhibit two $\Phi > 0$ states separated by gaps. When these suggested phenomena occur, it is important to drive the system with a finite R_L so that effects due to the NDC regions could manifest themselves. For example, since here the NDC regions are due to quasiparticle tunneling across a junction, they can be stabilized against circuit oscillations by a proper choice of the circuit damping parameter, $\sqrt{LC}/(R_0C)$, discussed in Chap. 2. Or, if stabilization cannot be achieved, the resulting circuit controlled oscillations alone could help quantify values of the important capacitive and inductive elements in the system.

In studying the systems shown in Figs. 6-15 and 6-16 great care must be taken to separate out effects due to applied and induced magnetic fields, heating, rf effects, and noise, all of which will act to alter the $I_c'(\Phi)$ characteristic. We can shield out external fields, and the current-induced fields will act to alter I_c in a manner we can predict. If we assume isothermal conditions for the moment, then in order to appreciate the influence of noise on the $I_c(\Phi)$ characteristics, it is useful to first understand how an imposed ac signal affects the $I(\Phi)$ characteristic of a Josephson tunnel junction. Although a constant voltage source is not a

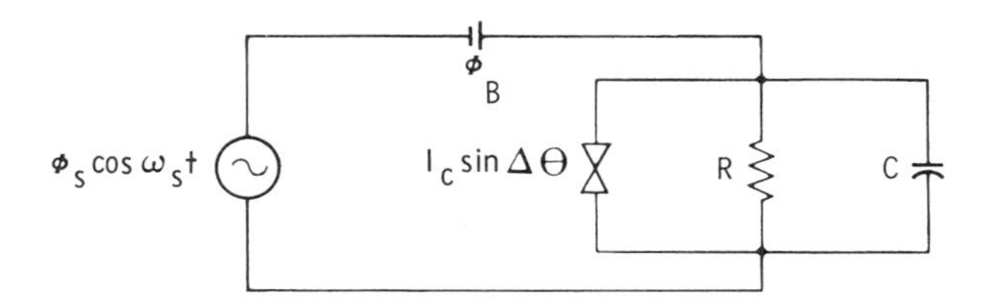

FIGURE 6-17. Equivalent circuit for a Josephson device connected to dc and ac voltage sources (after Van Duzer and Turner, 1981).

realistic configuration, it is easiest to see the effects of the ac signal by considering this mode of operation. Figure 6-17 shows the idealized circuit under analysis.

The current through the Josephson element is

$$I_J(t) = I_c \sin\left[\int_0^t \frac{2e\Phi(t')}{\hbar}\,dt' + \Delta\theta_0\right] \tag{6-53}$$

where $\Phi(t') = \Phi_B + \Phi_S \cos \omega_s t$. ω_s is the angular frequency of the rf signal. By integrating, using trigonometric identities and Bessel function relations, we arrive at

$$I_J(t) = I_c \sum_{n=-\infty}^{\infty} (-1)^n J_n\left(\frac{2e\Phi_s}{\hbar\omega_s}\right) \sin[(\omega_J - n\omega_s)t + \Delta\theta_0] \tag{6-54}$$

where J_n is a Bessel function of the first kind of order n and $\omega_J = 2e\Phi_B/\hbar$ the Josephson frequency. At values of voltage such that $\omega_J = n\omega_s$, i.e., when $\Phi_B = n\hbar\omega_s/2e$, the current will exhibit a spike whose height is given by $I_c J_n(2e\Phi_s/\hbar\omega_s) \sin \Delta\theta_0$. At all values of Φ_B there will also be a dc current equal to $G\Phi$ flowing through the resistor $R \equiv G^{-1}$. Hence, the $I(\Phi)$ characteristic will have the form shown in Fig. 6-18. Since the junctions have impedances much lower than typical driving systems, rather than spikes, a series of steps is usually observed, as sketched by the dashed lines in the figure. Indeed, a more accurate model employs a current source rather than a voltage source (Auracher and Van Duzer, 1973; Stancampiano, 1980). Numerical calculations for the small signal impedance as a function of dc current yield results as shown in Fig. 6-19. We see

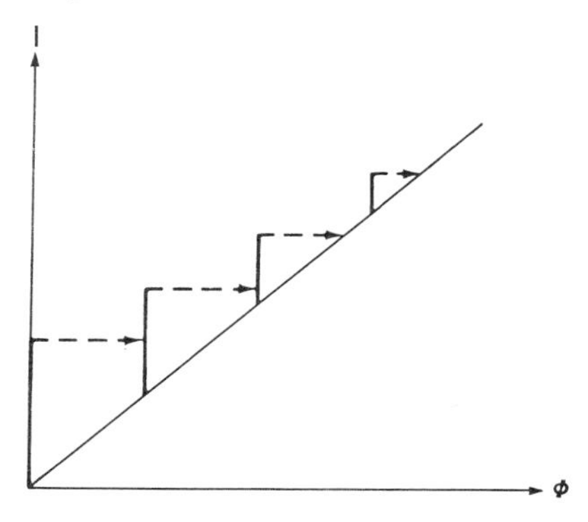

FIGURE 6-18. $I(\Phi)$ characteristics for the circuit of Fig. 6-17. In the voltage-source model, spikes in the dc current occur at voltages $\Phi = n\hbar\omega_s/2e$. Measurements are usually made with sources that do not have zero impedances; steps, rather than spikes are observed (broken lines). (After Van Duzer and Turner, 1981.)

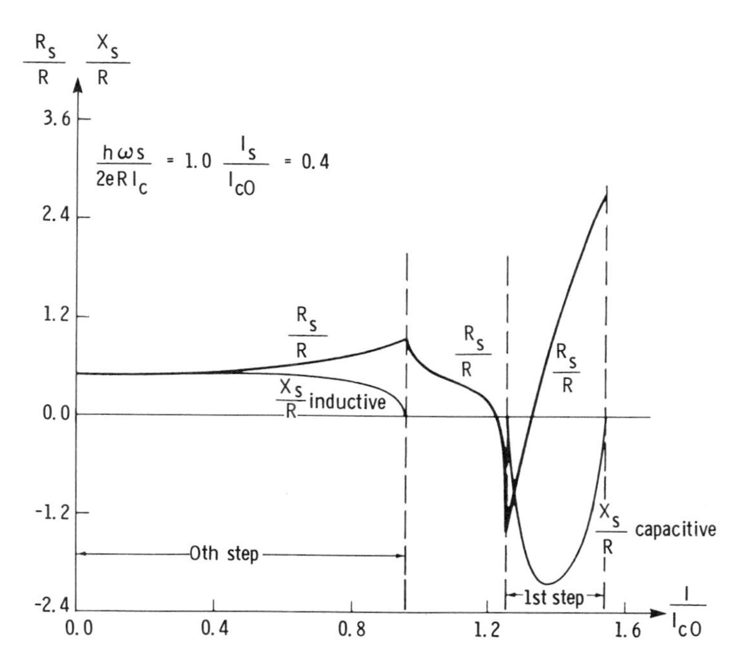

FIGURE 6-19. Dependence of impedance components at the signal frequency on the dc current. Impedance components are normalized to the value of the resistor in the equivalent circuit and current is normalized to the maximum zero-voltage current in the absence of applied rf current I_{c0}. (After Van Duzer and Turner, 1981.)

that the imaginary component vanishes between the current steps and the rf resistance is negative at the bottom of the step; the junction can oscillate and/or amplify under these conditions, as expected (see, e.g., McGrath et al., 1981). It is also interesting to note that for vanishingly small signal amplitudes and when Φ_B and C both are zero, we can find the impedance analytically. Here we can differentiate Eq. (6-31) to yield

$$\frac{d\Delta\theta}{dt} = \frac{1}{(I_c \cos \Delta\theta)} \frac{dI}{dt} \tag{6-55}$$

substitute this into Eq. (6-33), and, noting that $\Phi = L\, dI/dt$, obtain

$$L_J = \frac{\hbar}{2eI_c \cos \Delta\theta} \tag{6-56}$$

Since L_J is time dependent, parametric oscillations can occur. The equivalent circuit is this Josephson inductance L_J in parallel with R. Thus, although we originally neglected the L_{ip} term in Fig. 6-13, the Josephson junction is, in fact, inductive under certain ac conditions and it has been possible in some situations to identify L_J with L_{ip} (Langenberg, 1980).

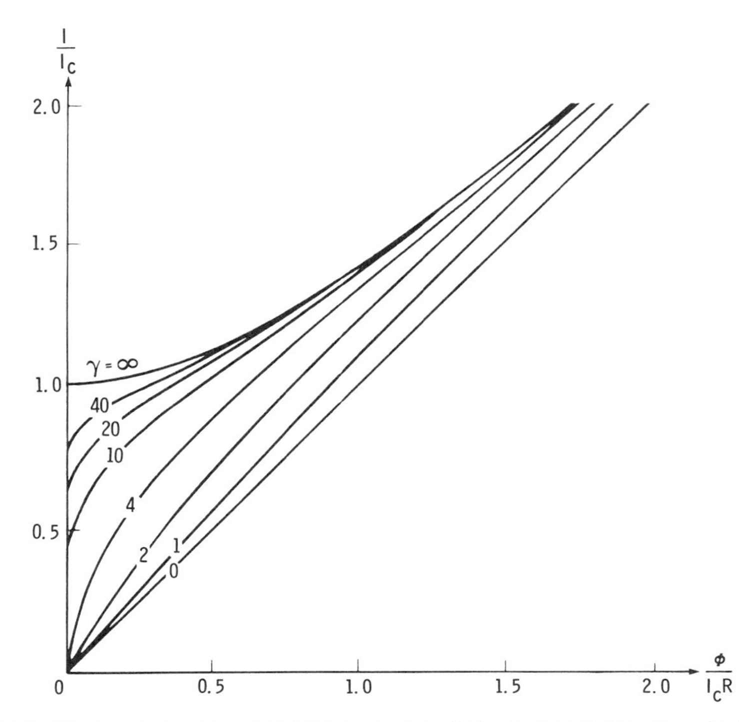

FIGURE 6-20. Effects of thermal fluctuations on the $I(\Phi)$ characteristics of a highly damped Josephson junction. The parameter γ is proportional to the ratio of the Josephson coupling energy to the thermal energy. (After C. M. Falco et al., 1974.)

The effects of thermal noise fluctuations on the $I(\Phi)$ characteristics were first analyzed by Ambegaokar and Halperin (1969). They employed a constant current source, a parallel noise–current source, and the conditions $\beta_c \ll 1$ and $e\Phi \ll k_B T$. They define a parameter γ that is proportional to the ratio of the Josephson coupling energy to the thermal energy: $\gamma = \hbar I_c / e k_B T$. Figure 6-20 shows how the $I(\Phi)$ characteristics are affected. Further studies for higher values of β_c were reported by Falco et al. (1974). Note that when thermal noise is negligible ($\gamma = \infty$) the $I(\Phi)$ characteristics are similar to the $\beta_c = 0$ case shown in Fig. 6-15. As the noise increases we see that the effect of the displacement current component in adding current to the system is diminished. When the noise is substantial ($\gamma \cong 0$), the Josephson component is essentially washed out.

More recent noise studies have emphasized rf driven systems and treated the transition to deterministic chaos, which appears to be coupled to a "noise rise" in the junction (see Sec 6.5, below).

We have already seen how a Josephson junction can convert dc to high-frequency ac. Let us now examine this phenomenon in a bit more detail. The simplest situation to analyze is where the capacitance can be neglected ($\beta_c = 0$), the junction is driven by a constant current source, and the parallel load resistor R_L is matched to the junction resistance R. When $I > I_c$, the voltage, $\Phi(t)$, across the junction is

$$\Phi(t) = \frac{RI[1 - (I_c/I)^2]}{2[1 + (I_c/I)\sin(\omega_J t + \Delta\theta_0)]} \tag{6-57}$$

By adjusting $\Delta\theta_0$, Eq. (6-57) can be expanded in a Fourier series of the form

$$\Phi(t) = \Phi_0\left(1 + \sum_{m=1}^{\infty} a_m \cos(m\,\omega_J t)\right) \tag{6-58}$$

where $\Phi_0 = (R/2)(I^2 - I_c^2)^{1/2}$ and the Fourier coefficient

$$a_m = 2\left\{\frac{I}{I_c} - \left[\left(\frac{I}{I_c}\right)^2 - 1\right]^m\right\} \tag{6-59}$$

For illustrative purposes let us examine the $m = 1$ component. At this frequency (ω_J) the power in the load $P_{L1} = \Phi_1^2/2R$. Typical operation dictates $I_cR \cong 1$ mV and $I_c \cong 1$ mA, with the junction biased at $\Phi = \frac{1}{2}I_cR$. The fundamental frequency is 242 GHz. Thus $P_{L1} \cong 10^{-7}$ W. Experiments usually produce powers two orders of magnitude lower, so circuit techniques have been developed to raise the power. The phase locking of a string of junctions provides higher output power than a single junction if the resistance of the string is matched to the load. For N junctions we choose $R = R_L/N$. Since I_cR can be made approximately the same for each junction, then I_c for N junctions is N times I_c for one junction. Under these conditions the voltage Φ_1 across each junction is the same as if we only employed one junction; the load power is then N^2 times that if only one junction were used to match the same R_L. Further, the phase-locking of N identical junctions narrows the spectral linewidth by a factor of N^2.

In the last section we showed that I_c is a strong function of magnetic field, **B**. A wide variety of switching, memory, and logic circuits are based on this phenomenon. For example, Fig. 6-21 shows a typical $I(\Phi)$ characteristic for a Josephson junction system switch suitable for such applications. If $I_c(0)$ is exceeded in the absence of **B** but with a current overdrive, switching can take place from points A to B. However, if a constant current source is utilized and I_c is suppressed by a magnetic field, switching from points A′ to B′ will occur. Switching times here can be as small as tens of picoseconds.

It is also common to employ parallel arrays of two or more Josephson junctions as switching or memory elements. These devices are called interferometers in that their $I_c(B)$ profiles are similar to the optical interference pattern observed in a double slit experiment. The physics of their operation is similar to what we have already discussed for the influence of **B** on a single junction, and their $I(\Phi)$ characteristics are often essentially the same as that shown in Fig. 6-21. Note in this figure that the device switches back to the zero voltage state when the current is below $I_{\min}$. This can occur here because the average voltage at this current is sufficiently low so that the superimposed oscillating component drives

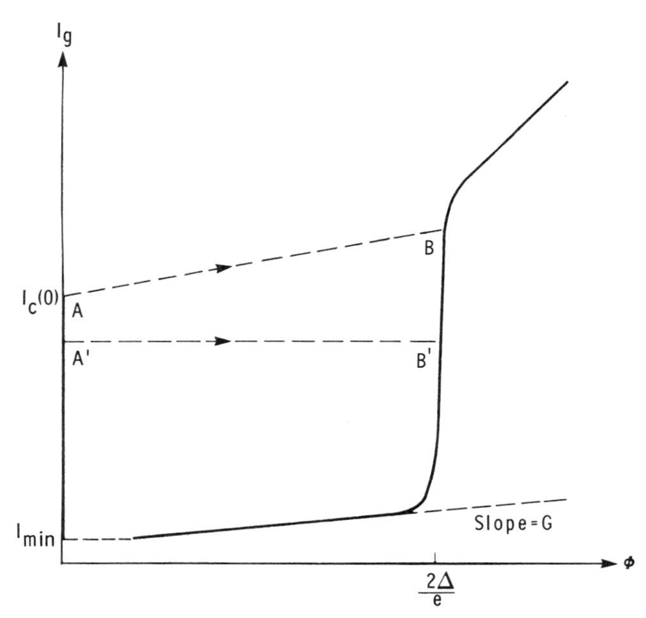

FIGURE 6-21. A typical hysteretic $I(\Phi)$ characteristic for a Josephson junction or interferometer suitable for switching applications. The broken line *A–B* represents the transition from the $\Phi = 0$ state to the $\Phi \neq 0$ state with gate-current overdrive. The *A′–B′* line shows the switching trajectory with a constant-current source and magnetic depression of the critical current. (After Van Duzer and Turner, 1981.)

the total voltage instantaneously to zero. The junction system may then return to the zero voltage state by this mechanism. On the other hand, the same phenomenon would be observed if the $I(\Phi)$ curve developed an NNDC region because of the low capacitance of the structure.

It is also important to realize that in some cases thermal effects can have great influence on the $I(\Phi)$ characteristics. Although tunnel junctions are not usually influenced by heating because of their extremely high resistances, weak-link microbridges have $I(\Phi)$ characteristics that are often dominated by thermal affects (Skocpol et al., 1974). In Chap. 8 we will treat thermal effects in detail, with emphasis on the development of SNDC characteristics. In the case of microbridges, however, thermal effects lead to observable NNDC characteristics. In particular, Skocpol et al. (1974) have shown that heating effects are important in both long and short microbridges, and the general $I(\Phi)$ characteristic can be understood via a simple model based on a localized hotspot maintained at $T > T_c$ (a normal region) by Joule heating. They show that the hot spot produces NNDC curves and is the dominant cause of the hysteresis observed in the low-temperature $I(\Phi)$ curve. The model they employ treats "long" one-dimensional bridges of length L ($\gg\eta$, the thermal heating length), width W, and thickness d. The normal region is of length $2X_0$, has resistivity ρ, and is symmetrically disposed in the bridge. Jackson and Shaw (1974) have also analyzed such a system, in an idealized fashion, and have obtained similar results (see Chap. 8 for details). The temperature distribution $T(x)$ satisfies

$$
\begin{aligned}
-\kappa_N \frac{d^2T}{dx^2} + \frac{\alpha}{d}(T - T_b) &= \left(\frac{I}{Wd}\right)^2 \rho, \qquad (|x| < X_0) \\
-\kappa_S \frac{d^2T}{dx^2} + \frac{\alpha}{d}(T - T_b) &= 0, \qquad (|x| > X_0)
\end{aligned}
\tag{6-60}
$$

where $\kappa_{N,S}$ are the thermal conductivities, α the heat transfer coefficient per unit area, and I the current. N denotes normal and S superconducting. The boundary conditions are $T(\pm\frac{1}{2}L) = T_b$, the ambient temperature. T and $\kappa_{N,S}(dT/dx)$ are matched at the N/S interface ($x = X_0$), where $T(\pm X_0) = T_c$. The temperature distribution is found, and then the self-consistent current

$$I(X_0) = I_1\left[1 + \left(\frac{\kappa_S}{\kappa_N}\right)^{1/2} \coth\left(\frac{X_0}{\eta_N}\right) \coth\left(\frac{\frac{1}{2}L}{\eta_S} - \frac{X_0}{\eta_S}\right)\right]^{1/2} \tag{6-61}$$

where $I_1 \equiv [\alpha W^2 d(T_c - T_b)/\rho]^{1/2}$. The corresponding voltage is

$$\Phi(X_0) = I(X_0) 2X_0\rho/Wd \tag{6-62}$$

As shown in Fig. 6-22, an NNDC region is observed at low voltages. One reason for its appearance is that once X_0 becomes less than η, the power dissipation, $I\Phi$, required to maintain $T > T_c$ in the center decreases slowly with a further decrease in X_0; Φ must vary as I^{-1} in this regime. Further discussion of the details of such thermally induced NDC curves is deferred to Chap. 8.

Skocpol et al. (1974) extend their model to short microbridges and also include the effects of microwave heating when rf excitation is present. Comparison between theory and experiment is shown in Fig. 6-22; we see that excellent agreement is obtained.

Other interesting instability phenomena occur in Josephson tunnel junctions when they are made wide enough so that transmission-line-type behavior is observed (see Van Duzer and Turner, 1981, Chap. 4) in a direction perpendicular to the current (Scott, 1976; Scott, in Bishop and Schneider, 1978). If we assume that a TEM mode can propagate in the x direction between two closely spaced superconductors carrying current in the z direction, the dependent variables of the curve are the transverse voltage, $\Phi(x, t)$, across the insulator and the longitudinal current, $I(x, t)$, flowing parallel to the barrier. Neglecting dissipation, the magnetic flux associated with the wave, $\phi = \int \Phi \, dt$, is governed by

$$\frac{\partial^2 \phi}{\partial x^2} - \frac{1}{\bar{C}^2}\frac{\partial^2 \phi}{\partial t^2} = \frac{1}{\lambda_J^2} \sin\left(\frac{2\pi\phi}{\phi_0}\right) \tag{6-63}$$

where $\bar{C}$ is the velocity of the wave front of the TEM wave, λ_J the Josephson penetration depth, and ϕ_0 the magnetic flux quantum $\pi\hbar/e$. Normalization of Eq. (6-63) results in the sine-Gordon equation [see Eq. (1-69)], whose "kink"

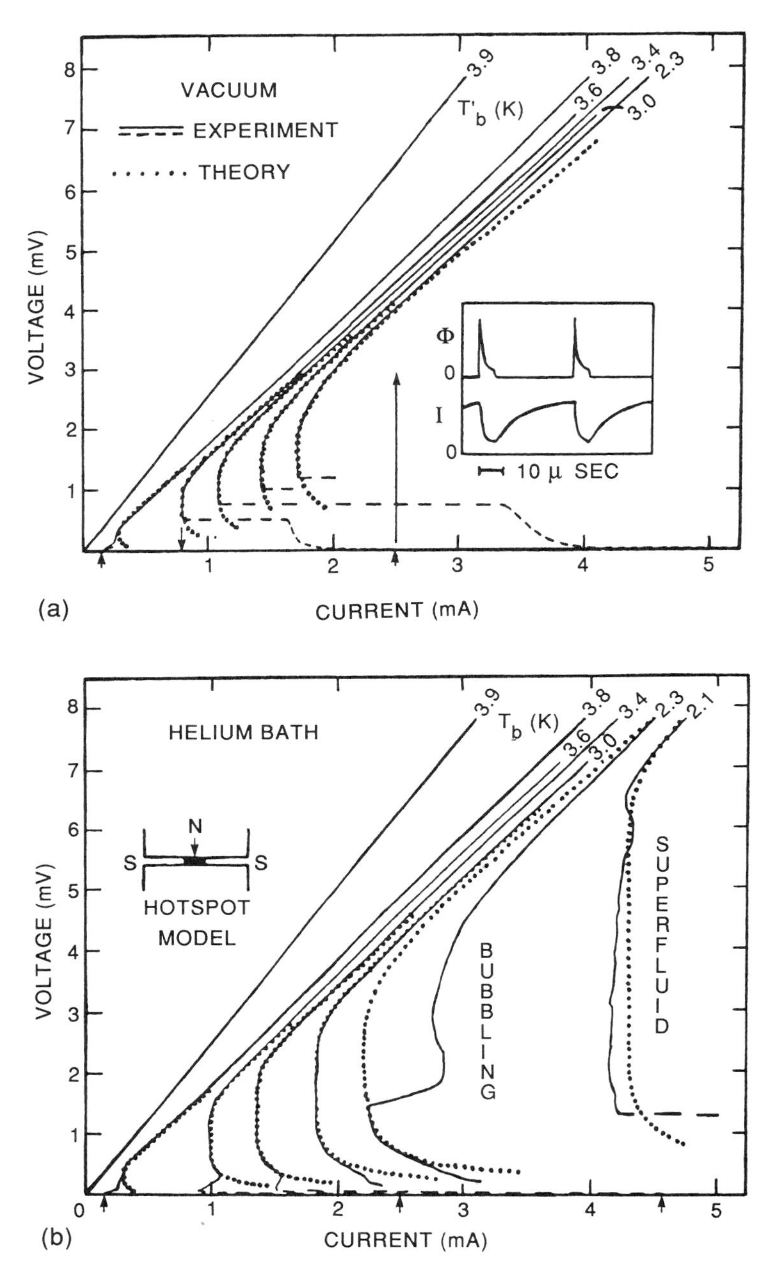

FIGURE 6-22. Voltage-biased $I(\Phi)$ characteristics of a typical long microbridge in various thermal environments. The solid and dashed curves are experimental data. The inset of the top figure shows a typical waveform of relaxation oscillations that follow the hysteresis loop indicated by the vertical arrows for 3.6 K; its dc average leads to the dashed curves. The dotted curves are the hot spot theory based on the model shown in the inset of the bottom figure. The arrows along the current axis indicate the superconducting critical current, showing the rapid development of hysteresis. (After Skocpol et al., 1974.)

solutions

$$\phi_{\pm}(x, t; u, x_0) = 4 \tan^{-1}\left\{\exp\left[\pm \frac{x - ut - x_0}{(1 - u^2)^{1/2}}\right]\right\} \tag{6-64}$$

represent either a fluxon (ϕ_+) traveling in the $+x$ direction or an antifluxon (ϕ_-) traveling in the $-x$ direction. These are soliton solutions. If the voltage is in units of $\phi_0 \bar{C}/2\pi\lambda_J$, then

$$\frac{\partial \phi}{\partial t} = \Phi \tag{6-65}$$

if the longitudinal current is in units of $J_0\lambda_J$, where J_0 is the maximum Josephson current per unit length, then

$$\frac{\partial \phi}{\partial x} = I \tag{6-66}$$

The moving fluxons can be used to process information and to generate electromagnetic radiation. For the radiation problem we must include perturbations to the sine-Gordon equation that permit input and dissipation of energy. The distributed bias current, γ, exerts a Lorentz force which accelerates a fluxon in the $+x$ direction or antifluxon in the $-x$ direction. The dissipation of energy can be represented phenomenologically by a term, ζ, of the form $\Gamma |\partial\phi/\partial t|\,(\partial\phi/\partial t)$ or $\alpha(\partial\phi/\partial t)$. Equation (6-63) then becomes

$$\frac{\partial^2 \phi}{\partial x^2} - \frac{\partial^2 \phi}{\partial t^2} - \sin\phi = \gamma + \zeta \tag{6-67}$$

Solutions of these equations (see Scott, in Bishop and Schneider, 1978) show three major types of time-dependent behavior: plasma oscillations; bound state oscillations of fluxon–antifluxon pairs ("breather" modes or "bions"); fluxon oscillations. The latter mode is simply a fluxon that moves to one end of the junction, is reflected as an antifluxon and moves back, etc. The $I(\Phi)$ characteristic of this fluxon mode depends upon the damping coefficient of the dissipative term, ζ and the ratio $L/2\pi\lambda_J$. In a medium-size sample ($L/2\pi\lambda_J \sim 1$) with negligible loss and in the absence of an external magnetic field, a fluxon can travel with the speed of electromagnetic waves and reflect back and forth with a frequency almost equal to the resonant frequency of the fundamental mode in the junction. As shown in Fig. 6-23, this resonant vortex mode leads to a series of current steps at constant voltages (Chen et al., 1971; and Fulton and Dynes, 1973). In a long junction ($L/2\pi\lambda_J \gg 1$) with large damping, the vortex speed as well as its reflection frequency in general cannot reach those of electromagnetic waves. This nonresonant vortex mode gives rise to a set of resistive branches in the $I(\Phi)$ characteristic of the junction shown in Fig. 6-24 (Rajeevakumar et al., 1980a). In the presence of an external magnetic field greater than B_{C1}, fluxons

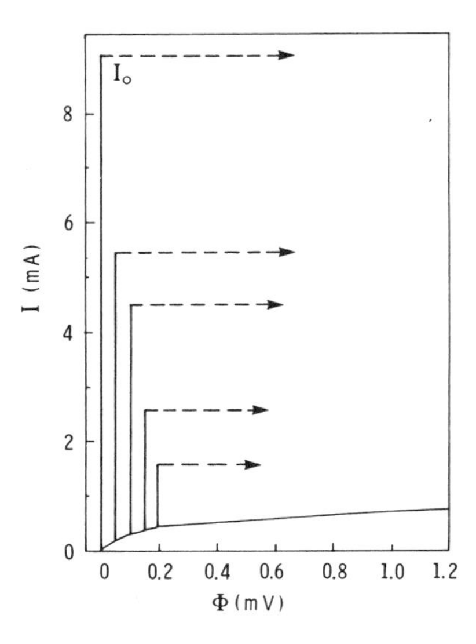

FIGURE 6-23. The $I(\Phi)$ curve of a junction carrying a resonant vortex mode, traced by repeated cycling without passage through zero current. The current singularities at nonzero voltage (in order of increasing voltage) are identified by different mode numbers. (After Chen et al., 1971.)

can not be reflected at the ends of a junction but can propagate only in one direction as determined by the Lorentz force produced by the current. This one-directional fluxon motion, called flux flow, leads to a single ohmic-like branch as shown in Fig. 6-25 (Rajeevakumar et al., 1980b; Yoshida et al., 1978). Unlike the other fluxon modes, the voltage induced by the flux–flow mode is tunable by an external magnetic field. Rajeevakumar et al. have demonstrated a fluxon device based upon this property. The external magnetic field can be varied by a control current flowing parallel to the length of a junction.

As mentioned earlier, it is possible for a moving fluxon to emit electromagnetic radiation. Experimentally, it was first done by dc biasing a Josephson tunnel junction on one of its resonant current steps (Chen and Langenberg, 1973). The detected radiation spectra are generally very complex. For example, a radiation peak occurs below the current peak. A simple explanation for the observed complex spectra and the radiation peak (shown in Fig. 6-26) has not yet been given. Chen has suggested that the rapid decrease in radiation before the current

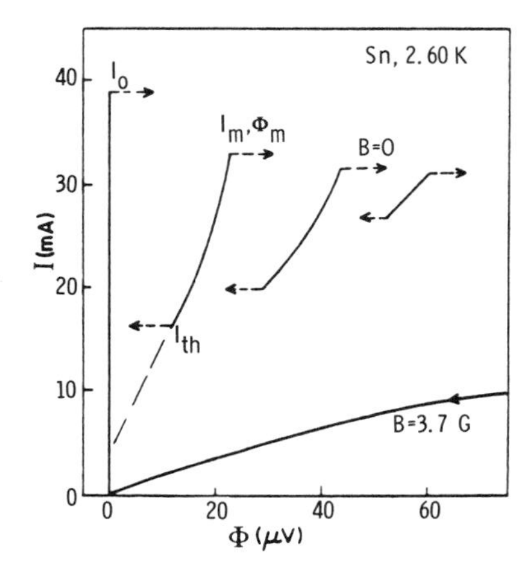

FIGURE 6-24. The $I(\Phi)$ characteristic of a long Sn–Sn oxide–Sn junction showing three resistive branches. Junction parameters are: length $L = 1.0$ mm, width $W = 0.1$ mm, and $L/\lambda_J = 31$ at 2.60 K. [λ_J, 0.032 mm, was calculated from $I_0 = 4\lambda_J W j_0$ and $\lambda_J = (\Phi_0/2\pi\mu_0 j_0 d)^{1/2}$, where j_0 is the maximum Josephson current density and d is twice the London penetration depth, about 850 Å.] (After Rajeevakumar et al., 1980a.)

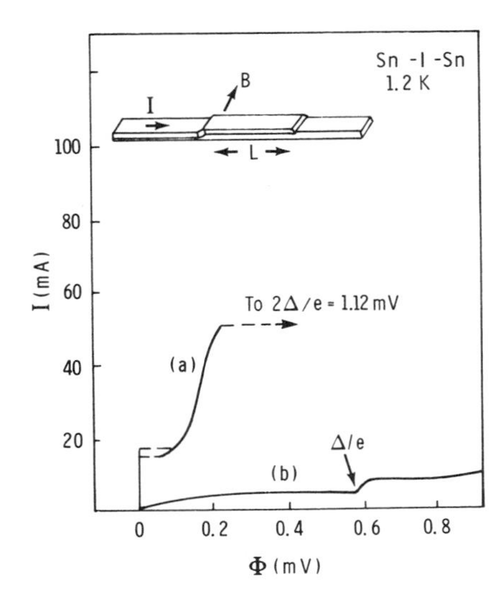

FIGURE 6-25. The $I(\Phi)$ characteristic of a Sn–Sn-oxide–Sn tunnel junction in various magnetic fields. The inset shows the junction geometry and the direction of the current and the applied magnetic field. (After Rajeevakumar et al., 1980b.)

reaches its peak is due to the increasing dissipation when the fluxon speed approaches the speed of electromagnetic waves. Near the speed of these waves a fluxon will experience a relativistic contraction (Fulton and Dynes, 1973) resulting in a voltage pulse with magnitude exceeding the pair-breaking value $2\Delta/e$. Other possible explanations include the nonlinear transfer of radiation energy to other modes and the creation of multiple fluxons upon reflection.

6.5. CHAOS IN THE rf-BIASED JOSEPHSON JUNCTION

The basic equation of the resistively shunted junction model, Eq. (6-48), which we studied in the preceding section, can be written as a two-variable

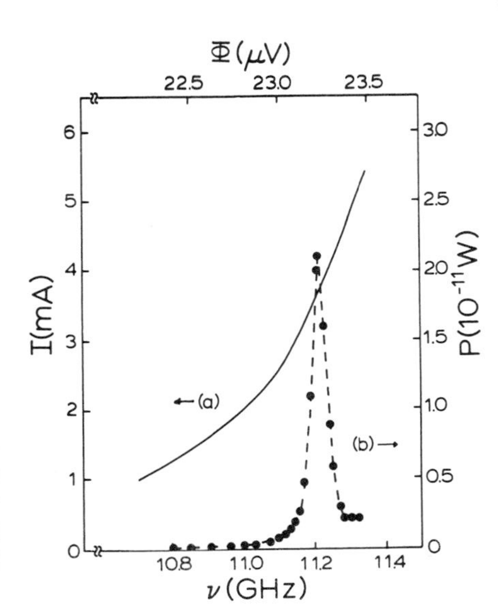

FIGURE 6-26. (a) Fine structure of the $I(\Phi)$ characteristic (solid line, left and upper scale) and (b) radiation power spectrum (dashed, right and upper scale) of a singularity associated with a moving fluxon in a Josephson junction. (After Chen and Langenberg, 1973.)

autonomous nonlinear dynamic system, and thus has two dynamic degrees of freedom. As we have seen in Chap. 1, at least three dynamic degrees of freedom are necessary to allow for chaotic time-dependent behavior. This can be provided by driving the Josephson junction with, in addition to the dc current I_{dc}, an ac current $I_1 \sin(\Omega_1 t')$ modulated with a radiofrequency (rf) ω_1:

$$\beta_c \Delta\ddot{\theta} + \Delta\dot{\theta} + \sin \Delta\theta = i_0 + i_1 \sin(\Omega_1 t') \tag{6-68}$$

with the normalized currents $i_0 = I_0/I_c$, $i_1 = I_1/I_c$, and the normalized frequency $\Omega_1 = \omega_1 \hbar/(2eI_cR)$. The dot denotes the time derivative with respect to the normalized time $t' = t(2eI_cR/\hbar)$. The resulting nonautonomous dynamic system can then be represented as a three-variable autonomous dynamic system, as we have demonstrated in Chap. 1, thus allowing for the possibility of chaotic solutions. Chaotic behavior was first revealed in the rf-biased Josephson junction in simulations by Huberman et al. (1980), Braiman et al. (1980), Kautz (1981), and Pedersen and Davidson (1981). Since that time, chaos in this system has been studied in both analog and digital simulations by many authors (see, e.g., reviews by Kautz and Monaco, 1985; Pedersen, 1986; Kurkijärvi, 1991). In these studies, the rf-biased junction has often served as a test case in the study of chaos because its equation of motion is one of the simplest equations to reveal chaotic behavior. It is identical to the equation of a damped driven pendulum, and also to an equation used to describe charge density waves in semiconductors. The rf-driven junction is also of practical importance in that the ac Josephson effect defines a standard of voltage. Thus, an understanding of how chaos can be avoided in this system is of significance to the design of Josephson voltage standards.

From Eq. (6-68) it is apparent that the dynamical behavior of the rf-biased junction is covered by four parameters: β_c, i_0, i_1, and Ω_1. Of these parameters, only β_c is necessarily positive, but we may assume without loss of generality that i_0, i_1, and Ω_1 are also positive. In order to simplify our search for chaos we will fix Ω_1 at 0.01. This value of Ω_1 approximates the situation in either of two types of voltage standards. In a conventional standard, the drive frequency is commonly around 10 GHz and the junction is biased above the energy gap so that I_cR is roughly 2 mV. In the zero-bias standard the drive frequency tends to be higher, say 50 GHz, and the relevant resistance is the subgap resistance so that I_cR is on the order of 10 mV. In either of these cases the reduced frequency Ω_1 is about 0.01. Because our model assumes a fixed resistance, it is most clearly applicable to the zero-bias standard and attention will be focused on this case. With regard to the other three parameters, our survey spans values of β_c between 1 and 20 000, values of i_0 between 0 and 1, and values of i_1 between 0.1 and 200.

Chaos does not occur for $\beta_c \ll 1$ (strong damping) and for $\Omega_1 \gg \beta_c^{-1/2}$. The latter relation means that the driving frequency is much larger than the plasma frequency of the junction, which is the resonant oscillation frequency of the corresponding linearized circuit. By "linearized" we mean that the Josephson junction is replaced by an inductor with $L = \hbar/(2eI_c)$, i.e., the Josephson current $\sin \Delta\theta$ is approximated by $\Delta\theta$, leading to a resonance frequency $\omega_1 = (LC)^{-1/2}$.

Because Eq. (6-48) represents a dissipative system driven at a single frequency, it might be expected that the steady-state solutions would be periodic, having a period equal to the drive frequency, $2\pi/\Omega_1$, or possibly some multiple of this period. That is, we expect that the steady state solutions will obey the relation

$$\Delta\theta(t' + 2\pi m/\Omega_1) = \Delta\theta(t') + 2\pi n \qquad (6\text{-}69)$$

for all t', where m and n are integers. For this solution the phase advances by exactly $2\pi n$ over m drive cycles, and the average voltage in units of $I_c R$ is

$$\langle\Phi\rangle \equiv \frac{\Omega_1}{2\pi m}\int_0^{2\pi m/\Omega_1} \frac{d}{dt'}(\Delta\theta)\, dt' = \frac{n}{m}\Omega_1 \qquad (6\text{-}70)$$

The smallest value of m for which (6-69) holds is the period of the solution measured in drive cycles.

While periodic solutions are commonly observed in the rf-biased junction, there are, in addition, steady state solutions having a chaotic nature that are not periodic. It is useful to distinguish three classes of solutions based on their period in

1. Harmonic phase-locked motion at frequency $\Omega = \Omega_1$, i.e., $m = 1$;
2. Subharmonic phase-locked motion at frequencies $\Omega = \Omega_1/m$, with $m = 2, 3, 4, \ldots$;
3. Aperiodic motion (quasiperiodic or chaotic, i.e, formally $m \to \infty$).

As we shall show, each of these solution classes is associated primarily with certain regions of the current–voltage characteristic of the junction.

The most easily identifiable part of the $I(\langle\Phi\rangle)$ characteristic is that associated with the harmonic solutions, namely, the principal rf-induced steps. The principal steps are regions of the characteristic where the average voltage is constant at the value $\langle\Phi\rangle = n\Omega_1$ over some range of dc bias i_0. The integer n is called the order of the step. Over the range of the step, the junction is phase-locked to the drive such that $\Delta\theta$ increases on the average $2n\pi$ during each drive cycle. By definition, all harmonic solutions fall on principal steps. For some ranges of dc bias there appear many small-amplitude constant-voltage steps at voltages of the form

$$\langle\Phi\rangle = \frac{n}{m}\Omega_1$$

where n/m is *not* an integer. They belong to subharmonic solutions. For other ranges of dc bias there exist chaotic solutions, but they are not apparent in the $I(\langle\Phi\rangle)$ characteristics. Instead, they may be identified by the calculation of Poincaré sections, power spectra, and Lyapunov exponents, as discussed in Chap. 1.

Let us consider first the Poincaré section of the state-space trajectory. At any given time, the state of the system described by Eq. (6-68) is completely determined if $\Delta\theta(t')$ and $\Delta\dot{\theta}(t')$ are specified. A Poincaré section of the trajectory can be plotted by recording the location of the system in the $\Delta\theta$–$\Delta\dot{\theta}$

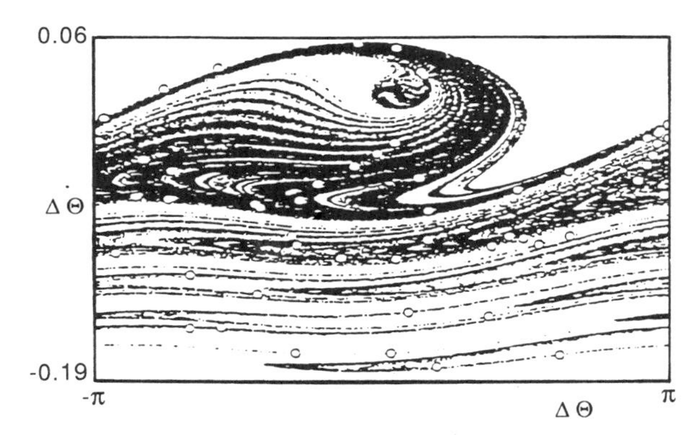

FIGURE 6-27. Poincaré section of a chaotic solution. The points $(\Delta\Theta(T_k), \Delta\dot{\Theta}(T_k))$ at the beginning of successive driving cycles are plotted in the $(\Delta\Theta, \Delta\dot{\Theta})$ phase plane. Circles indicate the locations of unstable period 1 solutions. The numerical parameters are $\beta_c = 1000$, $\Omega_1 = 0.01$, $i_1 = 10$, $i_0 = 0.1$. (After Kautz and Monaco, 1985.)

plane at the beginning of many successive rf cycles. Figure 6.27 shows a Poincaré section of the steady state solution for a bias on the chaotic part of the $I(\Phi)$ curve. For a chaotic solution, the successive values of $(\Delta\theta(t_k'), \Delta\dot{\theta}(t_k'))$, where $t_k' = 2\pi k/\Omega_1$, $k = 1, 2, 3, \ldots$, jump from one region of state space to another in an apparently random fashion, producing a geometrically complex Poincaré section.

Experimentally, chaotic behavior can be observed most easily as broadband noise in the power spectrum of the voltage. In a sense, a broadband noise spectrum is no more than a byproduct of chaotic motion. A concept more central to the nature of the chaotic state is that of the Lyapunov characteristic exponents introduced in Chap. 1. The Lyapunov exponents determine the local stability of a state-space trajectory. If one or more of the exponents are positive then the trajectory is locally unstable. A chaotic trajectory is globally stable in that it is confined to a bounded region of state space but locally unstable in that at least one Lyapunov exponent is positive. As can be shown (Kautz and Monaco, 1985), the Lyapunov exponents of this system obey the sum rule

$$\lambda_1 + \lambda_2 = -1/\beta_c \tag{6-71}$$

Because β_c is positive, one of the exponents is necessarily negative and it is sufficient to compute the larger of the two exponents, denoted by λ, to determine the stability of a trajectory. If λ is negative (positive) then the trajectory is locally stable (unstable). On the principal steps λ is typically $-(2\beta_c)^{-1}$, the minimum possible value according to Eq. (6-71). This negative value indicates the presence of stable periodic solutions. On the resistive part of the $I(\Phi)$ curve, λ is zero to within the accuracy of the calculations. If the solutions in this region are subharmonic solutions, as is probable, the fact that λ is nearly zero suggests that the tendency to phase lock is very weak. On the portion of the $I(\Phi)$ curve deemed to be chaotic based on Poincaré sections and power spectra, the Lyapunov exponent is positive, confirming the presence of chaos. The maximum Lyapunov exponent can thus be used to distinguish between the principal steps (λ

negative), the resistive part (λ near zero), and the chaotic part (λ positive) of the $I(\Phi)$ curve.

It is interesting to note that the Lyapunov characteristic exponents are related to the fractal dimension of the attractor in phase space (Schuster, 1987). For the present system one expects the dimension of the Poincaré section to be given by

$$d = \begin{cases} 0, & \lambda < 0 \\ 1 + \lambda/(\lambda + 1/\beta_c), & \lambda \geq 0 \end{cases}$$

For the Poincaré section shown in Fig. 6-27, we have $\lambda = 4.75 \times 10^{-3} > 0$. The solution is chaotic and the Poincaré section has a fractal dimension of 1.83. If $\lambda < 0$ then the solution is both stable and periodic, the Poincaré section consists of a finite number of points, and its dimension is zero. If $\lambda = 0$ then the solution is neutrally stable, $\langle \Phi \rangle / \Omega_1$ is irrational, the Poincaré section consists of an infinite number of points, and its dimension is 1.

In order to provide a structure for the discussion of Eq. (6-68), it is useful to distinguish two separate cases according to whether there is dc drive current present or not. In the case where there is no dc drive, we often see control parameter diagrams of different solutions in the i_1 vs. Ω_1 plane, such as Fig. 6-28, where quasiperiodic and chaotic areas are marked by hatching. Obviously, when β_c grows, the complicated behavior begins at smaller values of i_1. Experimentally, however, the case in the presence of a dc drive may be more interesting, since this tends to bring about a dc voltage that displays discernible effects in the presence of, e.g., chaotic states. Bak et al. (1984) have shown that dc and ac driven Josephson junctions can be described in terms of the circle map that we have discussed in Sec. 1.2.1. The circle map again seems capable of clearly explaining the subharmonic steps in the current–voltage relationship. At a point of operation that is on the verge of a chaotic domain, the circle map, in fact, predicts a complete Devil's staircase of steps below the first harmonic step at the frequency of the drive. Finally, the circle map predicts chaotic behavior whenever its slope goes negative.

With the advent of cell mapping methods it has become possible to find the

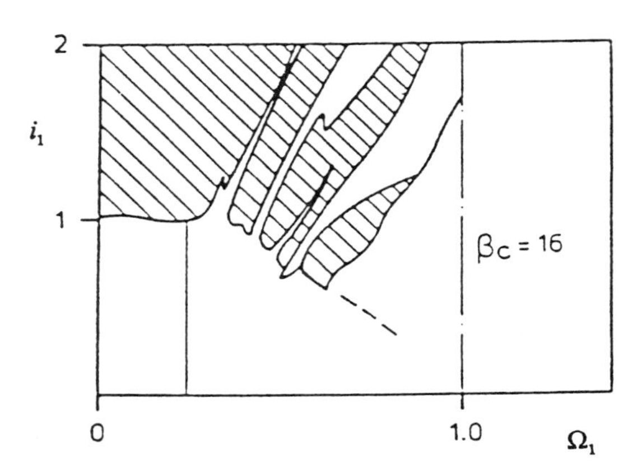

FIGURE 6-28. Regions of quasiperiodic and chaotic motion in the $i_1 - \Omega_1$ control parameter plane (hatched) (after Kurkijärvi, 1991).

different attractors of Eq. (6-68) at given parameter values and locate their basins of attraction. In this way quite detailed information about the nature of features that appear chaotic can be obtained. Soerensen et al. (1988), for example, have identified the end of a downward excursion from an harmonic step as a "boundary crisis," where an attractor in phase space collides with the boundary of its basis of attraction, and the beginning of the next, increasing the dc drive, as an "interior crisis."

To conclude this section, we mention that Josephson parametric amplifiers (externally pumped Josephson junctions) and dc-SQUIDS (*s*uperconducting *q*uantum *i*nterference *d*evices) are Josephson devices that also exhibit strongly nonlinear and chaotic behavior (Kurkijärvi, 1991).

6.6. SUMMARY

The pseudo-NNDC element Josephson superconducting junction was examined in the chapter. We first presented a simple phenomenological discussion of superconductivity, followed by a more detailed analytical description. Tunneling effects were emphasized, with a focus on both quasiparticle and pair tunneling, with the latter producing the Josephson effect. A major portion of the chapter was devoted to circuits and devices, and NNDC instability phenomena. Some of these were thermally induced; they will be discussed in more detail in Chap. 8. Finally, chaotic time-dependent behavior of the rf-biased Josephson junction was discussed.

As a concluding remark, it should be mentioned that the discovery of high-temperature superconductivity (Bednorz and Müller, 1986), with critical temperatures T_c now above the liquid nitrogen range, has given a new stimulus to research on Josephson junction device applications.

7

SNDC Multilayer Semiconductor Structures

7.1. INTRODUCTION

SNDC effects occur in a wide variety of semiconductor junction devices. In particular, we will discuss important cases such as the unijunction transistor (UJT) (Shockley et al., 1949); switchback in a bipolar junction transistor (BJT) (see, e.g., Grove, 1967); the *p-n-p-n* diode (thyristor or Shockley diode) (see, e.g., Blicher, 1976); the *p-i-n* diode (see, e.g., Weber, 1970 and Weber and Ford, 1970); the heterojunction hot-electron diode (Belyantsev et al., 1986; Emanuel et al., 1988). All of these have SNDC-type $I(\Phi)$ characteristics and are most useful at the lower end of the frequency spectrum. Further, they all involve some form of charge injection. The most useful, interesting, and important of these structures are the *p-n-p-n* and *p-i-n* diodes; the *p-i-n* device will be analyzed in some detail. We will, however, briefly discuss the UJT, BJT, and *p-n-p-n* structures first. The circuit response for uniform SNDC elements was discussed in detail in Chap. 2. However, when nonuniform current distributions develop a somewhat more complex problem must be solved. We will develop a "two subelement model" approach when we discuss the *p-i-n* diode; this model applies to devices where single high current density filaments usually dominate, such as some of the electrothermal-type devices discussed in Chap. 8. Shaw et al. (1979) and Solomon et al. (1972) discussed a two subelement model for NNDC devices that form high field domains.

7.2. THE UNIJUNCTION TRANSISTOR

The UJT is a member of the thyristor family (Sze, 1981). It has an emitter junction and two base low-resistance (ohmic) contacts, as shown in Fig. 7-1a. Figure 7-1b is the equivalent circuit. Under normal bias conditions for an *n*-type semiconductor bar having a *p*-type emitter, B_1 is grounded and a positive bias Φ_{BB} is put at B_2. Electrons flow from B_1 to B_2, with resistance to their flow denoted by R_{BB}. The emitter, biased at Φ_E, is at point A, and the resistance between B_2 and A is R_{B2}, and that between B_1 and A is R_{B_1} ($R_{BB} = R_{B_2} + R_{B_1}$). In Fig. 7.1b G_p and G_n represent the excess hole and electron conductances between B_1 and A. That fraction of Φ_{BB} that is dropped from A to B_1 is $R_{B_1}/(R_{B_1} + R_{B_2}) = R_{B_1}/R_{BB} \equiv \eta$, the "stand-off" ratio.

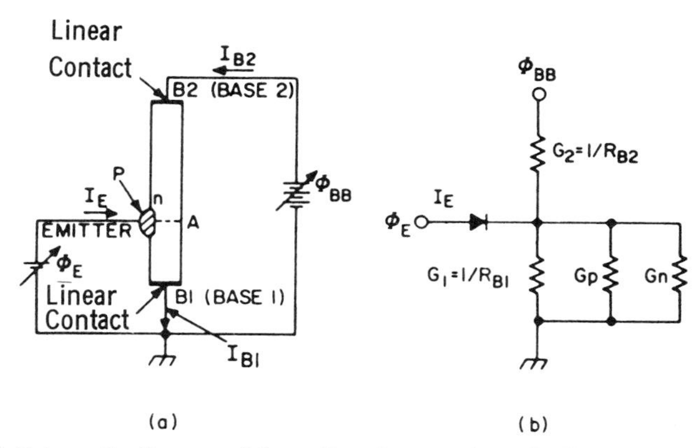

FIGURE 7-1. (a) Schematic diagram of the unijunction transistor (UJT). (b) Equivalent circuit of the UJT. (After Sze, 1981.)

For $\Phi_E < n\Phi_{BB}$ the emitter junction is under reverse bias; a small saturation current flows in the emitter circuit. For $\Phi_E > \eta\Phi_{BB} + \Phi_{FE}$, where Φ_{FE} is the forward voltage dropped across the emitter junction, holes are injected into the n-type bar. The holes move towards B_1 and increase the conductivity of the bar between A and B_1. As the forward I_E increases the voltage at the emitter will decrease as the conductivity of the region between A and B_1 increases; an SNDC characteristic will develop, as shown in Fig. 7-2.

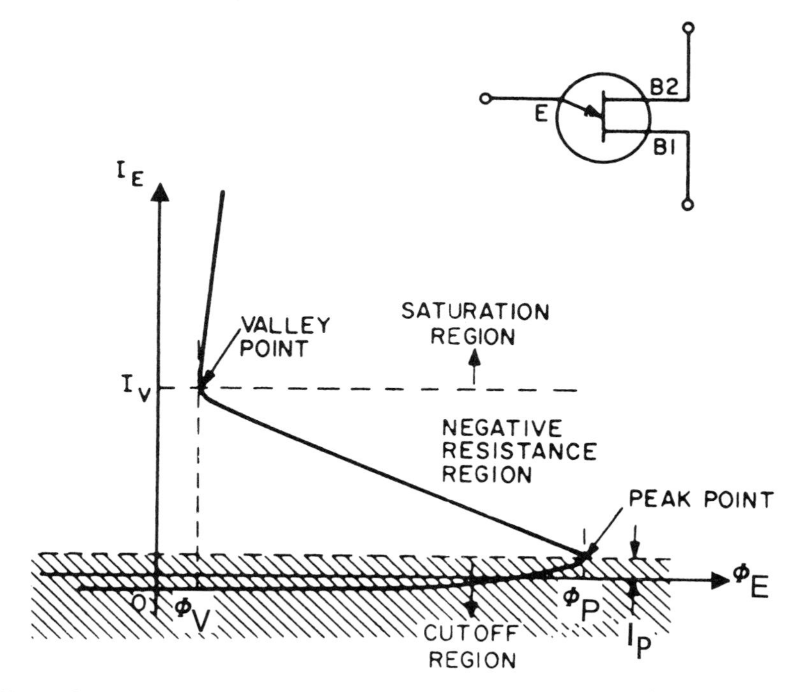

FIGURE 7-2. Current–voltage characteristics and device symbol of the UJT (after Sze, 1981).

7.3. THE BIPOLAR JUNCTION TRANSISTOR

In the common emitter mode of operation of a BJT the characteristics in the forward active region exhibit a "switchback" SNDC appearance under high enough bias for breakdown to occur, as shown in Fig. 7-3 (Grove, 1967). Of significance here is the maximum, or breakdown, voltage that can be reached when the base lead is open, $BV_{\rm CEO}$. When $\Phi_{\rm CE}$ is applied to the collector with the emitter grounded, some of $\Phi_{\rm CE}$ will drop across the base–emitter junction and put it under slight forward bias. In this case the collector current, I_C, will be the sum of the generation current in the reverse biased collector-base junction, $I_{\rm gen}$, plus the current carried by the carriers injected into the base from the emitter, $\gamma\alpha_T I_E$, where γ is the emitter injection efficiency and α_T is the base transport factor (Grove, 1967). Current continuity requires

$$I_E = I_C = \gamma\alpha_T I_E + I_{\rm gen} \tag{7-1}$$

But $I_{\rm gen}$ is the reverse bias leakage current of the collector–base junction under open emitter conditions; $I_{\rm gen} = I_{\rm CBO}$. Hence Eq. (7-1) leads to

$$I_{\rm CEO} = \gamma\alpha_T I_{\rm CEO} + I_{\rm CBO} \tag{7-2}$$

Equation (7-2) assumes that carrier generation or multiplication is absent. Under high bias, however, near breakdown, carrier multiplication will ensue; the normal current will be multiplied by a factor M to yield

$$I_{\rm CEO} = (\gamma\alpha_T I_{\rm CEO} + I_{\rm CBO})M \tag{7-3}$$

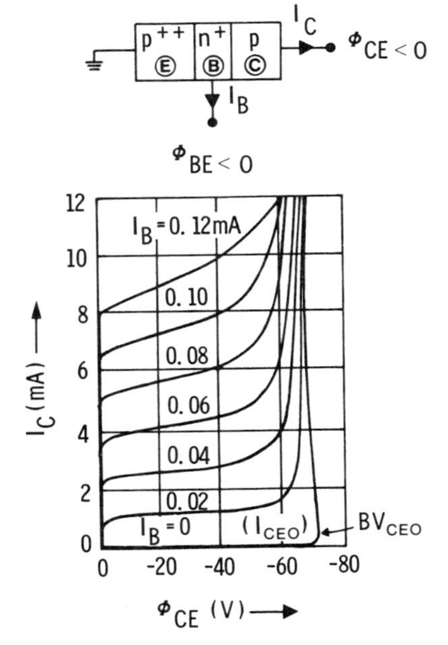

FIGURE 7-3. Current–voltage characteristics of a silicon *p-n-p* transistor in the common emitter configuration (after Grove, 1967).

which results in

$$I_{\text{CEO}} = \frac{I_{\text{CBO}} M}{1 - \gamma \alpha_T M} \tag{7-4}$$

that is, the leakage current under common emitter conditions is greater than that under common base conditions by the factor $1/(1 - \gamma\alpha_T M)$. The current will start to increase rapidly when $\gamma\alpha_T M \rightarrow 1$, rather than when $M \rightarrow \infty$, which is the case in a normal reverse-biased diode or in a BJT in the common base mode. Therefore, the breakdown voltage will be lower in the common emitter mode than in the common base mode. To determine how much this reduction is, we make use of the empirical relation

$$M = \left[1 - \left(\frac{\Phi_{VB}}{BV_{\text{CBO}}}\right)^n\right]^{-1} \tag{7-5}$$

where $3 \leq n \leq 6$ for most semiconductors. At $\Phi_{\text{CB}} = BV_{\text{CEO}}$, $\gamma\alpha_T M = 1$. Substituting Eq. (7-5) for M under these conditions results in

$$\frac{BV_{\text{CEO}}}{BV_{\text{CBO}}} = (1 - \gamma\alpha_T)^{1/n} \cong (h_{\text{FE}})^{-1/n} \tag{7-6}$$

where h_{FE} is the dc common emitter current gain;

$$h_{FE} = \frac{\gamma\alpha_T}{1 - \gamma\alpha_T} \tag{7-7}$$

It is usually the case that as I_c is increased, the dominant component of current carried by diffusion in the base increases, and this tends to increase the emitter efficiency when recombination in the emitter–base space charge region is considered (Grove, 1967). Both theory and experiment show that over a wide range of I_c, h_{FE} is an increasing function of I_c. Therefore, Eq. (7-6) tells us that under breakdown conditions, as the current increases the ratio $BV_{\text{CEO}}/BV_{\text{CBO}}$ will decrease. That is, an SNDC region will develop in the breakdown characteristic, as shown in Fig. 7-3b.

7.4. THE *p-n-p-n* DIODE

A thyristor (Blicher, 1976) is defined as any semiconductor switching device whose behavior depends upon the regenerative feedback available in a *p-n-p-n* structure. A silicon controlled rectifier (SCR) is a three-terminal thyristor. For the details of the operation of three-terminal thyristor structures, the reader is referred to the book *Thyristor Physics* (Blicher, 1976). In this chapter we concentrate on a two-terminal structure–the *p-n-p-n* or Shockley diode. However, for generality we will include a gate terminal where a gate current, I_g, can be injected into the structure.

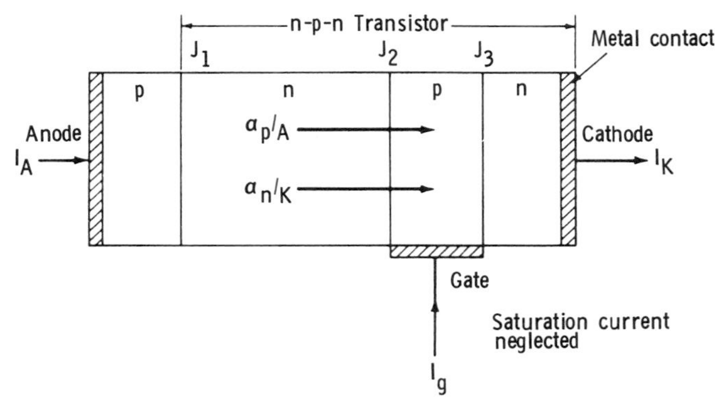

FIGURE 7-4. Current flow in an SCR (after Blicher, 1976).

Figure 7-4 shows the four-layer structure under consideration. It can be considered as a *n-p-n* transistor with an added *p* layer. The *n* emitter is the cathode and the added *p* layer is the anode hole emitter. The *p* base of the *n-p-n* transistor is the gate or control electrode. The device has three junctions and is often treated as a combination of a *n-p-n* and *p-n-p* transistor with a common collector junction J_2 in Fig. 7-4. Each transistor is supplied with a base current by the collector of the other transistor. Figure 7-5 shows this scheme. The device is understood in the following manner. First, let us apply a positive voltage to the anode, putting J_2 under reverse bias and J_1 and J_3 under forward bias. A small saturation current will flow as determined by the current-limiting properties of J_2

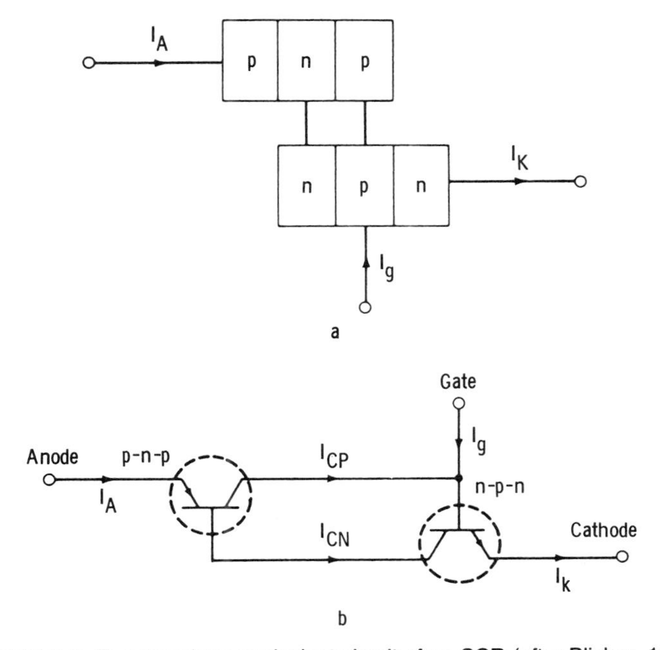

FIGURE 7-5. Two-transistor equivalent circuit of an SCR (after Blicher, 1976).

under reverse bias. The device is now in the forward-blocking OFF-state. Upon reversal of the polarity J_1 and J_3 are under reverse bias and they act to limit the current to low saturation values; we are now in the reverse-blocking OFF-state. These blocking OFF-states exist as long as the applied bias voltages are below the breakdown voltages of J_2 or the sum of the breakdown voltages of J_1 and J_3, and the gate is unbiased.

Next, under forward-blocking conditions let us forward bias the gate by external means (positive voltage applied). The *n*-type cathode will start emitting at a sufficient magnitude to put the *n-p-n* transistor into its conducting state. This in turn provides base current to the *p-n-p* transistor and it, too, is pushed into its conducting state. In the two-transistor model the total loop-gain exceeds unity at this point and regenerative feedback occurs, driving both transistors into saturation after a short delay time. All three junctions are in forward bias; the impedance is minimized and the device is now switched to the ON state as shown in Fig. 7-6. The gate current is usually applied in a pulsed form so that the device can be switched from the OFF to ON states by a short gate pulse. If the ON-state current is kept above a holding value, the device remains ON after the gate pulse is removed. Further, the switching (breakdown) point is a function of the magnitude of the gate current, so that a family of forward $I_A(\Phi_{AK})$ characteristics result, as shown in Fig. 7-6. Note also that an SNDC characteristic also occurs for $I_g = 0$. Here J_2 is avalanched by the application of sufficiently high applied positive bias to the anode; J_2 then ends up in a forward biased state. To understand how the device operates we must first consider the mechanisms

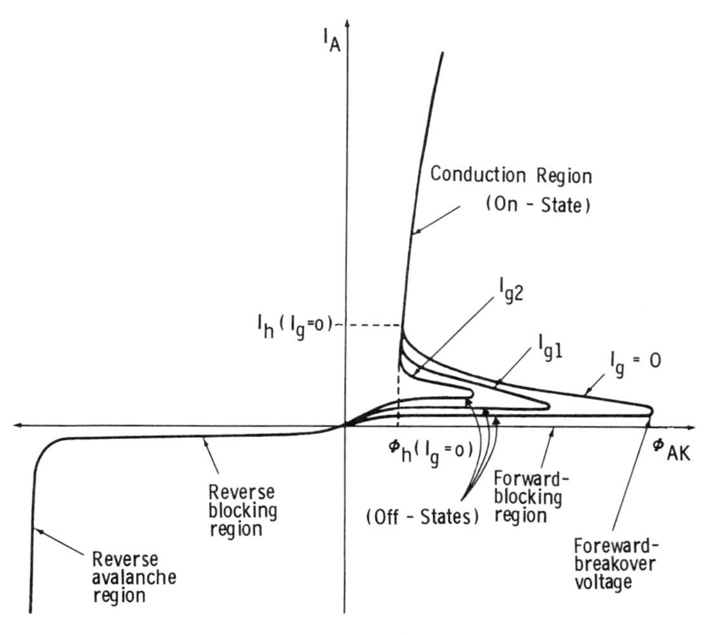

FIGURE 7-6. Family of SCR $I(\Phi)$ characteristics (after Blicher, 1976).

involved when the switching process is triggered by a forward gate pulse rather than by reaching a sufficiently high bias voltage.

The dc common base current gain ($\alpha \equiv h_{FB}$) is defined via the relation

$$I_c = \alpha I_E + I_{CBO} \tag{7-8}$$

The small signal current gain (α_1) is obtained by differentiating Eq. (7-8) with respect to I_E, keeping the collector voltage constant,

$$\alpha_1 = \alpha + I_E \frac{d\alpha}{dI_E} \tag{7-9}$$

Neglecting the saturation current, we can write the anode current in Fig. 7-7 as the sum of the two currents that reach the collector J_2:

$$I_A = \alpha_p I_A + \alpha_N I_K \tag{7-10}$$

where $I_K = I_A + I_g$ and $\alpha_p(\alpha_n)$ is the current gain of the *p-n-p* (*n-p-n*) transistor. We may write

$$I_A = \frac{\alpha_N I_g}{1 - (\alpha_p + \alpha_N)} \tag{7-11}$$

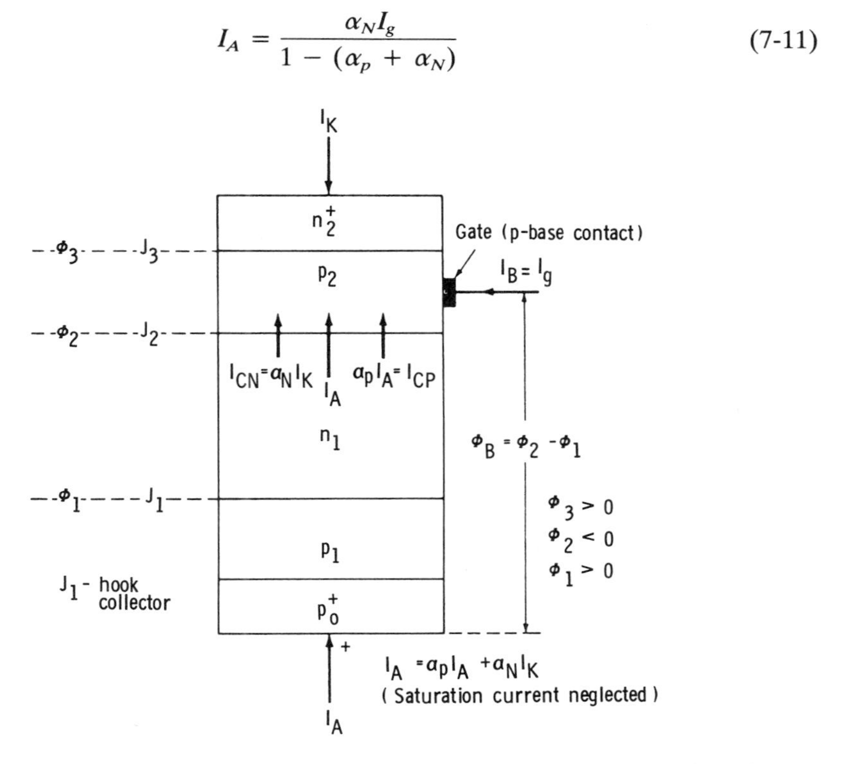

FIGURE 7-7. SCR schematic structure at low current levels, in the OFF condition (after Blicher, 1976).

Switching will occur when the breakdown point is reached, i.e., at

$$\left.\frac{dI_A}{dI_g}\right|_{\Phi=\text{const}} \rightarrow \infty \tag{7-12}$$

Equations (7-9) and (7-11) yield

$$\frac{dI_A}{dI_g} = \frac{\alpha_2}{1 - (\alpha_1 + \alpha_2)} \tag{7-13}$$

where $\alpha_1(\alpha_2)$ is the small signal gain of the *p-n-p* (*n-p-n*) transistor. Hence, the device turns ON when $\alpha_1 + \alpha_2 \rightarrow 1$.

Current gains are of course voltage dependent. Near the condition for avalanche in J_2, carrier multiplication must be considered so that the actual switching criterion should be

$$M_p\alpha_1 + M_n\alpha_2 = 1 \tag{7-14}$$

with M_p and M_n functions of the voltage across J_2. Equation (7-5) denotes the empirical relationship most commonly found.

Although the *p-n-p-n* diode is a junction device and does not possess bulk NDC properties, it can still develop current filamentation due to the development of a lateral instability in devices of sufficiently large cross-sectional area (Varlamov and Osipov, 1970). (Indeed, if the *p-n-p-n* structure is inhomogeneous and a local region possesses a lowered turnover or threshold voltage, current filamentation at this point is expected.) Varlamov and Osipov (1970) treat a system where the SNDC characteristics are associated with recombination in the emitter junctions and with leakage current uniformly distributed across the collector junction. They demonstrate that a uniform distribution of current is unstable against filamentation in the SNDC region for a sufficiently wide device and, for the boundary condition employed, the most stable configuration is a current filament adjoining a side surface.

To experimentally investigate current filamentation in *p-n-p-n* structures Varlamov et al. (1970) studied a distributed four-layer structure by combining several discrete *p-n-p-n* diodes of narrow width in parallel (connected by one base), as shown in Fig. 7-8. This structure is used to model a real *p-n-p-n* device.

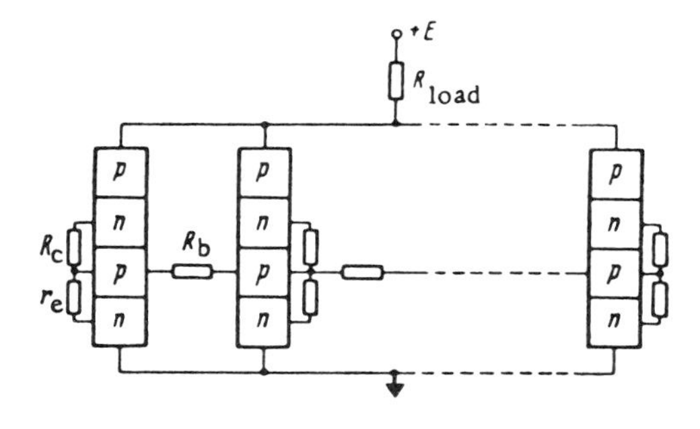

FIGURE 7-8. Diagram of distributed four-layer structure. R_c and r_e are the collector and emitter junction leakage resistance. R_b is the "longitudinal" resistance of the base. (After Varlamov et al., 1978.)

In fact, it can be considered to be a four-subelement model; we will discuss a two-subelement model with regard to circuit effects in Sec. 7.7. With this structure inhomogeneities can be simulated, the base resistance and sample size can be varied, and different boundary conditions can be produced. Typical $I(\Phi)$ characteristics are shown in Fig. 7-9. Each discrete device that comprised the parallel array had uniform, stable, SNDC $I(\Phi)$ characteristics that did not exhibit current filamentation. When put in parallel, however, Fig. 7-9 shows that "filamentary" characteristics are observed in that only one or a few of the devices will turn ON once threshold is reached. However, as the current is increased further in the ON-state, the filament expands, i.e., more of the subelements enter the ON-state. Eventually a change to a uniform distribution occurs when all the subelements are turned ON. Here the device voltage exhibits the $3 \rightarrow 4$ discontinuity shown in Fig. 7-9. With a reduction in current the transition $5 \rightarrow 6$ denotes a switchback to a nonuniform filamentary distribution; the transition $7 \rightarrow 8$ signifies the switch back to the uniform OFF-state.

These results hold for the case when the longitudinal component of current vanishes at the boundary. Here the filament lies at one boundary of the sample (Fig. 7-9b); shunting the emitter junction at the boundary moves the filament to the opposite wall. To obtain a filament at the center of the structure the emitter junctions at both boundaries are shorted. Connecting the first and last elements through a resistance R_b (periodic boundary conditions) produces an $I(\Phi)$ curve similar to that shown in Fig. 7-9.

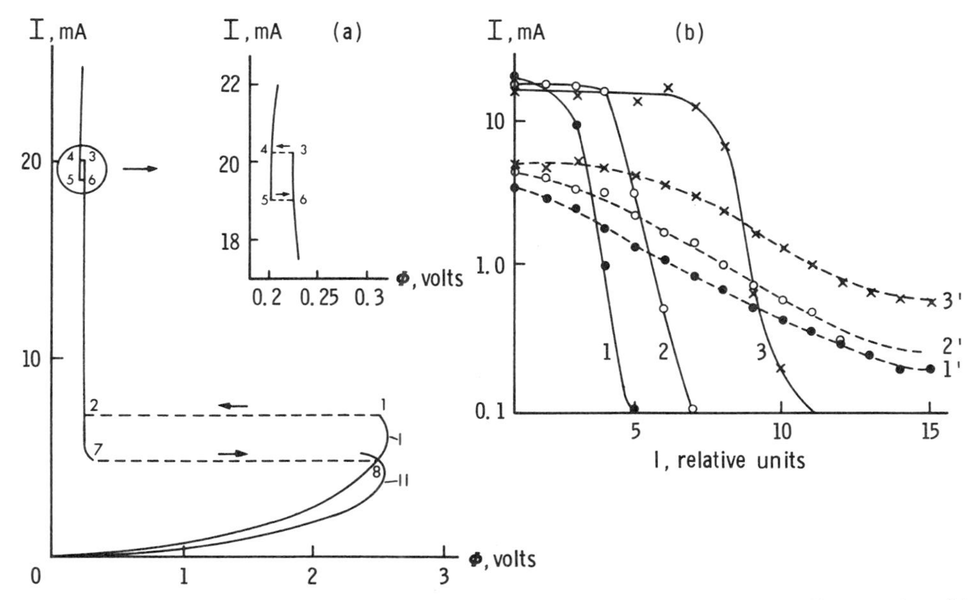

FIGURE 7-9. (a) Current–voltage characteristics of a *p-n-p-n* structure exhibiting filamentation; (b) current distributions in emitter (dashed curves) and collector (continuous curves) *p-n* junctions. (a) I, II, Current–voltage characteristics with 20 and 13 elements in the structure. (b) 1–3, 1′–3′ Filamentary current distributions in collector and emitter junctions for various values of the total current I through the structure ($I_1 < I_2 < I_3$). (After Varlamov et al., 1978.)

The above results show that the $I(\Phi)$ characteristics of a *p-n-p-n* structure under essentially isothermal conditions has two regions of hysteresis when the longitudinal component of the current vanishes at the boundaries. Discontinuities and hysteresis are related to the formation and disappearance of current filaments, and for these boundary conditions a filament adjoining one of the lateral surfaces is stable. If a filament is established at the center of the sample as a result of e.g., illumination, heating, or shunting, it will move back to its stable position near the edge of the sample once the local "inhomogeneity" is removed.

7.5. THE p-i-n DIODE

7.5.1. Current Injection in Solids—Injection of One Type of Carrier

The *p-i-n* diode exhibits a region of SNDC because of the injection of both electrons and holes into a material containing traps (see Chap. 1). In order to understand this process in detail, we will begin with a phenomenological analysis of the injection of one type of carrier into a semiconductor (Lampert and Mark, 1970). The details of the analytic and approximate $I(\Phi)$ solutions will be set forth when we discuss the problem of double injection.

To begin with, we choose cathode boundary conditions so that the cathode contact provides all the electrons demanded by the bulk. It is an accumulated majority carrier injecting contact (Shaw, 1981). The simplest system to analyze is the case of injection into the perfect trap-free insulator containing a negligible concentration of thermally induced free carriers. All the injected electrons will remain free, in the conduction band, and will contribute to space charge. Neglecting diffusion, the current density is

$$J = \rho v \tag{7-15a}$$

or

$$J = Q/t \tag{7-15b}$$

where ρ is the average injected free charge density, v the average velocity, Q the total injected charge per unit area, and t the transit time between electrodes. If L is the length of the sample (origin at the anode) then

$$t = L/v \tag{7-16a}$$

and

$$Q = \rho L \tag{7-16b}$$

To obtain the $J(\Phi)$ characteristic we must determine how Q, or ρ, and t, or v, depend on Φ. Analogously to the parallel plate condenser of capacitance C_0 $(=\epsilon/L)$, we might expect that $C = Q\Phi$, and, if the injected charge were uniformly distributed between the electrodes, its average distance would be at $L/2$; the capacitance would be twice the "geometric" capacitance of the parallel plate condenser with charge only on the plates. But, since the charge is injected from the cathode, we might expect a nonuniform distribution with the average distance somewhere between L and $L/2$. (The case of no injection puts the average distance at L.) Hence we expect that $C_0 < C < 2C$. Within a factor of 2 then, we can write

$$Q \cong C_0\Phi = (\epsilon/L)\Phi \tag{7-17}$$

whence, from Eqs. (7-15b), (7-16a), and (7-17) we obtain

$$J \cong C_0 v\Phi/L = \epsilon v\Phi/L^2 \tag{7-18}$$

At low uniform electric fields $v = \mu F = \mu(\Phi/L)$, so Eq. (7-18) becomes

$$J \cong \epsilon\mu\Phi^2/L^3 \tag{7-19}$$

To see how this result compares with the exact calculation, let us do the latter. Writing $J = e\mu n(x)F(x)$ and combining this with the Poisson equation, $\epsilon(dF/dx) = en(x)$ yields $J = \mu\epsilon F(x)(dF/dx)$, which is independent of position. integration yields

$$F(x) = \left(\frac{2Jx}{\epsilon\mu}\right)^{1/2} \tag{7-20}$$

To obtain the $J(\Phi)$ characteristic we make use of $\Phi = \int_0^L F(x)\,dx$, which yields

$$J = \frac{9}{8}\epsilon\mu\frac{\Phi^2}{L^3} \tag{7-21}$$

The exact result thus differs from the phenomenological result by a numerical factor of 9/8.

Equation (7-19) [or (7-21)] is a very useful result because it not only represents an ideal situation, but also that of an imperfect insulator under conditions where the number of injected carriers substantially exceeds the number of initially empty traps, a common case we shall shortly study. Further, the phenomenological equations can be used for the case of thermionic injection into vacuum. In this collision-free case $v \cong (e\Phi/m)^{1/2}$ and Eq. (7-18) becomes

$$J \cong \epsilon_0\left(\frac{e}{m}\right)^{1/2}\frac{\Phi^{3/4}}{L^3} \tag{7-22}$$

Child's law. When an exact derivation of Child's law is performed, a factor of $4\sqrt{2}/9$ appears.

We next include the presence of some thermally generated free carriers of concentration n_0, whose origins are not effective as electron traps. At low voltages, where the number of injected electrons is negligibly small, we expect

Ohm's law to hold:

$$J = en_0\mu\left(\frac{\Phi}{L}\right) \tag{7-23}$$

Once the average number of injected electrons, n_i, becomes comparable to n_0 we expect a crossover from the linear (neutral) to square law (space-charge-limited) regimes to occur. This should happen at a voltage Φ_x determined by

$$en_0L = Q = C\Phi_x \cong \left(\frac{\epsilon}{L}\right)\Phi_x \tag{7-24}$$

i.e., at

$$\Phi_x \cong \frac{en_0L^2}{\epsilon} \tag{7-25}$$

Now, Eq. (7-16a) for uniform fields yields for the transit time between electrodes $t_t = L^2/\mu\Phi_x$. The dielectric relaxation time $\tau_D = \epsilon/en_0\mu$. Equation (7-25) is therefore a statement that the crossover voltage occurs when $\tau_D \cong t_t$. Another way of looking at this equality involves the realization that at any value of Φ injected charge is being continuously drawn into the volume, and continuously being neutralized by dielectric relaxation. Let us for the moment imagine these mechanisms to operate sequentially rather than simultaneously. Consider a volume element near the anode. First, turn on the drift mechanism for a period equal to t_t. Charge will flow from the cathode reservoir and the volume element will charge up to some charge density ρ_0. Now turn off the drift mechanism and turn on the dielectric relaxation mechanism for an equal period t_t. Since $\rho = \rho_0 \times \exp(-t/\tau_d)$, we see that ρ_0 is diminished by a fraction $\exp(-t_t/\tau_d)$. For $t_t \gg \tau_d$ transport occurs essentially without space charge; we are in the Ohm's law regime. At $t_t = \tau_d$ we are in the space-charge-limited regime.

Suppose now that we increase the applied voltage even further, making t_t even shorter. What happens is that as the carriers are drawn more rapidly into the interior, their density increases (conductivity modulation). This raises the conductivity, which in turn shortens the dielectric relaxation time, exactly in proportion to the decreased transit time. Carriers are neutralized as fast as they arrive and they arrive as fast as they are neutralized; $t_t \cong \tau_d$ not only marks the onset of the space-charge-limited regime, it roughly defines the regime itself. Figure 7-10 is a plot of t vs. Φ; the two different regimes are separated at Φ_x.

When traps are also present in the material the current will be reduced at low injection levels since some of the injected charge is immobilized. But the amount of total excess charge in the material at any value of Φ will remain unchanged at $C\Phi$. Now, however, a fraction of the injected charge will be trapped. Equation (7-16b) then becomes

$$Q = (\rho + \rho_t)L \cong C_0\Phi = (\epsilon/L)\Phi \tag{7-26}$$

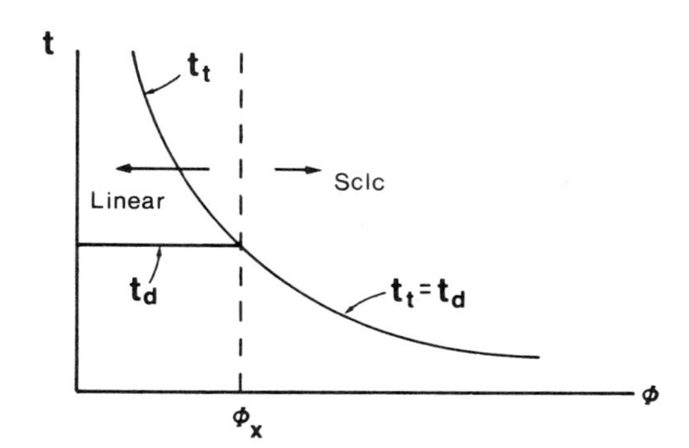

FIGURE 7-10. Time of transit, t_t, vs. applied bias Φ. At a sufficiently high voltage, Φ_x, the transit time equals the dielectric relaxation time and the two remain equal for all voltages greater than Φ_x.

where ρ_t is the average injected trapped charge density. (Note that we cannot write $J = Q/t$ anymore.)

In thermal equilibrium the free electron concentration is given by

$$n_0 = N_c \exp[(E_F - E_c)/kT] \tag{7-27}$$

The concentration, $n_{t,0}$, of filled electron traps at level E_t is

$$\begin{aligned} n_{t,0} &= \frac{N_t}{1 + (1/g)\exp[(E_t - E_F)/kT]} = \frac{N_t}{1 + (1/g)(N/n_0)} \\ N &\equiv N_c \exp[(E_t - E_c)/kT] \end{aligned} \tag{7-28}$$

where N_t is the trap concentration and g the degeneracy factor for the traps. The equilibrium trap occupancy results from a balance between the trapping of electrons at E_t and their thermal reemission, a process that we assume to be field independent for the moderate electric fields involved in the injection process. That is, the balance between free and trapped electrons can be changed only through the change in free carrier concentration induced by injection. Under injecting conditions the electron quasi-Fermi level, E_{Fn}, is defined via the total free electron concentration:

$$n = n_1 + n_0 = N_c \exp[(E_{Fn} - E_c)/kT] \tag{7-29}$$

where n_1 is the average excess injected free electron concentration. The trapped electron concentration is then given by

$$n_t = n_{t,i} + n_{t,0} = \frac{N_t}{1 + (1/g)\exp[(E_t - E_{Fn})/kT]} = \frac{N_t}{1 + (1/g)(N/n)} \tag{7-30}$$

where $n_{t,i}$ is the average injected excess trapped electron concentration.

A single level of traps is considered to be "shallow" if $E_{Fn} < E_t$. Here $E_F < E_t$ must be the case. Equation (7-28) then yields $[(E_t - E_{Fn})/kT > 1]$

$$\frac{n}{n_t} = \frac{\rho}{\rho_t} \cong \frac{N}{gN_t} = \frac{N_c}{gN_t}\exp\left(\frac{E_t - E_c}{kT}\right) = \theta \tag{7-31}$$

where θ is a constant if the trap remains shallow, independent of injection level.

If $\theta \ll 1$, the shallow trap concentration will affect the SCL injection in a major way. If $n \ll n_t$, Eq. (7-26) yields

$$Q \cong \rho_t L \cong \rho L/\theta \cong \epsilon\Phi/L \tag{7-32}$$

whence Eqs. (7-32) and (7-15b) yield

$$J \cong \theta\epsilon\mu\frac{\Phi^2}{L^3} \tag{7-33}$$

As for the trap free case, the analytically derived result (Lampert and Mark, 1970) again has the 9/8 factor present. Further, it is easily shown that the crossover voltage Φ_x in this shallow-trap region law regime is θ^{-1} times the crossover voltage for the trap free case.

It is now useful to define the concept of an effective drift mobility μ_e as

$$\mu_e = \left(\frac{n}{n + n_t}\right)\mu \tag{7-34}$$

For $n \ll n_t$ this becomes

$$\mu_e \cong \left(\frac{n}{n_t}\right)\mu = \theta\mu \tag{7-35}$$

In terms of μ_e we have $t_{x,e} \cong \tau_d$ with $t_{x,e} = L^2/\mu_e\Phi_x$, where $t_{x,e}$ is the effective cathode-to-anode transit time of the entire body of injected charge at voltage Φ_x.

When the single level of traps is "deep," i.e., for $(E_{Fn} - E_t)/kT > 1$, in thermal equilibrium the concentration of unoccupied traps, i.e., the hole occupancy of the traps, is

$$p_{t,0} = N_t - n_{t,0} = \frac{N_t}{1 + g\exp[(E_F - E_t)/kT]} \cong \frac{N_t}{g}\exp\left(\frac{E_t - E_F}{kT}\right) \tag{7-36}$$

a valid approximation when $(E_F - E_t)/kT > 1$. The $J(\Phi)$ curve will again be linear up to Φ_x, when $n_i \cong n_0$ will hold. Here the total thermal population of free carriers will be doubled, which corresponds to the movement of E_{Fn} toward E_c by about kT. This motion will be sufficient to essentially fill the deep traps. Here Φ_x will be coincident with the voltage Φ_{TFL} required to fill the deep traps; since $Q = C_0\Phi$, we expect that

$$\Phi_{\text{TFL}} \cong \frac{Q_{\text{TFL}}}{C_0} = \frac{ep_{t,0}L}{C_0} \cong \frac{ep_{t,0}L^2}{\epsilon} \tag{7-37}$$

(TFL denotes trap filled limit.)

We ask how we might expect the $J(\Phi)$ characteristic to behave above the voltage Φ_{TFL}. Let us imagine that $\Phi = 2\Phi_{\text{TFL}}$. Since the injected charge is proportional to Φ, we expect that $Q(2\Phi_{\mathrm{TFL}}) = 2Q_{\mathrm{TFL}}$. But since the traps were filled at Φ_{TFL}, the additional injected charge, $Q(2\Phi_{\mathrm{TFL}}) - Q_{\mathrm{TFL}} = Q_{\mathrm{TFL}} = ep_{t,0}L$, will all appear in the conduction band. The ratio of the currents will be

$$\frac{J(2\Phi_{\mathrm{TFL}})}{J(\Phi_{\mathrm{TFL}})} = \frac{2n(2\Phi_{\mathrm{TFL}})}{n(\Phi_{\mathrm{TFL}})} \cong \frac{p_{t,0}}{n_0} \tag{7-38}$$

In many insulators $p_{t,0}/n_0$ can be several orders of magnitude; the expected $J(\Phi)$ curve is shown in Fig. 7-11a, labelled I. We see that the current rises vary steeply at Φ_{TFL}, and shortly beyond $2\Phi_{\mathrm{TFL}}$ it merges with the trap-free square law since the injected free charge then dominates the injected trapped charge.

A family of $J(\Phi)$ characteristics for a single set of traps of density N_t results for different values of E_t. The family is contained in a triangle in the $\log J$–$\log \Phi$ plane as shown in Fig. 7-11a. The triangle is bounded by Ohm's law, the trap-free square law (TFSL), and the TFL vertical leg. This last line is the unphysical case where all traps are assumed filled at the outset, before Φ is applied. Since Φ_{TFL} is required to support the excess injected charge, no current will flow in this case until Φ_{TFL} is reached.

The family of $J(\Phi)$ curves consists of two subfamilies, for $E_t \lesseqgtr E_F$. For $E_t < E_F$, deep traps, a typical member is curve I in Fig. 7-11a, with the vertical

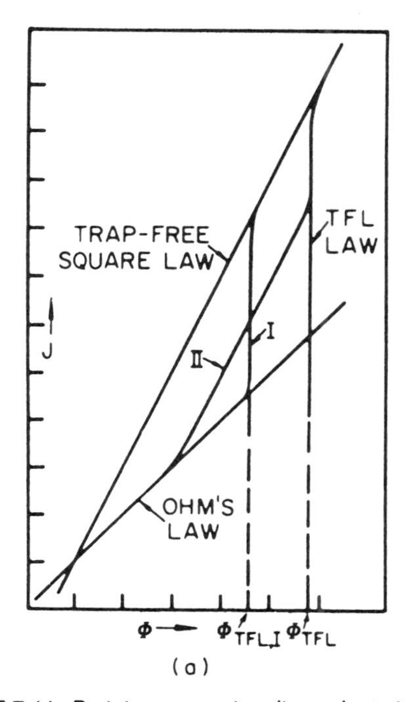

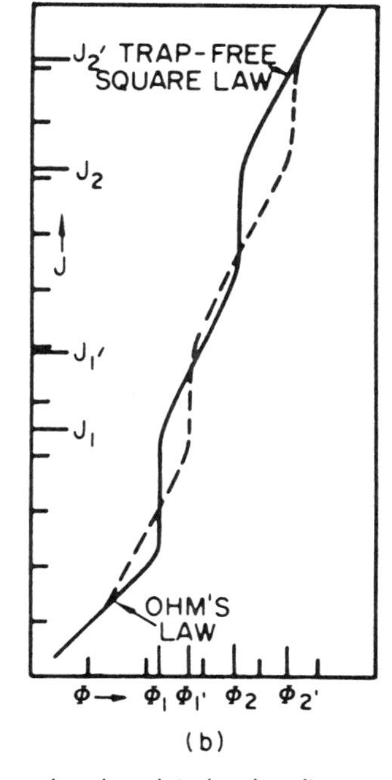

FIGURE 7-11. Prototype current–voltage characteristics on log–log plots (each unit corresponding to a decade): (a) prototype curves for a single set of traps; I corresponds to E_F lying above E_t, to E_F lying below E_t; (b) prototype curves for two sets of traps; the solid curve corresponds to E_F lying above one set of traps and below the other set; the dashed curve to E_F lying below both sets of traps. (After Lampert and Mark, 1970.)

section occurring at $\Phi_{TFL,I} \cong ep_{t,0}L^2/\epsilon$. When the traps are so deep that $p_{t,0} \cong n_0$ the vertical portion disappears and the traps no longer affect the current flow. For $E_t > E_F$ the traps are shallow; the current follows the square law with a θ factor after departing from linearity at Φ_x. Near Φ_{TFL}, E_F crosses E_t and the $J(\Phi)$ curve merges with the TFL law. A typical curve, labeled II, is shown in Fig. 7-11a. When $E_t \to E_c$ so that $\theta \cong 1$, the affect of the traps diminishes and only the TFSL is observed.

Figure 7-11b shows a representative case for two sets of traps. The argument for its shape follows the above line of reasoning (Lampert and Mark, 1970).

The assumption of one or more discrete-type trapping levels is valid for many single-crystal systems having a moderate to low density of impurity atoms. When the density of these impurities produces impurity bands, the trapping distribution may become Gaussian, e.g., $N_t(E) = N_1 \exp[-(E - E_t)^2/\Gamma^2]$, where N_t is the trap concentration per unit energy and the Γ factor represents a measure of the smearing out of the level at E_t. Complete basic structural disorder produces an exponential distribution of localized states (Mott and Davis, 1979) of the form

$$N_t(E) = N_0 \exp\left(\frac{E - E_c}{kT_t}\right) = N_n \exp\left(\frac{E - E_F}{kT_t}\right)$$
$$N_n = N_0 \exp\left(\frac{E_F - E_c}{kT_t}\right) \tag{7-39}$$

with T_t a temperature descriptive of the trap distribution. Many amorphous solids (Madan and Shaw, 1988) probably have a density of states in the "mobility" gap that is a superposition of exponential tail states and discrete or Gaussian levels associated with defects in the random structure (Mott and Davis, 1979; Kastner et al., 1976; Adler, 1978). For a simple exponential distribution Lampert and Mark (1970) show that for $T_t > T$

$$J \sim \Phi^{(l+1)}/L^{(2l+1)} \tag{7-40}$$

with $l = T_t/T$.

7.5.2. Injection of Two Types of Carriers

When one contact to the insulator or semiconductor is made electron injecting and the other hole injecting, we can obtain double injection. Since the injected carriers can largely neutralize one another, a two-carrier injection current will be larger than either single carrier current in a given sample. Further, a new limitation on the current now appears—carrier loss via recombination. This is usually a two-step process through a localized recombination center.

The vacuum analog of the above process will not lead to a total current substantially larger than the electron SCL current given by Child's law [Eq. (7-22)]. This occurs because in the vacuum diode the drift velocity depends upon the potential difference that the carrier has traversed. Where carriers of one sign are drifting slowly and have a high concentration, carriers of the opposite sign are at high velocity and low concentration. In the solid we note that scattering causes the drift velocity to depend upon the local field and both species drift slowly when the field intensity is low, and rapidly at high fields; maximum opportunity for mutual neutralization results. Much greater enhancement of the current can occur here. The difference between collision dominated flow and collision free flow, therefore, has a more interesting consequence for double injection than for single injection in the SCL current regime. Indeed, double injection is a richer phenomenon in general. In particular, contact effects are more important, since we find that contacts that are injecting for one type of carrier are generally blocking for the other type. Contact constraints and the diffusion currents associated with them must now be treated when the sample size is less than about the ambipolar diffusion length. For longer samples, however, a simplified theory can be usefully applied, and we shall emphasize that situation here.

For the single injection case we found that space charge limited the current flow. For double injection we will find that another mechanism, recombination, also plays a major role. In the single carrier case the presence of space change forced dF/dx to be nonvanishing, so that F will be nonvanishing over the bulk of the semiconductor. A given J will then be associated with a given Φ. In the double injection problem recombination plays a similar role; it constitutes a "disappearance" of injected carriers. The continuity equations require that J_n and J_p have nonvanishing divergences when recombination is present. Since $J \sim F$, then nonvanishing divergences imply nonvanishing dF/dx, a role similar to that of space charge.

In general double injection theory we have two equations involving dF/dx. In some problems, however, it is reasonable as a first approximation to ignore the Poisson equation in obtaining a solution; dF/dx is determined solely by recombination, and the space charge is accommodated by small perturbations in the electron and hole distributions, which themselves are determined by recombination. This type of situation occurs in the case of double injection into a semiconductor, where the thermally generated free carriers are sufficient in number to relax any injected space charge. For example, let us consider an n-type semiconductor with equilibrium population n_0 and p_0, where $n_0 - p_0 = N_D - N_A$. Assume that the excess thermal electron concentration, $n_0 - p_0$, is sufficiently large so that the excess injected charge is dissipated by dielectric relaxation in a time shorter than any other time involved in the problem; we have local charge neutrality everywhere:

$$n - p = n_0 - p_0 = N_D - N_A \tag{7-41}$$

If we create, by means such as optical excitation, an electron–hole plasma in this system, the ambipolar drift mobility is (Lampert and Mark, 1970)

$$\mu_a = \frac{n - p}{n/\mu_p + p/\mu_n} \tag{7-42}$$

For the case of averages, with $n \gg n_0$, $p \gg n_0$ and $\bar{n} \cong \bar{p}$, we have from Eqs. (7-41) and (7-42)

$$\bar{\mu}_a = \frac{n_0 - p_0}{\bar{n}} \frac{1}{1/\mu_n + 1/\mu_p} \tag{7-43}$$

If $\bar{\tau}$ is the common average lifetime for the injected carriers, $|\bar{\mu}_a|$ the average pair drift mobility, and $\bar{t}_a$ the average pair drift transit time, Lampert and Mark (1970) show that

$$\bar{t}_a \cong \bar{\tau} \qquad \text{with } \bar{t}_a = L^2/(|\bar{\mu}_a|\,\Phi) \tag{7-44}$$

Equations (7-43) and (7-44) yield

$$\bar{n} \cong \frac{(n_0 - p_0)\mu_n\mu_p\bar{\tau}}{\mu_n + \mu_p}\frac{\Phi}{L^2} \tag{7-45}$$

whence

$$J \cong e(n_0 - p_0)\mu_n\mu_p\bar{\tau}\left(\frac{\Phi^2}{L^3}\right) \tag{7-46}$$

In the above analysis the use of Eq. (7-44) actually means that plasma injected into semiconductors is recombination limited; the assumption of local neutrality precludes any role for space charge.

When traps are present in a double injection system the possibility of SNDC arises, and this is, of course, the focal point of our treatise. Indeed, we have already shown how SNDC arises in this case in detail in Sec. 1.3.4. The SNDC usually results from situations where the free carrier lifetime increases substantially with injection level; usually the lifetime of one species increases while the other decreases. For example, consider the case where there is a single set of N_R recombination centers that lie well below E_F and are therefore completely occupied by electrons in equilibrium. At low levels of injection the occupancy of the N_R states is negligibly perturbed; the electron lifetime $\tau_{n,\text{low}}$ is essentially infinite since no empty recombination centers are available to capture the electrons. The hole lifetime $\tau_{p,\text{low}} = 1/(N_R\langle v\sigma_p\rangle)$, where σ is the capture cross section, is relatively short. Hence we have a "recombination barrier" to the passage of holes, but none to the passage of electrons. The injected current will be like a trap free electron SCL current that recombines with injected holes within about a diffusion length from the anode. When Φ is sufficient to drive the holes across the bulk, i.e., when Φ_{th} is reached such that $t_{p,\text{th}} = L^2/(\mu_p\Phi_{\text{th}}) \cong \tau_{p,\text{low}}$, the situation changes. Because $\sigma_p \gg \sigma_n$ the center captures holes preferentially, filling up with them. This causes the hole lifetime to increase, and, since both carriers are being injected into the bulk, the bulk will be mostly neutral. When the injected carrier levels substantially exceed N_R, they are essentially free and equal. With $n \cong p$, the steady state requirement is $\tau_{n,\text{high}} \cong \tau_{p,\text{high}}$. But since the centers are all now almost completely occupied by holes, $\tau_{n,\text{high}} \cong 1/(N_R\langle v\sigma_n\rangle)$. Therefore, between the low and high level injection regimes the hole lifetime increases the ratio σ_p/σ_n, which can be many orders of magnitude. The recombination barrier decreases with increasing injection level; the greater the number of injected holes, the easier it is for them to traverse the sample. Indeed, it can become so much easier that the voltage required to drive them across can actually decrease, resulting in a region of NDC (Stafeev, 1959), as shown in Fig. 1-37 and discussed in some detail in the next section.

It is interesting to estimate the minimum voltage Φ_m and corresponding current I_m reached at the low-voltage end of the NDC segment shown in Fig. 1-37. Note that above I_m high-level injection conditions obtain. Here the injected plasma has $\tau_{n,\mathrm{high}}$ and $\tau_{p,\mathrm{high}}$; electrons initially in recombination centers are moved into the conduction band and the insulator behaves like an n-type semiconductor having N_R thermally generated electrons. Lampert and Mark (1970) derive the $I(\Phi)$ characteristic for such a situation. Φ_m is the lower limit of validity of the semiconductor square law:

$$t_{p,m} = L^2/(\mu_p \Phi_m) \cong \tau_{p,\mathrm{high}} = \frac{1}{N_R \langle v\sigma_n \rangle}$$

The corresponding current is

$$J_m \cong eN_R \tau_{p,\mathrm{high}} \mu_n \mu_p \Phi_m^2 / L^3 \cong eN_R Lb / \tau_{p,\mathrm{high}}$$

where b is the mobility ratio ($b = \mu_n/\mu_p$). Note that $\Phi_{\mathrm{th}}/\Phi_m = \tau_{p,\mathrm{high}}/\tau_{p,\mathrm{low}}$. All of the features just discussed are displayed in Fig. 1-37.

7.5.3. Double Injection in Long Silicon p-i-n Diodes Containing Deep Double Acceptors

Weber (1970), Weber and Ford (1970) and Dudeck and Kassing (1977) have discussed this particular case in great detail. Since it is perhaps the most common *p-i-n* structure we will examine it with some care. We must solve the continuity and Poisson equations in one dimension under steady state conditions. The recombination kinetics are treated via the Shockley–Read technique (1952). The continuity equations for electrons and holes are

$$-\frac{d}{dx}(\mu_n nF) = (e_n^- - c_n^- n)N^- - c_n n N^0 + e_n^= N^= \tag{7-47}$$

$$\frac{d}{dx}(\mu_p pF) = (e_p^- - c_p^- p)N^- - c_p^= p N^= + e_p N^0 \tag{7-48}$$

where N^0, N^-, and $N^=$ are the densities of neutral, singly charged, and doubly charged centers, $n(p)$ is the number density of free electrons (holes), and $e^i_{n(p)}$ and $c^i_{n(p)}$ are the emission and capture coefficients for electrons (holes) in the ith level. In steady state we must have

$$(e_p^- + c_n^- n)(N_R - N^0 - N^=) = (e_n^= + c_p^= p)N^= \quad (7\text{-}49)$$

$$(e_n^- + c_p^- p)(N_R - N^= - N^0) = (e_p + c_n n)N^0 \quad (7\text{-}50)$$

where $N_R = N^0 + N^- + N^=$.

Poisson's equation is

$$\frac{dF}{dx} = \left(\frac{e}{\epsilon}\right)(p - n - N^- - 2N^= + N_D^+ - N_A^-) \quad (7\text{-}51)$$

The appropriate boundary conditions are those for perfectly injecting hole (anode) and electron (cathode) contacts: $F(0) = F(L) = 0$. To obtain the exact $I(\Phi)$ characteristics Eq. (7-47)–(7-51) must be solved numerically. A quasi-neutrality approximation, however, provides solutions that are representative of the exact solutions. To this end we set Eq. (7-51) equal to zero, treating it as a constraint on the carrier density relationships. (The approximate solutions will not satisfy the boundary conditions at one contact, but will be good approximations across the rest of the sample.)

In equilibrium E_F is constrained to be between the two acceptor levels by setting $N_D^+ \cong N_R$ and $N_A^- = 0$. Further, we assume a sufficiently low T and sufficiently high injection level such that the trapped carrier density

$$n_t, p_t \ll n, p \ll N_R \quad (7\text{-}52)$$

Under these conditions the free carrier space charge is negligible in Eq. (7-51) and the emission coefficients are small. The neutrality condition is

$$N^- + 2N^= = N_D^+ \quad (7\text{-}53)$$

Using Eqs. (7-49) and (7-50), keeping only those terms linear in the emission coefficients, Eq. (7-53) can be written

$$\delta c_p^- c_p^= p^2 - 2c_n^- c_n(1 - \tfrac{1}{2}\delta)n^2 - c_n c_p^=(1 - \delta)np = 2(e_p c_n^- + e_p^- c_n)(1 - \tfrac{1}{2}\delta)n$$
$$- \delta(e_n^- c_p^= + e_n^= c_p^-)p + c_p^= e_p(1 - \delta)p + c_n e_n^=(1 - \delta)n \quad (7\text{-}54)$$

where $\delta = N_D^+/N_R$.

Since the e's are small, a first-order solution of Eq. (7-54) occurs when the right-hand side vanishes; we obtain

$$\frac{p}{n} = \frac{c_n(1 - \delta)}{2c_p^- \delta} + \frac{1}{2}\left[\frac{c_n^2(1 - \delta)^2}{(c_p^- \delta)^2} + \frac{8c_n^- c_n(1 - \frac{1}{2}\delta)}{c_p^- c_p^= \delta}\right]^{1/2} \quad (7\text{-}55)$$

If we neglect the e's in Eqs. (7-49) and (7-50) we obtain values for N^- and

$N^{=}$, which provide us with the lifetimes

$$\tau_n = \frac{c_n^- c_n + c_n c_p^=(p/n) + c_p^= c_p^-(p/n)^2}{N_R c_n c_p^=(p/n)[c_n^- + c_p^-(p/n)]}$$

$$\tau_p = (p/n)\tau_n \tag{7-56}$$

We see that in this approximation p/n and τ's are constants independent of position. In the next approximation the right-hand side of Eq. (7-54) is treated as a small quantity, yielding

$$\begin{aligned} p\tau_n - n\tau_p = {} & [(2-\delta)(e_p^- c_n + e_p c_n^-)\tau_n + (1-\delta)(e_p c_p^= \tau_p + e_n^= c_n \tau_n) \\ & - \delta(e_n^= c_p^- + e_n^- c_p^=)\tau_p][c_p^{=2} c_n^2 (1-\delta)^2 \\ & + 8c_n^- c_n c_p^= c_p^- \delta(1 - \tfrac{1}{2}\delta)]^{-1/2} \end{aligned} \tag{7-57}$$

Since we have assumed small e's, the right-hand side of Eqs. (7-47) and (7-48) are approximately equal to $-n/\tau_n$. If we then multiply Eq. (7-47) by τ_p/μ_n and Eq. (7-48) by τ_n/μ_p and add the resulting equations, we obtain

$$\frac{d}{dx}(p\tau_n - n\tau_p)F = -\frac{\mu_n\tau_n + \mu_p\tau_p}{\mu_n\mu_p}\frac{n}{\tau_n} \tag{7-58}$$

Noting that $p\tau_n - n\tau_p$ is a constant, and that

$$J = e(\mu_n n + \mu_p p)F \cong e(\mu_n\tau_n + \mu_p\tau_p)(n/\tau_n)F \tag{7-59}$$

Eq. (7-58) becomes

$$F\frac{dF}{dx} = -\frac{J}{e\mu_n\mu_p(p\tau_n - n\tau_p)} \tag{7-60}$$

Two spatial integrations then yield the $J(\Phi)$ characteristics:

$$J = \frac{9}{8}e\mu_n\mu_p\,|p\tau_n - n\tau_p|\frac{\Phi^2}{L^3} \tag{7-61}$$

Note that the dominant temperature dependence of the current comes from the emission coefficients contained in τ_n and τ_p; one emission coefficient is usually dominant. For example, in Zn-doped Si the e_p of holes from neutral impurities is much larger than the other e's. Here we expect the T dependence to be proportional to $\exp[(E_v - E_a^-)/kT]$, where E_a^- is the energy of the lower acceptor level. Experiments (Blouke, 1969) show a T dependence in the low injection regime of the form $\exp(-0.35\,\mathrm{eV/kT})$, which is different from that observed in the ohmic regime. The position of the lower acceptor level, 0.31 eV above the top of the valence band, agrees with this behavior, although the same experiments do not show an extended square law region.

In the high injection square law region, where $p, n \gg N_R$, the space charge is dominated by free carriers; $p \cong n$, with the neutrality condition

$$p - n = N^- + 2N^= - N_D^+ \tag{7-62}$$

The trapped charge densities are determined in this case by neglecting the e's in Eqs. (7-49) and (7-50) and taking $p = n$. The τ's are obtained by setting them equal and using $p = n$ in Eq. (7-56). We then use the arguments employed above for the low injection case, with $\tau_n = \tau_p = \tau$, to obtain

$$J = \frac{9}{8} e\mu_n\mu_p |p - n| \tau\Phi^2/L^3 \tag{7-63}$$

the high-level square law regime.

The threshold voltage for SNDC, often termed the breakdown voltage Φ_b for a forward-biased *p-i-n* structure, is determined by a technique developed by Ashley (1963). We omit thermal emission and calculate the voltage in the limit of zero current. The result is

$$\Phi_b = \frac{L^2}{2}\left[\frac{e(\mu_n\tau_n - \mu_p\tau_p - 4c_n^-\mu_p\tau_n/c_p^=)}{\pi\epsilon\mu_n\mu_p\tau_n\tau_p(c_n^-\tau_n + c_p^-\tau_p)}\right]^{1/2} \tag{7-64}$$

Figure 7-12 shows the SNDC $J(\Phi)$ curve calculated using the known parameters of Si at 200 K, the known position of the Zn impurity level, $N_R = 10^{16}\,\text{cm}^{-3}$, $N_D^+ = 10^{16}\,\text{cm}^{-3}$, $L = 10^{-2}$ and the following estimated values: $c_n = 10^{-9}\,\text{cm}^3/\text{s}$, $c_n^- = 5 \times 10^{-10}\,\text{cm}^3/\text{s}$, $c_p^- = 5 \times 10^{-8}\,\text{cm}^3/\text{s}$, and $c_p^= = 10^{-7}\,\text{cm}^3/\text{s}$. Φ_{min} is obtained by setting the hole transit time equal to its high injection lifetime. In Fig. 7-12 the dashed line indicates the $J(\Phi)$ curve that corresponds to the transition from high to low injection that would result if thermal emission were neglected; it tends to Φ_b as $J \to 0$. The threshold voltage should occur near the intersection point of the lower square law with the dashed curve. Φ_b as calculated from Eq. (7-64) is therefore an upper limit that is valid only at low temperatures. At higher T's, Φ_b tends to decrease with T until the NDC region vanishes completely.

In Chap. 2 we learned that SNDC points can be unstable under specific circuit conditions. In particular, if $C\,|d\Phi_c(I)/dI| > L/R$, where L is the intrinsic plus pack inductance, C the package capacitance, R the load resistance, and $|d\Phi_c(I)/dI|$ the slope of the NDC region, an SNDC point can be unstable against relaxation-type circuit oscillations. The stability analysis we performed was for the case of a uniform current distribution. The case of a centrally symmetric high current density filament will be treated in the next section when we discuss the "two-subelement model" of an SNDC element. In *p-i-n* diodes, however, before the SNDC region is reached, another type of current oscillation is often observed (Holonyak and Bevacqua, 1963). These spontaneous oscillations are associated with space-charge wave instabilities (Weber and Ford, 1971). To understand how they develop, we first rewrite Eq. (7-61) using zero subscripts

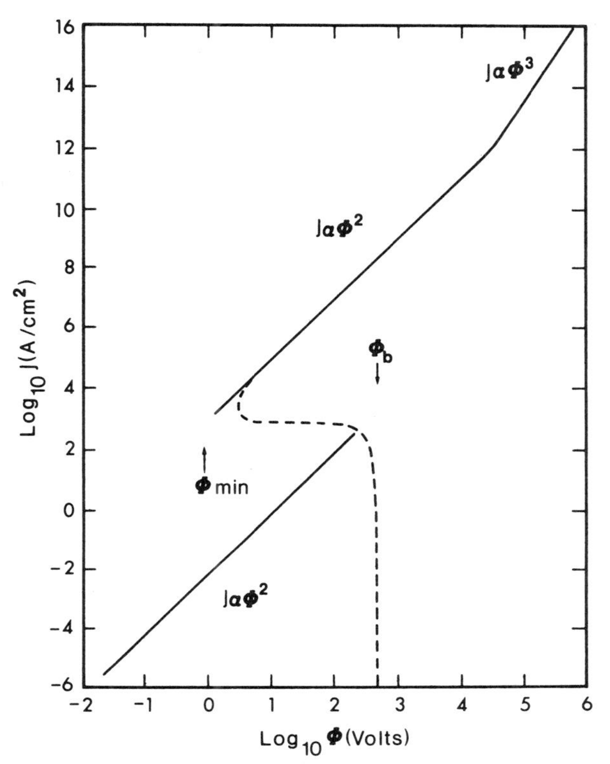

FIGURE 7-12. Calculated current–voltage characteristic for a Zn-doped Si *p-i-n* diode at 200 K using the parameters given by Weber (1971).

to denote steady state conditions:

$$J_0 = \tfrac{9}{8}\, e\mu_n\mu_p(p_0\tau_n - n_0\tau_p)(\Phi_0^2/L^3) \tag{7-65}$$

Further, the resulting field distribution is (Weber and Ford, 1970)

$$F_0 = \left(\frac{3\Phi_0}{2L}\right)\left(1 - \frac{x}{L}\right)^{1/2} \tag{7-66}$$

In a standard small signal analysis we assume that the time-dependent variables vary as e^{st} and are small compared to their dc values. They thus satisfy the following continuity equations and Poisson equation, where we have assumed τ's for the low injection square law regime and neglected thermal emission rates:

$$sn - \frac{1}{e}\frac{dJ_n}{dx} = -\frac{n}{\tau_n} \tag{7-67}$$

$$sp + \frac{1}{e}\frac{dJ_p}{dx} = -\frac{p}{\tau_p} \tag{7-68}$$

$$s(N^- - N^+) = \frac{p}{\tau_p} - \frac{n}{\tau_n} \tag{7-69}$$

$$\epsilon \frac{dF}{dx} = e(N^- - N^+ + p - n) \tag{7-70}$$

An immediate integral of these equations is the total time-dependent current

$$J = J_n(x) + J_p(x) + \epsilon s F(x) \tag{7-71}$$

Equations (7-67) through (7-71) are solved for $F(x)$ and the equations for $F(x)$ must then be solved subject to the standard boundary conditions $F(0) = F(L) = 0$. The impedence is then found via

$$Z(s) = \frac{1}{J} \int_0^L F(x)\, dx \tag{7-72}$$

and Nyquist's criterion for stability is examined: instabilities are associated with zeros or poles of $Z(s)$ in the right half s plane. (The details of a typical small signal stability analysis for nonuniform fields can be found in Shaw et al., 1979). The solutions must be obtained numerically; Weber and Ford (1972) found that $Z(s)$ has poles that move into the right half-plane when Φ_0 exceeds a threshold value. Similar results obtain when F_0 is approximated as uniform $(= \Phi_0/L)$, whence the equation for $F(x)$ becomes a linear second-order differential equation and the impedance has the form

$$Z(s) = \frac{L}{CJ_0}\left[1 - \frac{D_+ - D_-}{LD_+D_-} \frac{(e^{D_+L} - 1)(e^{D_-L} - 1)}{e^{D_+L} - e^{D_-L}}\right] \tag{7-73}$$

where the D's are functions of μ, τ, Φ_0, L, s, and τ_d.

Equation (7-73) has zeros at $s = \tau_n^{-1}$ and τ_p^{-1}, which correspond to the decay modes observed in transient experiments, and a pole at τ_d^{-1}, which corresponds to the decaying RC mode. There are also poles at

$$(D_+ - D_-)L = \pm 2n\pi i, \qquad n = 1, 2, \ldots \tag{7-74}$$

which are the poles that correspond to the observed space charge wave instabilities. In Fig. 7-13 we plot the observed oscillation frequency as a function of bias voltage and compare it with experimental results. In Fig. 7-14 the calculated temperature dependencies of frequency and voltage at the onset of the instability are plotted and compared with experiment.

Weber and Ford (1971) also found the electric field associated with a pole, and show that it has the form of a running wave modulated by a factor

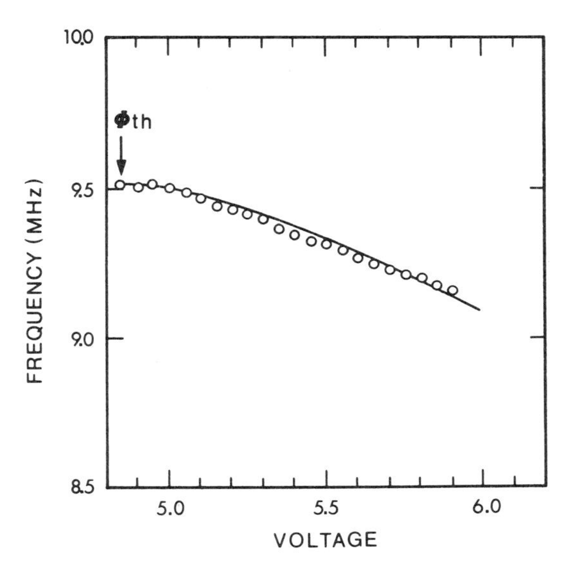

FIGURE 7-13. Oscillation frequency versus applied voltage for a Au-doped Si *p-i-n* diode. The solid line is the theoretical curve, where the threshold data is a fit using the parameters $\tau_n = 80$ ns, $\tau_p = 2.5$ ns, $L = 70\ \mu$m, $\tau_d = 0.47$ ns. The device was fabricated on 3-Ω cm compensated *n*-type Si having a carrier concentration of $\sim 10^{16}\ \text{cm}^{-3}$ Au impurities. $T = 300$ K. (After Weber and Ford, 1971.)

$\sin n\pi(x/L)$ (Weber and Ford, 1972). Using this the densities of free carriers and charged traps can be found; these also appear as waves running in the direction of the field wave. The electron and hole densities are found to be 180° out of phase with the field, with electrons dominating. The trapped charge density lags the electron density wave by ≈90°. With further increases in bias the SNDC regime is reached; other types of current controlled oscillations are then generated, making the *p-i-n* structure one of the richest systems we have for investigating instabilities in semiconductors.

Since a *p-i-n* diode is a bulk SNDC element, then we expect current filamentation to occur. The direct spatially resolved observation of the static and

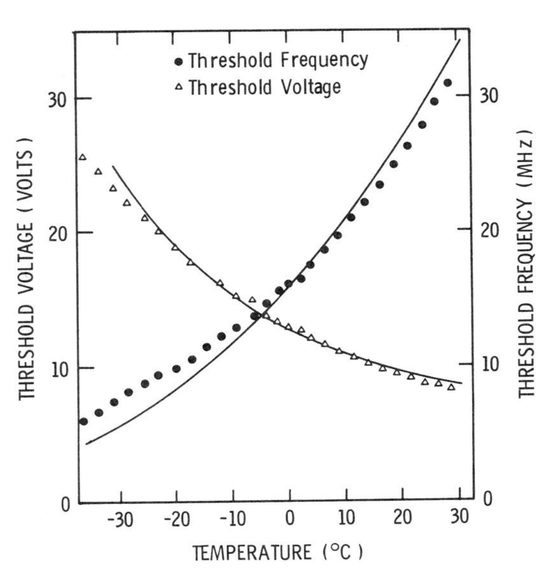

FIGURE 7-14. Temperature dependence of the threshold frequency and voltage for a Au-doped Si *p-i-n* diode. The solid lines are theoretical curves; the experimental data were fit at $T = 5°$C using $\tau_n = 60$ ns, $t_p = 1$ ns, $L = 93\ \mu$m, and $\tau_d = 0.224$ ns. The device was fabricated from 3-Ω cm *n*-type Si compensated with $\sim 10^{16}\ \text{cm}^{-3}$ Au impurities. (After Weber and Ford, 1971.)

dynamic behavior of current filaments in Si and GaAs *p-i-n* diodes has recently received much attention (Jäger et al., 1986; Baumann et al., 1987; Symanczyk et al., 1990, 1991; Jäger, 1991). These authors have demonstrated the inhomogeneous current flow by measuring the spatially resolved recombination radiation, the electric surface potential distribution, the temperature distribution, the infrared absorption, and the electron-beam-induced voltage by scanning electron microscopy. This complements earlier work in which the infrared (Barnett, 1969) or visible (Kerner and Sinkevich, 1982) recombination radiation was observed.

7.6. THE HETEROJUNCTION HOT-ELECTRON DIODE

As discussed in Chap. 3, a variety of new tunneling-type devices depend on multilayer structures that are superlattices. That is, they are repetitive heterojunctions that are closely spaced, usually less than a couple of thousand angstroms. Recently, a simple two-layer junction of *n*-GaAs and *n*-AlGaAs has been shown to exhibit SNDC characteristics, as originally suggested by Gribnikov and Mel'nikov (1966). It has been named the heterojunction hot electron diode (H^2ED) (Belyantsev et al., 1986, Emanuel et al., 1988, Kastalsky et al., 1989; Kolodzey et al., 1988; Hess et al., 1986; Tolstikhin, 1986; Mezrin and Troshkov, 1986; Alferov et al., 1987; Wacker and Schöll, 1991). The structure is shown in Fig. 7-15. It consists of a heavily doped contact, lightly *n*-doped GaAs (narrow gap), lightly *n*-doped AlGaAs (wide gap), and a heavily doped GaAs contact region. Electrons are made to drift from the left, where they encounter a barrier at the GaAs/AlGaAs interface (Fig. 7-15b). The small current here is controlled by the tunneling of electrons through this region. Under these low-current conditions most of the voltage is dropped across the interfacial barrier.

Increasing the bias thins the barrier and the tunneling current increases. There is a concomitant increase in voltage drop and hence electric field in the narrow gap GaAs region. When this field becomes sufficiently large to heat up the electron gas in the GaAs region, the hot electrons can emit over the barrier in a thermionic mode (Fig. 7-15c). The current will increase. However, this is an unstable condition since the higher current will induce an even higher field in the

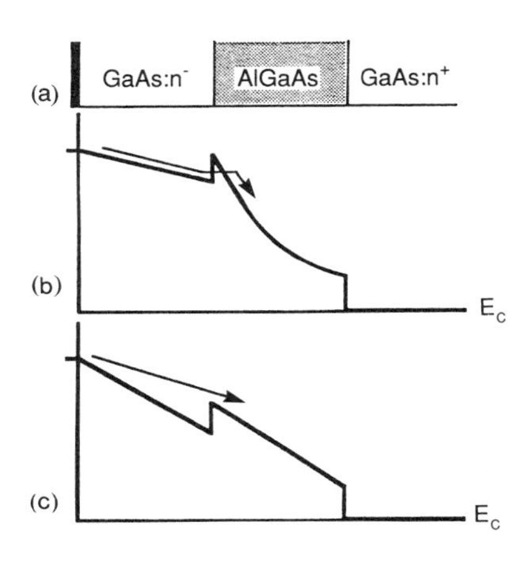

FIGURE 7-15. H^2ED (a) structure; (b) energy diagram for tunneling current mode; (c) energy diagram for thermionic emission current mode. E_c denotes the conduction band edge. (After Emanuel et al., 1988.)

narrow gap GaAs region. The $I(\Phi)$ characteristic often shows a low-current tunnel regime and a high-current thermionic emission regime, with an unstable region that separates them (Emanuel et al., 1986; Belyantsev et al., 1986, 1988).

As we have discussed in Chap. 2, this two-layer device, as well as multiple layers of the same structure (Belyantsev et al., 1988), will exhibit all the switching and oscillation processes that might be expected for such a system. Belyantsev et al. (1988) have calculated specific conditions for relaxation oscillations; near sinusoidal oscillations; damped oscillations.

7.7. THE TWO-SUBELEMENT MODEL OF CURRENT FILAMENTS

7.7.1. Introduction

In Sec. 2.3.3 we analyzed a uniform SNDC element in a circuit containing reactive elements. When current filamentation occurs, we must solve simultaneously for both the circuit response and the current density distribution, a process that is usually performed with the aid of numerical analysis. In general, we ask two questions: (1) how does filamentation influence the circuit response; and (2) how does the circuit response influence filamentation? We consider the second question first. The results of the uniform current case (see Sec. 2.3.3) indicated that the $\Phi(I)$ trajectory was determined primarily by the parameters $A = Z_0/R_0(=\sqrt{LC}/(R_0C))$, R_0, I_p, Φ_p, and Φ_s, and was relatively insensitive to the slope of the NDC region. Therefore, if we assume that the major effect of filamentation when sustained circuit-controlled oscillations are present is to change the slope of the NDC region, then the circuit will simply control the extent of filamentation. As we shall see, this is in fact the case. To answer the first question we note that when uniform currents flow and circuit-controlled oscillations occur, a specific current minimum is reached once each cycle. When filamentation occurs during sustained circuit-controlled oscillations we require filament quenching once each cycle. If we assume that there is a minimum sustaining current for filamentation, then the filament quenching criterion will impose an additional limitation on the range of circuit parameters for which circuit-controlled oscillations will occur.

We illustrate both conclusions below where we draw upon an approximate scheme for computing $\Phi_c(I)$ and obtaining a quenching criterion. The model neglects the skin effect, filament formation times, and spatial derivatives.

7.7.2. Computation of $\Phi_c(I)$

As in the NNDC element analysis (Solomon et al., 1972; Shaw et al., 1979), we divide the cylindrical SNDC element into two subelements: (1) a core cylinder of radius a_i: and (2) a surrounding cylindrical shell of inner radius a_i and outer radius, a_0, as shown in Fig. 7-16. The subelements have different $J(F)$ relations and within each subelement J and F are uniform. For the configuration shown in

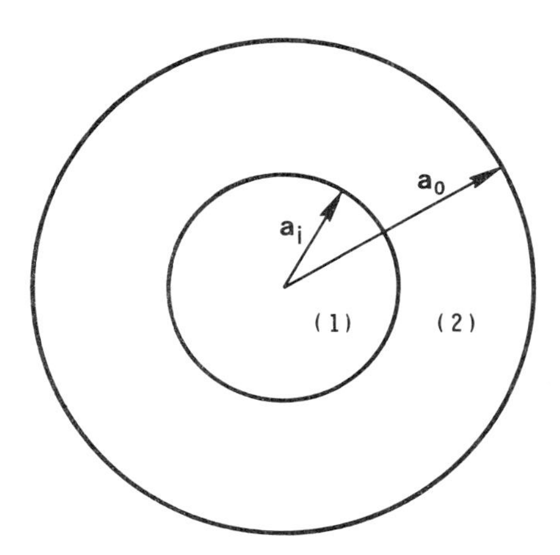

FIGURE 7-16. Two subelement model of a cylindrical current filament (radius a_i) in a device of radius a_0.

Fig. 7-16, the considerations in Chap. 2 (Shaw et al., 1973) yield

$$\Phi_c(\bar{I}) = \frac{1}{I}(J_1F_1S_1l + J_2F_2S_2l) \tag{7-75}$$

and the intrinsic inductance

$$L_i = \frac{\mu_0 l}{2\pi I^2}\left\{\frac{I^2}{4} + \left(I_1 - \frac{a_i^2 I_2}{a_0^2 - a_i^2}\right)\left[\frac{I_2}{2} + \left(I_1 - \frac{a_i^2 I_2}{a_0^2 - a_i^2}\right)\ln\frac{a_0}{a_i}\right]\right\} \tag{7-76}$$

where $I = I_1 + I_2$, $I_1 = J_1S_1$, and $I_2 = J_2S_2$ (S denotes area). The computation of $\Phi_c(I)$ for a relaxation oscillation with $L_p = 0$ [here $\Phi(I) = \Phi_0(I)$] is illustrated in Fig. 7-17 where, for ease in plotting, we compute $\Phi_c(\bar{J})$ rather than $\Phi_c(\bar{I})$. Here $J = I/S$, where $S = S_1 + S_2$. We consider the case $S_1 = S_2$, thus $a_0^2 = 2a_i^2$.

$$\bar{J} = \tfrac{1}{2}(J_1 + J_2) \tag{7-77}$$

and

$$\Phi_c(\bar{J}) = \frac{J_1}{J_1 + J_2}\Phi_c(J_1) + \frac{J_2}{J_1 + J_2}\Phi_c(J_2) \tag{7-78}$$

where $\Phi_c(J_1) = F_1l$ and $\Phi_c(J_2) = F_2l$. To determine $\Phi_c(\bar{J})$ we first obtain the $\Phi_0(\bar{J})$ trajectory from A, Φ_p, I_p, and Φ_s using the results of the circuit analysis for the uniform cylinder. Next, we consider the response of each subelement and determine $\Phi_c(\bar{J})$ by making use of Eqs. (7-77) and (7-78). From Eq. (2-186) we see that the induced voltage drop in the cylinder is just the difference $\Phi_0(J) - \Phi_c(J)$.

In Fig. 7-17a we show $\Phi_0(\bar{J})$ for a circuit-controlled relaxation oscillation, $\Phi_c(J_1)$, $\Phi_c(J_2)$, and $\Phi_c = \frac{1}{2}[\Phi_c(J_1) + \Phi_c(J_2)]$. In Fig. 7-17b we show computed values of $\Phi_c(\bar{J})$ (crosses) at four instants of time: a, b, c, and d. The conductive voltages in subelement 1 (solid circles) and subelement 2 (open circles) are also shown. At each instant of time an assumption is made about the time rate of change of current or voltage. At time a, $\Phi_0(\bar{J})$ has reached its maximum. If at this point we make the reasonable assumption of neglecting the inductive voltage

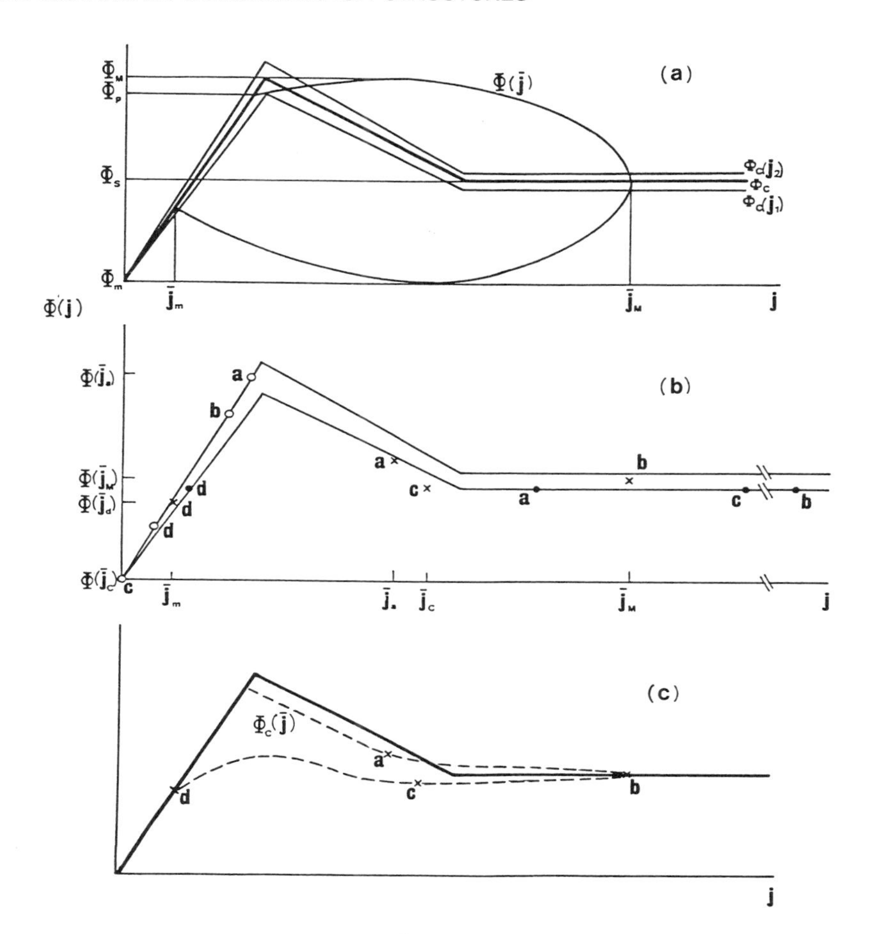

FIGURE 7-17. (a) Illustration of the way $\Phi_c(\bar{j})$ for a relaxation oscillation is obtained using the two subelement model. At time a, $\bar{j} = \bar{j}_a$ and $\Phi(\bar{j}) = \Phi(\bar{j}_a)$, etc. For clarity we have dropped the subscript 0 on Φ. In (b) the position of subelement 1 is designated by the closed circles, the position of subelement 2 by the open circles and the values of $\Phi_c(\bar{j})$ by the crosses. In (c) we plot a complete $\Phi_c(\bar{j})$ curve estimated point by point throughout one cycle of $\Phi_c(\bar{j})$.

drop in subelement 2, then $\Phi_0(\bar{J})_M = \Phi_c(J_2)$ and J_2 is determined. From Eq. (7-77) we obtain J_1, which in turn yields $\Phi_c(\bar{J})$ from Eq. (7-78). $\Phi_c(\bar{J})$ at time a is indicated in Fig. 7-17b as the cross a. At time b, $\bar{J}$ is a maximum, hence here $\Phi_0(\bar{J}) = \Phi_c(\bar{J})$. This point is the cross b. Similar determinations are made at c and d. A sketch of $\Phi_c(\bar{J})$ throughout the cycle is shown in Fig. 7-17c.

Important situations that occur often in the laboratory are the simple cases of either no filamentation (narrow base *p-n-p-n* diodes) or a narrow well-defined axial filament [*p-i-n* diode or amorphous threshold switch (see Chap. 8 and Adler et al., 1980 and Madan and Shaw, 1988)]. It is therefore of interest to examine the system in the two limiting cases: (1) $a_0 \gg a_i$ and (2) $a_0 \sim a_i$. If, for $a_0 \gg a_i$, we also consider the situation where the narrow central region has a much larger conductivity than the rest of the SNDC element, then the situation presumably corresponds quite closely to an amorphous threshold switch. Here we see from Eq. (7-76) that the $\ln(a_0/a_i)$ term of L_i becomes quite important. The term has no upper limit and is produced by the flux in subelement 2 produced by current in subelement 1. For the case $a_0 \sim a_i$, as I_2 vanishes Eqs. (7-75) and (7-76) produce the results required of a uniform cylinder where $L_i = \mu_0 l/8\pi$. Performing computations on these systems similar to those discussed with regard to Fig. 7-17 reveals several general features of the problem. We find that: (1) the $\Phi_c(\bar{J})$ curves can be regarded as members of a family of curves characterized by similar values of I_p, Φ_p, and Φ_s; (2) the $\Phi_c(\bar{J})$ trajectory reflects the current density evolution; and (3) when inhomogeneous current density distributions occur and sustained circuit oscillations are maintained, the major effect of filament formation will be to alter the shape of the region of negative slope. Since the circuit response is relatively insensitive to this parameter, the waveforms will be essentially indistinguishable from the uniform current density case discussed in Chap. 2.

7.7.3. Effect of Filamentation on Circuit Control of the Oscillations

In Chap. 2 we described the circuit oscillation in terms of the parameter $A \equiv \sqrt{LC}/(R_0C) \equiv Z_0/R_0$. When filamentation is present A may be regarded as circuit dependent insofar as L_i depends on the current distribution. As shown by Eq. (7-76), filamentation yields values of L_i greater than its value for uniform current densities. Therefore, A is larger above than below threshold. From the arguments for the uniform current density case, a larger A would yield higher values of I_m (minimum current). The implications of filamentation are therefore clear: if the SNDC element is part of a circuit that for the uniform field case yields a value of I_m substantially below a "filament sustaining current" I_f, the effect of filamentation on the circuit oscillations will be negligible. On the other hand, if $I_m \leq I_f$, then small increases in L_i due to filamentation may well increase I_m so that it exceeds I_f. In this case the circuit oscillations will damp and switching to an ON state will occur.

7.7.4. The Growth of Current Filaments

The above calculations illustrated filamentation when at threshold only one subelement entered its region of negative slope. The presence of relatively uniform current densities implies that at some time during the cycle both subelements are in their regions of negative slope. For the example considered above, when subelement 1 enters the NDC region its total voltage drop ceases to rise significantly and the voltage drop across subelement 2 soon begins to decrease. If we imagine a situation where $\Phi_0(\bar{J})$ rises significantly after threshold, then subelement 2 can also be pushed into its NDC region. The current distribution will initially be relatively uniform. Therefore, an important parameter in the growth of current filaments is the maximum voltage Φ_M reached during the cycle. From Eq. (2-191) we see that Φ_M is a strong function of A. For relaxation oscillations (small A) Φ_M barely exceeds the threshold voltage for the subelement with the highest carrier concentration. This leads to the highly nonuniform situation where a filament appears immediately upon reaching threshold. For near sinusoidal oscillations (large A) Φ_M may be much higher than the threshold voltage, producing a more uniform initial current density distribution.

7.7.5. The Quenching of Current Filaments

In order for circuit-controlled oscillations to be sustained it is necessary that all current density nonuniformities be quenched at the end of each cycle. When this occurs the $\Phi_0(\bar{J})$ trajectory is closed and completed on the low-current line of positive slope of $\Phi_0(\bar{J})$. If, however, the current density minimum J_m is not sufficiently low ($J_m > J_i$), the trajectory is open and the nonuniformities are enhanced. The circuit oscillation damps and a filament remains as a steady-state solution. This behavior is shown schematically in Fig. 7-18.

It is possible to make some qualitative predictions about how small I_m must be to ensure filament quenching. Certainly $\Phi_c(\bar{I})$ will not form a closed trajectory unless I_m is sufficiently below I_p. In general we expect that I_m must typically be below I_s to ensure circuit-controlled oscillations. According to Eq. (2-191) the elliptical part of the $\Phi_c(\bar{I})$ trajectory will pass through the line of positive slope at

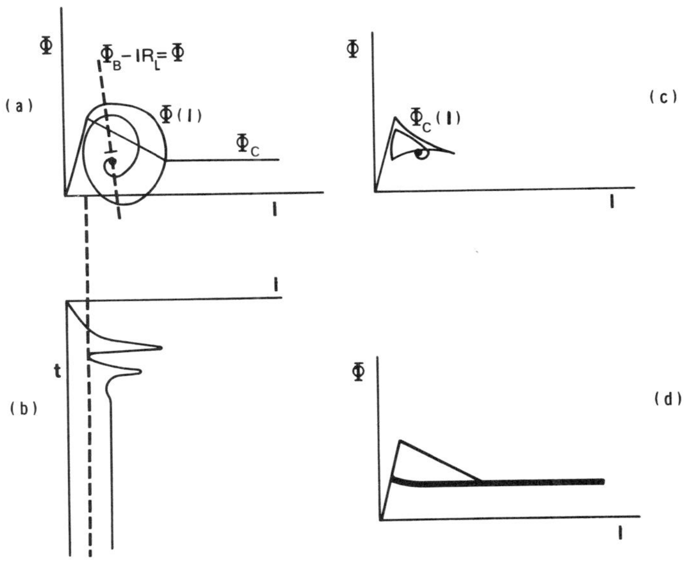

FIGURE 7-18. Estimated $\Phi(I)$, $\Phi_c(I)$, and $I(t)$ curves [(a), (b) and (c), respectively] for the case where I_m does not fall low enough to quench the inhomogeneities. The circuit oscillations damp, a filament forms, and switching occurs to the ON state. The ON state is the heavy dot. As Φ_B is varied the dots fall on a "filament characteristic," which is sketched (dark) in (d). When the load line intersects the SNDC element at points $I > I_v$, where I_v marks the onset of the region where $d\Phi/dI = 0$, a switch will occur to that point without relaxation oscillations (unless for some reason the stable point at $I > I_v$ is never reached). The switch will also take place, however, in a damped oscillatory fashion. A "spiraling in" to the ON state is fundamental to the switching process in general. In the current–time profile this appears as a damped ringing oscillation at a frequency determined in part by $(LC)^{1/2}$.

I_s when

$$A = \left[\frac{I_p - I_s}{2\Phi_B/R - (I_p + I_s)}\right]^{1/2}$$

For A less than this critical value, filaments will quench. Consider the case where $I_p \cong 2I_s$. Here, since $\Phi_B/R > I_p$, a necessary condition for quenching is $A \leq 1$. For a realistic bias $\Phi_B/R \sim 2I_p$ and $A \cong \frac{1}{2}$ is the critical value. Note that for a given A, as Φ_B increases, a transition from a relaxation oscillation to the switched ON state can occur as the critical calue of A is exceeded. Furthermore, if we design the SNDC element such that A is significantly greater than unity, filament quenching can be avoided. It is, therefore, possible to construct a filament forming SNDC element that will only turn ON and not exhibit relaxation oscillations for any Φ_B and R. To achieve this we require $\sqrt{LC} > R_0C$.

Throughout this discussion we have assumed that the package inductance L_p was zero. For a finite L_p, $L = L_i + L_p$ and some inductive voltage will be drained off the SNDC element. For a given $\Phi(\bar{J})$ trajectory, therefore, the maximum value of $\Phi_0(\bar{J})$ will be below its $L_p = 0$ value. Since a large Φ_M is necessary for relatively uniform current densities, increasing the package inductance enhances the possibility of filamentation. The two circuit factors that control filamentation are thus L_p and A, with filamentation optimized for large L_p and small A.

We now consider in more detail the inclusion of a primary filament nucleation site. If, for example, we were to model a thermal filament, the primary site would be at the center of the cylinder (see Chap. 8). The *p-n-p-n* diode experiments of Varlamov et al. (1970) show that the lateral edges of the SNDC element play a major role, and if the primary site is at an edge, the filament may tend to remain there rather than move towards the center. Thus, the role of, e.g., surface states on the sheath of the cylinder may be important. Wherever the site is, its role may be likened to that of a subelement with a high carrier concentration compared to the other subelements. Its presence will thus act to decrease $\Phi_p - \Phi_s$, which from Eq. (2-191) can be seen to reduce the amplitude of $\Phi(\bar{I})$. This will increase I_m and lead to the damping of the relaxation oscillations and the domination of a filament. Thus, if a primary site exists, a filament will not quench unless $A \ll 1$. Switching to the ON state will dominate. Amorphous threshold switches (Chap. 8) almost always exhibit relaxation oscillations for sufficiently large R (Shaw and Gastman, 1971) and also often have a high-conductivity central region. At first glance these two features seem incompatible. However, the high resistances of these devices ($R_0 \sim 10^6$–$10^7\Omega$) and values of $C_p \cong 10^{-12}F$ produce values of A much less than unity even for reasonably large estimates of the inductance. Hence, filament quenching and relaxation oscillations are still favored.

7.8. SUMMARY

In this chapter we discussed a wide variety of SNDC devices that developed their NDC regions because of specific properties of semiconductor junctions: the

UJT; BJT; *p-n-p-n* diode; *p-i-n* diode; and H^2ED. They were all analyzed in some detail, with specific emphasis on the *p-n-p-n* and *p-i-n* devices. The circuit response was analyzed for cases when current filamentation developed. In the next, and final chapter, we focus on SNDC devices that develop their NDC regions via thermal or electrothermal means.

8

Thermal and Electrothermal Instabilities

8.1. THE THERMISTOR

8.1.1. Introduction—Average Global Heating

Thermal effects in solids have been treated in great detail over the past 60 years (see, e.g., Carslaw and Jaeger, 1959). Of particular interest has been the myriad of phenomena associated with thermal runaway induced by Joule heating and the associated breakdown or switching processes often observed (see e.g., Fock, 1927; Lueder and Spenke, 1935; Becker, 1936; Franz, 1956; Skanavi, 1958; Böer et al., 1961; Stocker et al., 1970; Shousha, 1971; Altcheh et al., 1972; Thoma, 1976). Reviews have been given, e.g., by Klein (1969, 1978, 1983), Shaw and Yildirim (1983), and Madan and Shaw (1988). These instabilities often result in regions of NDC appearing in the $I(\Phi)$ characteristics of a variety of materials. Indeed, it is now well known that NDC can appear in the static and dynamic characteristics of common materials and devices in which the current level is determined not only by the applied voltage, but also by the temperature. One reason for this is that the Joule heating of the sample often causes the average temperature to rise above that of the ambient temperatures, T_a. Figure 8-1 shows how this might arise. Linear $I(\Phi)$ characteristics are sketched for isothermal cases where the ambient temperature $T_{a4} > T_{a3} > T_{a2} > T_{a1}$. These are the characteristics that would result were the heat sinking sufficient to maintain the system at the ambient levels shown. However, when the heat sinking is insufficient to remove heat fast enough, then, e.g., if the ambient is T_{a1}, it is possible that the steady state average T can correspond to a point on the T_{a4} line. The actual $I(\Phi)$ characteristic might then appear as the thick solid line; a region of SNDC could occur. A device in which NDC is induced in this manner is called a thermistor. Note that every nonlinear point on the thermistor characteristic corresponds to a different average steady state T distribution. The slope of the NDC characteristic will depend primarily upon the heat sinking, heating rate, pulse repetition frequency, and pulse width. Hence, its detailed form is a variable that depends on the way in which the measurement is performed, and a major feature of its shape is the question of the existence and position of a "turnover" or threshold voltage, Φ_T. Other types of thermal SNDC elements exhibit $I(\Phi)$ characteristics that are determined by critical electric fields or critical temperatures at which appreciable changes in conductivity occur. We will discuss them in later sections; first, we concentrate on the thermistor.

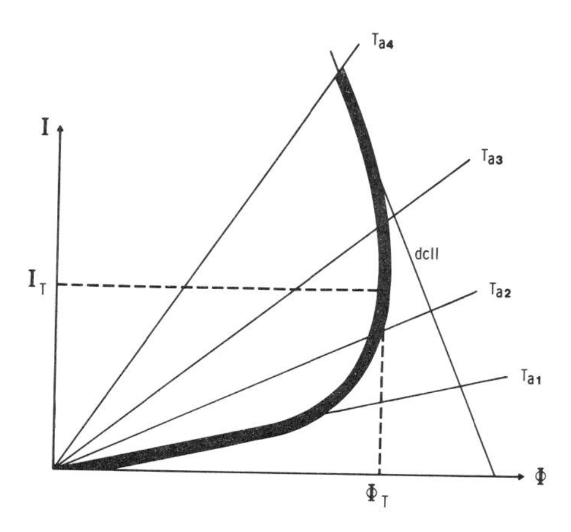

FIGURE 8-1. Nonlinear thermistor characteristics (heavy line) that might arise in a material where linear $I(\Phi)$ curves at different ambient temperatures T_a are shown at $T_4 > T_3 > T_2 > T_1$ (after Shaw and Yildirim, 1983).

Burgess (1955a, b, c, 1960) has made an extensive contributory study of thermistor behavior in materials having a conductance of the form $G = G_0 \exp(-b/T)$, where G_0 and b are constants, which encompasses a large number of important semiconductor materials and devices.

Let us first derive the turnover point of the current–voltage characteristic quite generally, independent of the particular form of the conductance. If the current is given by $I = \tilde{I}(\Phi, T)$ and the temperature is determined by the dissipated electric power $I\Phi = P(T)$, then the inverse slope of the current–voltage characteristic $I(\Phi)$ is given by

$$\frac{d\Phi}{dI} = R\frac{1-y}{x+y} \tag{8-1}$$

where

$$R = \Phi/I, \qquad x \equiv R(\partial\tilde{I}/\partial\Phi)_T, \qquad y \equiv (\Phi/B)(\partial\tilde{I}/\partial T)_\Phi$$

and

$$B = \frac{dP}{dT}$$

To derive Eq. (8-1), we write the total differentials of $I = \tilde{I}(\Phi, T)$ and $P = \Phi\tilde{I}(\Phi, T)$, and divide by $d\Phi$ and dT, respectively. From the resulting two differential relations we can eliminate $dT/d\Phi$, i.e. the differential temperature increase along the actual current–voltage characteristic. This leads directly to Eq. (8-1).

It is clear from Eq. (8-1) that the turnover point occurs at $y = 1$, with NDC setting in for larger values of y. Thus, at turnover we have

$$1 = \left(\frac{\partial\tilde{I}}{\partial T}\right)_\Phi \frac{dT}{dP} \tag{8-2}$$

In order to achieve the condition $y = 1$ the conductivity of the semiconductor must have the proper temperature dependence. In particular, the common form

$$I = G_0\Phi \exp(-b/T) \tag{8-3}$$

where G_0 and b are constants, is sufficient to achieve the turnover condition. In order to obtain an explicit current–voltage characteristic taking into account the heating as I is increased, we require additionally a relation between P and T. Often the dissipated power P is related to the excess temperature by a linear relationship

$$P = B(T - T_a) \quad \text{for} \quad T - T_a = aP \tag{8-4}$$

where T_a is the ambient temperature and $B = 1/a$ is a constant relating the excess temperature to the power supplied to the material or device.

Equations (8-2)–(8-4) yield for the temperature of the element at turnover

$$T_T = \tfrac{1}{2}b - (\tfrac{1}{4}b^2 - bT_a)^{1/2} \tag{8-5}$$

which shows that a requirement for turnover is $b > 4T_a$. The power at turnover is

$$P_T = a^{-1}[\tfrac{1}{2}b - T_a - (\tfrac{1}{4}b^2 - bT_a)^{1/2}] \tag{8-6}$$

The power at turnover increases with increasing ambient temperature. This feature is characteristic of a semiconducting thermistor having the property shown in Eq. (8-2). The voltage at turnover Φ_T is determined by $P_T = \Phi_T I$ with Eqs. (8-3), (8-5), (8-6). Φ_T for a given thermistor is a function only of T_a. For $b \gg T_a$, a good approximation of Eq. (8-6) is

$$P_T \cong T_a^2/ab \tag{8-7}$$

and the conductance at turnover is

$$\frac{I_T}{\Phi_T} = G_T \cong G_0 \exp\left(1 + \frac{T_a}{b}\right) \tag{8-8}$$

i.e., the conductance at turnover is enhanced by about a factor of e from its isothermal value, G_0, at the same T_a (Burgess, 1955a).

The potential of thermistor-type devices often hinges on their response times. Can the structure be heated and/or cooled fast enough for use as a high-speed switching device? The direction to take along these lines is to make the structure thin and small; thin film technology then becomes crucial. It is important to examine materials having desirable thermistor properties in thin film configurations (Hayes, 1974; Hayes and Thornburg, 1973). Hence, in what follows we will emphasize that particular geometry.

8.1.2. Heat Flow in Semiconductors

In the preceding section we treated the thermistor at a global average temperature. In reality, however, nonuniform temperature distributions often occur. Therefore, we will briefly review the problem of heat conduction in solids (Carslaw and Jaeger, 1959; Kittel, 1976; Kroemer and Kittel, 1980). Rather than be very detailed, we will emphasize the specific problem at hand with reference to Fig. 8-2. The semiconductor has an electrical conductivity $\sigma_1(T) = \sigma_0 e^{-\Delta E/kT}$, where ΔE is the thermal activation energy and k the Boltzmann constant. For later use we also include the presence of inhomogeneities with an electrical conductivity σ_2. For metallic inhomogeneities we take $\sigma_2 \gg \sigma_1$ and assume for simplicity that σ_2 is either constant or a slowly decreasing function of T. We also take the thermal conductivities as $K_1(T) = \alpha_1 T + \beta_1$, with $\alpha_1 \equiv dK_1/dT > 0$ and $K_2(T) = \alpha_2 T + \beta_2$, with $\alpha_2 \equiv dK_2/dT < 0$. The object of the exericse is to determine the $I(\Phi)$ characteristics of the device and study the effects of the: (1) temperature-dependent electrical conductivity; (2) temperature-dependent thermal conductivity; (3) thermal boundary conditions; (4) presence and morphology of inhomogeneities; (5) critical electric fields for the onset of impact ionization and carrier multiplication; (6) thermally induced phase changes and latent heats. To do the general analysis we must solve equations for the flow of both electric and thermal currents. For this we require knowledge of the boundary conditions. In prior chapters we have emphasized the electrical boundary conditions. Here we also need thermal boundary conditions, such as Eq. (8-31) below. Since in this chapter we are emphasizing thermal effects, we will first concentrate primarily on the heat flow equation. Later in the chapter we will discuss the coupled thermal and electronic equations for a general semiconductor device containing both holes and electrons.

We will first consider the simplest system and search for general criteria for (1) thermistor behavior; (2) switching effects. To do this we remove the inhomogeneities, provide a constant current source by setting $R_L = \infty$ and treat the general case where $d\sigma/dT > 0$.

For the total electronic current $\mathbf{J}_e$ with no sources or sinks we have the

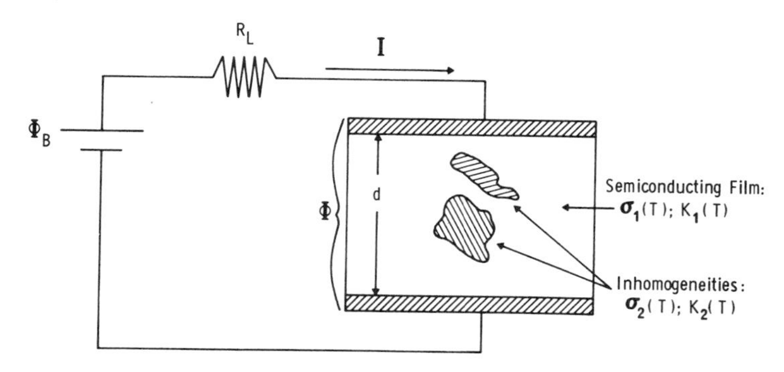

FIGURE 8-2. A thin-disk sample of semiconductor containing inhomogeneities placed in a resistive circuit. The symbols are defined in the text. (After Shaw and Yildirim, 1983.)

continuity equation

$$\operatorname{div} \mathbf{J}_e + \frac{\partial}{\partial t} \sum_i n_i q_i = 0 \tag{8-9a}$$

where $\sum_i n_i q_i$ is the total mobile charge density. In the steady state

$$\operatorname{div} \mathbf{J}_e = 0. \tag{8-9b}$$

In a homogeneous isotropic medium the constitutive equation for the electronic current density is in the simplest case given by

$$\mathbf{J}_e = -\sigma \operatorname{grad} \Phi \tag{8-10}$$

where σ is the electric conductivity and Φ the electrostatic potential. Hence we have for the steady state

$$\operatorname{div}(\sigma \operatorname{grad} \Phi) = 0 \tag{8-11}$$

Analogously, the constitutive equation for the heat current density is

$$\mathbf{J}_h = -K \operatorname{grad} T \tag{8-12}$$

where K is the thermal conductivity.

In general systems with potential, temperature, and carrier density gradients the linear constitutive equations also involve diffusion and thermodiffusion terms, an example of which is given by Eqs. (8-60) and (8-61) at the end of this chapter.

We shall now derive a continuity equation for the heat current density, using Eq. (8-9b) and the continuity equation for the total energy flux density which describes the conversion of electrical energy into heat:

$$-\operatorname{div}[\Phi \mathbf{J}_e + \mathbf{J}_h] = \frac{\partial w}{\partial t} \tag{8-13}$$

where w is the energy density and $\partial w / \partial t$ the power density. Here we have assumed that the product of the electronic current and enthalpy per unit charge carrier per unit length is independent of position and that thermoelectric power and Peltier heating effects are negligibly small (Carslaw and Jaeger, 1959).

Applying the divergence theorem to Eq. (8-13) yields

$$-\oint_s (\Phi \mathbf{J}_e) \cdot \mathbf{ds} - \oint_s \mathbf{J}_h \cdot \mathbf{ds} = \frac{\partial E}{\partial t} \tag{8-14}$$

where $E = \int_v w \, dv$ (joules) is the energy stored in the volume v enclosed by the surface s.

The first term on the left-hand side of Eq. (8-14) describes the electrical power flowing into the closed surface s:

$$P_e \equiv -\oint_s (\Phi \mathbf{J}_e) \cdot \mathbf{ds} \tag{8-15}$$

The second term on the left-hand side (without the minus sign) is the heat power flowing out of s:

$$P_h \equiv \oint \mathbf{J}_h \cdot \mathbf{ds} \tag{8-16}$$

The right-hand side is equal to the difference $P_e - P_h$; i.e., dE/dt denotes the rate of increase of the energy stored in v:

$$P_e - P_h = \frac{dE}{dt} \tag{8-17}$$

The energy stored in v is related to the heat capacity of v,

$$C = \frac{dE}{dT} \tag{8-18}$$

or

$$dE = C\,dT$$

Thus, Eq. (8-17) becomes

$$P_e - P_h = C\frac{dT}{dt} \tag{8-19}$$

Equation (8-19) states that the electrical input power P_e is used in two ways: Part of the input power flows out of S in the form of heat current with density $\mathbf{J}_h$; the rest is used to increase the temperature of the system.

The terms in Eq. (8-13) can also be interpreted in a similar way. Defining

$$\begin{aligned} p_e &= -\text{div}[\Phi \mathbf{J}_e] = -(\text{grad } \Phi) \cdot \mathbf{J}_e - \Phi(\text{div } \mathbf{J}_e) = -\mathbf{J}_e \cdot \text{grad } \Phi \\ &= \sigma\,|\nabla\Phi|^2 = \frac{|\mathbf{J}_e|^2}{\sigma} \end{aligned} \tag{8-20}$$

which is the electrical input power density at a point, and

$$p_h = \text{div } \mathbf{J}_h = -\text{div}(K \text{ grad } T) = -K\nabla^2 T - \text{grad } K \cdot \text{grad } T \tag{8-21}$$

which is the heat power density flowing out of that point, Eq. (8-13) becomes

$$p_e - p_h = \frac{dw}{dt} \tag{8-22}$$

The right-hand side of Eq. (8-22) is the power density required to alter the local temperature. The change of energy density dw is related to the change of temperature dT in a small volume v by

$$dw = \frac{dE}{v} = \frac{C\,dT}{v} \tag{8-23}$$

Using the definitions:

$$\rho_m = m_g/v = \text{mass density} = \text{specific mass} \tag{8-24}$$

where m_g is the mass in the volume v, and

$$c_v = \frac{C}{m_g} \tag{8-25}$$

is the heat capacity per unit mass (specific heat at constant volume), Eq. (8-22) becomes

$$p_e - p_h = \rho_m c \frac{\partial T}{\partial t} \tag{8-26}$$

Thus, we have two power continuity equations. One is in integral form, which can be used to relate the total input power, total heat power (efflux), and the rate of change of T:

$$C\frac{\partial T}{\partial t} + P_h = P_e \tag{8-27}$$

This equation can be used to study the system as a whole.

The other equation is a differential relation which can be used to study local regions of the material:

$$\rho_m c \frac{\partial T}{\partial t} + p_h = p_e \tag{8-28}$$

Both of these equations reemphasize the simple fact that the difference in input power and heat efflux goes into increasing the temperature of the system.

When the external source is switched on, $\partial T/\partial t$ will initially be greater than zero. That is, the temperature of the material will start to rise. The rise will continue until a sufficient temperature gradient is reached, whereby, neglecting radiative losses, all the incoming electrical power will flow out as heat. When steady state is reached $\partial T/\partial t = 0$, $P_e = P_h$ and $p_e = p_h$. The differential Eq. (8-28) is a forced diffusion equation, an inhomogeneous, parabolic partial differential equation which can be rewritten in terms of T, K, σ, and $\mathbf{J}_e$ using (8-20), (8-21) as

$$\rho_m c \frac{\partial T}{\partial t} - \operatorname{div}[K(T) \operatorname{grad} T] = \frac{|J_e|^2}{\sigma} \tag{8-29}$$

For a temperature-independent K it becomes

$$\frac{\partial T}{\partial t} - \frac{K}{\rho_m c} \nabla^2 T = \frac{|J_e|^2}{\sigma \rho_m c} \tag{8-30}$$

which defines a heat diffusion constant $D_h = K/(\rho_m c)$. Equation (8-30) has to be supplemented by suitable boundary conditions. If we assume that the heat flow at the surface of the semiconductor is dissipated at a rate proportional to the excess temperature, we obtain Newton's law of cooling

$$-K \operatorname{grad} T \cdot \hat{n} = G(T - T_a) \tag{8-31}$$

where $\hat{n}$ is the normal vector of the surface, T_a is the ambient temperature, and the constant G may be visualized as the thermal conductivity of the electrode divided by its thickness. In the limit of infinite G, i.e., an infinite heat sink at the surface, Eq. (8-31) reduces to the boundary condition of fixed temperature $T = T_a$. In the other extreme, $G = 0$, i.e., a thermally insulated surface, the gradient of the temperature distribution vanishes at the surface:

$$\operatorname{grad} T \cdot \hat{n} = 0$$

Since the heat flow process is diffusive, let us first discuss the transient state (where $\partial T/\partial t \neq 0$) in a qualitative manner. The most important feature is that the right-hand side of Eq. (8-30) is the heat power generated at a point. Considering that point alone (we use the superposition principle to study the other points in a similar way), it is seen that if D_h is large (good thermal conductivity, low ρ_m and c) the temperature wave will move rapidly (higher diffusion velocity) and a local thermal pulse or disturbance will propagate over a relatively long distance before being attenuated. A simple way of visualizing this process is offered next via the behavior of an analogous distributed RC network (Shousha, 1971).

8.1.3. An RC Network Analog of the Heating Process

Heat flow in a solid is a diffusive phenomenon. The differential equation and solutions for $T(t, \mathbf{r})$ are very similar to the solution describing current and voltage waves in a distributed RC network. Figure 8-3 shows a simple one-dimensional

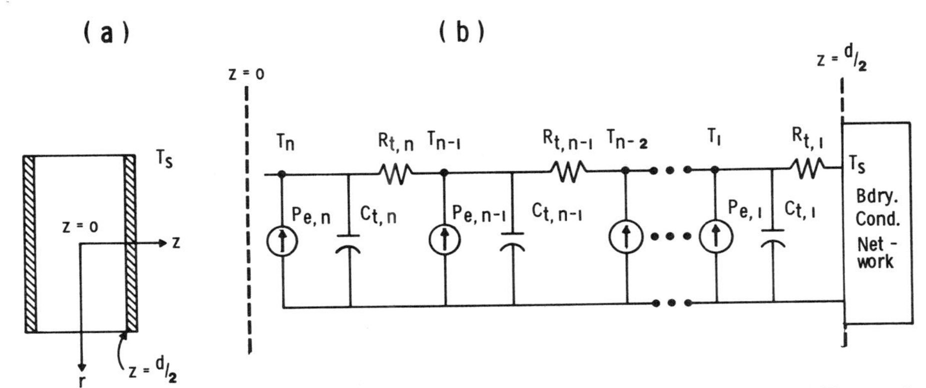

FIGURE 8-3. (a) Thin-disk sample of thickness *d*; (b) distributed *RC* analog network. The symbols are defined in the text. (After Shaw and Yildirim, 1983.)

analog-distributed *RC* circuit in which the voltage is analogous to temperature and current is analogous to heat current (Shaw and Yildirim, 1983). For simplicity and compactness we have considered heat flow only in one dimension (z). A similar model readily accounts for heat flow in the radial direction.

The R_t's are thermal resistances per unit length, which are inversely proportional to $K(T)$. The C_t's are thermal capacitances per unit length, which are proportional to ρ_m, c, and m_g. The $p_{e,n}$'s are the "heat current sources", the Joule heating at each point. The T_n's are the node temperatures (voltages developed across the capacitors). The energy stored in a capacitor is analogous to the thermal energy stored in the thermal capacity of the system. During the transient state, part of the current in R_t is used to charge C_t and the rest flows to the load network, which simulates the thermal boundary conditions. At steady state all the capacitors are charged to their final values (determined by R_t and p_e) and the input power flows to the load network.

The temperature at the zeroth node represents the surface temperature, T_s. A constant temperature boundary condition (infinite heat sink) can be simulated by connecting an ideal voltage (temperature) source with a voltage (temperature) equal to the ambient temperature T_a. Figure 8-4a shows the boundary condition.

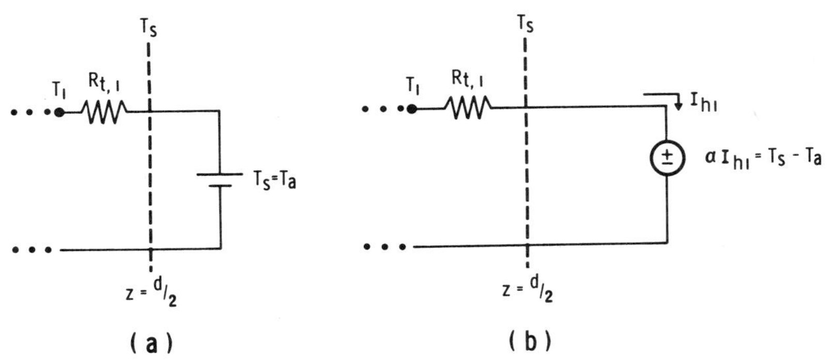

FIGURE 8-4. (a) *RC* network analog of a constant-temperature boundary condition; (b) *RC* network analog of Newton's law of cooling boundary condition (after Shaw and Yildirim, 1983).

A boundary condition of the type

$$\mathbf{I}_h \cdot \hat{n} = G(T_s - T_a) \tag{8-31'}$$

at interface, Newton's law of cooling, can be represented by an ideal current-dependent voltage source,

$$T_s - T_a = \alpha I_h \tag{8-32}$$

as shown in Fig. 8-4b, where I_h is the heat current.

Let us first consider an infinite heat sink and suddenly impose an electrical power source on the system. Heat will be generated everywhere in accordance with

$$p_e = |J_e|^2/\sigma(T) \tag{8-33}$$

For a temperature-independent σ, p_e will be highest at points where the electric current density is highest. However, since $\sigma(T)$ also increases with temperature, the variation of p_e with T must be inspected further, and we shall do this shortly.

If we treat one current source at a time, we see that the heat power will diffuse towards the short circuit load (electrodes) with a diffusion constant

$$D_h = \frac{K}{\rho_m c} \tag{8-34}$$

and with a diffusion velocity proportional to K. Considering just the source $p_{e,n}$ for the moment, at $t = 0$, $C_{t,n}$ acts as a short circuit and all the heat current flows through it. As $C_{t,n}$ is charged, the temperature $T_n(t)$ will start to increase, which will in turn cause part of the heat current to flow to the neighboring circuit $(R_{t,n-1}C_{t,n-1})$. This transient process will stop when a sufficient temperature gradient $(T_n, T_{n-1}, T_{n-2}, \ldots, T_s)$ is developed such that all the heat power will flow through the R_t's towards the boundary. That is, the C_t's are charged to their limits and draw no more current. Thus, the T_n's will not increase further. The limiting temperature (voltage) for each C_t is determined by the source strength (p_e), R_t and the boundary conditions.

The heating process, which we just discussed for a single heat source, occurs for all sources simultaneously. Since there are sources at each point in the material, the steady state may be established in a time shorter than for the single-source case. Because of the symmetry, at steady state the heat (current) flows only to the right toward the load. Therefore, we have

$$T_{\max} > T_n > T_{n-1} > T_{n-2} > \cdots > T_s \qquad \text{at steady state}$$

It is also interesting to note that during the transient states that may arise due to changes in some heat sources, the direction of heat flow may reverse. For example, let us assume that all the sources except one are dead at an instant $t = 0$. Also assume that the capacitors have initial temperatures $T_n(0) >$

$T_{n-1}(0) > \cdots > T_0$. Let the strength of $p_{e,h}$ suddenly increase, and let us kill all the other sources at $t = 0$. The source $p_{e,h}$ will start to charge the nearest capacitors toward the new steady state value, and all the other capacitors start to discharge (or charge) toward their new steady state values as determined by the new steady state network.

In the above we treated the case of an infinite-heat-sink boundary condition. Similar behavior will occur for a Newton-type boundary condition, which we will discuss further in the next section. First, however, we must examine the role of a temperature-dependent electrical conductivity.

Since σ increases with T, the inner region of the sample, which has the highest T, will have the highest σ. Hence, this region will draw the most current; a schematic is sketched in Fig. 8-5. The moderate "current-crowding" process shown in the figure is not self-accelerating or divergent. That is, a larger $\mathbf{J}_e$ does not necessarily mean more heat generation in this region. In fact, inspection shows that *less* heat is generated in the crowded region because $p_e = |J_e|^2/\sigma(T)$. Here an increase in $\mathbf{J}_e$ at a point is the result of an increase in $\sigma(T)$. Currents prefer to flow through the high-conductivity region to dissipate less power; the crowding process is a self-stabilizing one.

The underlying process can be explained by considering two conducting plates connected to each other by two parallel resistors R and $r(T)$ where $r(T)$ is a decreasing function of T. Initially we let $R = r(T_0)$; equal currents will pass through R and $r(T_0)$. The total power drawn will be

$$P_0 = I^2(r \parallel R) = I^2R/2 \tag{8-35}$$

As T increases r will decrease and draw more of the current. Let $r(T_1) \ll R$; almost all of I will pass through $r(T_1)$ and the power drawn from a constant current source will decrease to

$$P_1 = I^2r(T_1) \ll P_0 \tag{8-36}$$

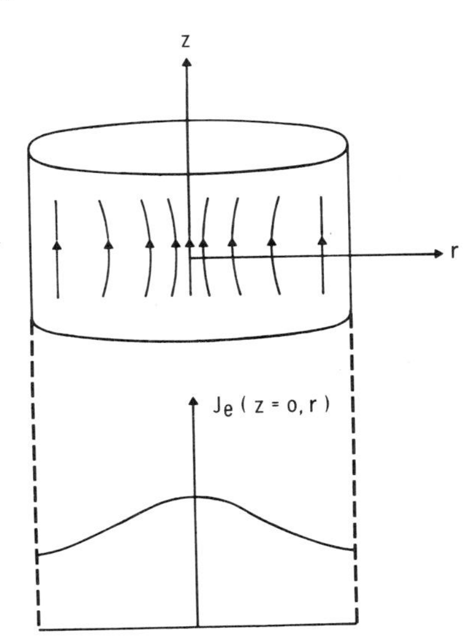

FIGURE 8-5. Lines of electric current (top) and current profile (bottom) sketched for a sample with a "hot spot" near its geometrical center (after Shaw and Yildirim, 1983).

Therefore, it is not correct to assume that an increase in local temperature, which increases the local conductivity, will increase the local current density and lead to a further increase in temperature. That is, thermal runaway will not occur because of this process alone.

It is important to note one further aspect of the effect of current crowding induced by the locally heated region. The channeled current will flow through the region in the z direction producing higher p_e's there. Hence, there will be a tendency for the hot region to expand in the direction of the electrodes. This expansion will stop when a sufficient temperature gradient is established to allow all the heat power to flow out of the film, and this brings us back to the questions of "turnover", the possible presence of NDC, and thermal runaway. All of these effects are related to the ability of the electrical resistance of the system to be reduced to a sufficiently low value by Joule heating. Ultimately, we must therefore find a mechanism by which the layers of the material adjacent to the electrodes can have their resistance lowered to sufficiently small values. The infinite-heat-sink boundary condition used above is unable to account for such effects since the material adjacent to the electrodes will remain cool and at a relatively high value of resistance. If turnover and NDC is to occur, we must have access to a mechanism that, e.g., will account for the fact that in the NDC region an increase in the current could lead to a decrease in the input power density, even though the temperature increases. To achieve this we can let the electrodes be heated (Newton's law of cooling) or realize that the electric fields adjacent to the electrodes might be raised to sufficiently large values to induce the field stripping of carriers, carrier multiplication, or avalanche breakdown effects (Shaw et al., 1973b).

8.2. THERMALLY INDUCED NDC

8.2.1. The Influence of Thermal Boundary Conditions

The development of an NDC region will lead to switching and oscillatory effects for a sufficiently lightly loaded system. It could also result in thermally induced current filamentation (Stocker et al., 1970; Altcheh et al., 1972), thermal runaway, and permanent or alterable breakdown type phenomena (memory). In order to understand how the above possibilities might occur, we return to the basic diffusion-free relation $\mathbf{J}_e = -\sigma(T)\mathrm{grad}\,\Phi$. We see that $\mathbf{J}_e$ can grow when either σ, grad Φ, or both increase. Let us imagine that we have established conditions for an infinite-heat-sink boundary condition on all surfaces, where substantial heating of the bulk material has occurred. The resulting schematic current (a), temperature (b), potential (c), and electrical conductivity (d) profiles are shown in Figs. 8-6a–8-6d. With regard to Fig. 8-6, the following important points should be noted:

1. There are large temperature gradients near the electrodes.
2. There are large electric fields near the electrodes caused by the low temperature of the electrodes (ambient) and nearby regions, and these produce "low-conductivity" layers.

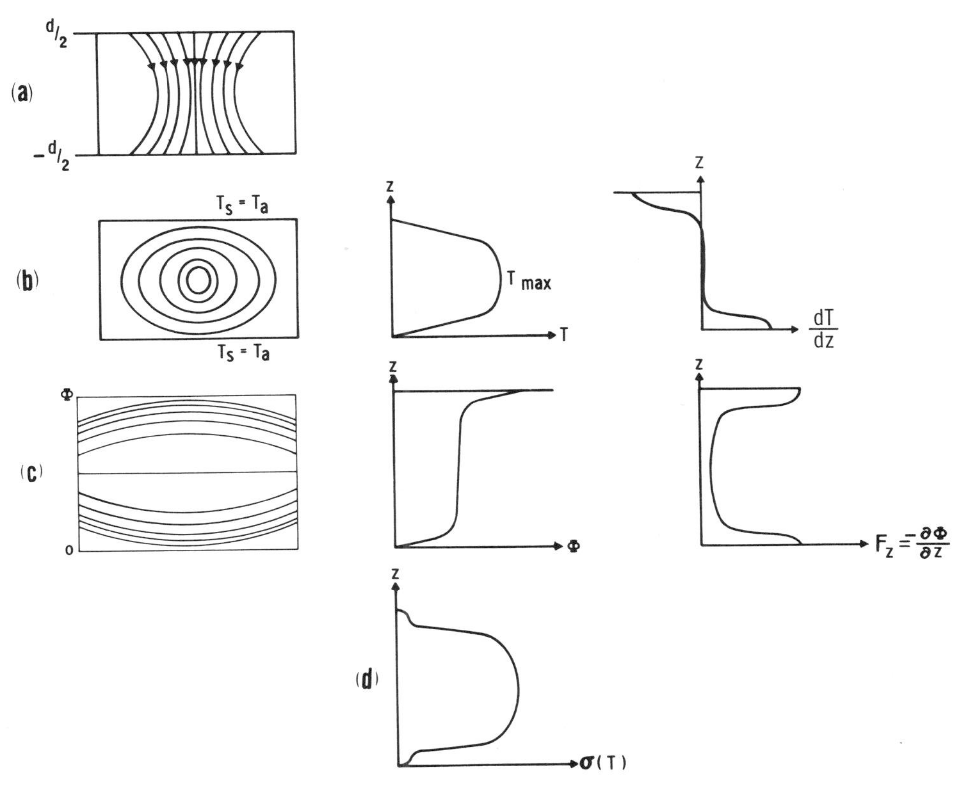

FIGURE 8-6. (a) Electric current lines; (b) isothermal lines (left), temperature profile (center), temperature gradient profile (right); (c) equipotential lines (left), potential distribution (center), electric field distribution (right); and (d) electrical conductivity profile (after Shaw and Yildirim, 1983).

3. The major voltage drop across the film is caused by the narrow low-temperature, low-conductivity layers.
4. If no other factor is included, as the current increases further these narrow regions near the contacts continue to shrink, but they never vanish. Therefore, although the resistance of the structure continues to decrease, the voltage across the device can continue to increase (PDC).

The situation outlined above cannot continue indefinitely. Eventually, at sufficiently high bias, the fields near the electrodes will cause a change in the transport properties of the semiconductor. Indeed, these important field effects will be treated shortly. Now, however, we will consider the effect that a more realistic boundary condition has on the profiles sketched in Fig. 8-6.

Instead of keeping the electrodes at a constant ambient temperature, we employ a more realistic Newton-type thermal boundary condition, which allows the electrodes to heat up and thereby decreases the resistance of the thin layer near the electrodes. We still keep the lateral surfaces at the constant ambient temperature T_a. The appearance of the isothermal lines will now have the form shown in Fig. 8-7. We see that the electrode surface near the z axis will be heated

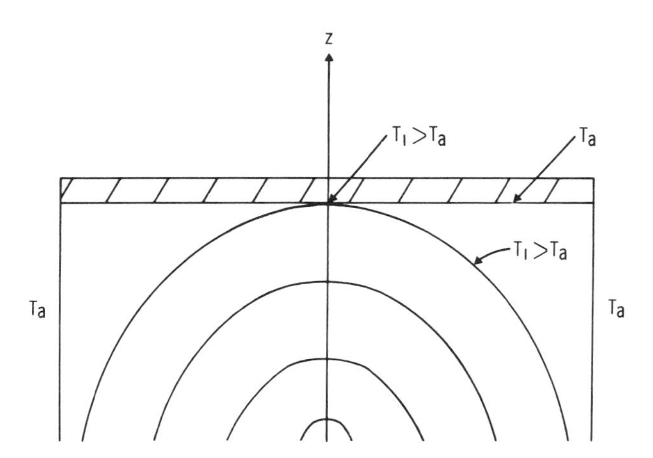

FIGURE 8-7. Schematic drawing of isothermal curves near the electrodes using Newton's law of cooling boundary condition (after Shaw and Yildirim, 1983).

first. This process can lead to current filamentation because the resistive barrier will be lowered in this region. The effect of this boundary condition is similar to the effect of a temperature-dependent electrical conductivity. That is, the electronic and heat current densities are redistributed in accordance with the new boundary condition. The development of current filamentation will be a smooth, continuous process, with filamentation growing as the current increases. (This is to be contrasted with the critical field case, where the transition might be sudden). Figure 8-8 shows how the filamentation develops.

As discussed in Sec. 8.1.1, the presence of a conductivity of the form $\sigma = \sigma_0 e^{-b/T}$ plus the realistic Newton boundary condition leads to turnover and NDC. However, for a sufficiently large load and proper circuit conditions (Shaw et al., 1973), these thermally induced NDC points are stable, and the entire $I(\Phi)$ curve can be, in principle, mapped out (Jackson and Shaw, 1974). Of course, if the load is made sufficiently light, switching will occur from the point I_T, Φ_T (see Fig. 8-1) to another point at higher current and lower voltage, as determined by the load line. For very lightly loaded systems the final state may not be stable because of thermally induced phase changes in the material. Here melting and shorting or opens may occur, and we describe this event as a thermally induced breakdown or memory phenomenon (Kotz and Shaw, 1982, 1983, 1984).

A word about stability is in order here. In Chap. 2 it was shown via Maxwell's equations (Shaw et al., 1973, 1979; Schöll, 1987) that NDC points produced electronically are intrinsically unstable against the formation of both inhomogeneous field and current density (Madan and Shaw, 1988) distributions. When the system evolves into these inhomogeneous states, the domain or filamentary characteristics so produced often exhibit regions of NDC (conduct*ance*) themselves. These NDC points *can be stabilized.* The same situation holds for thermally induced NDC points: any uniform, homogeneous NDC region is unstable; filamentary NDC regions can be stabilized. The latter comprise essentially all the known thermally controlled conditionally unstable systems.

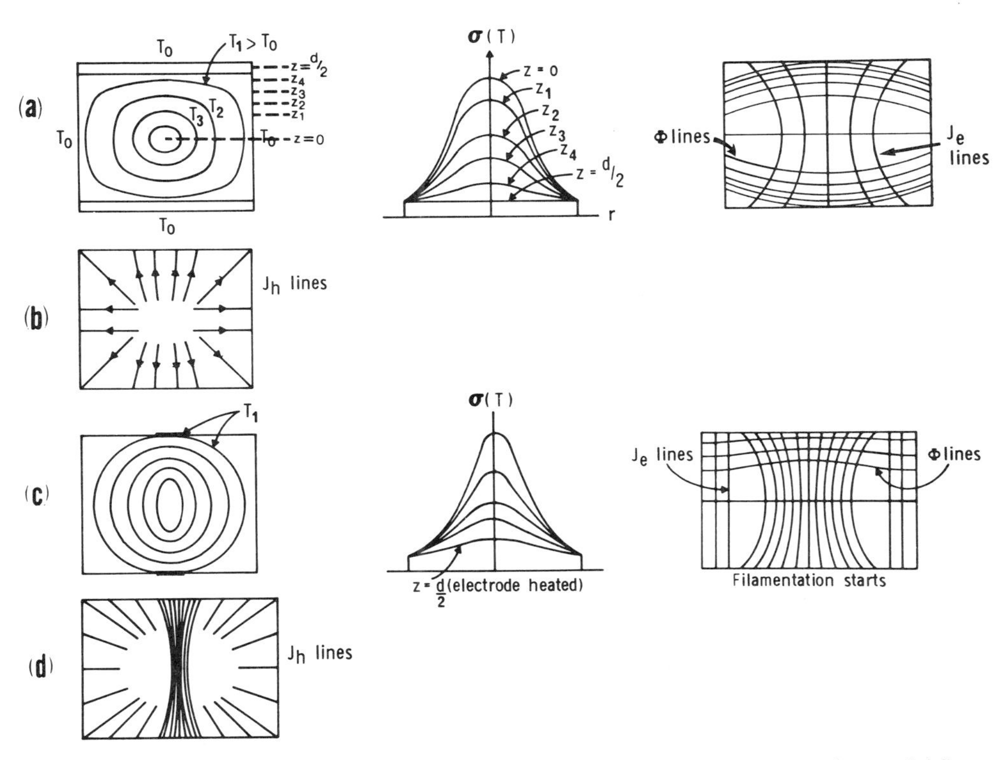

FIGURE 8-8. (a) Temperature (left), conductivity (center), and electric current and potential lines (right) for a constant-temperature boundary condition. (b) Heat current lines for a constant temperature boundary condition. (c) The same as (a) for a Newton's law of cooling boundary condition. (d) The same as (b) for a Newton's law of cooling boundary condition. (After Shaw and Yildirim, 1983.)

Let us now review some major features of the development so far. The main principle that governs the behavior of electric current lines is that these lines prefer to traverse the easiest path from source to sink. An increase in T is the cause of current concentration, not the result. Therefore, this process is self-stabilizing, not self-accelerating. The major role of the Newton boundary condition is basically that, since the electrode is heated at its center ($r = 0$, $z = d$), the electrical conductivity here will be increased and lead to current concentration. One effect of the increased conductivity is that the field here will not be as high as it was when the electrode was kept at ambient temperature. Therefore, if we combine a critical electric field concept (in the next section) with a Newton-type boundary condition, the boundary condition will inhibit the field, **F**, at the electrodes from reaching its critical value. Of further consequence is that if the thermal conductivity is temperature dependent and if $dK/dT > 0$, then the thermal resistance to the flow of heat (at the electrodes) will be reduced as the electrode is heated. This will cause a reduction in the overall temperature level of the system, making local heating more difficult to achieve. The thermal conductivity of the inner regions will increase and lead to a thermally short-circuited region. Thus, the heat generated in the inner regions will flow to the boundaries quite readily and the thermal resistances at the boundaries will be more effective in shaping the thermal behavior of the system. In the equivalent RC network model the situation can be described by reducing the R_t's in the middle of the system compared to the R_t's near the electrodes (see Fig. 8-3). At steady state, $R_{t,n} < R_{t,n-1} < R_{t,n-2}$ will be very small compared to $R_{t,1} > R_{t,2} > R_{t,3}$. We can exaggerate the situation by making $R_{t,n} = R_{t,n-1} = R_{t,n-2} = 0$. However, the $R_{t,n}$'s cannot decrease too much, otherwise the temperatures T_n, T_{n-1}, T_{n-2} will decrease because of the loss of heat. This in turn decreases the thermal conductivity, thus increasing $R_{t,n}$, $R_{t,n-1}$. That is, there is a lower limit for the $R_{t,n}$'s, which provides a sufficient T_n level to keep $K(T_n)$ high. The property $dK/dT > 0$ has a stabilizing effect on T. It prevents an excessive increase of T by allowing easier heat flow; when T tends to decrease, $K(T)$ will decrease and prevent excessive heat loss.

Since we have been emphasizing thin film configurations, the most important boundary conditions are those at the metal electrodes; the axial boundaries have been fixed at the ambient temperature. It is important to realize, however, that when the geometry becomes that of, e.g., a long cylinder, the axial boundary condition also becomes important. An excellent example of its importance is the problem of the Ballast Resistor (Busch, 1921; Bedeaux et al., 1977a, b; Landauer, 1978). This device is simply a long metallic wire immersed in a gas kept at an externally controlled temperature T_G, and is a useful example of a quasi-one-dimensional system exhibiting a variety of spatial and temporal thermal instabilities. The state of the wire at time t is described by only one variable, $T(x, t)$, the temperature field along the length x. Under the proper set of conditions the system exhibits NNDC-type instabilities which depend sensitively upon the boundary conditions. The reader is referred to the articles by Bedeaux et al. (1977, 1981) and Mazur et al. (1981) for further details.

8.2.2. The Effect of Inhomogeneities

The presence of a metalliclike inhomogeneity inside the semiconducting film (see Fig. 8-2) will influence the thermal and electrical transport processes and alter the conditions required for either turnover or switching. First, the inhomogeneity will cause current crowding. The highly conductive region draws the most current; therefore heat will be generated around the periphery of the inhomogeneity. Further, any sharp corners on the inhomogeneity will create large **F** fields leading to higher current densities and enhanced heating at these points. Although the heating process and concentration of current in the middle part of the film will also continue in the manner discussed above, the existence of a metallic inhomogeneity can affect the course of events in a number of ways, for example:

1. Local melting may occur at points surrounding the inhomogeneities, especially at sharp corners where the current density is high.

2. The points where the critical field, $\mathbf{F}_c$, is exceeded may differ from the normal configuration at the electrodes. That is, if there were no inhomogeneity, $\mathbf{F}_c$ would be exceeded at $r = 0$, $z = 0$, d, on the electrode–film boundaries. The

existence of an inhomogeneity will cause distortion of both the shape of the equipotential lines and their densities.

3. The thermal conductivity of the inhomogeneity may be important. If we denote the thermal conductivity of the metallic inhomogeneity $K_2(T)$, then a wide range of temperatures will exist for which $dK_2(T)/dT < 0$, whereas for the semiconductor $dK_1/dT > 0$ is typical. Since K_1 increases with T, the generated heat will flow out of the homogeneous material readily and inhibit excessive increases in temperature. With an inhomogeneity present, however, one must consider the system in somewhat more detail.

Let us consider the region in the vicinity of the inhomogeneity. As the input power increases, the local T will increase for two reasons: (1) heat will be produced inside the inhomogeneity; (2) heat will be produced in the region surrounding the inhomogeneity. This will act to raise the T of the inhomogeneity via the thermal boundary conditions. Consider the situation sketched in Fig. 8-9. As the temperature inside the inhomogeneity increases, $K_2(T)$ will decrease. The heat produced inside the inhomogeneity will see an increased thermal resistance to its flow out of this region. That is, a larger portion of the heat produced here will stay inside the region, which increases T further. This in turn leads to a further growth in T, which will reach a steady state at a higher level and in a longer time than if the thermal conductivity were independent of T. Further discussion of the role of inhomogeneities, including the case where several are present, is deferred to the next section, after we discuss the concept of a critical electric field.

8.2.3. Critical Electric Field Induced Thermally Based Switching Effects

The presence of a critical electric field, $\mathbf{F}_c$, at which a precipitous increase in conductivity occurs, will have a profound influence on the ability of a specific

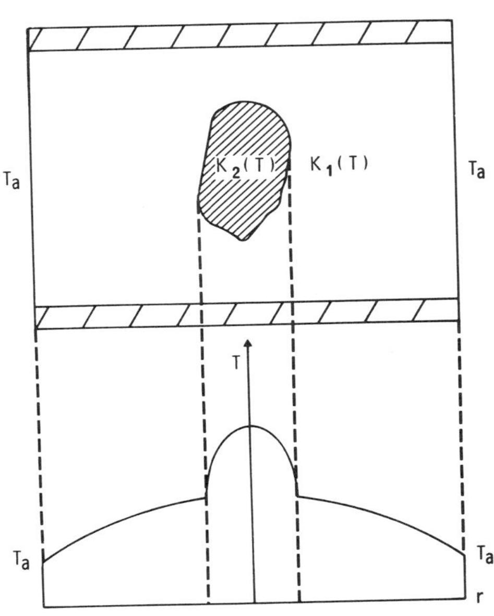

FIGURE 8-9. Thin-disk sample (top) containing an inhomogeneity and the associated temperature distribution (bottom) induced in the steady state by Joule heating (after Shaw and Yildirim, 1983).

system to undergo a switching transition (Shaw et al., 1973b; Kotz and Shaw, 1984). $\mathbf{F}_c$ will play a role for either one or both of the following reasons: (1) it can act to short out the coolest region near the electrodes; (2) it can cause large current densities to flow that produce local heating and melting. As we have shown in Chap. 2, and will reemphasize in Sec. 8.4, where we discuss VO_2, a sudden change in σ due either to $\mathbf{F}_c$ or a critical temperature, T_c, will produce filamentary NDC points that can be stabilized. Systems that exhibit NDC due solely to thermistor-type effects can also be stabilized in their NDC regions by a proper choice of inductance, capacitance, and load resistance (see Sec. 2.2 and Shaw et al., 1973a).

When a critical electric field is present we may write for the conductivity

$$\sigma(T) = \begin{cases} \sigma_0 e^{-\Delta E/kT}, & \mathbf{F} < \mathbf{F}_c \\ \sigma_h(\gg \sigma_0 e^{-\Delta E/kT}), & \mathbf{F} \geq \mathbf{F}_c \end{cases} \tag{8-37}$$

In any thermistor-type system, as the turnover condition is neared the field near the electrodes becomes large, with the largest field occurring just outside the metallic electrodes near the center of the sample. If $\mathbf{F}$ is made to exceed $\mathbf{F}_c$ at that point, then the region adjacent to the electrodes will become highly conducting. In essence this is similar to the sudden motion of a virtual electrode into the material over the warmest region, a shorting out of the highly resistive contact region (Kaplan and Adler, 1972) in this vicinity. This is a self-accelerating process. The expansion of the virtual electrode into the film increases the field just in front of it, which causes further penetration of the virtual electrode (under infinite heat sink boundary conditions, for example). Thus, the device resistance will decrease and the current will crowd into this region and nucleate a filament. When filamentation starts, *less* power will be drawn from the source. But since this power is dissipated in a narrower region, higher local temperatures will readily be produced. (Effects due to the variation of the thermal conductivity in this region will be discussed shortly.) We see that $\mathbf{F}_c$ can lead to the occurrence of a switching event prior to the turnover condition being achieved.

The above critical field effect is electronic in nature, and is invoked as a mechanism by which a thermally based switching event can be electronically *sustained.* When metallic inhomogeneities are present, however, fields can be produced near sharp points that can lead to high local current densities and initiate a local melting event. Indeed, the two types of effects might work together to produce a switching event. A substantial increase in temperature (or melting) might occur near an inhomogeneity, which would lead to a field redistribution and the subsequent attainment of $\mathbf{F}_c$ near an electrode; switching or breakdown would then occur. Other possible sequences and/or simultaneities are easy to visualize and categorize. The order of occurrence is not critical; a self-accelerating switching event will be induced.

The above considerations will be modified somewhat when the variation of the local power density, $p_e(T)$, with temperature is considered. As T increases, $\sigma(T)$ will increase and $p_e(T)$ could decrease. If $p_e(T)$ decreases at a sufficiently rapid rate as T increases, the steady state condition may be reached in a shorter time period and be at a lower T level. However, we have previously seen that $p_e(T)$ can be affected by things other than the presence of an inhomogeneity ($\mathbf{F}_c$, T_c, $dK_2/dT < 0$, etc.). these features can make $p_e(T)$ *increase* with T. Because of this the projected steady state T distribution may not be reached; a switching event will occur first. For example, a switching event can be launched prior to turnover in a system where $dK_2/dT < 0$ and an $\mathbf{F}_c$ is present.

Many common systems will not only have an $\mathbf{F}_c$, but will also contain many inhomogeneities. It is important, therefore, to consider the effect that multiple metallic inhomogeneities might have on the above conclusions. First, it is clear that current crowding effects will be more pronounced in regions that contain more inhomogeneities per unit volume. Therefore, local heating effects will be prevalent here. Further, at the sharpest corners of the inhomogeneities the fields and current densities will be largest. If critical values are exceeded and melting occurs, the presence of a nearby inhomogeneity will aid in the development of the potential instability. This case can be seen simply by realizing that the presence of nearby inhomogeneities will lead to more current crowding about the specific inhomogeneity we are heating. Once the major hot spot is nucleated, the high conductance region will expand toward neighboring inhomogeneities and raise the local fields, for example, above $\mathbf{F}_c$, thus reducing the time it takes to switch the sample, along with the voltage at switching.

Next, it is very possible that the switching event may also start at regions near the electrode interface where the inhomogeneities are densest. Further, since $dK_2/dT < 0$, the power generated inside the inhomogeneities will tend to remain there. Although negligible power should develop there because of the high values of $\sigma(T)$, $p_e(T)$ $[= |J_e|^2/\sigma(T)]$ may become large due to substantial increases in current density. The existence of regions with high thermal resistances (metallic inhomogeneities at elevated temperatures) will also narrow the path of the flow of heat current, which will then act to raise the average temperature of the inhomogeneous region.

Finally, the heating of the electrodes (Newton's law of cooling) may be enhanced in regions where $\mathbf{F}_c$ is reached, and this will also tend to accelerate the switching process.

It is useful to conclude this section with a general description of the possible types of thermal behavior that might occur in a system where the electrical conductivity is a function of both temperature and electric field. A qualitative-analytical approach, with support from numerical calculations, leads to the following observations and conclusions.

Consider the nonlinear eigenvalue problem

$$\Delta u = -\lambda f(x, u), \qquad x \in D \tag{8-38a}$$

$$\alpha u + \beta\, \partial u/\partial n = 0, \qquad x \in S \tag{8-38b}$$

where $D \cup S$ is the active region, S the boundary surface, and Eq. (8-38b) a general boundary condition. λf is the source of u. Some well-known results on the existence, uniqueness, and stability of the solutions of Eq. (8-38) are as follows

(Joseph, 1965; Joseph and Sparrow, 1970; Keller and Cohen, 1967; Simpson and Cohen, 1970):

(A) If $\partial f/\partial u \equiv f_u \geq 0$ and $f(x, u) < F(x)$ ($\equiv$ a function independent of u) for $0 \leq u \leq \infty$, then a finite unique and stable solution $u > 0$ exists for all λ in the interval $0 \leq \lambda \leq \infty$.

(B) If $f_u > 0$ and unbounded as $u \to \infty$, then a solution $u > 0$ exists only in a range $0 < \lambda < \lambda^*$, where λ^* is a limit determined by $f(u)$, the boundary conditions and geometry of the system. This solution is unique if $f(u)$ is concave $[f_u(u_2) < f_u(u_1)$ for $u_2 > u_1]$ and nonunique if $f(u)$ is convex $[f_u(u_2) > f_u(u_1)]$.

Equation (8-38) is in a form similar to Eq. (8-30) under steady state conditions and with $J_e = \sigma(T, F)F$:

$$K\nabla^2 T = -F^2\sigma(T, F) \tag{8-39}$$

where T is analogous to u, $\sigma(T, F)$ to $f(u)$, and F^2 to λ. Therefore, the well-known features of Eq. (8-38) may be used to study the solutions of Eq. (8-39). Examples are the following:

(a) If the conductivity σ has no F dependence, then Eq. (8-39) has the same form as Eq. (8-38); for a conductivity of the form

$$\sigma(T) = \sigma_0 e^{-a/T}$$

(or any other form which is monotonic and remains finite for all T), Eq. (8-39) has a unique solution because $\sigma(T)$ has the property (A) stated above. Under these conditions an SNDC region can be generated by using Newton's law of cooling as the boundary condition for T. The filamentary solution can be stabilized in the NDC region using a source with a sufficiently high resistance and the proper local circuit environment (Shaw et al., 1973).

(b) For a conductivity of the form

$$\lim_{T\to\infty} \sigma(T) \propto T^n \qquad (n > 0)$$

which goes to infinity as $T \to \infty$, property (B) above holds and solutions do not exist beyond a certain local electric field F_c (analogous to λ_c). Also, for $0 < n < 1$, the existing T solutions are unique and stable; for $n > 1$, the solutions are nonunique and only one (the lowest) is stable.

(c) for an electric-field-dependent conductivity $\sigma(T, F)$, the problem is slightly more complicated because the two coupled Eqs. (8-39) and (8-11) for the thermal and the electric current must be considered together in determining the existence, uniqueness, and stability of the solutions.

$$\mathrm{div}[K(T)\mathrm{grad}\, T] = -\sigma(T, F)F^2 \tag{8-39'}$$

$$\mathrm{div}[\sigma(T, F)F] = 0 \tag{8-11'}$$

Experimentally, it is known that in materials such as the chalcogenides the electrical conductivity increases with electric field,

$$d\sigma/dF > 0$$

This property has a limiting effect on the maximum field value, $F_{\max}$, which occurs at the electrode–film interface. As the material is heated, $F_{\max}$ steadily increases. However, this increase is decelerated by the increase in conductivity, which tends not to support high F fields. Thus, the right-hand side of Eq. (8-39) remains finite, and property (A) is expected to hold again, yielding a unique T solution for all possible values of field.

The best fits between numerical and experimental results are obtained with such conductivities. For example, Kaplan and Adler (1972) have obtained a switching phenomenon with a conductivity of the form $\sigma(T, F) = \sigma_0 e^{-a/T} e^{F/F_0}$. Shaw and Subhani (1981) have obtained a better fit by assuming a discontinuous $\sigma(F)$ variation given by

$$\sigma(T, F) = \begin{cases} \sigma_a = \sigma_0 \exp[-(\Delta E - \beta F)/kT - F_0/F], & F < F_c \\ \sigma_h \gg \sigma_a, & F > F_c \end{cases} \tag{8-40}$$

where ΔE is the thermal activation energy, β is a constant representing a field-dependent decrease in activation energy, and F_0 is a constant associated with carrier multiplication. Both conductivity functions satisfy the condition $d\sigma/dF > 0$, lead to current filamentation, and cause a discontinuous jump from a low conductance to a high conductance state under sufficiently high bias, independently of the external circuit. With such conductivity functions no stabilizable uniform NDC region apparently exists. We have an inherently nonhysteretic system performing a jump in conductance independently of the external circuit.

Considering all of these features and the numerical results, electrothermal switching phenomena in semiconductors can be described, in general, as follows.

As the applied voltage is increased, the temperature of the central region grows faster than the regions close to the boundaries. At low heating levels the electric field is not very effective in controlling the conductivity, hence the conductivity can be approximated by $\sigma_0 e^{-a/T}$. Therefore, the conductivity of the central region will be larger than the regions close to the electrodes. This produces a low F field in the central region and a high F field near the electrodes. As the voltage is increased further, if σ were not a function of F, very high F fields would be produced near the electrodes. Since the heat generation rate is σF^2, increasingly large power will be generated near the electrodes. To keep the power density at the electrode–material interface finite, the conductivity there will increase by some mechanism, which in turn will decrease the local field. This can be accomplished in a number of ways. For example, Newton's law of cooling is a realistic thermal boundary condition. It allows a certain amount of heating at the interface, thus increasing the conductivity and lowering the field. Newton's law of cooling introduces a higher thermal resistance at the electrodes, which elevates the temperature level of the material, thus increasing the conductivity everywhere. However, the electrodes are still cooler than the central region, resulting in higher F fields at the electrodes. Therefore, most of the heat generation again occurs near the electrodes. The temperature distribution over the electrode surface will be such that it will be maximum at the center of the electrode if there is no inhomogeneity. However, if a metallic inhomogeneity exists, T will be maximum at the point nearest to the inhomogeneity. At that point, where the low conductivity barrier is reduced, current filamentation starts. This confines the dominant heat generation near the electrodes to these weak points, which reduces the electrical resistance further.

The dependence of σ on F with $d\sigma/dF > 0$ can also lead to a limitation on the high power generation near the electrodes. For example, we have pointed out that as the Joule heating increases, the $e^{-a/T}$ dependence produces a higher conductivity in the central regions compared to the regions near the electrodes, leading to low F's in the central region and high F's near the electrodes. The e^{F/F_0} or F_c type dependence starts to be effective after a certain field level, thus increasing the conductivity more at a point on the electrode where F is maximum. In the presence of a metallic inhomogeneity, this point will be nearest the inhomogeneity. Here is where the filamentation begins, leading to higher F fields, which in turn produces a higher local conductance. This process is equivalent to the expansion of the electrode into the material at the point of interest. This is a self-accelerating event, because as the virtual electrode pushes into the material (Kaplan and Adler, 1972) the entire voltage drops across a shorter distance, which means a higher F field, and further expansion of the virtual electrode into the material.

It is clear from what we have discussed so far that the switching process in thin films is a very rich and complex subject, and the presence of inhomogeneities can play a very important role. Indeed, in an experimental study, Thoma (1976) provided evidence that bias-induced reversible switching transitions in a wide variety of thin insulating and semiconducting films between 2 and 100 μm thick occurred when a critical amount of power per unit volume was dissipated in the samples. The materials investigated—crystalline and polymeric, as well as amorphous—included ZnS, mica, Al_2O_3, anthracene crystals, Mylar, polystyrole foils, crystalline LiF, ZnO, Cds, Si, and GeAsTl glasses. He concluded that in order to explain this ubiquitous phenomenon, it must be assumed that many real insulating materials contain defects or inhomogeneities arranged in chainlike patterns which give rise to higher mobility and/or higher carrier concentration paths through the films, a view taken and exploited analytically for inhomogeneous multicomponent chalcogenide films by Popescu (1975). Under bias, these inhogeneities lead to very narrow current filaments that extend throughout the thickness of the films. Since the switching effect in chalcogenide films has been explored in great detail, we treat it first, although, strictly speaking, the switching events observed in these films are electronically initiated and sustained. However, the thermal aspects are so important with regard to the delay-time phenomenon, forming and memory, that this popular device structure is emphasized here. Our second example will be vanadium dioxide, which exhibits a purely thermal switching phenomenon.

8.3. ELECTROTHERMAL SWITCHING IN THIN CHALCOGENIDE FILMS

8.3.1. Introduction

8.3.1.1. Scope of the Problem

As we have already pointed out, the application of sufficiently high electric fields to any material sandwiched between metal contacts almost always results in departures from linearity in the observed current–voltage characteristics (Shaw, 1981). With further increases in bias in the nonlinear regime either breakdown, switching, or oscillatory events eventually occur (Shaw and Subhani, 1981; Thoma, 1976; Madan and Shaw, 1988). Breakdown usually results in local "opens" or "shorts," whereas switching is often involved with local changes in morphology that are not as catastrophic as those that result from a breakdown event; here reversible changes in conductance are induced (Kotz and Shaw, 1983, 1984). In many thin films, after a switching event from an "OFF" to an "ON" state occurs, when the ON state is maintained for a sufficiently long time, a "setting" or memory can occur such that when the bias is reduced the sample will not switch back to the OFF state until it is subjected to further treatment such as the application of high current pulses.

There are two classes of explanations for the above array of complex phenomena: thermal and electronic. In general, both effects must be considered in any quantitative analysis, and the two can produce a coupled response called "electrothermal". In a discussion of the physical mechanisms involved with a particular specimen, the major parameters controlling its operation must be identified and separated out from the less significant features. In this section we do this for bias-induced switching effects in amorphous chalcogenide films (De Wald et al., 1962; Ovshinsky, 1968; Pearson and Miller, 1969) typically 0.5–10.0 μm thick. Reviews have been given, e.g., by Adler et al. (1978, 1980), Shaw (1985, 1988), and Madan and Shaw (1988). It is the purpose of this section to review the major experimental features of these phenomena; present the results of numerical calculations that model the first-fire event in homogeneous films and compare favorably with experiment; discuss switching in inhomogeneous films that have become so because of the morphological changes induced by prior switching events—a process known as "forming." Here, we suggest that specific inhomogeneous films can show an electronic switching transition initiated by an instability that nucleates at a critical local power density (Thoma, 1976; Shaw, 1979). On the other hand, specific homogeneous films can be induced into a breakdown-type event at sufficiently high electric fields (Shaw et al., 1973b), but rather than resulting in an open or short circuit, intermediate (inhomogeneous) states are formed which serve as basis states for subsequent switching events (Kotz and Shaw, 1983, 1984). The differences and similarities between virgin and formed films and their electronic behavior will be emphasized throughout the section.

It has been common practice among some investigators to separate switching in thin amorphous chalcogenide films into two classes, threshold and memory

(Adler et al., 1978, 1980), according to whether the OFF state can be resuscitated after the ON-state has been maintained for a given length of time. In fact, in most of these materials a memory effect will occur when the ON-state is held by direct current for sufficiently long times. Specimens called memory switches usually are of relatively high conductivity (e.g., $Ge_{17}Te_{79}Sb_2S_2$ (Buckley and Holmberg, 1975; Kotz and Shaw, 1984), where the room temperature conductivity is about $10^{-5}\ \Omega^{-1}\ cm^{-1}$) and "set" in a matter of milliseconds. Specimens called threshold switches usually are of relatively low conductivity [e.g., $Te_{39}As_{36}Si_{17}Ge_7P_1$ (Petersen and Adler, 1976), where the room temperature conductivity is about $10^{-7}\ \Omega^{-1}\ cm^{-1}$) and require times much longer than a millisecond to set. On some occasions relatively low conductivity films can still be returned to the OFF-state (without additional treatment) by reducing the applied bias after being kept ON for times on the order of 10 h.

It is probably the case that the major difference between memory and threshold switches is the time required to set when excited by pulsed or continuous direct current. However, in what follows we will delineate between the two in deference to common practice. Since our emphasis here is on threshold switching, the memory effect will only be discussed when its understanding will help elucidate the threshold effect. The electronic mechanism for the initiation of the switching event in both cases in the same (Kotz and Shaw, 1984).

8.3.1.2. Switching Parameters

The $I(\Phi)$ characteristics of a typical threshold switch are shown in Fig. 8-10a. At low currents a high resistance is observed (approximately $10^7\ \Omega$); this is called the OFF-state regime. When a threshold voltage, Φ_T, typically 10–100 V, is exceeded, the sample switches to a low-resistance operating point on the load line, with the dynamic resistance falling to about 1–100 Ω; this regime is known as the ON state. As long as a minimum current, I_h, called the holding current, is maintained, the sample remains in the ON state. However, if the current falls below I_h, the sample either switches back to an OFF-state operating point on the load line or undergoes relaxation oscillations between the ON and OFF states, depending on the value of the load resistance (Shaw et al., 1973). Since I_h depends on the circuit conditions, it is better to treat an essentially circuit-independent parameter, the holding voltage, Φ_h, as more fundamental; Φ_h is usually about 1–2 V.

The corresponding $I(\Phi)$ characteristic of a typical memory switch is shown in Fig. 8-10b. Here, once the sample is switched to the ON state it will remain there even when the current is reduced to zero. To return the sample to the OFF state a reset pulse of sufficient magnitude and duration is required (Ovshinsky, 1968).

Voltage-pulse experiments (Buckley and Holmberg, 1975; Petersen and Adler, 1976; Pryor and Henisch, 1972; Shaw et al., 1973b; Kotz and Shaw, 1984) provide a major source of information about threshold switching and lead to the introduction of other parameters of interest. After a voltage pulse is applied, a delay time, t_d, typically less than 10 μs, elapses before the onset of switching. The switching time, t_0, has proven to be faster than any means found of measuring it,

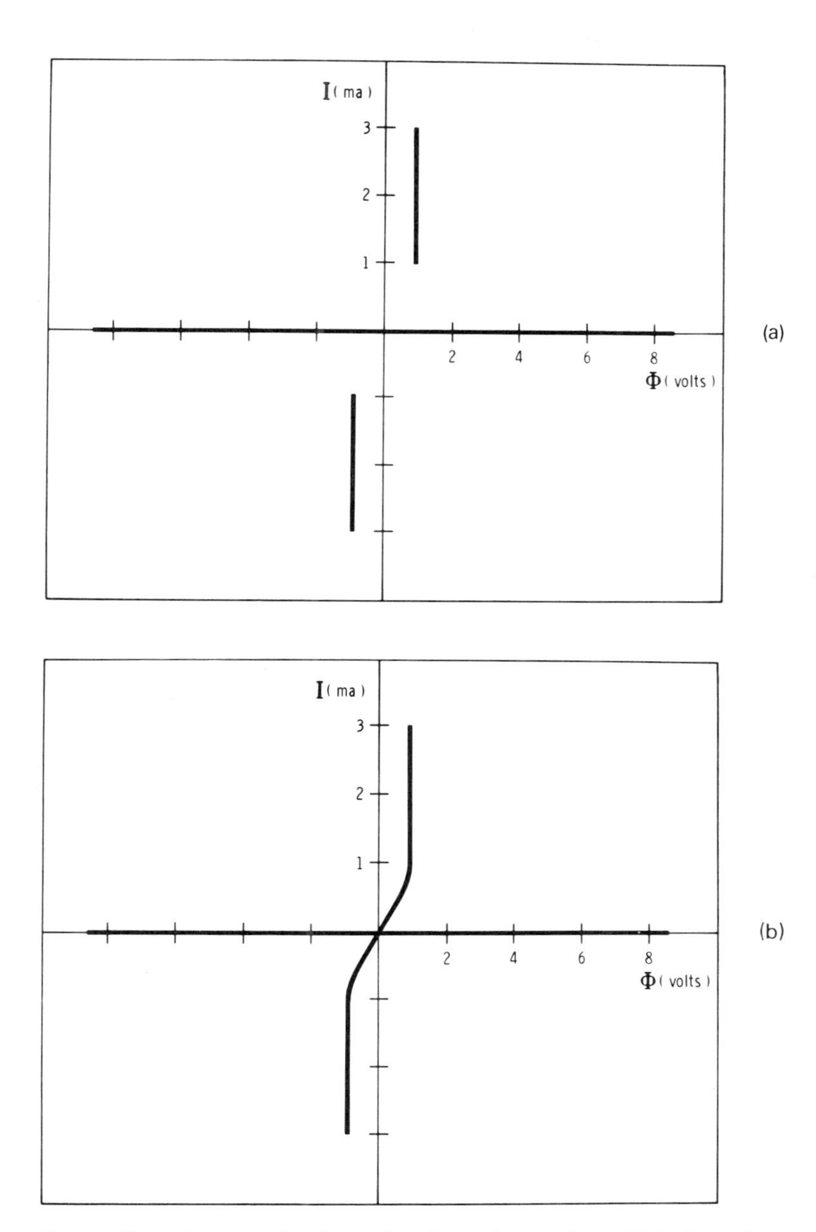

FIGURE 8-10. (a) Current as a function of voltage for a 1-μm-thick film of amorphous $Te_{39}As_{36}Si_{17}Ge_7P$ sandwiched beween Mo electrodes. This is a trace from a Tektronix curve-tracer oscilloscope, which implies a 60-Hz ac signal. (b) Same as (a) except this material is an alloy where memory switching readily occurs. (After Petersen and Adler, 1976.)

but is known to be less than 1.5×10^{-10} s. It has also been convenient to define a pulse interruption time (Pryor and Henisch, 1972), t_s, as the time between removal of Φ_h and application of an ensuing pulse with $\Phi > \Phi_h$. After Φ_h is removed for a time t_{sm}, the maximum benign interruption time, only Φ_h is required to restore the ON state. (t_{sm} is typically about 250 ns, but varies with the original ON-state operating point.) For longer values of t_s, the voltage required to reswitch the sample approaches the original threshold, Φ_T, the latter being completely restored in a recovery time, t_r. Figure 8-11 shows some of these parameters for two pulses, each of width t_p.

8.3.1.3. OFF-State Characteristics of a Homogeneous Film

For conciseness, we will emphasize a typical switching material, $Te_{39}As_{36}Si_{17}Ge_7P_1$, which has been perhaps the most thoroughly studied sample (Petersen and Adler, 1976). At low fields, less than about 10^3 V cm^{-1}, the $I(\Phi)$ characteristics are linear and the resistivity varies with temperature as $\rho(T) = 5 \times 10^3 \exp(0.5\,[\text{eV}]/kT)\Omega$ cm. This yields a room temperature resistivity of the order of $10^7\ \Omega$ cm. The optical energy gap is approximately 1.1. eV, or about twice the thermal activation energy, a result typical for amorphous chalcogenide semiconductors, which are usually p-type in nature (see, e.g., Tauc, 1974).

When, e.g., Mo electrodes are put in contact with the chalogenide, the bands in the latter bend upward by approximately 0.15 eV. Under extremely low applied bias the characteristics are linear, but as the applied bias is increased into the field range 10^3–10^5 V cm^{-1}, the Schottky barrier manifests itself and the current becomes contact limited; it is controlled by various tunneling contributions from field and thermionic field emission (Shaw, 1981). However, in the 10^5-V cm^{-1} range a "high field" characteristic appears in which the conduction is bulk-limited and of the form $\sigma = A \exp(F/F_a)$, where A and F_a are constants. In fact, in the field region above 10^4 V cm^{-1}, we shall show that the OFF-state $I(\Phi)$ characteristics of a virgin, homogeneous film can be fitted rather well by using an expression for the conductivity given by $\sigma = \sigma_0 \exp[-(\Delta E - \beta F)/kT - F_0/F]$,

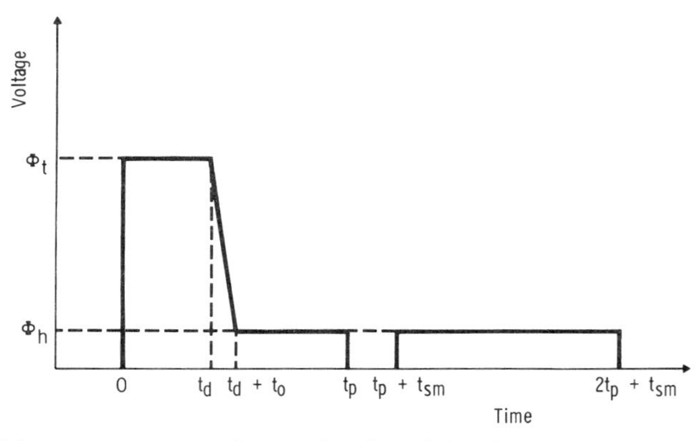

FIGURE 8-11. Voltage across a sample as a function of time for two pulses of width t_p separated in time by t_{sm}. All parameters are defined in the text. (After Shaw and Yildirim, 1983.)

where β is a constant representing a field-dependent decrease in activation energy, and F_0 is a constant associated with carrier multiplication. σ_0 is the conductivity as $T \to \infty$ in the absence of a field effect (Reinhard et al., 1973).

8.3.1.4. The Switching Transition

When thin amorphous chalcogenide films containing tellurium are homogeneous, uniform, and virgin, the initial (first-fire) switching process is an electronic event (Shaw et al., 1973b; Buckley and Holmberg, 1975). Although first-fire is classified as a switching phenomenon, it is perhaps also useful to treat it as a breakdown-type process, since it is often, but not always, the case that the voltage at threshold is substantially lower after the first few firing events. Furthermore, after further firings the voltage at threshold often continues to drop slowly until, after a sufficient number of firings, in many cases it stabilizes at a "running" value.

The drop in threshold voltage upon firing is associated in most cases with a forming process (Shaw et al., 1973c; Kotz and Shaw, 1984) wherein either crystalline (Bosnell and Thomas, 1972), morphological (Allinson et al., 1979), or amorphous imperfections are produced locally. In a formed or inhomogeneous sample the instability that develops at a critical value of local power density has some features that are somewhat different from the first-fire event in homogeneous films. In the latter case, for short enough pulses a critical electric field is reached isothermally over the entire sample, independent of the thickness (Buckley and Holmberg, 1975). As we shall discuss, we expect that this field strips trapped carriers off local defects, and then the significantly increased Joule heating, often dominated by the capacitive discharge (Kotz and Shaw, 1984), causes a thermal breakdown at the weakest point in the film.

For pulse widths greater than about 10^{-9} s the common delay-time mode is observed in formed samples of all thicknesses. (To our knowledge, single-shot data showing the existence of a delay-time event occurring during first fire in a thin homogeneous sample is not available.) There is abundant experimental evidence (Thoma, 1976; Balberg, 1970; Reinhard, 1977) that the delay-time mode produces a switching event at a critical local power input. It was first shown (Balberg, 1970) that intimate double inverted pulses produced identical delay times and later (Reinhard, 1977) that a critical rms voltage switched samples after identical times in a study of their response to pulse burst waveforms. These results are evidence that t_d is associated with the time it takes for a local hot spot to grow and perhaps propagate through the sample (Shaw et al., 1973c; Homma, 1971).

As we have emphasized above, voltage pulse measurements have been very useful in elucidating several important aspects of t_d. In fact, for sufficiently short pulses (Shaw et al., 1973b) t_d can be made comparable to the time it takes for the voltage across the sample to collapse from Φ_T to Φ_h, the switching time, t_0. Whereas t_d is thermal in origin, t_0 is due to an electronic process in both virgin and formed films. Models for both will be discussed shortly.

8.3.1.5. ON-State Characteristics

The forming process produces a local inhomogeneous region typically 1–5 μm in diameter. It is through this relatively high conductance region that the major portion of the current flows when the sample is in the ON state. This filamentary current-carrying path has a radius r_f, and its major features have been described for samples that did not exhibit forming (Petersen and Adler, 1976) (the first-fire and running threshold voltage were the same). These samples had quite low values of threshold voltage ($\cong 17$ V). Therefore, the films could have contained large numbers of inhomogeneities even when virgin (perhaps crystallites of various sizes), so that the hot spot nucleated at a site present in the sample when it was prepared.

In the studies of Petersen and Adler (1976) the current dependence of r_f was experimentally exposed by several independent methods. The results are shown in Fig. 8-12. First, a study of velocity saturation in crystalline-Si/amorphous-chalcogenide heterojunctions provided a means of determining the current density in the chalcogenide in the 2–9-mA range (prior to avalanche breakdown

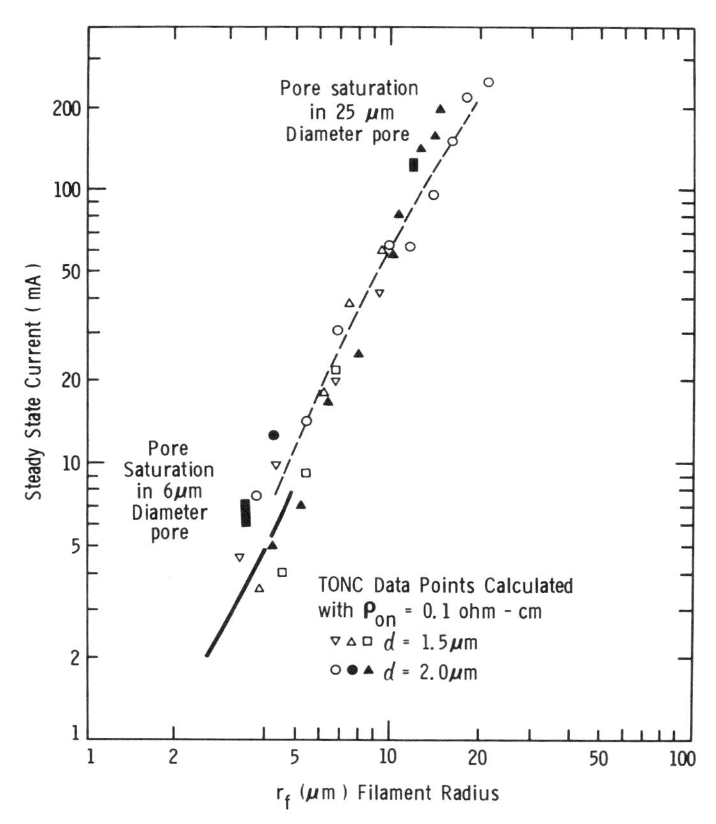

FIGURE 8-12. Steady state current as a function of filament radius determined by four methods. The solid line represents the results of the velocity-saturation analysis, the data points are the TONC results, and the dashed line is calculated from Shanks' (1970) carbon/chalcogenide/carbon results. Two pore-saturation points are also indicated. (After Petersen and Adler, 1976.)

in the Si depletion region). It was found that the area of the current filament, A_f, increased more or less proportionally to the increase in current, indicating that the current density remains constant in the filament over a wide range of current. Second, the transient-ON-$I(\Phi)$ characteristic (TONC) technique (Pryor and Henisch, 1972) was used to analyze the ON-state behavior (see Fig. 8-13). It was found that the TONC was stable for only about 50 ns, after which the response gradually relaxed to the steady ON-state $I(\Phi)$ characteristic. Therefore, for TONC pulses less than 50 ns (Kotz and Shaw, 1984), we expect that the area of the current filament remains the same as in the steady state; the shape of the TONC should then depend upon the value of the operating steady ON-state current for which a particular TONC is taken. This is in fact the case.

In general we expect three contributions to the voltage drop across the sample in the ON state: the resistance of the ON state material; the contact resistance, R_c; the interfacial barrier, Φ_B. The TONC curves should obey

$$\Phi_{\mathrm{TONC}} = \Phi_B + I[R_c + \rho_{\mathrm{ON}} d/A_f(I_{\mathrm{dc}})] \tag{8-41}$$

where ρ_{ON} is the ON-state resistivity, d the thickness of the film, and $A_f(I_{\mathrm{dc}})$ the area of the current filament at the steady state operating point. Extrapolation of the sub-50-nsec TONC curves should yield the same value for Φ_B, and this value should be the same as the metal/amorphous-chalcogenide barrier measured by other means. The agreement is good. The TONC slopes then determine the variation of A_f with steady state current.

Further, the steady ON-state voltage is

$$\Phi_{\mathrm{dc}} = \Phi_B + IR_c + J(I)\rho_{\mathrm{ON}} d \tag{8-42}$$

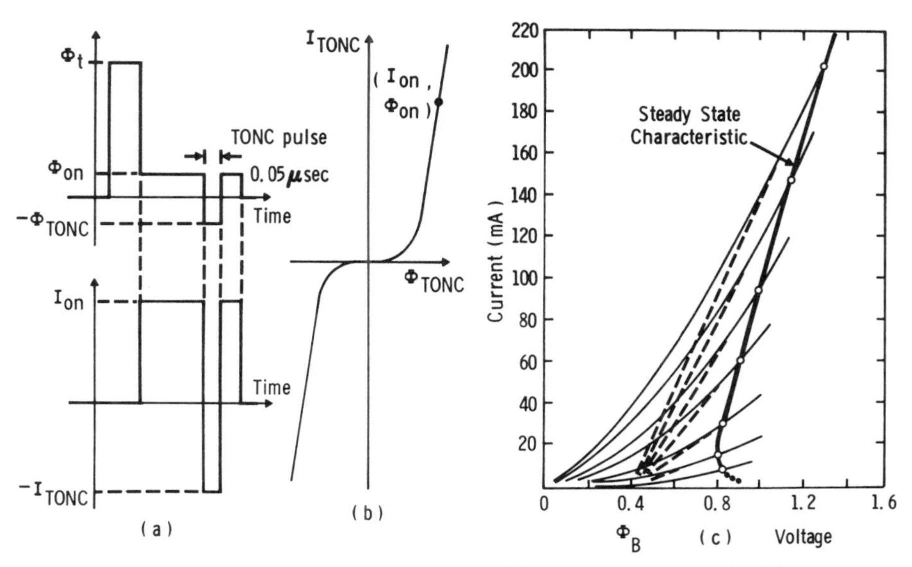

FIGURE 8-13. (a) Transient ON-state characteristic (TONC) measurement. (b) A typical result. (c) Different Mo/amorphous/Mo sample TONC curves taken from different (I_{on}, V_{on}) points. The sample is 50 μm in diameter. (After Petersen and Adler, 1976.)

If, as expected, the current density J is independent of l, extrapolation of the steady ON-STATE characteristics to $l = 0$ will yield an effective barrier voltage

$$\Phi_{B(\mathrm{eff})} = \Phi_B + J\rho_{\mathrm{ON}}d$$

and this should vary linearly with the thickness of the chalcogenide film. Experiments show this correlation rather well. Furthermore, extrapolation of $\Phi_{B(\mathrm{eff})}$ to $l = 0$ should yield Φ_B. Again, this is the experimentally observed situation. These results yield $\rho_{\mathrm{ON}} \cong 0.07\ \Omega\ \mathrm{cm}$.

Shanks' result (1970) for the ON-state $I(\Phi)$ characteristics of a formed chalcogenide film having pyrolytic graphite electrodes were also used to determine the area of the current filament as a function of current. The results are shown as the dashed line in Fig. 8-12.

Analysis of the gain observed in an n(ON-state)-p-n amorphous-crystalline heterojunction transistor (Petersen et al., 1976) as a function of the crystalline-Si base doping concentration showed that the free-carrier concentration in the ON state is of the order of $10^{19}\ \mathrm{cm}^{-3}$. This implies that the ON-state carrier mobility is about $10\ \mathrm{cm}^2/\mathrm{V\ s}$.

The total of the cathode and anode interface barriers for Mo/amorphous/Mo samples is 0.4 eV. If this is distributed evenly between cathode and anode, and if in the ON-state fields above $10^5\ \mathrm{V\ cm}^{-1}$ must be maintained near both electrodes, the band bending will then extend about 30–70 Å into the amorphous material. This is sufficently narrow that it is possible that the ON state can be maintained by either strong-field emission or thermionic field emission tunneling through the electrode barriers. However, if the barriers are asymmetric the depletion regions can be larger in extent, and tunneling processes become less likely. Alternatively, the ON state can be maintained from carrier generation in the high field regions themselves. Since the potential drop in these regions is less than E_g, such generation would have to be from localized states rather than from across-the-gap excitation. In either event, it is likely that both electrons and holes contribute to the ON-state (Petersen et al., 1976). A generation–recombination model incorporating these features has been proposed by Schöll (1978, 1987) and elaborated by Landsberg et al. (1978), Schöll and Landsberg (1979), and Robbins et al. (1981). It is based upon the simultaneous impact ionization of trapped electrons and trapped holes from localized states at about midgap. For electric fields in the range between a holding field F_h and a threshold field F_T there exist two stable steady states. In one of these—the low conductivity (OFF) state—the localized states are largely occupied by electrons, and the current is carried by holes. In the other—the high-conductivity, electron-dominated (ON) state—the localized states are mainly occupied by holes. The threshold and holding fields of the switching transistion scale with the field dependence of the electron and hole impact ionization coefficients, which have been assumed to be of the simple form $\alpha \exp(-F_0/F)$.

8.3.1.6. Recovery Properties

When the current is reduced below I_h, the sample switches back to the OFF state. One possible mechanism for the initiation of this transition suggests that there might exist a minimum r_f for which radial diffusion would break the filament. This would set an absolute minimum value for the current that can be maintained in the ON-state, I_{hm}. However, observation of I_{hm} is normally difficult to achieve because of the reactive components in the circuit. If we define I_h as that current below which circuit controlled relaxation oscillations occur (Shaw et al., 1973), then for most sample configurations there will always be a range of currents between I_h and I_{hm} that are unstable against relaxation oscillations. The package capacitance, C, and intrinsic plus package inductance, L, will always produce $I_h > I_{hm}$. On the other hand, if an intimate double-pulse technique is employed (Hughes et al., 1975), where the sample is forced to remain in the ON state after switching by first rapidly reducing the applied bias, then by minimizing C and maximizing the load resistor R_L, values of I_h as low as 10 μA can be observed. (These are currents that would produce relaxation oscillation were the ON state not "held" by the second pulse.) For current densities in the filament in the range of 10^3 A cm^{-2}, such low values of current imply that filament radii in the 0.5-μm regime can be stabilized. In this case, Φ_h is rather high since the ON-state (filamentary) characteristic itself exhibits a long, stabilizable NDC region for these currents. The fields are therefore sufficiently high so that the ON state is maintained in an extremely narrow filament; as $\Phi_h \rightarrow \Phi_T$, r_f approaches its minimum value. There is evidence that the minimum radius of the current filament may, in fact, be in the fractional-micrometer regime; thus there is a

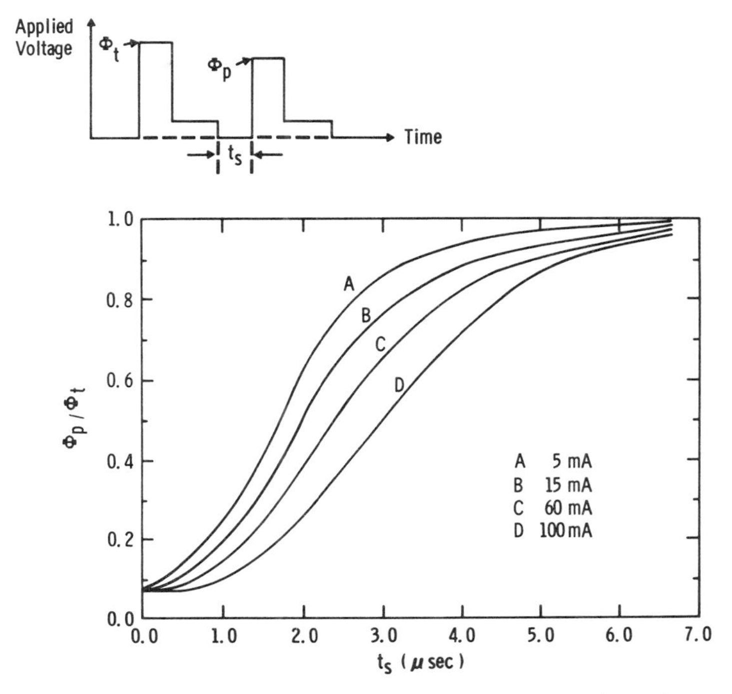

FIGURE 8-14. Recovery of threshold voltage as a function of interruption time t_s for several values of ON-state current. The double pulse sketched on top is as shown, but with $t_s > t_{sm}$ so that $\Phi_p > \Phi_h$. Curves: A, 5 mA; B, 15 mA; C, 60 mA; D, 100 mA. (After Petersen and Adler, 1976.)

possibility that I_{hm} exists. However, for essentially all circumstances where the battery voltage is kept constant after switching occurs, I_h should be treated as a completely circuit-controlled parameter.

Once the voltage across the sample is removed, the recovery curve can be studied. As shown in Fig. 8-14, the recovery process depends upon the steady state operating point. One explanation of the data is that after the voltage is removed, the field at the anode adjusts almost instantaneously but the cathode field decays slowly (Frye et al., 1980), maintaining carrier generation or tunneling near that contact. Since the applied voltage is now zero, a counter field will be built up near the anode within a dielectric relaxation time; this explains the symmetry of the TONC results shown in Fig. 8-13b. The limiting feature of the recovery process is then the ambipolar diffusion of carriers radially out of the conducting filament. As the diffusion proceeds, the radius of the filament decreases. As long as any filament remains, only Φ_h is required to resuscitate the ON state. However, after a time which depends on the original r_f (and thus I_{ON}), the filament shrinks to zero radius and the contact barriers begin to decay (this is the origin of the parameter t_{sm} discussed in Sec. 8.3.1.2). Once the equilibrium contact barriers are reestablished, the original Φ_T is completely restored.

An alternative explanation of the recovery data is that t_{sm} is the time it takes for the contact barriers to widen to lengths too great to sustain the tunneling motion of large numbers of carriers through them. At this point a sharp increase in resistance occurs as the contact-to-contact path is broken; the remainder of the recovery process involves diffusion of heat out of the filament.

8.3.2. Numerical Calculations of the First-Fire Event in Homogeneous Films

8.3.2.1. Introduction

Shaw et al. (1973b) and Buckley and Holmberg (1975) have presented experimental data on the first-fire event in both virgin threshold and memory material, with the latter work being more extensive in that, among other things, a range of samples thicknesses were explored. Both sets of experiments showed that for short enough pulses a critical electric field exists that initiates a breakdownlike switching process which, in these experiments, leads to forming and a substantial drop in threshold voltage upon subsequent firings. Although the workers cited above did not explore the formed filamentary region via scanning electron microscopy, others (see, e.g., Bosnell and Thomas, 1972) have done so in great detail, and different types of inhomogeneities have been shown to be present. We will discuss one type shortly.

In this section we present the results of extensive numerical electrothermal calculations for both threshold and memory type material, and compare our results with the experiments discussed above. The details of the calculations can be found in Subhani (1977). We study both time-independent and time-dependent processes, incorporating a critical electric field into the model in order to obtain agreement with experiment. It is important to note in what follows that the only difference between the two types of samples involves the setting or lock-on in the memory ON-state. The mechanism for the initiation of the ON-state (prior to memory lock-on if it occurs) is the same for both.

8.3.2.2. Memory-Type Samples—Calculations of the Steady State

We first solve an electrothermal model for the steady state $I(\Phi)$ and $T(r)$ characteristics of a virgin memory sample, and then compare our calculated $I(\Phi)$ characteristics with the experimental results of Buckley and Holmberg (1975). By electrothermal we mean solutions of the heat balance equation explicitly including nonohmic contributions such as a field-dependent conductivity and/or a critical electric field.

The geometry of the sample under analysis is that of a homogeneous thin circular disk of radius R and thickness d sandwiched between metallic electrodes (see inset, Fig. 8-15). Because switching occurs primarily along a central axial path, the temperature far from the center of the sample remains at ambient. Hence, rather than apply the boundary condition $(\partial T/\partial r)|_{r=R} = 0$, the radial surface of the sample is kept at a temperature $T(r = R) = T_a$, where T_a is the ambient temperature. However, the axial surfaces of the sample (the amorphous/electrode interfaces) have finite heat losses and are modeled using the

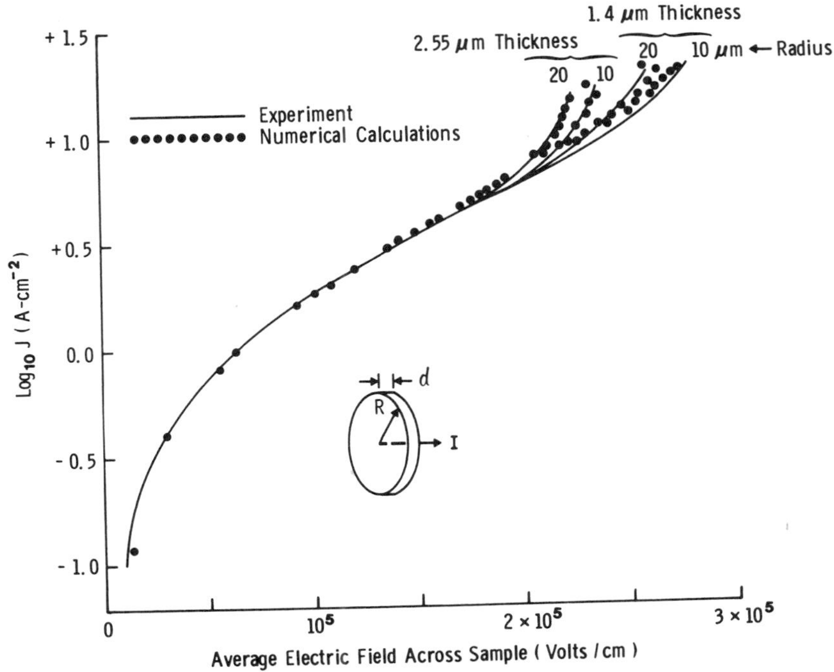

FIGURE 8-15. Comparison of the experimental and calculated dc OFF-state characteristics for a memory-type sample whose parameters are given in Table 8-1. Four different sizes are modeled. The insert shows the geometry of the sample. Experiment (solid); numerical calculations (dashed). (After Shaw and Subhani, 1981.)

electrode boundary condition previously described,

$$K_\alpha \text{ grad } T \cdot \hat{n} = -G_\alpha(T - T_a) \tag{8-31}$$

Newton's law of cooling, where $\alpha = e$ for the electrodes and $\alpha = a$ for the chalcogenide material; $\hat{n}$ is a unit vector normal to the boundary. Here both G_e and K_e are taken as finite and independent of T (G_e is the Newton coefficient). For simplicity K_a is also taken as a constant, although it is a slowly increasing function of T. The electrical conductivity of the amorphous material is taken as thermally activated and of the form

$$\sigma = \sigma_0 \exp[-(\Delta E - \beta F)/kT - F_0/F] \tag{8-40}$$

where the term in the parentheses represents a field-induced decrease of thermal activation energy and the last term in the bracket represents carrier multiplication effects [in the actual numerical calculation, this term is written as $F_0/(F + F')$ in order to yield a finite conductivity at zero field].

For the steady-state calculations the inhomogeneous elliptic partial differential equation of heat conduction [Eq. (8-30) with $\partial T/\partial t = 0$],

$$K\nabla^2 T + J_e^2/\sigma = 0 \tag{8-39$''$}$$

together with Eq. (8-9b) are solved on a grid in the finite-difference approximation. It is important to appreciate that for well heat sunk electrodes the solution of this equation for a temperature-independent K cannot successfully account for the observed virgin $I(\Phi)$ characteristics of either memory or threshold type material unless a critical electric field, F_c, at which a precipitous increase in conductivity occurs, is included in the calculations. The OFF-state characteristics used for memory material are given in Table 8-1. To compare with experiment, calculations are performed for four cases: 20 μm and 40 μm pore diameter; 1.40 μm and 2.55 μm thickness.

Figure 8-15 shows a comparison of the numerical results with the experimental values reported by Buckley and Holmberg (1975). The agreement is quite good. We determined steady state values using a constant-current source. Starting a low current density in the OFF state, the current was incremented slowly until

TABLE 8-1. Parameters Used for Memory-Type Samples

Parameter
$F_0 = 7 \times 10^3$ V cm^{-1}
$F_c = 3.1 \times 10^5$ V cm^{-1}
$\Delta E = 0.43$ eV
$\beta = 1.5 \times 10^{-7}$ e cm
$\sigma(T_a, \epsilon = 0) = 1.1 \times 10^{-5}$ $(\Omega\text{ cm})^{-1}$
$T_a = 297$ K
$K_{a,\text{eff}} = 3.0$ mW/°C cm
$G_e/K_e = 3.5 \times 10^4$ cm^{-1}

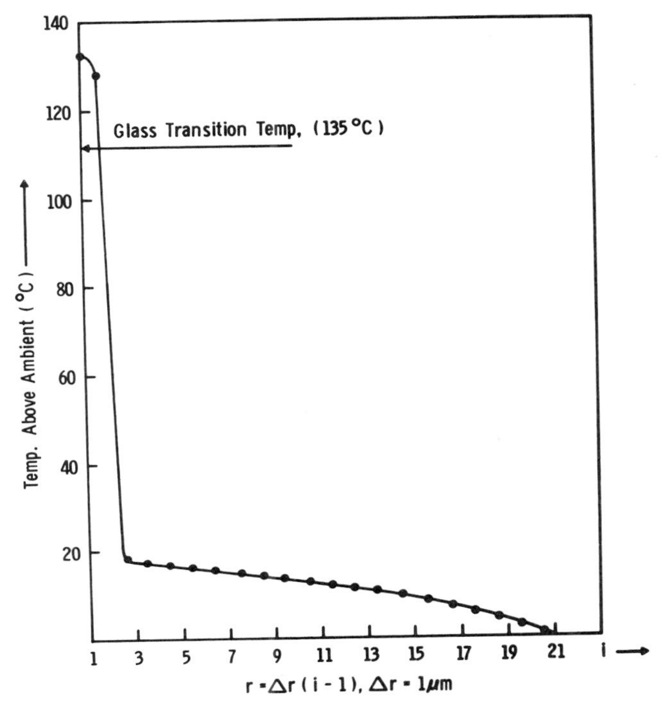

FIGURE 8-16. Temperature above ambient versus radial position for the ON state of a memory switch. Using the experimental fact that the microcrystalline ON-state filament is about 2–4 μm in diameter, and the glass transition temperature in the material is about 135°C, we inserted the filament diameter and found that the best fit to the $I(\Phi)$ characteristic was obtained for $K_{ON}/K_{OFF} = 10^2$. We also used $\sigma_{ON} = 10\ \Omega^{-1}\ \text{cm}^{-1}$. The sample is 2.55 μm thick and 40 μm in diameter; $R_L = 10\ k\Omega$. (After Shaw and Subhani, 1981.)

the sample underwent a large change in conductance. The steady-state voltage was recorded for each current density. Figure 8-16 shows the temperature as a function of radial position on a plane through the center of the memory sample just after breakdown, where the steady ON-state current is equivalent to that produced with a load resistor, $R = 10\ \text{k}\Omega$. Note that (1) in Fig. 8-15, the $J(\bar{F})$ characteristics exhibit a slight region of thermistor-type SNDC before the onset of switching, and (2) in Fig. 8-16 the temperature distribution and current density define a sharp filamentary conducting path after switching.

Prior to switching, the maximum temperature in the OFF state is calculated to be about 20°C above ambient. Just after switching, the maximum temperature at the center of the filament is 126°C above ambient, which is above the glass transition temperature for this material. However, to develop a crystalline filament substantially higher temperatures must be reached. It appears that the energy required to do this comes from the capacitive discharge during the switching transition (Shaw et al., 1973; Kotz and Shaw, 1984). The microcrystalline filament eventually formed is approximately 2–4 μm in diameter, in agreement with experimental observations. This demonstrates clearly that electrothermal numerical calculations, modified with a critical electric field, can quantitatively mimic the breakdown-type switching characteristics observed in *memory-type* chalcogenide films.

In general, our calculations yield for virgin memory-type samples:

(1) The calculated and experimental $I(\Phi)$ values are coincident at low fields for the range of film thicknesses and radii investigated. This indicates that a uniform field and current density distribution is present at low fields and thermal effects are unimportant here.

(2) The $I(\Phi)$ characteristics diverge from a common line near the threshold voltage, indicating that the electric field or the current density, or both, become nonuniform within the sample because of local heating.

(3) As the film thickness increases, the characteristics diverge from the common curve at lower fields and current densities, again because of local heating.

(4) As the sample diameter increases, the breakdown voltage decreases.

(5) The current density at breakdown is unaffected by variations in diameter for a given sample thickness (in the range investigated).

(6) As the film thickness decreases, the effects of the diameter variations are sharply diminished, thus making the $I(\Phi)$ characteristics less sensitive to diameter for very thin samples.

(7) For all the memory samples under investigation there is a tendency for the OFF-state characteristic to bend back upon itself near the onset of switching (thermistor-type behavior). The resulting effect is to decrease the average applied field because of the NDC region.

(8) Complete thermistor-type behavior (see Fig. 8-17) can be observed by

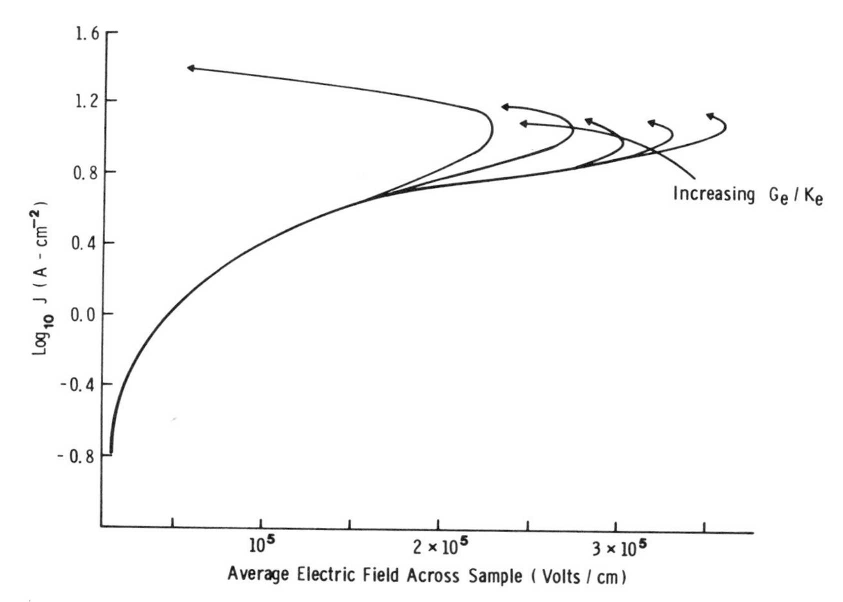

FIGURE 8-17. Calculated prethreshold $I(\Phi)$ characteristics of a typical memory sample having a 20-μm radius and a 1.40-μm thickness as a function of G_e/K_e, the heat-conducting properties of the electrodes (after Shaw and Subhani, 1981).

decreasing the thermal conductivity of the electrodes or eliminating F_c. When G_e/K_e is increased, thermistor behavior with a turnover voltage either above or below Φ_T can be induced. Here, in many cases F_c is never reached in the sample. For the case where F_c is removed from the calculations, a turnover voltage without breakdown occurs; the turnover voltage is much higher than Φ_T. (It is important to reemphasize here that for a sufficiently light load, thermistor characteristics produce a thermal runaway event.)

(9) The current density and temperature distributions define sharp filamentary conducting paths connecting the electrodes for currents above threshold.

Our calculations reveal the fact that for any one sample thickness, the switching current density is independent of radius. This is also evident from a current-density versus radius plot in the immediate preswitching region, where prethreshold heating is observed. In other words, any divergence from the common curve at high fields (near switching) is due to a nonuniform field distribution along the axis of the sample.

We can also expect a somewhat radially nonuniform current distribtion as a result of nonmetallic inhomogeneities (film imperfections). We have verified this expectation by simulating an inhomogeneous conductivity model wherein we allow for small conductivity perturbations that model imperfections having the same value of thermal conductivity as the amorphous material. Relatively small changes in the virgin $I(\Phi)$ characteristics are observed for conductivity variations across the sample of up to 30%. The general shape of the $I(\Phi)$ curve shown in Fig. 8-15 is maintained.

8.3.2.3. Threshold-Type Samples—Calculations of the Steady State

Now that is clear that the OFF-state conditions of a memory sample can be modeled, we turn to the threshold case (Shaw et al., 1973b), making use of those ON-state parameters that have emerged from the best fit for the memory behavior. We assume that the thermal properties of the memory and the threshold material are the same, and in our calculations use the same ratio of ON to OFF-state thermal conductivity for the threshold sample as for the memory sample. All other parameters are obtained directly from observations on the threshold material. The parameters are given in Table 8-2.

The major features of the $I(\Phi)$ and $T(r)$ calculations shown in Figs. 8-18 and 8-19 are similar to that of memory-type samples. However, note that in Fig. 8-18

TABLE 8-2. Parameters used for Threshold-Type Samples

$F_0 = 7 \times 10^3$ V cm^{-1}
$F_c = 7.0 \times 10^{-5}$ V cm^{-1}
$\Delta E = 0.55$ eV
$\beta = 1.5 \times 10^{-7}$ e cm
$\sigma(T_a, \epsilon = 0) = 2 \times 10^{-8}$ $(\Omega$ cm$)^{-1}$
$T_a = 297$ K
$K_{a,\text{eff}} = 3.0$ mW/°C cm
$G_e/K_e = 3.5 \times 10^4$ cm^{-1}

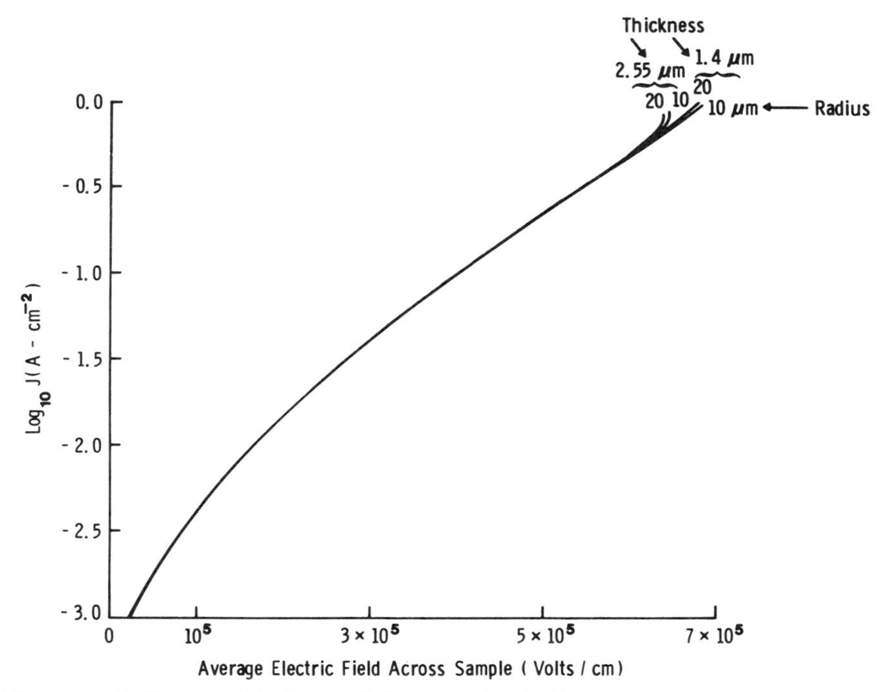

FIGURE 8-18. OFF-state $J(\Phi)$ characteristics for a threshold-type sample whose parameters are given in Table 8-2. Four different sizes are modeled. (After Shaw and Subhani, 1981.)

the departure from the common curve at high fields is very small compared to that of the memory samples. Thus, according to the calculation, there should be no phase change induced in threshold-type samples. Forming should be absent and the switching initiation and maintainance should be electronic processes with only minor thermal overtones. Experiments, however, tell us otherwise. First, scanning electron microscopy reveals the presence of both crystalline and morphological imperfections (Bosnell and Thomas, 1977; Allinson et al., 1979) in formed films, and these are by far the most common films encountered in practice. Hence, we suggest that F_c causes a switching event that produces high temperatures often because of the capacitive discharge in both memory and threshold samples, and also because of operation at high ON-state currents (Kotz and Shaw, 1983, 1984).

In order to explain the divergence from the common curve at high fields for a given (memory or threshold) film thickness with different diameters, we study the temperature profile of the sample in the OFF state. Investigation reveals that for low fields the power input is small, causing negligible heating effects, and the curves are coincident for all geometries. As the current density increases, heating effects are observed if the power input approaches the power dissipation capacity of the sample. Generally, the diameter of the sample is much larger than its thickness and heat is dissipated primarily along the axial direction. Thicker films will develop higher internal temperatures than thinner films under the same conditions because the heat transfer is limited primarily by the low thermal

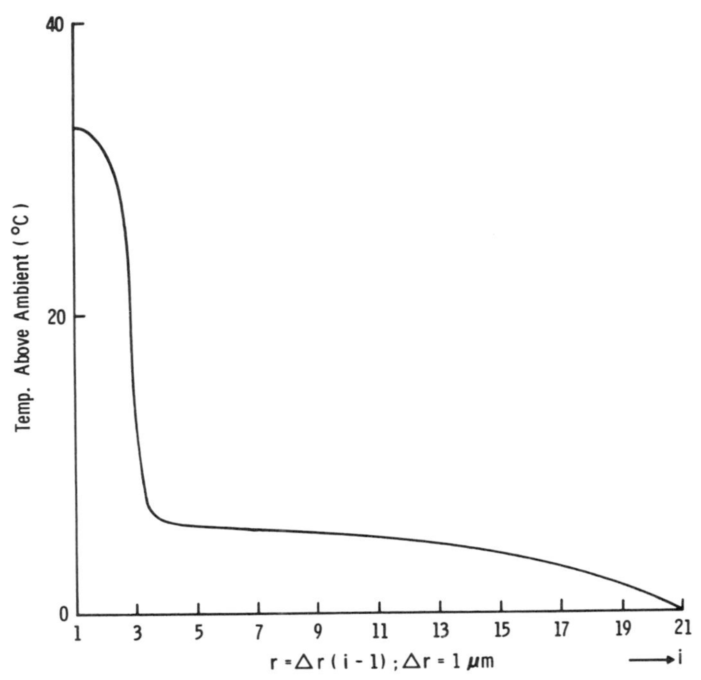

FIGURE 8-19. Temperature above ambient versus radial position for an electronic model of the ON state of a threshold switch. The sample is 2.55 μm thick and 40 μm in diameter (for a 1.40-μm-thick sample the maximum temperature is 18°C above ambient); $R_L = 10\ k\Omega$. No overvoltage is applied. These results assume that the carriers induced by F_c produce no additional Joule heating. (After Shaw and Subhani, 1981.)

conductivity of the material. Furthermore, the conductivity expression descriptive of the material is a temperature-activated type; a small change in temperature will result in a comparatively large change in conductivity. Hence, the temperature gradient will redistribute the applied voltage across the colder regions of the film, causing an axially nonuniform field distribution. Therefore, it is reemphasized that aside from electronic contact effects, the highest fields will occur next to the electrodes, where the film is coolest. When the local field exceeds F_c, the sample will switch. Our calculations indicate that for $R \gg d$, radial heat transfer is small. However, it is not negligible for $R \cong d$ (thick films or small diameters) because the ratio of diameter to thickness is reduced and the relative contribution of the radial heat transfer is increased. This can result in a lower internal temperature rise and higher calculated average breakdown field.

The experimental results as well as the calculations provide good evidence of internal heating in the immediate preswitching region for memory material. A sample will usually undergo breakdown at higher average fields and higher average current densities if heating effects are reduced by changing the sample geometry or material composition (conductivity) for a given set of thermal boundary conditions.

In comparing the memory-type virgin $I(\Phi)$ characteristics (Fig. 8-15) with the threshold-type characteristics (Fig. 8-18), we see that the threshold sample has (1) a lower OFF-state conductance; (2) a higher breakdown voltage; (3) almost no departure from the "common" curve; (4) essentially the same breakdown voltage for different diameters. We therefore suggest that prethreshold heating in virgin threshold type samples is not important in producing the breakdown-type switching event. Rather, it is the critical field that ultimately causes discharge-induced changes in the nature of the material and produces a formed filamentary region suitable for reversible switching events upon subsequent firings. We suggest how this might happen in Sec. 8.3.3.

8.3.2.4. Calculations of Time-Dependent Processes

Although to our knowledge no direct data exists on delay-time effects involved in switching events in virgin samples, we calculate the average field at breakdown as a function of pulse width, t_p, which can be compared with experiments of this type (Shaw et al., 1973b, Buckley and Holmberg, 1975). To do this, we solve the time-dependent heat equation [Eq. (8-30) with $\mathbf{J}_e = \sigma\mathbf{F}$]

$$c\rho_a \frac{\partial T}{\partial t} = K\nabla^2 T + \sigma F^2 \tag{8-43}$$

subject to the boundary conditions previously described [Eq. (8-13)]. The results of the calculations for memory-type samples are shown in Fig. 8-20. Pulse widths in the range $2 \times 10^{-9} \leq t_p \leq 10^{-4}$ s were investigated. Comparison of Fig. 8-20 with Fig. 8-21 shows that good qualitative agreement exists between experiment and the numerical calculations. We see that the threshold voltage saturates for $t_p \leq 10^{-6}$ s and $t \geq 10^{-5}$ s, in approximate agreement with the data. However,

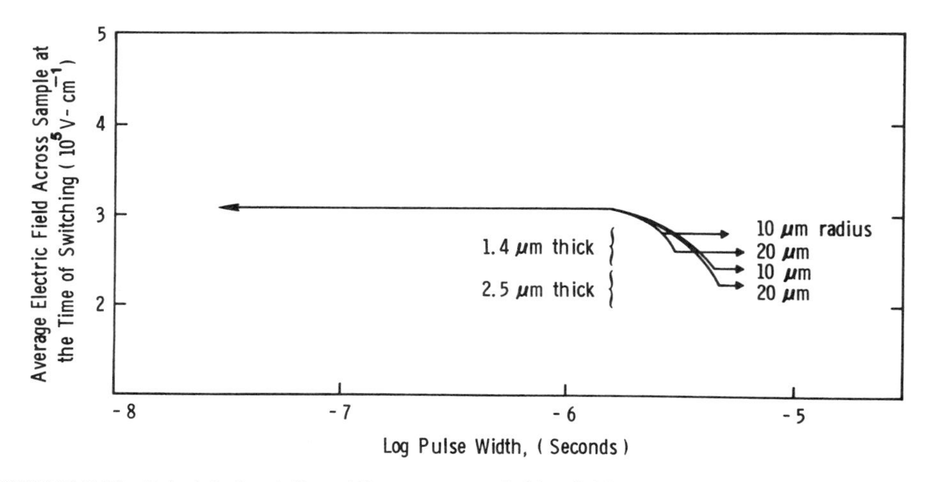

FIGURE 8-20. Calculated variation of the average switching field versus log pulse width (seconds) of a memory-type sample modeled after that producing the data shown in Fig. 8-17. The arrowheads denote that Φ_t was independent of t_p in both directions for all larger and smaller values of t_p. (After Shaw and Subhani, 1981.)

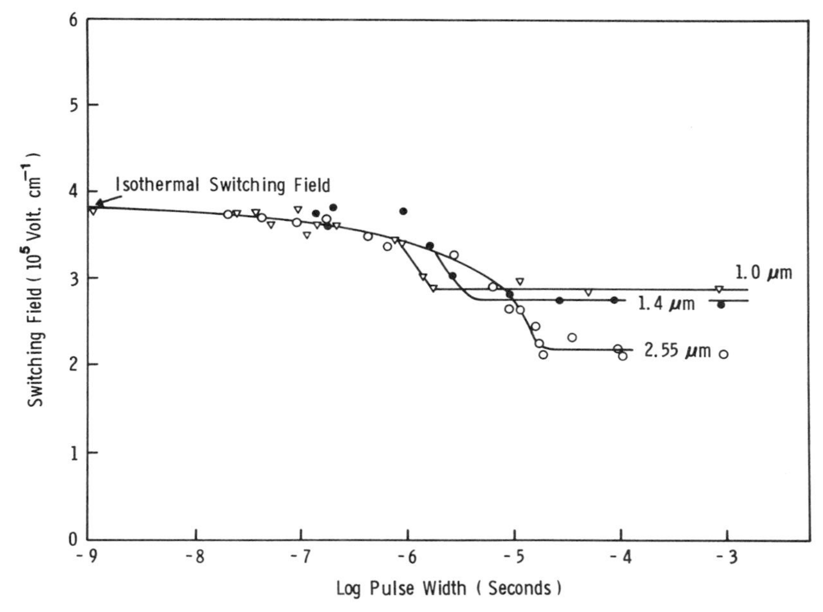

FIGURE 8-21. Variation of the average switching field with voltage pulse width for virgin samples of $Ge_{17}Te_{79}Sb_2S_2$ having a 20-μm pore diameter and three different thicknesses. Note that for the shortest pulse the switching field asymptotically approaches the same value independent of sample thickness. (After Buckley and Holmberg, 1975.)

the difference in the average fields at which the long- and short-pulse results saturate, which we call ΔF_T, is generally not as great in the numerical calculations as it is in the experimental data. Furthermore, the experimental value of F_T is about 20% higher than the value required to obtain the precise $I(\Phi)$ fit in the memory material shown in Fig. 8-15. The closeness of F_c in these two cases is, in fact, evidence that our model applies rather well to this memory-type virgin material. Finally, the typical t_d's of about 5 μs predicted from the calculations for virgin samples are sufficiently close to those observed experimentally in inhomogeneous "running" samples to support a model where t_d is thermally induced in formed memory-type samples for sufficiently long pulses. In this standard model, a "hot spot" nucleates in the center or high conductance region of the sample and the conductivity there increases, thereby reducing the voltage in the central region and increasing it near the electrodes where F_c is eventually reached and switching occurs. This model is basically the one we outlined in Sec. 8.2.3.

The agreement between experiment and numerical calculations is not nearly as good for the threshold-type virgin material (Shaw and Subbani, 1981) [here the data are sparse since only a single thickness was studied (Shaw et al., 1973b)]. The experiments revealed no clear-cut saturation for $t_p \leq 10^{-7}$ s. Furthermore, the experimental value of ΔF_T is substantially larger than the predicted value. Finally, the predicted t_d is less than 10^{-6} s, a value much below those experimentally observed in inhomogeneous samples, which is of the same order of magnitude as for formed memory-type samples, less than about 10^{-5} s. This is an important point. Experimental values of t_d are typically the same for both formed (inhomogeneous) threshold and memory-type material. The numerical calculations for virgin samples, however, show that t_d should be about an order of magnitude longer in memory-type material. This result is in harmony with the switching model we shall discuss in the next section, and is based on the precept that F_c initiates switching in both types of materials. This event causes forming in both, and the mechanism for the switching effect observed in subsequent firings is the same in both—an electronic instability that is thermally modified and electronically sustained.

One other important point, most clearly seen with reference to Fig. 8-21, is that as t_p decreases below about 10^{-7} s, the rate of rise of $\bar{F}_T$ first tends to diminish and then eventually saturates. These data are contrary to the behavior expected from a model of the first-fire event based solely on heating with weakly heat sunk electrodes. For long values of t_p ($\geq 10^{-5}$ s), $\bar{F}_T$ decreases with increasing film thickness and is almost independent of t_p for any one thickness. Furthermore, as previously discussed, the axial nonuniformity that causes $\bar{F}_T$ to diverge with thickness is consistent with our model over the range of studied t_p's. However, $\bar{F}_T$ is independent of film thickness for short t_p's. Calculated values deviate slightly in the sense that the increase of $\bar{F}_T$ with thickness disappears as t_p decreases, whereas a small slope is seen in the experimental data. This may be due to the fact that in our model the voltage drop near the electrodes is symmetric, which is probably not the case for actual samples. However, the fact that $\bar{F}_T$ is independent of film thickness in this time regime suggests that the field is only slightly nonuniform along the axial direction and the breakdown event is a bulk effect; i.e., the threshold voltage equals $F_c d$ for sufficiently short pulses.

We have seen that the first-fire event in homogenous films can be interpreted as an event induced by a critical electric field. In the next section we discuss how this event might lead to forming and how formed samples might act as reversible switches. In what follows we make use of the several experimental observations showing that differences exist between virgin and formed films; e.g., virgin films show a short-pulse critical field and a dc critical voltage, while formed films seem to show a critical local power density; virgin films show large prethreshold currents for short pulses, while formed films only do so when the temperature is lowered substantially.

8.3.3. Electrothermal Switching Mechanisms in Formed Chalcogenide Films

8.3.3.1. Introduction

In the next section of this chapter we discuss switching effects in vanadium dioxide (VO_2). In this material a structural phase change occurs at a critical temperature, T_c; at T_c the conductivity rises precipitously by several orders of magnitude. By modeling this phenomenon with a step function change in conductivity at T_c, we can treat the problem analytically and predict the observed $I(\Phi)$ curves successfully. A similar phenomenon occurs for inhomogeneous chalcogenide films. Here, however, it is again an F_c that causes σ to rise precipitously, and it is reached locally in many cases because of thermal effects.

In the last few sections we have discussed switching effects in uniform homogeneous chalcogenide films. Thermal theories attempting to explain this phenomenon have been presented by many authors (see, e.g., Stocker et al., 1970; Croitoru and Popescu, 1970; Shousha, 1971; Duchene et al., 1971; Altcheh et al., 1972; Kaplan and Adler, 1972; Warren, 1973; Kroll, 1974; Owen et al., 1979; Shaw and Subhani, 1981). From these studies, it has been made clear that for nonthermistor-type switching to occur an electronic mechanism must also be operative in order to short out the low-conductance regions adjacent to the cool electrodes. A critical electric field will certainly suffice, and it is this assumption that we have used to provide good agreement with the first-fire event data for memory samples discussed in the last section. It is our contention that during the first few firing events a breakdown-type process often occurs that is driven by the capacitive discharge and/or high operating currents (Kotz and Shaw, 1984). An open, and most often a short (memory) occurs if the sample is kept ON for a sufficiently long time. In general, the first-fire event produces an intermediate state that is a narrow ($\lesssim 5\ \mu m$ diameter) filamentary region containing crystalline or morphological imperfections. [Sometimes several firings are required to develop a formed state that is amenable to easy observation by scanning microscopy, but it is the first-fire event that often results in the largest change in threshold voltage (Allinson et al., 1979).] For a threshold switch the formed region is of higher conductance than the surrounding homogeneous film (Coward, 1971), but still of substantially lower conductance than the ON state. We can imagine the intermediate state as being formed in the following manner. Consider, for example, the case where sufficiently short pulses are applied such that F_c is reached isothermally over the entire sample, independent of its thickness (Buckley and Holmberg, 1975). In this region of pulse width ($\leq 10^{-8}$ s) the power, P, dissipated in the sample because of Joule heating is given by $P \cong \sigma_{RT} F^2$, where σ_{RT} denotes the room-temperature (ambient) conductivity. When F_c is reached the current increases by orders of magnitude (Buckley and Holmberg, 1975) at constant voltage. (This large increase in current at constant voltage prior to breakdown or switching has only been observed at room temperature in virgin samples for sufficiently short pulses. Formed samples show this effect at low temperatures.) In this region $P \cong \sigma_h F_c^2$, where $\sigma_h \gg \sigma_{RT}$. The large increase in conductivity induced by F_c can be due to either the field-stripping of trapped carriers and/or avalanching. Once these carriers are generated, because of the ensuing capacitive discharge the significantly increased Joule heating causes a breakdown at the weakest point in the film. The sequence leading to forming in virgin samples subjected to short pulses is first electronic, then thermal. As previously stated, the outcome of the switching event can be (1) an open; (2) a short (e.g., the memory state); and (3) an inhomogeneous formed region (threshold switch). In what follows we support the view that a formed or intrinsically inhomogeneous region is common in conventional threshold switches (Popescu, 1975) made from thin amorphous chalcogenide films. We also suggest that the mechanism for the switching event has features that are somewhat different from that of the first-fire switching event in a homogeneous film. In the latter case F_c is either reached isothermally over the entire sample for short pulses or, for longer pulses, some thermal modification allows for switching to occur when F_c is reached over only part of the sample. In either case the switch occurs very rapidly when a critical field or voltage is reached. Formed samples, however, show a switching transition, after a delay time t_d, when a critical local power density is reached (Balberg, 1970; Thoma, 1976; Reinhard, 1977; Shaw and Subhani, 1981). In the following section a model for these latter effects is presented.

8.3.3.2. An Electrothermal Model for Threshold Switching in Inhomogeneous Films

Popescu (1975) has provided a detailed analytical model of how switching can occur in inhomogeneous chalcogenide films. It is our view that his arguments center correctly on the properties of the formed region and the thermal nature of the current instabilities possible in the vicinity of such paths. In what follows we offer a simple supplement to Popescu's work by suggesting possible means by which an electronic instability can be encouraged in such systems.

Figure 8-22a shows the geometry under analysis. As in Fig. 8-15, the sample is a thin cylindrical disk of radius R composed of material having a thermal conductivity, $K_a(T)$, and electrical conductivity, $\sigma_a(T)$, that increase with increasing temperature. These are the conductivities associated with the homogeneous parts of the film. Now, however, we have imbedded in the material an array of inhomogeneities (shaded) confined to the region $r < R_I$ (these would commonly be near or attached to an electrode). For the specific, but common, case where the inhomogeneities are Te-rich crystallites, they have a thermal conductivity $K_i(T)$ that decreases with increasing temperature (over the temperature range of interest) and an electrical conductivity $\sigma_i(T)$, substantially higher than $\sigma_a(T)$ and weakly dependent on temperature. A bias voltage, Φ_B, is applied

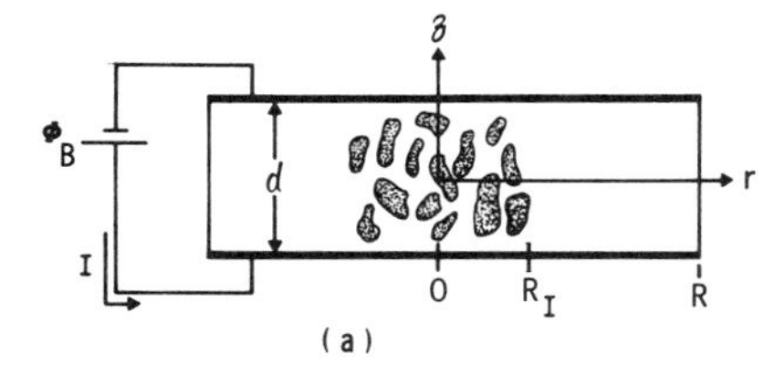

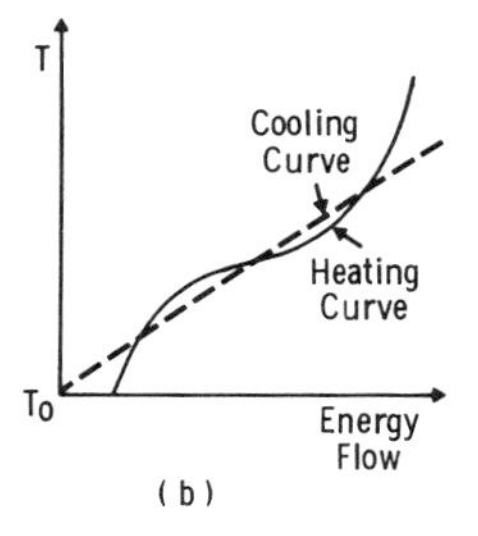

FIGURE 8-22. (a) Geometry under analysis: all parameters are defined in the text. (b) Heating and cooling curves as a function of temperature. T_0 is the ambient temperature. (After Shaw and Yildirim, 1983.)

across its thickness, d, and current, I, flows in the external circuit. Because of the properties of the system outlined above, the current density in the region $r < R_I$ is greater than in the surrounding homogeneous medium.

The conventional thermal instability (Landauer, 1978) that can occur at a critical value of local power density in such systems has been outlined by Landauer and Woo (1972), and treated in detail by Popescu (1975). It can be understood most simply by considering Fig. 8-22b. Here the cooling curve represents the rate at which heat can be taken away from the region $r < R_I$, $-(d/2) < z < (d/2)$, when it is excited by Joule heating. The heating curve is sketched for the case $\sigma(T) = \sigma_0 \exp(-\Delta E/kT)$. A stable solution exists at the lower intersection of the heating and cooling curves. As the input power is increased the heating curve shifts to the right and an instability results when no lower intersection point between the two curves is possible. The upper intersection point represents another stable state of the system and switching occurs between these two stable states; a sudden increase in local temperature can occur.

Experiments, however, do not suggest that this is the mechanism for switching at a critical local power density (Kotz and Shaw, 1984). Rather, the sample simply heats, a "hot-spot" grows, and the delay time, t_d, is the time it takes the hot front to spread through the formed region, approach both electrodes (Shaw et al., 1973c; Homma, 1971; Newland, 1975), and cause F_c to be reached near an electrode. In this model the delay time is a consequence of a completely thermal process; thus the critical parameter is the power density.

When the field near an electrode reaches a magnitude sufficient to sustain field stripping or avalanche within, or tunneling through the amorphous regions adjacent the contacts, switching occurs (Adler et al., 1980), along with the concomitant capacitive discharge. Under conditions where the entire formed region melts (Kotz and Shaw, 1984), this picture would be in harmony with the observations of Pearson and Miller (1969). Here, upon turning the switch off, the molten region could revitrify. The subsequent switching event could then initiate at a different spot; the conducting path could "jump around" from cycle to cycle. However, if partial crystallization occurred, then the same spot could initiate the switch upon consecutive firings.

The switching transition electronically stabilizes the filamentary region, which can sustain relatively high temperatures in its center. The electrode temperature, however, is cooler. The regions near the electrodes are amorphous, maintain a large temperature gradient, and have an average energy band gap that depends upon the temperature gradient. In the narrow amorphous regions carriers are being supplied by fields on the order of $10^5\ \mathrm{V\ cm^{-1}}$. Recombination radiation is being emitted near 0.5 eV (Walsh et al., 1978, 1979); it could be originating from either (1) defect transitions in the amorphous layers or (2) band-to-band transitions in the core of the filament. A black-body spectrum has not been observed in these studies, although it has been in others.

In the above switching model the width of the current filament in the ON state is largely constrained to the width of the formed region. The spatial extent of typical formed regions in threshold and memory-type samples has been measured by scanning microscopy and found to be 1–5 μm in diameter. We expect that formed threshold-type samples will generally have highly conducting

ON-state paths of this size. Thus, for a given load line that produces an ON state below current saturation of the formed region, the current filament will be smaller than the formed region. As the load is lightened or the current increased at fixed load, the current filament will widen until it fills the formed region. Further increases in current will result primarily in heating of the current-carrying path rather than its continued spatial expansion (Kotz and Shaw, 1984).

The model presented here satisfactorily explains the phenomenology of threshold switching. It is consistent with the experimental observation that the instability initiates at approximately zero time for any overvoltage (Shaw et al., 1973c) and the inference, taken from the data, that is like a convective instability. Furthermore, it explains the behavior of t_d in the "statistical" regime just at threshold. Here, very long t_d's can be observed, where the current is not observed to rise until within a microsecond or two prior to the switching event. We suggest that the instability is triggered by a fluctuation associated with the injection of carriers. Slightly past threshold, t_d is usually 1–2 μs in a 1-μm-thick film. This is the time it takes for the hot spot to grow. As this occurs, the current increases with time. In the statistical regime we must wait for the fluctuation that will trigger an event in a material that will be slightly different each cycle. There will be no rise in current while we wait.

The model also explains the results of Henisch et al. (1974) and Rodgers et al. (1976). The former group found that the voltage at threshold was insentitive to light intensity at low excitation levels, even though the current at threshold increased owing to the enhanced conductivity of the material. The latter group showed that the voltage at threshold decreased with intensity at high levels where the materials is heated by the optical pulse. A straightforward explanation can now be given for these effects. The Te-rich crystallites are essentially unaffected by the light. At low intensities the conductivity of the region surrounding the crystallites is increased, but the local field is thereby decreased, and the local power density remains essentially unaltered. The critical condition is local, and if the temperature of the surrounding medium is unchanged, the instability will occur at the same value of local power density. Once the temperature increases locally, however, the threshold power density will drop. The excess currents observed in these experiments with increasing light intensity are due primarily to the enhanced conductance in those (major) parts of the films that remain homogeneous.

8.4. THERMAL SWITCHING IN VANADIUM DIOXIDE

8.4.1. Introduction

Vanadium dioxide (VO_2) exhibits a first-order structural phase transition at 68°C from a high-temperature tetragonal structure to a low-temperature monoclinic structure (Anderson, 1954; Berglund, 1969; Duchene et al., 1971a,b,c, 1972; Fisher, 1973; Jelks et al., 1975). Accompanying this phase change is a change in conductivity by a factor of near 10^4. The high-temperature phase is metallic; the low-temperature phase is akin to an intrinsic semiconductor. When the high-

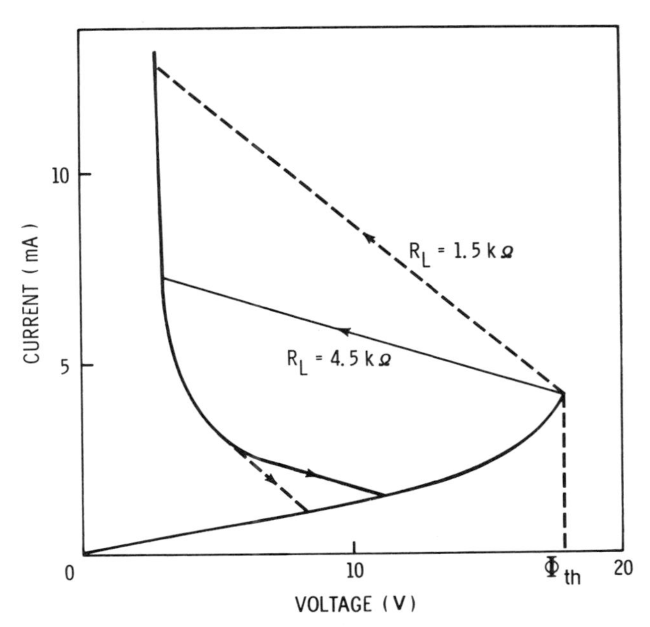

FIGURE 8-23. Static $I(\Phi)$ characteristic for two values of load resistor R_L for a coplanar VO_2 sample (after Duchene et al., 1971b).

temperature phase is induced locally by Joule heating, switching is observed in the $I(\Phi)$ characteristics as shown in Fig. 8-23. Further, narrow high-temperature filaments are easily produced and studied (Berglund, 1969; Duchene et al., 1971b). A substantial measure of the understanding of the high current density filament has come from a symbiosis between approximate calculation, numerical analysis, and experiment. In this section we will outline the results of *exact* calculations of the current–voltage characteristics and stability for ideal one-dimensional models (Jackson and Shaw, 1974). We solve, analytically, systems having (1) parallelepiped and cylindrical geometries, (2) heat flow J_h parallel and perpendicular to current flow I, and (3) abrupt conductivity increases and decreases at a critical temperature T_c.

We find that for a given direction of J_h, the steady state $I(\Phi)$ characteristics are completely determined by a single parameter, the conductivity ratio $\epsilon = \sigma_0/\sigma_s$, where σ_0 is the value of σ below T_c and σ_s the value of σ above T_c. The derived $I(\Phi)$ characteristics for the parallelepiped slab geometry of Fig. 8-24, where J_h is restricted to the plane of the slab, are shown in Fig. 8-25. The various characteristics for the parameters chosen are identical below the threshold ($\Phi < \Phi_T$, $I < I_T$) characteristic and NDR (*resistance*) only results for $\epsilon < 1$ with $J_h \perp I$ and for $\epsilon > 1$ with $J_h \parallel I$. We prove that a filamentary (high electric field domain) SNDR (NNDR) characteristic ($a[d]$) is stable only if $R_L + d\Phi/dI > 0$, (<0), in agreement with experiment. An important consequence of these criteria is that if there is only one possible intersection between the load line and $I(\Phi)$ curve, it must be stable if we neglect the effect of the reactive elements in the circuit.

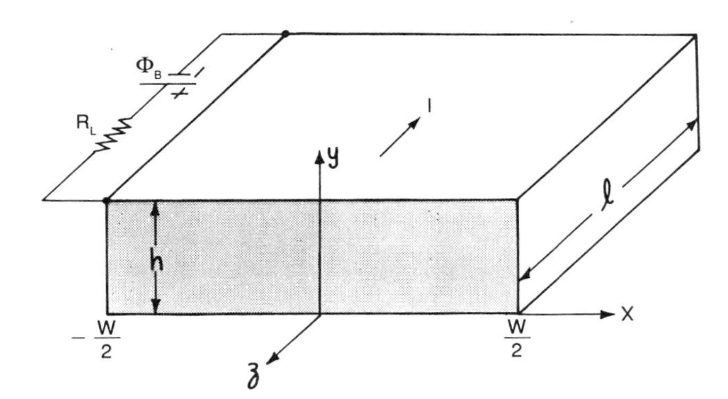

FIGURE 8-24. Sample geometry under analysis: $\Phi = \Phi_B - IR_L$ (after Jackson and Shaw, 1974).

Although the thermally induced NNDR case (*d*) has applications in the area of superconductivity (flux flow in the intermediate state, microbridges, etc.) we will not emphasize it in this section (see Chap. 6). By way of example, we choose the case where $\epsilon < 1$. For the geometry of Fig. 8-24 with $J_h \perp I$, the results are in harmony with the major experimental features of switching in VO_2 films (Duchene et al., 1971b).

8.4.2. An Ideal Model for Thermal Switching in Thin VO_2 Films

We first derive a closed-form expression for the steady state $I(\Phi)$ characteristics, and then outline the details of a full analysis of the stability of the sample when it is in series with R_L (Jackson and Shaw, 1974). Assuming that the thermal conductivity K is a constant and the electric field is uniform in the sample, the heat flow equation (8-29) now reads

$$\rho_C \frac{\partial T}{\partial t} = K \frac{\partial^2 T}{\partial x^2} + \sigma(T) \frac{\Phi_B^2}{l^2} \frac{R_S^2}{(R_L + R_S)^2} \tag{8-44}$$

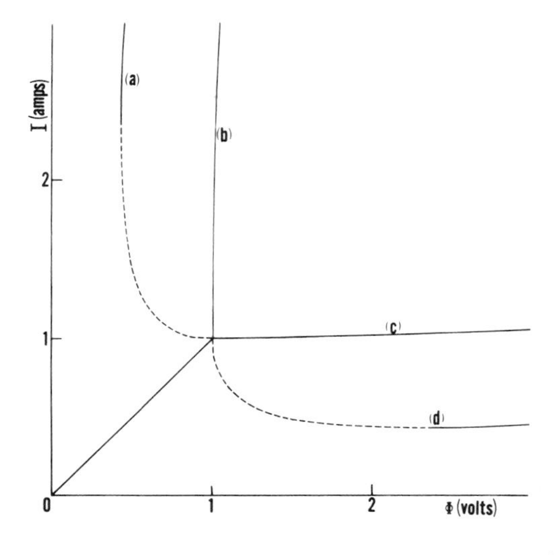

FIGURE 8-25. $I(\Phi)$ characteristic for the case $w = 1$. The threshold point $[\Phi_c, I_c] = (1, 1)$ is common to all curves. Solid lines, stable; dashed lines, conditionally stable. (a) $\epsilon < 1$, $J_h \perp I$. A high-current-density filament and SNDC results: NDC points stable only if $R_L + d\Phi/dI < 0$. (b) $\epsilon < 1$, $J_h \parallel I$. A low-electric-field domain and no NDC results; always stable. (c) $\epsilon > 1$, $J_h \perp I$. A low-current-density filament and no NDC results; always stable. (d) $\epsilon > 1$, $J_h \parallel I$. A high-electric field domain and NNDC results. NDC points stable only if $R_L + d\Phi/dI < 0$. $\epsilon = 0.1$ for curves (a) and (b) and $\epsilon = 10$ for curves (c) and (d). (After Jackson and Shaw, 1974.)

where Φ_B is the applied bias, and R_S, the nonlinear sample resistance, is given by

$$l\left[h \int_{-w/2}^{w/2} \sigma(T(x)\, dx\right]^{-1}$$

We do not include a latent heat of transformation at T_c; its inclusion will not alter the results. The boundary condition is that $T(\pm\frac{1}{2}w) = T_a$ (the ambient temperature). The electrodes, top, and bottom of the sample are perfectly insulated.

We first seek steady state solutions of Eq. (8-44). As long as the maximum temperature is less that T_c, there is a solution

$$T(x) = \frac{\sigma_0 \Phi^2}{K l^2}\left(\frac{w^2}{8} - \frac{x^2}{2}\right) + T_a \tag{8-45}$$

When Φ is increased to the point where Eq. (8-45) would yield a value greater than T_c at $x = 0$, the equation becomes invalid and we look for a "two-phase solution" with an internal hot filamentary region for which $T > T_c$ and $\sigma = \sigma_s$. The critical voltage Φ_T at which Eq. (8-45) becomes invalid is

$$\Phi_T = \frac{l}{w}\left[\frac{8K(T_c - T_a)}{\sigma_0}\right]^{1/2} \tag{8-46}$$

A two-phase solution satisfies

$$K\frac{\partial^2 T}{\partial x^2} = -\frac{\sigma_s \Phi^2}{l^2}, \qquad -\frac{f}{2} \leq x \leq \frac{f}{2} \tag{8-47a}$$

$$K\frac{\partial^2 T}{\partial x^2} = -\frac{\sigma_0 \Phi^2}{l^2}, \qquad \frac{f}{2} \leq |x| \leq \frac{w}{2} \tag{8-47b}$$

with $T = T_c$ at $x = \pm\frac{1}{2}f$. If these equations are integrated with the additional conditions that the temperature and heat current are continuous at $\frac{1}{2}f$, we obtain the width of the hot filamentary region as a function of Φ:

$$\frac{|f|}{w} = \frac{1-\epsilon}{2-\epsilon} \pm \frac{1}{2-\epsilon}\left[1 - \frac{\Phi_T^2}{\Phi^2}(2\epsilon - \epsilon^2)\right]^{1/2} \tag{8-48}$$

For $\Phi > \Phi_T$, only the plus sign in Eq. (8-48) applies and $T > T_c$ in all but a small strip near the surface; as $\Phi \to \infty$, $f \to w$.

We see from Eq. 8-48 that there are also two-phase solutions for $\Phi < \Phi_T$. Defining

$$\Phi_h = (2\epsilon - \epsilon^2)^{1/2}\Phi_T$$

there are two positive values of f for voltages in the region $\Phi_h < \Phi < \Phi_T$. The current in the two-phase state is

$$I = \frac{\Phi}{R_0}\left(1 + \frac{1-\epsilon}{\epsilon}\frac{|f|}{w}\right) \tag{8-49}$$

where $R_0 = L/\sigma_0 wh$.

A typical $I(\Phi)$ curve is shown in Fig. 8-25, curve a. Φ_h is the minimum voltage reached in the filamentary state. At Φ_h we have

$$\frac{|f|}{w} = \frac{1-\epsilon}{2-\epsilon} \tag{8-50}$$

Furthermore, independent of ϵ and the relative dimensions of the parallelepiped,

$$I(\Phi_h)\Phi_h = I(\Phi_T)\Phi_T$$

the powers dissipated at Φ_h and the low-current Φ_T are the same.

To investigate stability we consider a small perturbation about the time-independent solutions which we have found, and see whether Eq. (8-44) causes the perturbation to grow or decay. If $T_0(x)$ is a time-independent solution and $\eta(x, t)$ the perturbation, we write $T(x, t) = T_0(x) + \eta(x, t)$ insert it into Eq. (8-44) and linearize to terms of first order in $\eta(x, t)$. Since the resulting equation is symmetric with respect to inversion $x \to -x$, we can restrict ourselves to the domain $x \geq 0$ if we use appropriate boundary conditions at $x = 0$. The equation for $\eta(x, t)$ is then for $x \geq 0$

$$\begin{aligned}\rho c\frac{\partial\eta}{\partial t} &= K\frac{\partial^2\eta}{\partial x^2} + (\sigma_s - \sigma_0)\left(\left|\frac{\partial T}{\partial x}\right|_{x=f/2}\right)^{-1}\frac{\Phi_B^2}{l^2}\frac{R_S^2}{(R_L + R_S)^2} \\ &\quad \times \left[\delta\left(x - \frac{f}{2}\right) - \frac{2\sigma(T_0(x))}{\int_{-w/2}^{w/2}\sigma(T(x))\,dx}\frac{R_L}{(R_L + R_S)}\right]\eta\end{aligned} \tag{8-51}$$

We seek a solution of the form

$$\eta(x, t) = \exp(\alpha t)X(x) \tag{8-52}$$

If there is a solution of this form with a positive α, the perturbation grows and the time-independent $T_0(x)$ is unstable; if not, it is stable. Inserting Eq. (8-52) into (8-51) we find that there is no positive α (and hence stability) when

$$\frac{f}{w} \geq \frac{(1-\epsilon)R_0}{R_L + \epsilon(2-\epsilon)R_0} \tag{8-53}$$

For $R_L \to 0$, Eq. (8-53) becomes

$$\frac{f}{w} \geq \frac{1-\epsilon}{2-\epsilon}$$

Comparing this result with Eq. (8-50) we see that NDR points are unstable for the unloaded case. PDR points are always stable.

As R_L increases from zero, additional NDR states will stabilize and finally as $R_L \to \infty$, Eq. (8-53) shows us that $|f|/w \geq 0$. All NDR states can be stabilized with an infinite load. The condition for stability in the NDR region can be shown to be $R_L + d\Phi/dI > 0$. Identical results are obtained for a right circular cylinder of radius r_0, which attests to the general validity of the stability criterion. For the cylinder we find that in two-phase region

$$I = \frac{\sigma_0 \pi r_0^2}{l} \Phi\left(1 + \frac{f^2}{r_0^2}\frac{1-\epsilon}{\epsilon}\right) \tag{8-54}$$

and

$$\frac{\Phi_T^2}{\Phi^2} = 1 - \frac{f^2}{R_0^2}\left(1 + \frac{1-\epsilon}{\epsilon}\ln\frac{f^2}{r_0^2}\right) \tag{8-55}$$

where

$$\Phi_T = \frac{2l}{R_0}\left[\frac{K(T_c - T_a)}{\sigma_0}\right]^{1/2} \tag{8-56}$$

Although all NDR states can in principle be stabilized, characteristic (a) in Fig. 8-25 approaches the point (Φ_T, I_T) from the NDR region with zero slope [Eqs. (8-48) and (8-49)]. Prohibitively large resistive loads are therefore required to stabilize all the NDR points and an "open" region should always be present in the experimental characteristics. This behavior is in fact what is observed. The model also predicts the major features of the observed $I(\Phi)$ characteristics including the narrow filaments that are observed even for relatively low ($\ll \Phi_T$) voltages. The calculated filamentary $I(\Phi)$ characteristics produced by a thermally induced switch when a critical temperature is reached somewhere in the sample can also be compared with $I(\Phi)$ characteristics produced by a thermally induced switch when a critical electric field is reached somewhere in the sample, as for the amorphous chalcogenide films discussed in the last section (Hughes et al., 1975). The similarities are striking.

It is important to appreciate how these results fit into the general framework of the NDC stability problem. Two aspects are important: the role of reactive components and the stability of uniform isothermal bulk NDC points. In Chap. 2 we showed that uniform isothermal bulk NDC points are unstable both against (1) current filamentation (Shaw et al., 1973) and electric field domains (Shaw et al., 1979), and (2) circuit-controlled relaxation oscillations (Solomon et al., 1972, Shaw et al., 1973a) (if particular conditions are satisfied). Our solutions are the nonuniform filamentary and domain configurations; we need only inquire about their stability in a circuit containing reactive elements. It is easily shown that if the above stability criteria are met ($R_L + d\Phi/dI \geq 0$), then the nonuniform NDR points will be circuit stable if the circuit response allows those points to be reached (Shaw et al., 1979).

The above analysis is a simple model for an ideal case. In most real systems the change in σ is never as abrupt as we have made it (in most semiconductors it is thermally activated, a situation we have discussed in prior sections). Furthermore, the heat flow will be in all directions. For the case $\epsilon < 1$, the filamentary $I(\Phi)$ characteristic will be an admixture of curves a and b in Fig. 8-25. Nevertheless, the basic physics involved in a wide variety of switching phenomena is contained in this model. For example, if a local melting occurs due to Joule heating, and the molten region has a higher conductivity than the solid phase, then a switching event will occur for a thin film even with $J_h \parallel I$ if the cool regions near the electrodes are shorted by the effect of a critical electric field.

8.5. SECOND BREAKDOWN IN TRANSISTORS

When a sufficiently large reverse bias (Φ_{RB}) is applied to a diode or either one of the two or more junctions of a transistor-type structure, avalanche or Zener breakdown is induced (Sze, 1969). In this mode the device acts as a voltage limiter with little or no NDC observed. Once avalanche or Zener breakdown is achieved, the application of a sufficiently large current often causes a second breakdown (SB) to occur (Thornton and Simmons, 1958). In this mode there is a large "switchback" effect to a holding voltage $\Phi_h \ll \Phi_{RB}$. Typical transistor characteristics are shown in Fig. 8-26 (Schafft and French, 1966). In some cases SB is due to purely electronic effects, such as avalanche injection at the collector n–n^+ junction (Hower and Reddi, 1970). In this chapter we will concern ourselves with SB effects in transistors due solely to thermal instabilities (SB in diodes is discussed by Tauc and Abrahám (1957) and Oka and Oshima (1962)). As is the case for the devices discussed previously, once a sufficiently large input power is provided a switching effect (SB) will occur after a time t_d (Schafft and French, 1962); a triggering energy, or local power density of sufficient magnitude

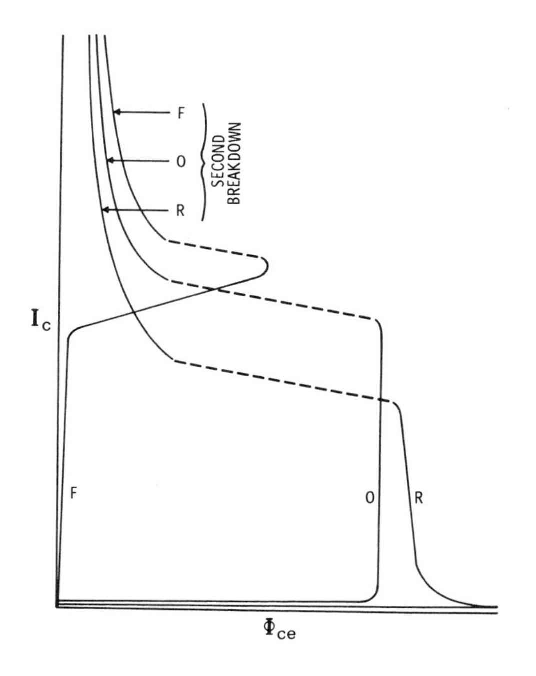

FIGURE 8-26. Swept $I_C(\Phi_{CE})$ characteristics of a transistor with forward (F), zero (0), and reverse (R) constant-base current drive. The characteristics are drawn for only the first half of the sweep cycle. The initiation of second breakdown is indicated by the abrupt drop Φ_{CE}. (After Alwin et al., 1977.)

must be present in the transistor before SB will occur. Some portion of the structure must therefore reach a critical temperature, T_c, before SB will occur. The switching phenomena here are similar to those previously discussed. Nucleation of the hot spot here can be caused by (1) material nonuniformities; (2) base width nonuniformities; (3) heat sinking irregularities; (4) localized regions of breakdown (microplasmas); (5) application of reverse base currents. Indeed, the current constriction that results from any of the above mechanisms (Bergmann and Gerstner, 1966) can cause local melting to occur and extend through the base to create a collector-to-emitter short circuit, the characteristic failure mechanism of SB.

Scarlett et al. (1963) and Bergmann and Gerstner (1963) were the first to study the thermal stability of a uniform current distribution in a transistor. The main feature of these linear models is that any transistor structure will develop a lateral current instability when its temperature rises above that of the heat sink by a sufficiently large amount. Later work indicated that a very important factor in initiating and sustaining SB is the distribution of minority carrier current injected into the base (Schafft and French, 1966b).

A symbiosis between experiment, simple analysis, and numerical modeling is an optimum way of understanding the behavior of semiconductor devices. Since the complexities involved with thermal switching (SB) effects often make it extremely difficult to obtain quantitative fits between experiment and the prediction of simple models, numerical analysis is often called for (see, e.g., Gaur and Navon, 1976; Alwin et al., 1977; Shaw and Subhani, 1981). Typical techniques involve the use of the finite difference method to solve the nonlinear partial differential equations of both heat [Eq. (8-29)] and current flow. The latter are

$$\nabla^2\Phi = -q(p - n + N_D^+ - N_A^-) \tag{8-57}$$

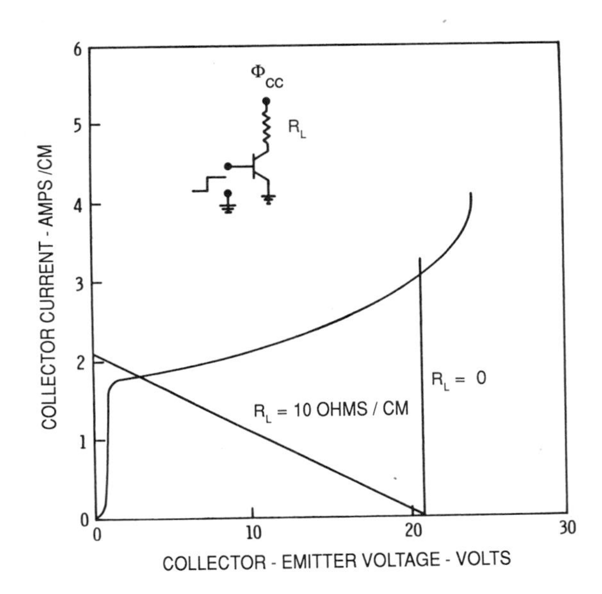

FIGURE 8-27. The transistor steady state collector current versus collector–emitter voltage characteristic at $\Phi_{be} = 0.9$ V. Also indicated are load lines corresponding to $R_L = 0$ and $R_L = 10\,\Omega\,\text{cm}^{-1}$, with a circuit battery voltage $\Phi_{CC} = 20.9$ V. (After Alwin et al., 1977.)

$$\operatorname{div} \mathbf{J}_p = -q\left(R - G + \frac{\partial p}{\partial t}\right) \tag{8-58}$$

$$\operatorname{div} \mathbf{J}_n = q\left(R - G + \frac{\partial n}{\partial t}\right) \tag{8-59}$$

Here Φ is the local electrostatic potential, p and n are the mobile hole and electron densities, N_D^+ and N_A^- are the number of ionized donors and acceptors, R is the net carrier recombination rate, and G is the carrier generation rate.

The constitutive equations for the hole and electron current densities are

$$\mathbf{J}_p = -\mu_p q n \operatorname{grad} \Phi - qD_p \operatorname{grad} p - qpD_p^T \operatorname{grad} T \tag{8-60}$$

$$\mathbf{J}_n = -\mu_n q n \operatorname{grad} \Phi + qD_n \operatorname{grad} n - qnD_n^T \operatorname{grad} T \tag{8-61}$$

$$\mathbf{J} = \mathbf{J}_p + \mathbf{J}_n \tag{8-62}$$

Here μ_p, μ_n are the respective mobilities, D_p, D_n are diffusion constants, and D_p^T, D_n^T are thermal diffusion constants.

The boundary conditions and the circuit interactions are then defined and the problem then reduces to solving the equations for Φ, T, and $\mathbf{J}$ as functions of time and position. [Often the use of quasi-Fermi potentials is useful (see e.g., Alwin et al., 1977).] Typical results are shown in Figs. 8-27 through 8-31. These figures do not demonstrate the postswitching effects. Rather, they display the role of heating in the pre-SB regime and characterize the current, voltage, and

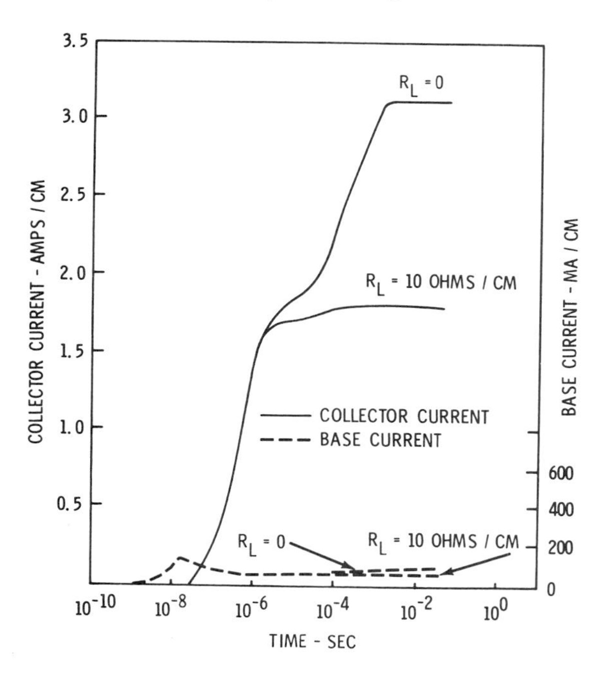

FIGURE 8-28. Curves showing the turn on of the collector current along a load line: $R_L = 0$ and $R_L = 10\ \Omega\ \text{cm}^{-1}$. In each case the voltage applied to the base contact is 0.9 V and the battery voltage is 20.9 V. The base currents are also shown. The rise in collector current beginning at 0.1 μs is the electronic rise time, whereas the rise beginning at about 0.02 ms corresponds to the current increase caused by thermal effects; collector current (solid), base current (dashed). (After Alwin et al., 1977.)

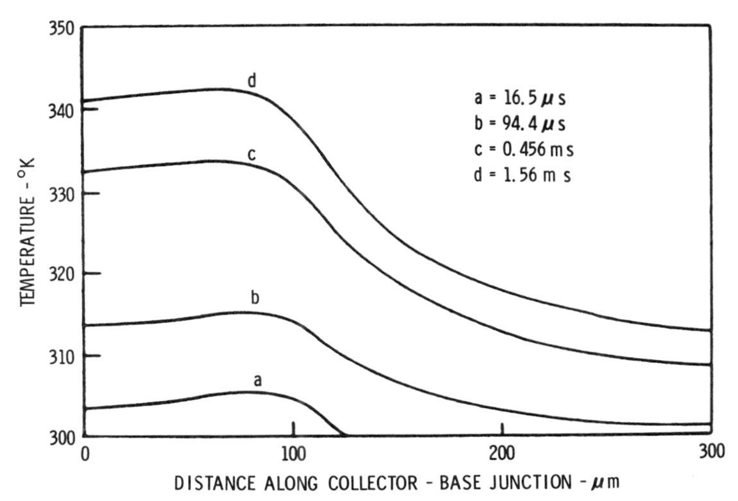

FIGURE 8-29. Temperature distribution along the collector–base junction at various points in time indicating the heat effect due to the flow of emitter current with a load resistance $R_L = 0$. Curves: (a) 16.5 μs; (b) 94.4 μs; (c) 0.456 ms; (d) 1.56 ms. (After Alwin et al., 1977.)

temperature profiles in the stable steady state. These predictions are similar to those presented in Sec. 8.3.2 for the preswitching characteristics of thin chalcogenide films. Both phenomena and their basic understanding are quite the same. Indeed, the manifestation and explanation of thermal instabilities in solids in general can be treated via the model outlined in Sec. 8.2 and idealized in Sec. 8.4.

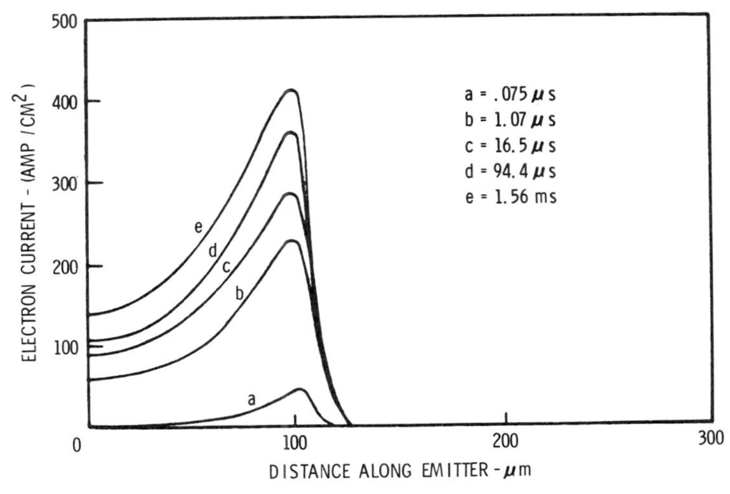

FIGURE 8-30. Longitudinal electron current density along the emitter–base junction showing the current buildup in time. Curves (a) (0.075 μs) and (b) (1.07 μs) reflect primarily the electronic buildup, whereas curves (c) (16.5 μs), (d) (94.46 μs), and (e) (1.56 ms) reflect the effect of heating. All curves show current crowding to the emitter edge. (After Alwin et al., 1977.)

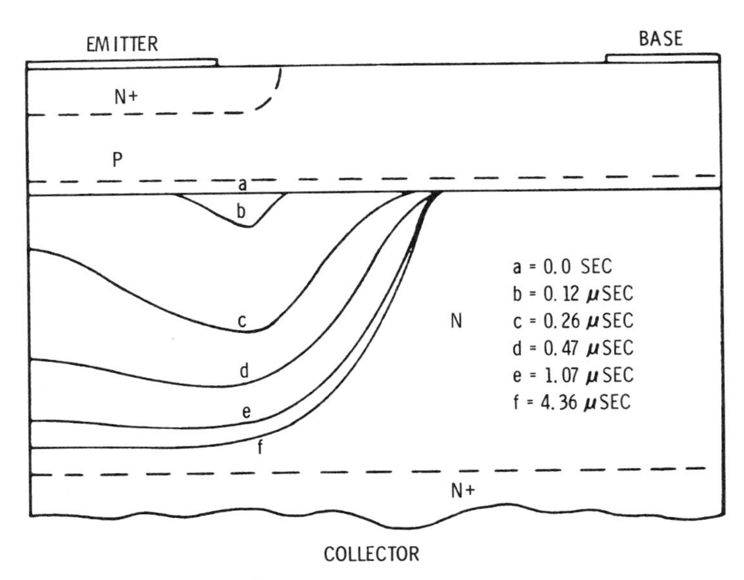

FIGURE 8-31. Plot of the electrostatic potential $\Phi = 20.5$ V for the transistor at $\Phi_{BE} = 0.9$ V and $\Phi_{CE} = 20.9$ V at various points in time, indicating the current buildup in the collector region for the transistor operating in the circuit with $R_L = 0$. (After Alwin et al., 1977.)

8.6. SUMMARY

We have reviewed the basic thermal and electrothermal instabilities that lead to breakdown, switching, or memory effects in semiconductor materials and devices. By way of introduction we first discussed the thermistor, a device that relies on the development of a temperature gradient induced by Joule heating in a material having the proper variation of electrical conductivity with temperature. The thermistor can be stabilized at all of its $I(\Phi)$ points for the proper set of circuit parameters. We then discussed the general problem of heat flow in semiconductors, with emphasis on thermally induced NDC, critical fields and temperatures, boundary conditions, and inhomogeneities. Our understanding was then applied to three specific configurations: thin chalcogenide films, where critical electric field effects are important; VO_2, where a critical temperature dominates; second breakdown in transistors. We compared theory with experiment as much as possible, relying primarily on numerical calculations and simple analytical models for special situations such as VO_2.

List of Pertinent Symbols

(Others are self-evident and coupled to the specific analysis, such as certain constants, etc.)

B	magnetic flux density
C	capacitance
D	diffusion coefficient
e	electron charge
E	energy
f	frequency
F	electric field
H	Hamiltonian, magnetic field
I	current
J	current density
k	wave number (vector), Boltzmann constant (scalar)
K	wave number (vector), thermal conductivity (scalar)
l	sample length
L	inductance, length
m	mass
M	mutual inductance
n	electron density
N	number density (particles, traps, etc.)
p	momentum (vector), hole density (scalar), power density (scalar)
P	power
q	displacement
r	recombination rate
R	resistance
t	time
T	temperature
v, V	average drift velocity, velocity, volume, volts
x, y, z	position
Z	impedance
ϵ	permittivity
μ	mobility
μ_0	permeability of free space
ξ	normalized displacement
ρ	charge density
σ	electrical conductivity, capture cross section

τ_D	dielectric relaxation time
τ_R	recombination lifetime
ϕ	magnetic flux
Φ	electric potential/voltage drop
Φ_b	battery voltage, breakdown voltage
Φ_V	battery voltage
ω	frequency (angular), energy density

References

Abe, Y. (Ed.) (1989). Special Issue on Nonlinear and Chaotic Transport in Semiconductors, *Appl. Phys.* **A48,** 93.

Adler, D. A. (1978). *Phys. Rev. Lett.* **41,** 1755.

Adler, D. A., H. K. Henisch, and N. Mott (1978). *Rev. Mod. Phys.* **50,** 209.

Adler, D. A., M. S. Shur, M. Silver, and S. R. Ovshinsky (1980). *J. Appl. Phys.* **51,** 3289.

Alferov, Zh. I., O. A. Mezrin, M. A. Sinitsyn, S. I. Troshkov, and B. S. Yavich (1987). *Sov. Phys. Semicond.* **21,** 304.

Allinson, D. L., T. I. Barry, D. J. Clinton, A. J. Hughes, A. H. Lettington, and J. A. Savage (1979). *J. Non-Cryst. Solids* **31,** 307.

Altcheh, L., N. Klein, and I. N. Katz (1972). *J. Appl. Phys.* **43,** 3258.

Alwin, V. C., D. H. Navon, and L. J. Turgeon (1977). *IEEE Trans. Electron Devices* **ED-24,** 1297.

Ambegaokar, V., and A. Baratoff (1963). *Phys. Rev. Lett.* **11,** 104, (Erratum of *Phys. Rev. Lett.* **10,** 486, 1963).

Ambegaokar, V., and B. I. Halperin (1969). *Phys. Rev. Lett.* **22,** 1364.

Amrachov, A. Sh., V. M. Ivastchenko, and V. V. Mitin (1986). *Ukr. Phys. J. (USSR)* **31,** 689.

Anderson, C. L., and C. R. Crowell (1972). *Phys. Rev. B* **5,** 2267.

Anderson, F. (1954). *Acta Chem. Scand.* **8,** 1599.

Anderson, P. W., and J. M. Rowell (1963). *Phys. Rev. Lett.* **10,** 230.

Ando, T., A. B. Fowler, and F. Stern (1982). *Rev. Mod. Phys.* **54,** 437.

Andronov, A. A., A. Witt, and S. E. Khaikin (1966). *Theory of Oscillations* (Pergamon Press, London).

Andronov, A. A., E. A. Leontovich, I. I. Gordon, and A. G. Maier (1971). *Theory of Bifurcations of Dynamic Systems on a Plane,* (Israel Program for Scientific Translations, Jerusalem).

Andronov, A. A., E. A. Leontovich, I. I. Gordon, and A. G. Maier (1973). *Qualitative Theory of Second-Order Dynamic Systems,* (Wiley, New York).

Aoki, K., and K. Yamamoto (1989). *Appl. Phys.* **A48,** 111.

Aoki, K., K. Miyamae, T. Kobayashi, and K. Yamamoto (1983b). *Physica* **117** & **118B** & **C,** 570.

Aoki, K., K. Yamamoto, N. Mugibayashi, and E. Schöll (1989). *Solid State Electron.* **32,** 1149.

Aoki, K., O. Ikezawa, and K. Yamamoto (1983a). *Phys. Lett.* **98A,** 217.

Aoki, K., O. Ikezawa, and K. Yamamoto (1984). *Phys. Lett.* **106A,** 343.

Aoki, K., O. Ikezawa, N. Mugibayashi, and K. Yamamoto (1985). *Physica* **134B,** 288.

Aoki, K., T. Kobayashi, and K. Yamamoto (1981). *J. Phys. Coll.* **C7,** 51.

Aoki, K., T. Kobayashi, and K. Yamamoto (1982). *J. Phys. Soc. Jpn.* **51,** 2373.

Asche, M., Z. S. Gribnikov, V. M. Ivastchenko, H. Kostial, and V. V. Mitin (1982). *Phys. Status Solidi (b)* **114,** 429.

Asche, M., Z. S. Gribnikov, V. V. Mitin, and O. G. Sarbey (1979). *Goryachie Elektrony v Mnogodolinikh Poluprovodnikakh* (Naukova Dumka, Kiev).

Ashley, K. L. (1963). Ph.D. dissertation, Carnegie Institute of Technology.

Auracher, F., and T. Van Duzer (1973). *J. Appl. Phys.* **14,** 848.

Avramenko, V. A., and M. V. Strikha (1986). *Sov. Phys. Semicond.* **20,** 1152.

Baccarani, G., M. Rudan, R. Guerrieri, and P. Ciampolini (1986). In *Advances in CAD for VSLI,* W. Engl (Ed.) (North-Holland, Amsterdam), Vol. 1, p. 107.

Bak, P., T. Bohr, M. H. Jensen, and P. V. Christiansen (1984). *Solid State Commun.* **51,** 231.

Balberg, I. (1970). *Appl. Phys. Lett.* **16,** 491.

Baraff, G. A. (1962). *Phys. Rev.* **128,** 2507.

Baraff, G. A. (1964). *Phys. Rev.* **A135,** 528.
Bardeen, J., L. N. Cooper, and J. R. Schrieffer (1957). *Phys. Rev.* **108,** 1175.
Barker, J. R. (1973). *J. Phys. C, Solid State Phys.* **6,** 2663.
Barnett, A. M. (1969). *IBM J. Res. Dev.* **13,** 522.
Barone, A., and G. Paterno (1982). *Physics and Applications of the Josephson Effect,* (Wiley, New York).
Bass, F. G., Yu. G. Gurevich, S. A. Kostylev, N. A. Terent'eva (1983). *Sov. Phys. Semicond.* **17,** 808.
Bauhahn, P., and G. I. Haddad (1977). *IEEE Trans.* **ED-24,** 634.
Baumann, H., R. Symanczyk, C. Radehaus, H. G. Purwins, and D. Jäger (1987). *Phys. Lett.* **123,** 421.
Beasley, M. R., and C. J. Kircher (1981). *Superconductor Materials Science: Metallurgy, Fabrication and Applications* , S. Foner and B. B. Schwartz (Eds.), Chap. 9 (Plenum Press, New York).
Becker, K. (1936). *Arch. Elektrotech. (Berlin)* **30,** 411.
Bedeaux, D., and P. Mazur (1981). *Physica* **105A,** 1.
Bedeaux, D., P. Mazur, and R. A. Pasmanter (1977a). *Statistical Mechanics and Statistical Methods in Theory and Application,* U. Landman (Ed.) (Plenum Press, New York).
Bedeaux, D., P. Mazur, and R. A. Pasmanter (1977b). *Physica* **86a,** 355.
Bednorz, J. G., and K. A. Müller (1986). *Z. Phys. B* **64,** 189.
Belyantsev, A. M., A. A. Ignatov, V. I. Piskarev, M. A. Sinitsyn, V. I. Shashkin, B. S. Yavich, and M. L. Yakovlev (1986). *JETP Lett.* **43,** 437.
Belyantsev, A. M., V. I. Gavrilenko, A. A. Ignatov, V. I. Piskarev, V. V. Shashkin, and A. A. Andronov (1988). *Solid State Electron* **31,** 379.
Berglund, C. N. (1969). *IEEE Trans.* **ED16,** 432.
Berglund, C. N., and N. Klein (1971). *Proc. IEEE* **59,** 1099.
Bergmann, F., and D. Gerstner (1963). *Arch. Elektr. Übertr.* **17,** 467.
Bergmann, F., and D. Gerstner (1966). *IEEE Trans. Electron Devices* **ED-13,** 630.
Bishop, A. R., and T. Schneider (Eds.) (1978). *Solitons and Condensed Matter Physics* (Springer-Verlag, Berlin).
Blicher, A. (1976). *Thyristor Physics* (Springer-Verlag, Berlin).
Blotekjaer, K., and E. B. Lunde (1969). *Solid State Phys.* **35,** 581.
Blotekjaer, K. (1970). *IEEE Trans. Electron Devices* **ED-17,** 38.
Blouke, M. M. (1969). Ph.D. dissertation, University of Illinois.
Boer, K. W., E. Jahne, and E. Neubauer (1969). *Phys. Status Solidi* **1,** 231.
Bonch-Bruevich, V. L. (1966). *Sov. Phys. Solid State* **8,** 1397.
Bonch-Bruevich, V. L. (1966). *Sov. Phys. Solid State* **8,** 290.
Bonch-Bruevich, V. L., and Sh. M. Kogan (1965). *Sov. Phys. Solid State* **7,** 15.
Bonch-Bruevich, V. L., I. P. Zvyagin, and A. G. Mironov (1975). *Domain Electrical Instabilities in Semiconductors* (Consultant Bureau, New York).
Bosch, B. G., and R. W. Engelmann (1975). *Gunn Effect Electronics,* (Halsted Press, Wiley, New York).
Bosch, R., and H. W. Thim (1974). *IEEE Trans. Electron Devices* **ED-21,** 16.
Bosnell, J. R., and C. B. Thomas (1972). *Solid State Electron.* **15,** 1261.
Bott, I. B., and W. Fawcett (1968). *Adv. Microwaves* **3,** 224.
Braiman, Y., E. Ben-Jacob, and Y. Imry (1980). In *SQUID 80,* H. D. Hahlbohm and H. Lubbig (Eds.), (de Gruyter, Berlin), p. 783.
Brandl, A., T. Geisel, and W. Prettl (1987). *Europhys. Lett.* **3,** 401.
Brandl, A., M. Völcker, and W. Prettl (1989). *Appl. Phys. Lett.* **55,** 238.
Brandl, A., W. Kröninger, W. Prettl, and G. Obermair (1990). *Phys. Rev. Lett.* **64,** 212.
Brown, E. R. (1989). *Solid State Electron.* **32,** 1179.
Brown, E. R., T. C. L. G. Sollner, W. D. Goodhue, and C. L. Chen (1988). *Proc. SPIE* **943,** 3.
Buckley, W. D., and S. H. Holmberg (1975). *Solid State Electron.* **18,** 127.
Bulman, P. J., G. S. Hobson, and B. C. Taylor (1972). *Transferred Electron Devices* (Academic Press, New York).
Burmeliene, S. B., Yu. K. Pozhela, K. A. Pyragas, and A. V. Tamaševičius (1985). *Physics* **134B,** 293.

Burgess, R. E. (1955a). *Proc. Phys. Soc.* **B68,** 706.
Burgess, R. E. (1955b). *Proc. Phys. Soc.* **B68,** 908.
Burgess, R. E. (1955c). *J. Electron.* **297,** 459.
Burgess, R. E. (1960). *Can. J. Phys.* **38,** 369.
Burstein, E., and S. Lundqvist (Eds.) (1969). *Tunneling Phenomena in Solids* (Plenum Press, New York).
Burt, M. G. (1985). *J. Phys. C, Solid State Phys.* **18,** L477.
Busch, H. (1921). *Ann. Phys. (N.Y.)* **64,** 401.
Butcher, P. N. (1967). *Rep. Prog. Phys.* **30,** 97.
Butcher, P. N., and C. J. Hearn (1968). *Electron. Lett.* **4,** 459.
Butcher, P. N., and W. Fawcett (1966). *Brit. J. Appl. Phys.* **17,** 1425.
Büttiker, M., and H. Thomas (1979). *Z. Phys. B* **34,** 301.
Büttiker, M., and H. Thomas (1981). *Phys. Rev. A* **24,** 2635.
Capasso, F. (1987). *Science* **235,** 172.
Capasso, F., K. Mohammed, and A. Y. Cho (1986). *IEEE Trans. Electron Devices* **ED-34,** 297, 1853.
Capasso, F., S. Sen, A. Y. Cho, and D. Sivco (1987). *IEEE Trans. Electron Devices* **EDL-8,** 297.
Carslaw, H. S., and J. C. Jaeger (1959). *Conduction of Heat in Solids* (Oxford, New York).
Chang, D. M., and J. G. Ruch (1968). *Appl. Phys. Lett.* **12,** 111.
Chen, J. T., and D. N. Langenberg (1973). *Proceedings of the International Conference on Low-Temperature Physics* LT-13, (Plenum, New York) Vol. 3, p. 298.
Chen, J. T., T. F. Finnegan, and D. N. Langenberg (1971). *Physica* **55,** 413.
Chen, Y., and T. Tang (1988). *IEEE Trans. Electron Devices* **ED-35,** 2180.
Cheung, P. S., and C. J. Hearn (1972). *Electron Lett.* **8,** 79.
Chevoir, F., and B. Vinter (1989). Workbook of 8th Conference on Electronic Properties of Two Dimensional Systems, Grenoble, 4–8 September, p. 65, and *Appl. Phys. Lett.* **55,** 1859.
Chwang, R., C. W. Kao, and C. R. Crowell (1979). *Solid State Electron.* **22,** 599.
Cole, J. D. (1968). *Perturbation Methods in Applied Mathematics,* (Ginn, Boston).
Coleman, P., J. Freeman, H. Morkoc, K. Hess, B. Streetman, and M. Keever (1982). *Appl. Phys. Lett.* **40,** 493.
Collet, P., and J. P. Eckmann (1980). *Iterated Maps on the Interval as Dynamical Systems* (Birkhäuser, Basel).
Collins, D. A., D. H. Chow, D. Z. Ting, E. T. Yu, J. R. Söderström, and T. C. McGill (1989). *Solid State Electron.* **32,** 1095.
Conwell, E. M. (1967). *High Field Transport in Semiconductors,* Solid State Phys. Suppl. 9, (Academic Press, New York).
Conwell, E. M. (1970). *IEEE Trans. Electron Devices* **ED-17,** 262.
Conwell, E. M., and M. O. Vassell (1966). *IEEE Trans. Electron Devices* **ED-13,** 22.
Cook, R. K., and J. Frey (1981). *IEEE Trans. Electron Devices* **ED-28,** 951.
Coon, D. D., S. N. Ma, and A. G. U. Perera (1987). *Phys. Rev. Lett.* **58,** 1139.
Cooper, L. N. (1956). *Phys. Rev.* **104,** 1189.
Coward, L. A. (1971). *J. Non-Cryst. Solids* **6,** 107.
Crandall, R. S. (1970). *J. Phys. Chem Solids* **31,** 2069; *Phys. Rev.* B **1,** 730.
Croitoru, N., and C. Popescu (1970). *Phys. Status Solidi* **3** (*a*), 1047.
Czekaj, J., M. P. Shaw, and H. L. Grubin (1988). *Microwave Opt. Technol. Lett.* **1,** 70.
Czekaj, J., M. P. Shaw, J. East, P. A. Blakey, and H. L. Grubin (1985). Proceedings of the 4th International Conference on Hot Electrons in Semiconductors, Innsbruck, Austria, July, *Physica* **134B,** 499.
Davies, R. A., M. J. Kelly, and T. M. Kerr (1985). *Phys. Rev. Lett.* **55,** 1114.
Davies, R. A., M. J. Kelly, and T. M. Kerr (1986). *Electron. Lett.* **22,** 131.
DeGroot, A. W., G. C. McGonigal, D. J. Thomson, and H. C. Card (1984). *J. Appl. Phys.* **55,** 312.
DeLoach, Jr., B. C. (1976). *IEEE Trans. Electron Devices* **ED-23,** 57.
Demikhovskii, V. Ya., and B. A. Tavger (1966). *Radiotekh. Elektron.* USSR **11,** 1147.
Demikhovskii, V. Ya., and B. A. Tavger (1963). Patent Disclosure No. 827216/26-9 of March 22, USSR.
Dennemeyer, R. (1968). *Introduction to Partial Differential Equations and Boundary Value Problems* (McGraw-Hill, New York), Secs. 1–5.

DeWald, J. F., A. D. Pearson, W. R. Northover, and W. F. Peck (1962). J. Electrochem. Soc. **109,** 243C.
Dmitriev, A. P., M. P. Mikhailova, and I. N. Yassievich (1987). *Phys. Status Solidi* (*b*) **140,** 9.
Doumbia, I., G. Salmer, and E. Constant (1975). *J. Appl. Phys.* **46,** 1831.
Dubitskij, A. L., B. S. Kerner, and V. V. Osipov (1986). *Sov. Phys. Semicond.* **20,** 755.
Duchene, J., G. Adam, and D. Augier (1971a). *Phys. Status Solidi* **8,** 459.
Duchene, J., M. Tervaillon, and M. Pailly (1972). *Thin Solid Films* **12,** 231.
Duchene, J., M. Tervaillon, M. Pailly, and G. Adam (1971b). *Appl. Phys. Lett.* **19,** 115.
Duchene, J., M. Tervaillon, M. Pailly, and G. Adam (1971c). *IEEE Trans. Electron Devices* **ED-18,** 1151.
Dudeck, I., and R. Kassing (1977). *J. Appl. Phys.* **48,** 4786.
Duke, C. B. (1969). *Tunneling in Solids* (Academic Press, New York).
East, J. R., and P. A. Blakey (1984). *Physics of Submicron Structures Proceedings—1982 Workshop,* H. L. Grubin (Ed.) (Plenum, New York), p. 287.
Eaves, L., F. W. Sheard, and G. A. Toombs (1989). In *Band Structure Engineering in Semiconductor Microstructures,* R. A. Abram and M. Jaros (Eds.) (Plenum Press, New York), p. 177.
Eaves, L., G. A. Toombs, F. W. Sheard, C. A. Payling, M. A. Leadbeater, E. S. Alves, T. J. Foster, P. E. Simmonds, M. Henini, O. H. Hughes, J. C. Portal, G. Hill, and M. A. Pate (1988). *Appl. Phys. Lett.* **52,** 212.
Eckmann, J. P. (1981). *Rev. Mod. Phys.* **53,** 643.
Emanuel, M. A., T. K. Higman, J. M. Higman, J. M. Kolodzey, J. J. Coleman, and K. Hess (1988). *Solid State Electron.* **31,** 589.
Esaki, L. (1958). *Phys. Rev.* **109,** 603.
Esaki, L. (1969). In *Tunneling Phenomena in Solids,* E. Burstein and S. Lundquist (Eds.) (Plenum Press, New York), p. 47.
Esaki, L., and R. Tsu (1970). *IBM J. Res. Dev.* **14,** 61.
Falco, C. M., W. H. Parker, S. E. Trullinger, and P. K. Hansma (1974). *Phys. Rev.* B **10,** 1865.
Fawcett, W., and D. C. Herbert (1972). *Electron. Lett.* **8,** 592.
Fawcett, W., and D. C. Herbert (1974). *Brit. J. Appl. Phys.* **C7,** 1641.
Feigenbaum, M. J. (1978). *J. Stat. Phys.* **19,** 25.
Feynman, R. P., R. B. Leighton, and M. Sands (1965). *Lectures on Physics,* Vol. 3 (Addison-Wesley, Reading, MA).
Feynman, R. P., R. B. Leighton, and M. Sands (1964). *Lectures on Physics,* Vol. 2, (Addison-Wesley, Reading, MA).
Fisher, B. (1975). *J. Phys. C* **8,** 2072.
Flasck, R., and H. K. Rockstad (1973). *J. Non-Cryst. Solids* **12,** 353.
Fock, W. A. (1927). *Arch. Electrotechn.* **19,** 71.
Foyt, A. G., and A. M. McWhorter (1966). *IEEE Trans. Electron Devices* **ED-13,** 79.
Franz, W. (1956). *Encyclopedia of Physics* Vol. 18, p. 166, S. Flügge (Ed.) (Springer-Verlag, Berlin).
Franz, W. (1969). In *Tunneling Phenomena in Solids,* E. Burstein and S. Lundquist (Eds.), (Plenum Press, New York), pp. 13 and 207.
Frensley, W. R. (1987). *Appl. Phys. Lett.* **51,** 448.
Frensley, W. R. (1989a). In *Band Structure Engineering in Semiconductor Microstructures,* R. A. Abram, and M. Jaros (Eds.) (Plenum Press, New York), p. 177.
Frensley, W. R. (1989b). *Solid State Electron.* **32,** 1235.
Frye, R., D. Adler, and M. P. Shaw (1980). *J. Non-Cryst. Solids* **35–36**(2), 1099.
Fulop, W. (1963). *IEEE Trans. Electron Devices* **ED-10,** 120.
Fulton, T. A., and R. C. Dynes (1973). *Solid State Commun.* **12,** 56.
Gaur, S. P., and D. H. Navon (1976). *IEEE Trans. Electron Devices* **ED-23,** 50.
Goldman, I. I., and V. D. Krivchenkov (1961). *Problems in Quantum Mechanics* (Pergamon, New York), Chap. 2.
Goldman, V. J., D. C. Tsui, and J. E. Cunningham (1987a). *Phys. Rev.* B **36,** 7635.
Goldman, V. J., D. C. Tsui, and J. E. Cunningham (1987b). *Phys. Rev.* B **35,** 9387.
Goldman, V. J., D. C. Tsui, and J. E. Cunningham (1987c). *Phys. Rev. Lett.* **58,** 1256.
Goldman, V. J., D. C. Tsui, and J. E. Cunningham (1988). *Solid State Electron.* **31,** 731.
Gram, N. O., (1972). *Phys. Lett.* **38A,** 235.

Gray, K. W., J. Pattison, H. D. Rees, B. A. Prew, R. Clarke, and L. D. Irving (1975). Proceedings of the Biannual Cornell University Conference, 5th, p. 215.
Gribnikov, Z. S. (1973). *Sov. Phys. Semicond.* **6,** 1204.
Gribnikov, Z. S., and V. I. Mel'nikov (1966). *Sov. Phys. Solid State* **7,** 2364.
Gribnikov, Z. S., (1978). *Sov. Phys. JETP* **47,** 1099.
Gribnikov, Z. S., V. M. Ivastchenko, and V. V. Mitin (1981). *Phys. Stat. Sol.* (*b*) **105,** 451.
Grove, A. S. (1967). *Physics and Technology of Semiconductor Devices* (Wiley, New York).
Grubin, H. L. (1972). *IEEE Trans. Electron Devices* **ED-19,** 110.
Grubin, H. L. (1976). *IEEE Trans. Electron Devices* **ED-23,** 1012.
Grubin, H. L. (1978). *IEEE Trans. Electron Devices* **ED-25,** 511.
Grubin, H. L., and J. P. Kreskovski (1983). *Surf. Sci.* **132,** 594.
Grubin, H. L., and J. P. Kreskovski (1985). *VLSI Electronics, Microstructure Science,* Vol. 10, N. G. Einspruch (Ed.), (Academic Press, New York).
Grubin, H. L., and R. D. Kaul (1975). *IEEE Trans. Electron Devices* **ED-20,** 63.
Grubin, H. L., D. K. Ferry, G. J. Iafrate, and J. B. Barker (1982). In *VLSI Electronics,* Vol. 3, N. G. Einspruch (Ed.) (Academic Press, New York), p. 198.
Grubin, H. L., D. K. Ferry, and J. R. Barker (1979b). *Proceedings of the IEDM* (IEEE Press, New York), p. 394.
Grubin, H. L., D. K. Ferry, J. R. Barker, M. A. Littlejohn, T. H. Glisson, and J. R. Hauser (1979). *Proceedings of the Biannual Electron Engineering Conference 7th* (Cornell University, Ithaca, NY).
Grubin, H. L., M. P. Shaw, and E. M. Conwell (1971). *Appl. Phys. Lett.* **18,** 211.
Grubin, H. L., M. P. Shaw, and P. R., Solomon (1973). *IEEE Trans. Electron Devices* **ED-20,** 63.
Guckenheimer, J., and P. Holmes (1983). *Nonlinear Oscillations, Dynamical Systems and Bifurcation of Vector Fields* (Springer-Verlag, Berlin).
Gunn, J. B. (1964). *IBM J. Res. Dev.* **8,** 141.
Gunn, J. B. (1966). *IBM J. Res. Dev.* **10,** 300.
Hahn, W. (1967). *Stability of Motion* (Springer-Verlag, Berlin).
Haken, H. (1970). *Light and Matter,* Encyclopedia of Physics XXV/2C, S. Flügge (Ed.) (Springer-Verlag, Berlin).
Haken, H. (1983a). *Synergetics, An Introduction,* 3rd ed., (Springer-Verlag, Berlin).
Haken, H. (1983b). *Advanced Synergetics* (Springer-Verlag, Berlin).
Hasty, T., R. Stratton, and E. L. Jones (1968). *J. Appl. Phys.* **39,** 4623.
Hauser, J. R., T. H. Glisson, and M. A. Littlejohn (1979). *Solid State Electron.* **22,** 487.
Hayes, T. M. (1974). *J. Phys.* **C7,** 371.
Hayes, T. M., and D. D. Thornberg (1973). *J. Phys.* **C6,** 450.
Heiblum, M., M. V. Fischetti, W. P. Dumke, D. J. Frank, I. M. Anderson, C. M. Knoedler, and L. Osterling (1987). *Phys. Rev. Lett.* **58,** 816.
Heiblum, M., U. Sivan, and M. V. Weckwerth (1989). Abstracts of 6th International Conference "Hot Carriers in Semiconductors," 23–28 July, Scottsdale, Arizona, p. 46.
Held, G. A., C. D. Jeffries, and E. E. Haller (1985). *Proceedings of the 17th International Conference on the Physics of Semiconductors,* San Francisco, J. D. Chadi and W. A. Harrison (Eds.) (Springer-Verlag, Berlin), p. 1289; and *Phys. Rev. Lett.* **52,** 1037 (1984).
Henisch, H. K., W. R. Smith, and W. Wohl (1974). *Amorphous and Liquid Semiconductors,* J. Stuke, and W. Brenig (Eds.) (Taylor & Francis, London), p. 567.
Hensel, F. (1977). *Amorphous and Liquid Semiconductor,* W. E. Spear (Ed.) (U. Edinburgh, Scotland), p. 815.
Herbert, D. C., and S. J. Till (1982). *J. Phys. C. Solid State Phys.* **15,** 5411.
Hess, K. (1988). *Advanced Theory of Semiconductor Devices* (Prentice Hall, Englewood Cliffs, New Jersey).
Hess, K. (1990). *Phys. Today,* February, 34.
Hess, K., and G. J. Iafrate (1985). *VLSI Electronics, Microstructure Science,* Vol. 9, N. G. Einspruch (Ed.), (Academic Press, New York).
Hess, K., H. Morkoc, H. Shichijo, and B. G. Streetman (1979). *Appl. Phys. Lett.* **35,** 469.
Hess, K., T. K. Higman, M. A. Emanuel, and J. J. Coleman (1986). *J. Appl. Phys.* **60,** 3775.
Higman, J. M., I. C. Kizilyalli, and K. Hess (1988). *IEEE Trans. Electron Devices* **EDL-9,** 399.

Hilsum, C. (1962). *Proc. IRE* **50,** 185.
Holonyak, N., and S. F. Bevacqua (1963). *Appl. Phys. Lett.* **2,** 71.
Homma, K. (1971). *Appl. Phys. Lett.* **18,** 198.
Hopf, E. (1942). *Ber. Math.-Phys. Klasse Sächs. Akad. Wiss. Leipzig* **XCIV,** 1.
Hower, P. L., and V. G. K. Reddi (1970). *IEEE Trans. Electron Devices* **ED-17,** 320.
Huang, C. I., M. J. Paulus, C. A. Bozada, S. C. Dudley, K. R. Evans, C. E. Stutz, R. L. Jones, and M. E. Cheney (1987). *Appl. Phys. Lett.* **51,** 121.
Huberman, B. A., J. P. Crutchfield, and N. H. Packard (1980). *Appl. Phys. Lett.* **37,** 750.
Hughes, A. J., P. A. Holland, and A. H. Lettington (1975). *J. Non-Cryst. Sol.* **17,** 89.
Hüpper, G., and E. Schöll (1991). *Phys. Rev. Lett.* **66,** 2372.
Hüpper, G., E. Schöll, and L. Reggiani (1989). *Solid State Electron.* **32,** 1787.
Hurwitz, C. E., and A. L. McWhorter (1964). *Phys. Rev.* **134,** A1033.
Iafrate, G. J., and K. Hess (1985). In *VLSI Electronics,* N. Einspruch (Ed.) Vol. 9, Chap. 6, (Academic, New York).
Imenkov, A. N., S. S. Meskin, D. N. Nasledov, V. N. Ravich, and B. V. Tsarenkov (1965). *Sov. Phys. Solid State* **6,** 1808.
Inata, T., S. Muto, Y. Nakata, S. Sasa, T. Fujii, and S. Hiyamizu (1987). *Jpn. J. Appl. Phys.* **26,** L1332.
Ivanov, I. L., and S. M. Ryvkin (1958). *Sov. Phys. Techn. Phys.* **3,** 722.
Ivastchenko, V. M., and V. V. Mitin (1990). *Modelirovanie kineticheskich javlenij v poluprovodnikach. Method Monte Carlo* (Naukova Dumka, Kiev).
Ivastchenko, V. M., R. V. Konakova, and V. V. Mitin (1984). *Phys. Status Solidi* (*a*) **84,** 669.
Jackson, J. L., and M. P. Shaw (1974). *Appl. Phys. Lett.* **25,** 666.
Jäger, D. (1991). In *Nonlinear Dynamics in Solid State Physics,* H. Thomas (Ed.) (Springer-Verlag, Berlin).
Jäger, D., H. Baumann, and R. Symanczyk (1986). *Phys. Lett.* **A117,** 141.
Jelks, E. C., G. M. Walser, R. W. Bené, and W. H. Neal II (1975). *Appl. Phys. Lett.* **26,** 355.
Johnston, R. L., B. C. DeLoach, Jr., and B. G. Cohen (1965). *Bell System Tech. J.* **44,** 369.
Jonson, M. (1989). *Phys. Rev.* B **39,** 5924.
Joseph, D. D. (1965). *Int. J. Heat Mass Transfer* **8,** 281.
Joseph, D. D., and E. M. Sparrow (1970). *Q. Appl. Math.* **28,** 327.
Josephson, B. D. (1962). *Phys. Lett.* **1,** 251.
Josephson, B. D. (1965). *Adv. Phys.* **14,** 419.
Kane, E. O. (1957). *J. Phys. Chem. Sol.* **1,** 249.
Kane, E. O. (1961). *J. Appl. Phys.* **32,** 83.
Kane, E. O. (1967). *Phys. Rev.* **159,** 624.
Kane, E. O. (1969). In *Tunneling Phenomena in Solids,* Burstein and S. Lundquist (Eds.) (Plenum Press, New York), p. 1.
Kane, E. O., and E. I. Blount (1969). In *Tunneling Phenomena in Solids,* E. Burstein, and S. Lundquist (Eds.) (Plenum Press, New York), p. 79.
Kao, K. C., and W. Hwang (1981). *Electronic Transport in Solids* (Pergamon, Oxford).
Kaplan, T., and D. Adler (1972). *J. Non-Cryst. Solids* **8–10,** 522.
Karlovsky, J. (1962). *Phys. Rev.* **127,** 49.
Kastal'skii, A. A. (1968). *Sov. Phys. Semicond.* **2,** 546.
Kastal'skii, A. A. (1973). *Phys. Status Solidi* (*a*) **15,** 599.
Kastal'skii, A. A., and S. M. Ryvkin (1968). *Sov. Phys. JETP Lett.* **7,** 350.
Kastalsky, A., S. Luryi (1983). *IEEE Trans. Electron Devices* **EDL-4,** 334.
Kastalsky, A., M. Milshtein, L. G. Shantharama, J. Harbison, and L. Florez (1989). *Appl. Phys. Lett.* **54,** 2452.
Kastner, M., D. Adler, and H. Fritzsche (1976). *Phys. Rev. Lett.* **37,** 1504.
Kaul, R. D., H. L. Grubin, G. O. Ladd, Jr., and J. M. Berak (1972). *IEEE Trans. Electron Devices* **ED-19,** 988.
Kautz, R. L., and R. Monaco (1985). *J. Appl. Phys.* **57,** 875.
Kautz, R. L. (1981). *J. Appl. Phys.* **52,** 6241.
Kazarinov, R. F., and R. A. Suris (1971). *Sov. Phys. Semicond.* **5,** 707.
Kazarinov, R. F., and R. A. Suris (1972). *Sov. Phys. Semicond.* **6,** 120.

Keever, M., H. Shichijo, K. Hess, S. Banerjee, L. Witkowski, and H. Morkoc (1981). *Appl. Phys. Lett.* **38,** 36.
Keldysh, L. V. (1958a). *Sov. Phys. JETP* **6,** 753.
Keldysh, L. V. (1958b). *Sov. Phys. JETP* **7,** 665.
Keldysh, L. V. (1960). *Sov. Phys. JETP* **10,** 509.
Keldysh, L. V. (1965). *Sov. Phys. JETP* **21,** 1135.
Keller, H. B., and D. S. Cohen (1967). *J. Math. Mech.* **16,** 1361.
Kelly, M. J., R. A. Davies, A. P. Long, N. R. Couch, and T. M. Kerr (1986). *GEC J. Res.* **4,** 157.
Kerner, B. S., and V. F. Sinkevich (1982). *Sov. Phys. JETP Lett.* **36,** 437.
Kerner, B. S., and V. V. Osipov (1980). *Sov. Phys. JETP* **52,** 112.
Khan, F. S., J. H. Davies, and J. W. Wilkins (1987). *Phys. Rev.* B **36,** 2578.
Kirchgässner, K. (1977). In *Synergetics*. H. Haken (Ed.) (Springer-Verlag, Berlin), p. 34.
Kittel, C. (1976). *Introduction to Solid State Physics* (Wiley, New York).
Klein, N. (1971). *Thin Solid Films* **7,** 149.
Klein, N. (1969). *Adv. Electron. Electron Phys.* **26,** 309.
Klein, N. (1978). *Thin Solid Films* **50,** 233.
Klein, N. (1983). *Thin Solid Films* **100,** 335.
Kleinman, L. (1969). In *Tunneling Phenomena in Solids,* E. Burstein and S. Lundquist (Eds.) (Plenum Press, New York), p. 181.
Knap, W., M. Jezewski, J. Lusakowski, and W. Kuszko (1988). *Solid State Electron.* **31,** 813.
Knight, B. W., and G. A. Peterson (1966). *Phys. Rev.* **147,** 617.
Knight, B. W., and G. A. Peterson (1967). *Phys. Rev.* **155,** 393.
Kolodzey, J., J. Laskar, T. K. Higman, M. A. Emanuel, J. J. Coleman, and K. Hess (1988). *IEEE Trans. Electron Devices* **EDL-9,** 272.
Kotz, J., and M. P. Shaw (1982). *Proceedings of the 16th International Conference on the Physics of Semiconductors,* Montpellier (North-Holland, Amsterdam), p. 986.
Kotz, J., and M. P. Shaw (1983). *Appl. Phys. Lett.* **42,** 199.
Kotz, J., and M. P. Shaw (1984). *J. Appl. Phys.* **55,** 427.
Kroemer, H., and C. Kittel (1980). *Thermal Physics* (Freeman, San Francisco).
Kroemer, H. (1966). *IEEE Trans. Electron Devices* **ED-13,** 27.
Kroemer, H. (1968a). *IEEE Spectrum* **5,** 47.
Kroemer, H. (1968b). *IEEE Trans. Electron Devices* **ED-15,** 819.
Kroemer, H. (1970). *Proc. IEEE* **58,** 1844.
Kroemer, H. (1971). *Proc. IEEE* **59,** 1844.
Kroemer, H. (1978). *Solid State Electron.* **21,** 60.
Kroll, D. M. (1974). *Phys. Rev.* **9,** 1669.
Kurkijärvi, J. (1991). In *Nonlinear Dynamics in Solid State Physics,* H. Thomas (Ed.) (Springer-Verlag, Berlin).
Lampert, M. A., and P. Mark (1970). *Current Injection in Solids,* (Academic Press, New York).
Landauer, R. (1978). *Phys. Today* **31,** 23.
Landauer, R., and J. W. F. Woo (1972). *Comments Solid State Phys.* **4,** 139.
Landsberg, P. T., D. J. Robbins, and E. Schöll (1978). *Phys. Status Solidl* (*a*) **50,** 423.
Landsberg, P. T., E. Schöll, and P. Shukla (1988). *Physica* **D30,** 235.
Langenberg, D. N. (1980). *Phys. Rev.* B **21,** 5432.
Larrabee, R. D., and M. C. Steele (1960). *J. Appl. Phys.* **31,** 1519.
Lebwohl, P. A., and R. Tsu (1970). *J. Appl. Phys.* **41,** 2664.
Lee, J., M. O. Vassel, and H. F. Lockwood (1984). In *The Physics of Submicron Structures,* H. L. Grubin and K. Hess (Ed.) (Plenum Press, New York), p. 33.
Lent, C. S. (1987). *Superlattices and Microstruct.* **3,** 387.
Levinson, L. I., and Ya. Yasevichyute (1972). *Sov. Phys. JETP* **35,** 991.
Likharev, K. K. (1986). *Dynamics of Josephson Junctions and Circuits* (Gordon and Breach, New York).
Lippens, D., and E. Constant (1981). *Electron. Lett.* **17,** 878.
Lippens, D., E. Constant, M. R. Friscourt, P. A. Rolland, and G. Salmer (1984). In *The Physics of Submicron Structures,* H. L. Grubin and K. Hess (Ed.) (Plenum Press, New York), p. 93.
Lippens, D., J. Nieruchalski, and E. Constant (1983). In *Noise in Physical Systems and 1/f Noise,* M. Savelli, G. Lecoy, and J. P. Nougier (Eds.) (Elsevier, New York), p. 227.

Littlejohn, M. A., J. R. Hauser, and T. H. Glissen (1977). *J. Appl. Phys.* **48,** 4587.
Liu, H. C., and D. D. Coon (1987). *Appl. Phys. Lett.* **50,** 1669.
Logan, R. A. (1969). in *Tunneling Phenomena in Solids,* E. Burstein and S. Lundquist (Eds.) (Plenum Press, New York), p. 149.
Lueder, H., and E. Spenke (1935). *Phys. Z.* **36,** 767.
Luryi, S. (1985). *Appl. Phys. Lett.* **47,** 490.
Luryi, S. (1987). *Heterojunction Band Discontinuities,* F. Capasso and G. Margaritondo (Eds.) (North-Holland, Amsterdam), p. 489.
Madan, A., and M. P. Shaw (1988). *The Physics and Technology of Amorphous Semiconductors* (Academic Press, New York).
Mahan, G. D. (1972). In *Polarons in Ionic Crystals and Polar Semiconductors,* J. T. Devreese (Eds.) (North-Holland, Amsterdam), p. 554.
Mahan, G. D. (1981). *Many-Particle Physics,* (Plenum, New York).
Mahrous, S., and P. N. Robson (1966). *Electron. Lett.* **2,** 108.
Maracas, G. N., W. Porod, D. A. Johnson, D. K. Ferry, and H. Goronkin (1985). *Physica* **134B,** 276.
Maracas, G. N., D. A. Johnson, R. A. Puechner, J. L. Edwards, S. Myhajlenko, H. Goronkin, and R. Tsui (1989). *Solid State Electron.* **32,** 1887.
Matisoo, J. (1980a). *IBM J. Res. Dev.* **24,** 113.
Matisoo, J. (1980b). *Sci. American* **May,** 50.
Mayer, K. M., J. Parisi, and R. P. Huebener (1988). *Z. Phys.* **B71,** 171.
Mazur, P., and D. Bedeaux (1981). *J. Stat. Phys.* **24,** 215.
McCumber, D. E. (1968). *J. Appl. Phys.* **39,** 3113.
McCumber, D. E., and A. G. Chynoweth (1966). *IEEE Trans. Electron Devices* **ED-13,** 4.
McGrath, W. R., P. L. Richards, A. D. Smith, H. van Kempen, R. A. Batchlor, D. E. Probor, and P. Santhanam (1981). *Appl. Phys. Lett.* **39,** 655.
McWhorter, A. L., and A. G. Foyt (1966). *IEEE Trans. Electron Devices* **ED-13,** 1979.
Mezrin, O. A., and S. I. Troshkov (1986). *Sov. Phys. Semicond.* **20,** 819.
Minorsky, N. (1962). *Nonlinear Oscillations* (Van Nostrand, Toronto).
Misawa, T. (1966). *IEEE Trans. Electron Devices* **ED-13,** 137.
Mitin, V. V. (1977). *Sov. Phys. Semicond.* **11,** 727.
Mitin, V. V. (1986). *Appl. Phys.* **A39,** 123.
Mitin, V. V. (1987). *Sov. Phys. Semicond.* **21,** 142.
Mizuta, H., T. Tanoue, and S. Takahashi (1988). *IEEE Trans. Electron Devices* **ED-35,** 1951.
Moon, F. C. (1987). *Chaotic Vibrations,* (Wiley, New York).
Morgan, T. N. (1966). *Phys. Rev.* **148,** 890.
Morkoc, H., J. Chen, U. K. Reddy, T. Henderson, and S. Luryi (1986). *Appl. Phys. Lett.* **49,** 70.
Mott, N., and E. A. Davis (1979). *Electronic Processes in Non-Crystalline Materials,* (Clarendon Press, Oxford).
Muto, S., A. Tackeuchi, T. Inata, E. Miyauchi, and T. Fujii, (1989). 4th International Conference on Modulated Semiconductor Structures, 17–21 July, Ann Arbor, MI, p. 575.
Naber, H., and E. Schöll (1990a). *Z. Phys. B* **78,** 301.
Naber, H., and E. Schöll (1990b). *Z. Phys. B* **78,** 305.
Nakamura, K. (1988). *Phys. Lett.* **134,** 173.
Newland, F. J. (1975). *Jpn. J. Appl. Phys.* **14,** 1.
Nicolis, G., and I. Prigogine (1977). *Self-Organization in Non-Equilibrium Systems* (Wiley, New York).
Novikov, S., S. V. Manakov, L. P. Pitaevskij, and V. I. Zakharov (1984). *Theory of Solitons—The Inverse Scattering Method,* (Consultant Burau, New York).
Oka, H., and S. Oshima (1962). *Mitsubishi Duiki Lab. Rep.* **3,** 165.
Okamoto, H. (1975). *IEEE Trans. Electron Devices* **ED-22,** 558.
Okuto, Y., and C. R. Crowell (1974). *Phys. Rev.* B **10,** 4284.
Ovshinsky, S. R. (1968). *Phys. Rev. Lett.* **21,** 1450.
Owen, A. E., J. M. Robertson, and C. Main (1979). *J. Non-Cryst. Solids* **32,** 29.
Owen, C. S., and D. J. Scalapino (1967). *Phys. Rev.* **164,** 144.
Pacha, F., and F. Paschke (1978). *Electron. Commun.* **32,** 235.
Pearsall, T. P., R. E. Nahory, and J. R. Chelikowsky (1977). *Phys. Rev. Lett.* **39,** 295.

Pearson, A. D., and C. F. Miller (1969). *Appl. Phys. Lett.* **14,** 280.
Pedersen, N. F. (1986). *Phys. Scripta* **T13,** 129.
Pedersen, N. F., and A. Davidson (1981). *Appl. Phys. Lett.* **39,** 830.
Peinke, J., A. Mühlbach, R. P. Huebener, and J. Parisi (1985). *Phys. Lett.* **108A,** 407.
Peinke, J., U. Rau, W. Clauss, R. Richter, and J. Parisi (1989). *Europhys. Lett.* **9,** 743.
Peregrine, D. W. (1966). *J. Fluid Mech.* **25,** 321.
Perlman, B. S., C. L. Upadhyayula, and R. E. Marx (1970). *IEEE Trans. Microwave Theory and Techniques* **MIT18,** 911.
Petersen, K. E., and D. Adler (1976). *J. Appl. Phys.* **47,** 256.
Petersen, K. E., D. Adler, and M. P. Shaw (1976). *IEEE Trans. Electron Devices* **ED-23,** 471.
Pickin, W. (1978). *Solid State Electron.* **21,** 309; 1299.
Piragas, K., Yu. Pozhela, A. Tamashyavichyus, and Yu. Ulbikas (1987). *Sov. Phys. Semicond.* **21,** 335.
Popescu, C. (1975). *Solid State Electron.* **18,** 671.
Pozhela, J. (1981). *Plasma and Current Instabilities in Semiconductors* (Pergamon, Oxford).
Price, P. J., and J. M. Radcliffe (1959). *IBM J. Res. Dev.* **3,** 364.
Prim, R. C. (1953). *Bell Syst. Tech. J.* **32,** 665.
Prince, M. B. (1956). *Bell. Syst. Tech. J.* **35,** 661.
Proctor, W. G., P. Lawaetz, Y. Marfaing, and R. Triboulet (1982). *Phys. Status Solidi* (*b*) **110,** 637.
Pryor, R. W., and H. K. Henisch (1972). *J. Non-Cryst. Solids* **7,** 181.
Purwins, H. G., G. Klempt, and J. Berkemeier (1987). *Festkörperprobleme* **27,** 27.
Quade, W., F. Rossi, and C. Jacoboni (1991). *Semicond. Science and Technol.*, to be published.
Rajakarunanayake, Y., and T. C. McGill (1989). *Appl. Phys. Lett.* **55,** 1537.
Rajeevakumar, T. V., J. X. Przybysz, and J. T. Chen (1980a). *Phys. Rev.* B **21,** 5432.
Rajeevakumar, T. V., L. M. Geppert, and J. T. Chen (1980b). *J. Appl. Phys.* **51,** 2744.
Rau, U., J. Peinke, J. Parisi, R. P. Huebener, and E. Schöll (1987). *Phys. Lett.* **124,** 335.
Rau, U., K. M. Mayer, J. Parisi, J. Peinke, W. Clauss, and R. P. Huebener (1989). *Solid State Electron.* **32,** 1365.
Read, W. T. (1958). *Bell System Tech. J.* **37,** 401.
Reed, M. A. (1986). *Superlattices and Microstruct.* **2,** 65.
Reggiani, L., and V. V. Mitin (1989). *Riv. Nuovo Cimento* **11**(12), 1.
Reggiani, L., P. Lugli, and A. P. Jauho (1987). *Phys. Rev.* B **36,** 6602.
Reggiani, L., P. Lugli, and A. P. Jauho (1988). *J. Appl. Phys.* **64,** 3072.
Reinhard, D. K. (1977). *Appl. Phys. Lett.* **31,** 527.
Reinhard, D. K., F. O. Arntz, and D. Adler (1973). *Appl. Phys. Lett.* **23,** 521.
Rhoderick, E. H., and R. H. Williams (1988). *Metal–Semiconductor Contacts* (Clarendon Press, Oxford).
Ricco, B., and M. Ya Azbel (1984). *Phys. Rev.* B **29,** 1970.
Ridley, B. K. (1963). *Proc. Phys. Soc.* **82,** 954.
Ridley, B. K. (1983). *J. Phys. C, Solid State Phys.* **16,** 3373.
Ridley, B. K. (1987). *Semicond. Sci. Technol.* **2,** 116.
Ridley, B. K., and T. B. Watkins (1961). *Proc. Phys. Soc. London* **78,** 293.
Ridley, B. K., and F. A. El-Ela (1989). *Solid State Electron.* **32,** 1393.
Rieger, M., and P. Vogl (1989). *Solid State Electron.* **32,** 1399.
Robbins, D. J. (1980). *Phys. Status Solidi*: (*b*) **97,** 9; **97,** 387; **98,** 11.
Robbins, D. J., P. T. Landsberg, and E. Schöll (1981). *Phys. Status Solidi*: (*a*) **65,** 353.
Rockstad, H. K., and M. P. Shaw (1973). *IEEE Trans. Electron Devices* **ED-20,** 593.
Rodgers, B. D., C. B. Thomas, and H. S. Reehal (1976). *Phil. Mag.* **31,** 1013.
Rolland, P. A., E. Constant, G. Salmer, and R. Ranquemberque (1979). *Electron. Lett.* **15,** 374.
Rosencher, E. (1981). 3rd International Conference on Hot Carriers in Semiconductors, *J. Phys.* **C-7,** 351.
Rössler, O. E. (1977). In *Synergetics, A Workshop*, H. Haken (Ed.), (Springer-Verlag, Berlin), p. 184.
Rousseau, K. V., J. L. Schulman, and K. L. Wang (1988). *Proc. SPIE* **943,** 30.
Roy, D. K. (1986). *Quantum Mechanical Tunneling and its Applications* (World Scientific, Singapore).

Ruch, J. G. (1972). *IEEE Trans. Electron Devices* **ED-19,** 652.
Ruch, J. G., and G. K. Kino (1967). *Appl. Phys. Lett.* **12,** 111.
Sah, C. T. (1969). In *Tunneling Phenomena in Solids,* E. Burstein and S. Lundquist (Eds.) (Plenum Press, New York), p. 193.
Sakamoto, R., K. Akai, M. Inoue (1989). *IEEE Trans. Electron Devices* **ED-36,** 2344.
Sattinger, D. H. (1973). *Topics in Stability and Bifurcation Theory* (Springer-Verlag, Berlin).
Sawaki, N., M. Susuki, E. Okuto, H. Goto, I. Akasaki, H. Kano, Y. Tanaka, M. Hashimoto (1988). *Solid State Electron.* **31,** 351.
Sawaki, N., R. Höpfel, and E. Gornik (1989). *Solid State Electron.* **32,** 1321.
Scanlan, J. O. (1966). *Analysis and Synthesis of Tunnel Diode Circuits* (Wiley, London).
Scarlett, R. M., W. Shockley, and R. K. Haitz (1963). *Physics of Failure in Electronics,* M. F. Goldberg and J. Vacaro (Eds.) (Spartan Books, Baltimore) Vol. 1, p. 194.
Schafft, H. A., and J. C. French (1962). *IEEE Trans. Electron Devices* **ED-9,** 129.
Schafft, H. A., and J. C. French (1966). *IEEE Trans. Electron Devices* **ED-13,** 613.
Schafft, H. A., and J. C. French (1966b). *Solid State Electron.* **9,** 681.
Scheuermann, M., J. R. Lhota, P. K. Kuo, and J. T. Chen (1983). *Phys. Rev. Lett.* **50,** 74.
Schlögl, F., and E. Schöll (1988). *Z. Phys.* **B71,** 231.
Schlögl, F. (1980). *Phys. Rep.* **62,** 267.
Schöll, E. (1978). Ph.D. thesis, University of Southampton.
Schöll, E. (1981). *J. Phys. (Paris) Colloque* **C7,** 57.
Schöll, E. (1982a). *Z. Phys. B* **46,** 23.
Schöll, E. (1982b). *Z. Phys. B* **48,** 153.
Schöll, E. (1983). *Z. Phys. B* **52,** 321.
Schöll, E. (1985a). *Proceedings of the 17th Conference on the Physics of Semiconductors,* San Francisco, J. D. Chadi and W. A. Harrison (Ed.) (Springer-Verlag, Berlin), p. 1353.
Schöll, E. (1985b). *Physica* **134B,** 271.
Schöll, E. (1986a). *Solid State Electron.* **29,** 687.
Schöll, E. (1986b). *Phys. Rev.* B **34,** 1395.
Schöll, E. (1987). *Nonequilibrium Phase Transitions in Semiconductors* (Springer-Verlag, Berlin).
Schöll, E. (1988). *Solid State Electron.* **31,** 539.
Schöll, E. (1989a). *Appl. Phys.* **A48,** 95.
Schöll, E. (1989b). *Phys. Scripta* **T29,** 152.
Schöll, E. (1989c). *Solid State Electron.* **32,** 1129.
Schöll, E. (1991). *Nonlinear Dynamics in Solid State Physics,* H. Thomas (Ed.) (Springer-Verlag, Berlin).
Schöll, E., and K. Aoki (1991). *Appl. Phys. Lett.* **58,** 1277.
Schöll, E., and D. Drasdo (1990). *Z. Phys. B State Physics,* **81,** 183.
Schöll, E., and P. T. Landsberg (1979). *Proc. R. Soc. London* **A365,** 495.
Schöll, E., and P. T. Landsberg (1983). *J. Opt. Soc. Am.* **73,** 1197.
Schöll, E., and P. T. Landsberg (1988). *Z. Phys. B* **72,** 515.
Schöll, E., and W. Quade (1987). *J. Phys.* **C20,** L861.
Schöll, E., J. Parisi, B. Röhricht, J. Peinke, and R. P. Huebener (1987). *Phys. Lett.* **A119,** 419.
Schuller, M., and W. W. Gartner (1961). *Proc. IEEE* **49,** 1268.
Schuster, H. G. (1987). *Deterministic Chaos,* (Physik-Verlag, Weinheim), 2nd ed.
Schwartz, B. B., and S. Frota-Passoa (1982). *Topics in Electromagnetism,* D. Talpitz (Eds.) (Plenum, New York).
Scott, A. C., F. Y. Chu, and S. A. Reible (1976). *J. Appl. Phys.* **47,** 3272.
Scott-Russell, J. (1845). Report of the 14th Meeting of the British Association, 311.
Seeger, A., and A. Kochendörfer (1950). *Z. Phys.* **130,** 321.
Seeger, A., H. Donth, and A. Kochendörfer (1953). *Z. Phys.* **134,** 173.
Seiler, D. G., C. L. Littler, R. J. Justice, and P. W. Milonni (1985). *Proceedings of the 17th International Conference on the Physics of Semiconductors,* San Francisco, J. D. Chadi and W. A. Harrison (Eds.) (Springer-Verlag, Berlin), p. 1385; and *Phys. Lett.* **108A,** 462.
Seminozhenko, V. P. (1982). *Phys. Rep.* **3,** 103.
Sen, S., F. Capasso, A. C. Gossard, R. A. Spah, A. L. Hutchinson, and S. N. G. Chu (1987). *Appl. Phys. Lett.* **51,** 1428.

Sen, S., F. Capasso, A. Y. Cho, and D. Sivco (1987b). *IEEE Trans. Electron Devices* **ED-34,** 2185.
Shanks, R. R. (1970). *J. Non-Cryst. Solids* **2,** 504.
Shapiro, S., P. H. Smith, J. Nicol, J. L. Miles, and P. F. Strong (1962). *IBM J. Res. Dev.* **6,** 34.
Shaw, M. P. (1979). *IEEE Trans. Electron Devices* **ED-26,** 1766.
Shaw, M. P. (1981). *Handbook on Semiconductors,* Vol. 4, Chap. 1, (North-Holland, Amsterdam).
Shaw, M. P. (1985). *Physics of Disordered Materials,* D. Adler, H. Fritzsche, and S. R. Ovshinsky (Eds.) (Plenum Press, New York), p. 793.
Shaw, M. P. (1988). *Disorder and Order in the Solid State,* R. W. Pryor, B. B. Schwartz, and S. R. Ovshinsky (Eds.) (Plenum Press, New York), p. 53.
Shaw, M. P., and I. J. Gastman (1971). *Appl. Phys. Lett.* **19,** 243.
Shaw, M. P., and I. J. Gastman (1972). Proceedings of the 4th International Conference Amorphous and Liquid Semiconductors, *J. Non-Cryst. Solids* **8–10,** 999.
Shaw, M. P., and K. F. Subhani (1981). *Solid State Electron.* **24,** 233.
Shaw, M. P., and N. Yildirim (1983). *Advances in Electronics and Electron Physics* Vol. 60, (Academic Press, New York), p. 307.
Shaw, M. P., H. L. Grubin, and I. J. Gastman (1973a). *IEEE Trans. Electron Devices* **ED-20,** 169.
Shaw, M. P., H. L. Grubin, and P. R. Solomon (1979). *The Gunn–Hilsum Effect* (Academic Press, New York).
Shaw, M. P., P. R. Solomon, and H. L. Grubin (1969). *IBM J. Res. Dev.* **13,** 587.
Shaw, M. P., S. H. Holmberg, and S. A. Kostylev (1973b). *Phys. Rev. Lett.* **23,** 521.
Shaw, M. P., S. C. Moss, S. A. Kostylev, and L. A. Slack (1973c). *Appl. Phys. Lett.* **22,** 114.
Sheard, F. W., and G. A. Toombs (1988). *Appl. Phys. Lett.* **52,** 1228.
Shichijo, H., and K. Hess (1981). *Phys. Rev.* B **23,** 4197.
Shichijo, H., K. Hess, and B. G. Streetman (1980). *Solid State Electron.* **23,** 817.
Shockley, W. (1954). *Bell. Syst. Tech. J.* **33,** 799.
Shockley, W. (1961). *Solid State Electron.* **2,** 35.
Shockley, W., and W. T. Read, Jr. (1952). *Phys. Rev.* **87,** 835.
Shockley, W., G. L. Pearson, and J. R. Haynes (1949). *Bell Syst. Tech. J.* **28,** 244.
Shousha, A. K. M. (1971). *J. Appl. Phys.* **42,** 5131.
Shuey, R. T. (1969). In *Tunneling Phenomena in Solids,* E. Burstein and S. Lundquist (Ed.) (Plenum Press, New York), p. 93.
Simpson, R. B., and D. S. Cohen (1970). *J. Math. Mech.* **19,** 895.
Skanavi, G. I. (1958). *Fizika Dielektrikov* (Gosudarstvennyi Izdatel'stvo Fiziko-Mathematicheskoj Literatury, Moscow).
Skocpol, W. J., M. R. Beasley, and M. Tinkham (1974). *J. Appl. Phys.* **45,** 4054.
Smith, R. A. (1978). *Semiconductors,* 2nd ed. (Cambridge Eng., Cambridge).
Soerensen, M. P., A. Davidson, N. F. Pedersen, and S. Pagano (1988). *Phys. Rev. A* **38,** 5384.
Solner, T. C. L. G. (1987). *Phys. Rev. Lett.* **59,** 1622.
Solner, T. C. L. G., E. R. Brown, W. D. Goodhue, and H. Q. Le (1987). *Appl. Phys. Lett.* **50,** 332.
Solner, T. C. L. G., W. D. Goodhue, P. E. Tannenwald, C. D. Parker, and D. D. Peck (1983). *Appl. Phys. Lett.* **43,** 588.
Solomon, P. R., M. P. Shaw, and H. L. Grubin (1972). *J. Appl. Phys.* **43,** 159.
Solomon, P. R., M. P. Shaw, H. L. Grubin, and R. Kaul (1975). *IEEE Trans. Electron Devices* **ED-22,** 127.
Sommerfeld, A. (1956). *Thermodynamics and Statistical Mechanics,* (Academic Press, New York).
Spinnewyn, J., H. Strauven, and O. Verbeke (1989). *Z. Phys. B* **75,** 159.
Spitalnik, R., M. P. Shaw, A. Rabier, and J. Magarshack (1973). *Appl. Phys. Lett.* **22,** 162.
Stafeev, V. I. (1959). *Sov. Phys. Solid State* **1,** 763.
Stancampiano, C. V. (1980). *IEEE Trans. Electron Devices* **ED-27,** 1934.
Stewart, W. C. (1968). *Appl. Phys. Lett.* **12,** 277.
Stocker, H. J., C. A. Barlow, Jr., and D. F. Weirauch (1970). *J. Non-Cryst. Solids* **4,** 523.
Stoker, J. J. (1950). *Nonlinear Vibrations* (Wiley Interscience, New York).
Stoker, J. J. (1980). *Commun. Pure Appl. Math.* **33,** 215.
Stratton, R. (1962). *Phys. Rev.* **126,** 2002.
Subhani, K. F. (1977). Ph.D. dissertation, Wayne State University, Detroit, MI.
Summers, C. J., and K. F. Brennan (1987a). *Appl. Phys. Lett.* **51,** 276.

Summers, C. J., K. F. Brennan, A. Torabi, and H. M. Harris (1987b). *Appl. Phys. Lett.* **52,** 132.
Symanczyk, R., E. Pieper, and D. Jäger (1990). *Phys. Lett.* **A143,** 337.
Symanczyk, R., D. Jäger, and E. Schöll (1991). *Appl. Phys. Lett.* **59,** 105.
Sze, S. M. (1981). *Physics of Semiconductor Devices* (Wiley, New York).
Sze, S. M., and G. Gibbons (1966). *Appl. Phys. Lett.* **8,** 111.
Sze, S. M. (1985). *Semiconductor Devices, Physics and Technology,* (Wiley, New York), 523 pp.
Tauc, J., and Abraham (1957). *Phys. Rev.* **108,** 936.
Tauc, J. (Ed.) (1974). *Amorphous and Liquid Semiconductors* (Plenum, New York).
Tavger, B. A., and V. Ya. Demikhovskii (1968). *Usp. Fiz. Nauk* **96,** 61; translation *Sov. Phys. Usp.* **11,** 644, 1969.
Teitsworth, S. W., and R. M. Westervelt (1984). *Phys. Rev. Lett.* **53,** 2587.
Teitsworth, S. W., R. M. Westervelt, and E. E. Haller (1983). *Phys. Rev. Lett.* **51,** 825.
Thim, H. W., and M. R. Barber (1966). *IEEE Trans. Electron Devices* **ED-13,** 110.
Thim, H. W., M. R. Barber, B. M. Hakki, S. Knight, and M. Uenohara (1965). *Appl. Phys. Lett.* **1,** 167.
Thom, R. (1975). *Structural Stability and Morphogenesis* (Benjamin, Reading, MA).
Thoma, P. (1976). *J. Appl. Phys.* **47,** 5304.
Thomas, H. (1973). In *Synergetics,* H. Haken (Ed.) (Teubner, Stuttgart), p. 87.
Thompson, G. H. B. (1980). *Physics of Semiconductor Laser Devices,* (Wiley, New York).
Thompson, J. M. T., and H. B. Stewart (1986). *Nonlinear Dynamics and Chaos* (Wiley, New York).
Thornton, C. G., and C. D. Simmons (1958). *IEEE Trans. Electron Devices* **ED-5,** 6.
Tolstikhin, V. I. (1986). *Sov. Phys. Semicond.* **20,** 1375.
Tomizawa, K., Y. Aweno, N. Hashizume, and M. Kawashima (1982). *IEEE Proc.* **129,** 131.
Tsu, R., and L. Esaki (1973). *Appl. Phys. Lett.* **22,** 562.
Val'd-Perlow, V. M., A. V. Krasilov, and A. S. Tager (1966). *Radio Eng. and Electron. Phys.* **Nov.,** 1764.
Van Duzer, T., and C. W. Turner (1981). *Principles of Superconducting Devices and Circuits* (Elsevier, New York).
Van Hove, M., C. Van Hoof, W. De Raedt, P. Jansen, I. Dobbelaere, J. Peeters, G. Borghs, and M. Van Rossum (1989). *ESSDERC'89, 19th European Solid State Device Research Conference,* Berlin, H. Heuberger, H. Ryssel, and P. Lange (Ed.) (Springer-Verlag, Berlin), p. 270.
Varlamov, I. V., and V. V. Osipov (1970). *Sov. Phys. Semicond.* **3,** 803.
Varlamov, I. V., V. V. Osipov, and E. A. Poltoratskii (1970). *Sov. Phys. Semicond.* **3,** 1978.
Vodjdani, N., F. Chevoir, D. Thomas, D. Cate, P. Bois, E. Costard, and S. Delaitre (1989). *Appl. Phys. Lett.* **55,** 1528.
Volkov, A. F., and Sh. M. Kogan (1969). *Sov. Phys. Usp.* **11,** 881.
Wacker, A., and E. Schöll (1991). *Appl. Phys. Lett.* **59,** 1702.
Walsh, P. J., D. Pooladdej, M. S. Thompson, and J. Allison (1979). *Appl. Phys. Lett.* **34,** 835.
Walsh, P. J., S. Ishioka, and D. Adler (1978). *Appl. Phys. Lett.* **33,** 593.
Wang, Y. (1979). *Int. J. Electron.* **47,** 49.
Warren, A. C. (1973). *IEEE Trans. Electron Devices* **ED-20,** 123.
Weber, W. H. (1970). *Appl. Phys. Lett.* **16,** 396.
Weber, W. H., and G. W. Ford (1970). *Solid State Electron.* **13,** 1333.
Weber, W. H., and G. W. Ford (1971). *Appl. Phys. Lett.* **18,** 241.
Weber, W. H., and G. W. Ford (1972). *Solid State Electron.* **15,** 1277.
Weil, T., and B. Vinter (1987). *Appl. Phys. Lett.* **50,** 1281.
Wersinger, J. M., J. M. Finn, and E. Ott (1980). *Phys. Fluids* **23,** 1142.
Westervelt, R. M., and S. W. Teitsworth (1985). *J. Appl. Phys.* **57,** 5457.
Wingreen, N. S., K. W. Jacobsen, and J. W. Wilkins (1988). *Phys. Rev. Lett.* **61,** 1396.
Wolak, E., K. L. Lear, P. M. Pitner, B. G. Park, E. S. Hellman, T. Weil, J. S. Harris Jr., and D. Thomas (1988). *Proc. SPIE* **943,** 36.
Wolff, P. A. (1954). *Phys. Rev.* **95,** 1415.
Woodward, T. K., T. C. McGill, H. F. Chung, and R. D. Burnham (1987). *Appl. Phys. Lett.* **51,** 1542.
Woodward, T. K., T. C. McGill, H. F. Chung, and R. D. Burnham (1988). *IEEE Trans. Electron*

Wu, J. S., C. Y. Chang, C. P. Lee, Y. H. Wang, and F. Kai (1989). *IEEE Trans. Electron Devices* **ED-10,** 301.
Wu, K. F., and M. P. Shaw (1989). *IEEE Trans. Electron Devices* **ED-36,** 603.
Wu, K. F., J. Czekaj, and M. P. Shaw (1991). *J. Appl. Phys.*, to be published.
Yadau, K. S., S. K. Sharma, C. Shekhar, and K. S. Balain (1976). *Indian J. Phys.* **50,** 986.
Yamada, K., N. Takara, H. Imada, N. Miura, and C. Hamaguchi (1988). *Solid State Electron.* **31,** 809.
Yang, C. H., J. M. Carlson-Swindle, S. A. Lyon, and J. M. Worlock (1985). *Phys. Rev. Lett.* **55,** 2359.
Yoshida, K., F. Irie, and K. Hamasaki (1978). *J. Appl. Phys.* **49,** 4468.
Young, J. F., B. M. Wood, H. C. Liu, M. Buchanan, D. Landheer, A. J. Springthorpe, and P. Mandeville (1988). *Appl. Phys. Lett.* **52,** 1398.
Zabrodskij, A. G., and I. S. Shlimak (1975). *Sov. Phys. Solid State* **16,** 1528.
Zaslavsky, A., V. J. Goldman, D. C. Tsui, and J. E. Cunningham (1988). *Appl. Phys. Lett.* **53,** 1408.
Zener, C. (1943). *Proc. R. Soc. (London)* **A145,** 523.
Zur, A., T. C. McGill, and D. L. Smith (1983). *Surf. Sci.* **132,** 456.

Index